PROGRESS IN CLINICAL AND BIOLOGICAL RESEARCH

RECENT TITLES

See pages following the index for previous titles in this series.

MECHANISMS OF SPECIATION

MECHANISMS OF SPECIATION

Proceedings from the
International Meeting on
Mechanisms of Speciation
Sponsored by the
Accademia Nazionale dei Lincei
May 4–8, 1981
Rome, Italy

Editor

Claudio Barigozzi

Institute of Genetics
University of Milan
Milan, Italy

ALAN R. LISS, INC., NEW YORK

Address all Inquiries to the Publisher
Alan R. Liss, Inc., 150 Fifth Avenue, New York, NY 10011

Copyright © 1982 Alan R. Liss, Inc.

Printed in the United States of America.

Library of Congress Cataloging in Publication Data

International Meeting on Mechanisms of Speciation
(1981 : Rome, Italy)
Mechanisms of speciation.

(Progress in clinical and biological research ;
96)
Bibliography: p.
Includes index.
1. Species—Congresses. 2. Evolution—Congresses.
I. Barigozzi, Claudio. II. Academia nazionale dei
Lincei. III. Title. IV. Series.
QH371.I53 1981 / 575   82-13014
ISBN 0-8451-0096-3

CONTENTS

CONTRIBUTORS

F.A. Abreu-Grobois [345]
Department of Genetics, University College of Swansea, Swansea SA2 8PP, U.K.

Francisco J. Ayala [51]
Department of Genetics, University of California, Davis, CA 95616

B. Battaglia [377]
Istituto di Biologia Animale, Universita di Padova, Padova, Italy, and Istituto di Biologia del Mare, C.N.R., Venezia, Italy

J.A. Beardmore [345]
Department of Genetics, University College of Swansea, Swansea SA2 8PP, U.K.

Mario Benazzi [307]
Department of Zoology, University of Pisa, Via Volta 4, 56100 Pisa, Italy

Luciano Bullini [241]
Institute of Genetics, University of Rome, P. le Aldo Moro 5, 00185 Rome, Italy

Ernesto Capanna [155]
Istituto di Anatomia Comparata, Rome University, Via Borelli 50, I-00161 Roma, Italy

Hampton L. Carson [411]
Department of Genetics, University of Hawaii at Manoa, Honolulu, HI 96822

Mario Coluzzi [143]
Institute of Zoology, University of Camerino, Italy

E.S. Dennis [123]
Division of Plant Industry, C.S.I.R.O., P.O. Box 1600, Canberra City, A.C.T. 2601, Australia

Gabriel Dover [435]
Department of Genetics, University of Cambridge, Cambridge CB2 3EH, U.K.

Friedrich Ehrendorfer [479]
Institute of Botany and Botanical Garden of the University, A-1030 Vienna, Rennweg 14, Austria

W.L. Gerlach [123]
Division of Plant Industry, C.S.I.R.O., P.O. Box 1600, Canberra City, A.C.T. 2601, Australia

The boldface number in brackets following each contributor's name indicates the opening page of that author's paper.

L.D. Gottlieb [179]
Department of Genetics, University of California, Davis, CA 95616

S.H. James [461]
Botany Department, University of Western Australia, Nedlands, 6009 Australia

Ernst Mayr [1]
Museum of Comparative Zoology, Harvard University, Cambridge, MA 02138

G. Montalenti [xi]
Accademia Dei Lincei, Via Della Lungara 10, Roma, Italy

Eviatar Nevo [191]
Institute of Evolution, University of Haifa, Haifa 31999, Israel

Georges Pasteur [511]
Ecole Pratique des Hautes Etudes and Centre de Recherches sur l'Evolution et ses Mécanismes, Place Eugène-Bataillon, 34100 Montpellier, France

W.J. Peacock [123]
Division of Plant Industry, C.S.I.R.O., P.O. 1600, Canberra City, A.C.T. 2601, Australia

Jeffrey R. Powell [67]
Department of Biology, Yale University, P.O. Box 6666, 260 Whitney, Ave., New Haven, CT 06511

Ralph Riley [471]
Agricultural Research Council, 160 Great Portland Street, London W1N 6DT, England

Valerio Sbordoni [219]
Institute of Zoology, University of Roma, Viale dell'Università 32, Roma, Italy

Valerio Scali [393]
Institute of Zoology, Bologna University, Via S. Giacomo 9, Bologna, Italy

Steven M. Stanley [41]
Department, of Earth and Planetary Sciences, Johns Hopkins University, Baltimore, MD 21218

G. Ledyard Stebbins [21]
Department of Genetics, University of California, Davis, CA 95616

Alan R. Templeton [105]
Department of Biology, Washington University, St. Louis, MO 63130

Bruce J. Turner [265]
Department of Biology, Virginia Polytechnic Institute and State University, Blacksburg, VA 24061

M.J.D. White [75]
Department of Population Biology, Research School of Biological Sciences, Australian National University, P.O. Box 475, Canberra City, A.C.T. 2601, Australia

Introductory Remarks

G. Montalenti

Distinguished colleagues, dear friends, and ladies and gentlemen: On behalf of the Accademia Nazionale dei Lincei, which is honored to offer its hospitality, it is my pleasant duty at the opening of this meeting to welcome all the participants to the symposium.

Concerning the subject of the meeting, Mechanisms of Speciation, let me tell a little story that occurred to me recently.

A few weeks ago a friend of mine came to visit me at a time when I was perusing M.J.D. White's book on Modes of Speciation. "What!—said he—are you still busy with that problem? I thought the whole matter had been settled more than a century ago by Darwin's book. Doesn't it bear the title The Origin of Species?"

I had some difficulty persuading him that the problem is still open to debate although genetics has thrown considerable light on it; that Darwin plowed a furrow which we are still tracing: the recent spectacular achievements of molecular biology should also be taken into account in considering the intimate mechanisms underlying speciation, et cetera.

The most difficult thing was to convince my friend—a cultivated man interested in biology, but, not a specialist in the field—that mechanisms of speciation should be a multi-dimensional concern because the problem of species formation cannot be solved by a single simple answer, as he believed, and as many people, including some biologists, still believe. The problem requires, as it were, as many solutions as there are species that exist in the animal and plant kingdoms—to say nothing of prokaryotes, where the concept of species itself is in some way disappearing.

However, I did not dare to refer my friend to R.C. Lewontin's (1974) statement: "It is an irony of evolutionary genetics that, although it is a fusion of Mendelism and Darwinism, it has made no direct contribution to what Darwin obviously saw as the fundamental problem: the origin of species" [1]. Otherwise my friend would have said: "What have you biologists been doing in the hundred and twenty years since the publication of Darwin's book!"

I believe that the majority of biologists—if not all of them—are aware both of the complexity of the problem and of its paramount importance for biology, as well as of its more general philosophical implications. In fact speciation is the fundamental process of biological evolution, as Darwin for the first time in history clearly understood.

For these two reasons—the still open status of the problem and its momentous significance for the whole theory of evolution, which in turn is one of the most important achievements of modern scientific thought—the Accademia Nazionale dei Lincei was glad to comply with the suggestion offered by our foreign fellow M.J.D. White to organize an international symposium on this subject. Michel White is thus the main person responsible for this conference and has given invaluable help to the organizing committee. On behalf of the committee and of the Accademia I want to thank him heartily.

The Accademia has always been interested in the problem of the process of evolution as is shown by the number of national and international meetings which it has organized in the last decades, and by the Seminar on Evolution and the great problems of biology which is sponsored by the Centro Linceo Interdisciplinare and has been held every year in February, since 1974. This initiative is intended to offer school teachers and the general public a series of lectures of a high scientific standard on evolutionary and other biological problems. In this way we think that the Accademia may accomplish an important task, that is to maintain contact with the public at large. I think that both tasks—discussion of problems at a highly specialized level (as will occur presently in this symposium) and dissemination of this information for the benefit of a larger public, hence adding to the general body of knowledge—are very important.

The field of biology is still haunted by some ghosts. I will mention two that are polar opposites, namely the Galilean and Newtonian concepts of nature, and the belief in a strict physicalist reductionism. The first means the persistence of the Aristotelian conception of species, which, adopting Ernst Mayr's terminology, we may call typological. A great number of people, and even several biologists, still have great difficulties in thoroughly adopting the concept that Mayr calls the populational concept of species, with all its corresponding implications. The other ghost is the attitude taken by those who adhere to a belief in strict reductionism, and which seems to many biologists the only alternative to the rejection of vitalism. But the reduction of all biological phemomena to the laws elaborated by physicists and chemists for their own purpose is not possible. I remember that time in the thirties when Fermi, Rasetti, and the physicists of their group became interested in biology, and were, just for fun, considering the alternative of biology eventually becoming a branch of physics, or vice versa. There was much discussion and many jokes between the physicists and the biologists, upon this admittedly paradoxical dilemma. But this was just a way of recognizing that, although physics and chemistry can sufficiently explain a great many phenomena basic to biological processes, they are unable to supply a full explanation of all the more complex processes.

This failure is due to several peculiarities pertaining to biological phenomena. I will mention only one bearing directly on our study of speciation: unlike the great majority of physical and chemical events, the biological processes are essentially historical in nature. I am glad to say that this has been explicitly stated by, among others, a distinguished physicist, Professor Mario Ageno, professor of biophysics in our faculty in Rome, who has devoted a great deal of attention to the demarcation between biology and physics. In a recent text book of biophysics he writes: "Tali differenze interspecifiche, come molti altri fatti biologici, non sono una conseguenza delle leggi della chimica e della fisica" (Such interspecific differences, like many other biological facts, are not a consequence of physical and chemical laws), and, further "la biologia è una scienza storica" [2] (biology is an historical science). And in another place: "Il vivente è dunque un sistema complessa, *non ripetitivo,* e i metodi generali con cui il fisico ha finora studiato correntemente i sistemi complessi, in questo caso non sono applicabili" [3] (The living being is a complex system, *nonrepetitive,* and the general methods by which the physicist has so far currently studied the complex system cannot be used in this case).

The processes we are going to consider today and in the next few days, ie the emergence and differentiation of species, are historical phenomena, and undoubtedly they will be considered under the populational perspective. Thus, the results of the work that the participants in this symposium will present and discuss will certainly be an important contribution not only to a limited sector of biology, but will have a wider scientific and philosophical relevance, contributing to the history of evolution and to an understanding of life in scientific terms.

Perhaps I should apologize for emphasizing a few concepts which are quite obvious and familiar to all the speakers invited to this symposium. But, since there are in attendance many young biologists and some laymen, I thought it important to show them the conceptual frame in which we are working.

Finally, let me mention, just in passing, the debate which recently exploded upon the occasion of the opening of an exhibition on "Dinosaurs and their living relatives" organized at the British Museum, Natural History, London. The argument is whether the "cladistic" mode of speciation is the only way by which new species arise. The problem of "phyletic evolution" versus "punctuated equilibria" is a matter of current polemics, as you know, and no doubt it will be discussed in the course of this meeting. But I do not agree with L.B. Halstead, who, in an article entitled "The Museum of Errors" brings into the dispute no less than the venerable figure of Karl Marx [4]. This just shows how the subject of speciation can be of concern to a wider public, with the obvious danger of allowing people to stray from the scientific track.

I will say nothing about the anti-evolutionary movements which are proliferating like mushrooms in many countries. The facts and hypotheses which will emerge from this symposium will no doubt help to counteract them.

In concluding my introductory remarks I want to express, on behalf of the Accademia dei Lincei and of the Italian biologists, the deepest appreciation and the warmest thanks to our foreign guests, who undertook a long journey to reach this place. Special thanks are due to those who were able to come to the aid of the exhausted finances of the Accademia by finding total or partial support for their travel expenses from other sources. We ask them to convey the expression of our gratitude to the institutions which supported them.

Unfortunately the finances of the Accademia dei Lincei are in quite a dramatic condition. The inflation which is accelerating at a dreadful rate in our country has seriously threatened the fulfillment of this meeting. Therefore, I want to appeal to your indulgence for anything you might find wrong or imperfect during the following days.

Before starting with Mayr's lecture, let me perform my duty as President of the Accademia by giving the membership badge of the Accademia to our foreign fellows, M.J.D. White, who was elected in 1978, and Ernst Mayr, who was elected in 1980.

REFERENCES

1. Lewontin RC: The Genetic Basis of Evolutionary Change. New York: Columbia University Press, 1974, p 159.
2. Ageno M: Lezionidi Biofisica. Bologna: Zanichelli, 1980, p 560.
3. Ibid., p 382.
4. Halstead LB: The museum of errors. Nature 288:208, 1980.

Mechanisms of Speciation, pages 1–19
© 1982 Alan R. Liss, Inc., 150 Fifth Avenue, New York, NY 10011

Processes of Speciation in Animals

Ernst Mayr

INTRODUCTION

Speciation, as M.J.D. White has recently stated quite rightly, now appears as the key problem of evolution. It is remarkable how many problems of evolution cannot be fully understood until speciation is understood. In view of this, it is somewhat puzzling why genetics has so long neglected a serious study of this process. This has now changed, and White's "Modes of Speciation" (1978) [1] provides a comprehensive summary of the current state of speciation research. The availability of this volume greatly facilitates my task. For many aspects of the speciation problem, I shall simply refer to the relevant chapters of White's book. There have been attempts by authors such as Bateson, de Vries, Morgan, and Goldschmidt to explain speciation by macromutations, or of others to describe it simplistically merely as a change of gene frequencies in populations, but these endeavors failed to do justice to the real situation in nature. To state it factually, speciation research did not become part of the tradition of genetics until very recently.

Some of the reasons for this neglect are obvious. Speciation, except for polyploidy and some other chromosomal processes, is too slow to be observed directly. Therefore, the method of speciation research must consist of an attempt to reconstruct the historical precedents, derive from this reconstruction certain deductive generalizations, and test their validity by proper comparative methods. Since the conditions prevailing at the time of the division of an ancient gene pool into two cannot be observed directly, they must be inferred. In order to arrive at the most probable course of events, one must have a great deal of experience with cases of incipient speciation, and with the genetic, chromosomal, ecological, and geographical circumstances of recently speciated populations and of those that seem just now to be undergoing the process of speciation. There are nearly always several possible scenarios, and it is not surprising that different authors may differ in the choice of their explanations. Owing to the slowness of the speciation process, it is not possible to study the same individual or population "just before" and "just after" speciation. By necessity there is some arbitrariness in the sequence of events one postulates to have occurred, and authors with different professional backgrounds—let us say a geneticist, a cytologist, a paleontologist, a zoologist, and a botanist—may select different scenarios as most probable.

I shall merely mention a second difficulty, because other speakers in this symposium will expand on this subject: I refer to the pluralism of speciation processes. Plants offer particularly striking illustrations of this, but even among animals different species groups within the same genus may have rather different speciation patterns, as shown by the speciation of Thomomys talpoides compared with other pocket gophers. This pluralism of speciation processes shows that it is not permissible to apply an answer that is correct in one case automatically to other situations. Speciation patterns in groups with high dispersal facility are different from those with low dispersal; speciation in groups in which premating isolating mechanisms are acquired first and postmating isolation only subsequently is different from that in groups for which the reverse is true.

A third difficulty, now largely overcome, was that many authors had only vague ideas as to what the term *speciation* really meant. It is now generally agreed that it signifies a multiplication of species—that is, the production of new, reproductively isolated individuals (allopolyploids) or populations.

Perhaps even more damaging and causing greater confusions was the failure to recognize that certain phenomena or processes occur simultaneously, rather than being "either-or" alternatives. Let me begin with *the chromosomal vs geographical alternative*. White, Bush, and other recent proponents of stasipatric speciation have repeatedly made statements approximately of this form: "Chromosomal rearrangements and not geographical isolation must have been responsible for speciation" in a given case. Such a statement implies that one must make a choice between either one or the other factor. By contrast I have long insisted that they are not mutually exclusive and therefore [2] that the two processes are independent of each other and therefore can, but need not necessarily, coincide with each other. Failure to recognize this means that an author will fail to make a distinction between necessary and sufficient conditions. Even if we were to make the extreme assumption that speciation would always require chromosomal reorganization, it could well be possible that such a reorganization could be completed successfully only in an isolated population of minimal size. I shall come back to this question in the discussion of stasipatric speciation.

A second erroneous alternative is that between the *reproductive* and the ecological aspects of speciation. Speciation is *not* either the acquisition of reproductive isolation *or* of the ability to compete successfully with other species. It is always primarily the acquisition of reproductive isolation, but sympatry cannot be achieved until sufficient niche segregation has evolved to prevent fatal competition. To this problem I shall also return later in my discussion.

GEOGRAPHIC OR ALLOPATRIC SPECIATION

The first authors in the 19th century to speculate about the origin of species were local naturalists who studied the variation of species in their home country and attempted to find such deviations from the type as could be considered incipient species. This nondimensional approach was particularly favored by the early botanists, but it has remained the approach of certain zoologists and botanists right up to the present time.

The application of multidimensional thinking to the speciation problem by L. von Buch (1825), Darwin (March 1837), and M. Wagner (1841) [3] was a conceptual breakthrough of the first order. By adopting a new sequence of events, namely by first splitting off a population from the parental species and then accumulating the genetic differences that permit speciation, most of the difficulties were avoided that had previously bedeviled the typological and nondimensional (single-population) approach.

The Dumbbell Model

The classical concept of the process of geographical speciation was that of a widespread species, the range of which was divided into two halves by a geographical barrier. It is illustrated in the so-called dumbbell diagram. It was this concept most authors had in mind when talking about allopatric speciation, and particularly geneticists and cytologists without personal acquaintance with actual speciation in the field. Actually there exist many forms of geographic speciation, depending primarily on the size of the isolated populations. One extreme is represented by the dumbbell model, in which two large, subequal portions of the species are separated from each other. No special name exists for this extreme, even though it is what most people had in mind when referring to geographic speciation. More interesting, and apparently far more important for speciation, is the other extreme, where a strong disparity in size exists between the isolated populations.

Peripatric Speciation

Zoologists, and some botanists, who analyzed actual speciation patterns among closely related species, however, have long been aware of the disparity situation. Already Darwin, when comparing South American with Galapagos species, knew that speciation is far more active among small island populations than among large continental species. That there is a roughly inverse relation between population size and rate of speciation had long been intuitively appreciated by many students of mammals, birds, fishes, and certain groups of insects, as is evident from the taxonomic literature. Ever since the publication in 1942 of my "Systematics and the Origin of Species," I puzzled over this difference of rates, but it was not until 1954 that I proposed an entirely new theory of allopatric speciation [4]. How drastically different from traditional geographic speciation this new theory was, is missed by all those who, like Michael White, lump the two models and still speak of "*the* allopatric model of speciation." Actually, the two allopatric models are worlds apart. To make this even clearer than it has been in the past, and in order to preclude the continuous confounding of my new model of speciation with traditional geographic speciation of large populations, I propose that my 1954 model be designated as "peripatric speciation."

Here, the gene pool of a small either founder or relict population is rapidly, and more or less drastically, reorganized, resulting in the quick acquisition of isolating mechanisms and usually also in drastic morphological modifications and ecological shifts. It involves populations that pass through a bottleneck in population size.

I illustrated this process by the distribution pattern of the Tanysiptera galathea species group of birds which shows hardly any geographic variation on the mainland of New Guinea over a distance of more than 1,000 kilometers and with a distribution over several climatic zones and across several geographical barriers, whereas all populations on islands off New Guinea (eg, Koffiao, Biak, Numfor) are so strikingly different that they were considered to be separate species. Each of these islands, none of which could have been in recent continental connection with New Guinea, was almost certainly colonized by a founding pair of birds, giving rise to a highly distinct population. My analysis of distribution patterns in the Indo-Pacific archipelagoes revealed an even more spectacular phenomenon. In many cases, when I traced a series of closely related allopatric species, I found that the most distant, the most peripheral species, was so distinct that ornithologists had either described it as a separate genus (Serresius derived from Ducula pacifica, Dicranostephes derived from Dicrurus hottentottus) or had, at least, failed to recognize its true relationship (Dicaeum tristrami as an allospecies of Dicaeum erythrothorax). In genus after genus I found the most peripheral species to be the most distinct.

It is on this *strictly observational* basis that I proposed in 1954 my theory of drastic speciation in peripheral founder populations—that is, my theory of peripatric speciation. Nothing has been discovered in the more than 25 years since my publication that would have weakened my case. On the contrary, it has been enormously strengthened by the brilliant researches of Carson on speciation of Drosophila in the Hawaiian Islands, which have put flesh on the skeleton of my theory.

When it came to provide a genetic interpretation for my observations, all I could offer were hypotheses, because no adequate genetic analysis of any peripherally isolated founder population was as yet available. Still, I saw a number of things rather clearly and emphasized them in my original paper: 1) that there is always a considerable if not drastic loss of genetic variability in a founder population; 2) that there is greatly increased homozygosity in the new population and that this will affect the selective value of many genes as well as the total internal balance of the genotype; 3) that conditions are provided for the occasional occurrence of a veritable genetic revolution; 4) that such a genetic revolution occurs only sometimes, but by no means in all cases of speciation in founder populations; 5) that such founder populations can pass quickly through a condition of heterozygosity in those cases in which the heterozygotes are of lowered fitness as is, for instance, often the case with chromosomal rearrangements; 6) that such genetically unbalanced populations may be ideally situated to shift into new niches; 7) that the genetic reorganization might sufficiently loosen up the genetic homeostasis to facilitate the acquisition of morphological innovations; and 8) that the genetic revolution is a population phenomenon, not like Goldschmidt's hopeful monsters, one of individuals, and that therefore the change, no matter how drastic and rapid it may be, is gradual and continuous; it is not a saltation (or transilience as Galton had called it) [5].

In 1954 geneticists thought that all DNA of the nucleus consisted of a single kind of traditional genes, and, not being a geneticist myself, I naturally based my own tentative explanation on this assumption. It is a source of great satisfaction to me that my interpretation is equally applicable to the newer genetic explanations that have been suggested in recent years. Indeed, any more drastic disturbance of the genotype, particularly one that produces a deleterious heterozygous condition, is coped with much more easily in a small, inbred, founder population than in any other kind of population.

A few years later the botanist Harlan Lewis [6] proposed independently a very similar theory of rapid speciation in peripherally isolated populations of plants and, as I had, emphasized strongly the importance of selection in such genetic revolutions.

One major addition must be made to the theory of peripatric speciation. Haffer [7] has clearly demonstrated relatively rapid speciation in Amazonian rain forest birds which had been isolated in forest refugia during Pleistocene drought periods. Even though these populations were subjected to a severe reduction of population size, they did not go through nearly as drastic a bottleneck as do founder populations. Nevertheless, there has been active speciation in these forest refugia, as confirmed by Vanzolini and Williams for reptiles and by Turner for butterflies. None of these neospecies are so strikingly different as those that I found among Pacific birds, produced by genetic revolutions in peripheral founder populations. Since these refuge populations consisted even at their smallest size presumably of hundreds if not thousands of individuals, their rapid change is probably due more to greatly increased selection pressure (particularly owing to the drastically altered biotic environment) than to the genetic consequences of inbreeding. Relatively rapid speciation occurred apparently also in some Pleistocene refuges in the Holarctic.

SPECIATION ON CONTINENTS

The two theories of allopatric speciation demand the existence of natural barriers than can isolate a population sufficiently long and effectively that it can pass through the bottleneck of deleterious heterozygosity and reorganize itself genetically. From the beginning some naturalists have had difficulties in conceiving the existence of such barriers on continents, even though these same authors freely adopted allopatric speciation for islands. Darwin himself accepted various schemes of sympatric speciation to explain speciation on continents, and so have other authors right up to the present. Mr. White's major reason for proposing stasipatric speciation on continents was that "there is no plausible geographic or paleographic reason for believing that the two populations were ever geographically isolated" [1 p 108]. The known facts thoroughly refute White's opinion. Vegetational, climatic, and other physiographic barriers abound on continents, particularly in the tropics, as demonstrated by literally scores of zoologists. These areas are rich in allopatric species of greater or lesser recency, separated from each other by physiographic barriers. During the Pleistocene and post-Pleistocene climatic fluctuations many, perhaps most, species ranges were fractured and the remnants were pushed into refugia. A series of splendid monographs has appeared in recent decades documenting the causative role of vegetational-climatic barriers for the speciation of Australian birds (Keast), African birds (Moreau-Hall), South American birds (Haffer), South American reptiles (Vanzolini and Williams), and South American butterflies (Turner), to mention only a few papers of a rich literature. Ken Key, Michael White's collaborator in the work on the Australian grasshoppers [8], has adopted the same interpretation, and the allopatric distribution pattern of nearly all closely related species of morabine grasshoppers makes this interpretation, of course, almost inevitable. Ironically, one of the best listing of cases of allopatric speciation on continents is provided by M. White himself, labeled, however, as cases of stasipatric speciation.

FATE OF NEOSPECIES AND INCIPIENT NEOSPECIES

The distribution pattern of neospecies raises the question, What happens during the isolation? We must clearly distinguish here two very different topics: 1) phenotypic manifestations of species status, and 2) the nature of the genetic mechanisms responsible for these manifestations. Let me begin with the phenotypic manifestations. When such an isolated population survives and expands its range after the obliteration of the barrier, it will sooner or later encounter the range either of its parental species or that of a sister neospecies. The fate of such an encounter depends on two sets of factors: 1) the nature and extent of the isolating mechanisms acquired during the spatial isolation, and 2) the degree of ecological compatibility acquired. If complete reproductive isolation as well as ecological compatibility were acquired, the two neospecies can widely overlap but usually do so only partly, owing to a previous acquisition of different habitat preferences.

If complete postmating reproductive isolation was acquired, but not premating isolation, either no hybrids at all or sterile hybrids will be produced in the zone of contact. Single invaders into the range of the alien species will therefore leave no offspring, as is documented by most parapatric species pairs of morabine grasshoppers. A parapatric pattern, however, can also develop through competitive exclusion when reproductive isolation but no ecological compatibility had developed during the isolation. This seems to be the major cause for parapatric distribution in birds.

Finally, if neither premating nor effective postmating isolating mechanisms had developed during the isolation, a wide or narrow hybrid belt will develop in the zone of contact, the width of the belt depending both on ecological and genetic factors.

In this analysis I have repeatedly used the term *parapatric pattern*. Let me explain this in more detail.

PARAPATRIC PATTERN OF DISTRIBUTION

A pattern of distribution in which the borders of closely related species touch, often for considerable distances, without major overlap or massive hybridization, is called parapatric distribution. Some cases of this pattern had long been known, but only the careful mapping of tropical and Australian species has revealed how frequent parapatry is. The problem we have is to explain the reasons for the existence of this pattern. We are facing here an explanatory problem that applies to all sciences and which deals with historical narratives. Our task is, as I pointed out in the introduction, to infer from the present pattern through what past events and processes it had originated.

There are two possible interpretations of the causation of parapatry. According to the traditional explanation, which I have just presented, the two parapatric species speciated in isolation, expanded their ranges subsequently, and finally met in the zone of parapatric contact. Mutual overlap of their ranges is prevented either for ecological reasons (ie, competition) or owing to the production of sterile hybrids resulting from a lack of premating isolating mechanisms.

White and other authors have advanced an alternative theory according to which parapatric patterns are due to in situ speciation by which the previously continuous range of an ancestral species is disrupted along a step in an environmental gradient. No doubt this explanation will be fully discussed in the course of this symposium. Let me merely state that I find it singularly unconvincing [9].

The fact that for many species pairs the parapatric zones coincide, in the absence of any physiographic or vegetational line, is particularly convincing evidence for the secondary origin of these zones [10]. I must confess that in all the cases of postulated parapatric speciation listed by White in Chapter 5 of his book, there is not a single one in which the nonallopatric mode of origin would seem more probable to me than the allopatric one. Parapatric patterns of distribution, thus, as I see it, are always caused by the fact that the development of the isolating mechanisms between the two neospecies had not yet reached perfection.

A sterility barrier, caused by chromosomal restructuring, is only one among the known kinds of isolating mechanisms between species. The true nature of isolating mechanisms was appreciated by most evolutionary biologists remarkably late. Dobzhansky in 1941 and H. J. Muller in 1942 still included geographical barriers among the isolating mechanisms. The true intrinsic isolating mechanisms can be divided into premating (mostly behavioral or ecological) mechanisms, and postmating mechanisms. Because it was found in birds and some other groups of animals that often there were behavioral (premating) but no sterility barriers between closely related species, it was long assumed that in animals the evolution of premating barriers precedes that of postmating (sterility) barriers, whereas in plants speciation would always begin with the development of sterility barriers. However, we now also know several groups of animals in which the evolution of sterility barriers preceded that of behavioral barriers. This is, for instance, true for the Australian grasshoppers on which White and Key are working, and apparently also for some groups of amphibians. The literature reveals that authors who work with groups of animals or plants in which sterility is the primary isolating mechanism, find it very difficult to accept speciation primarily based on the origin of behavioral barriers. Yet, such a process is now abundantly established for numerous animal groups. Unfortunately, I do not have the time to discuss various aspects of behavior that effect patterns of speciation in animals.

THE GENETICS OF ALLOPATRIC KINDS OF SPECIATION

The traditional presentation of speciation in genetics textbooks was that of a continuous piling up of gene differences, leading to a gradual divergence of populations separated by a barrier. The discovery of the method of enzyme electrophoresis raised the hope that this method would reveal the nature of the genetic changes occurring during speciation. As surely other speakers, particularly F. Ayala, will describe in detail, this hope was not fulfilled. Nevertheless this method supplied the extremely important evidence that in speciation there is no major involvement, if any at all, of the classical enzyme genes. There are multiple proofs for this conclusion, perhaps the most convincing being that the rate of isozyme replacement is no more rapid in actively speciating phyletic lines than in conservative ones. With the enzyme genes removed from consideration, we must ask what other kinds of genetic factors are responsible for speciation.

Let me begin with a consideration of the chromosome as a whole. Since White and others will deal with this subject in detail, I will restrict myself to a few comments. Since in the eukaryotes virtually all the genetic material is located on the chromosomes, no one will question that the chromosomes are important in speciation—the only question being, in what way? To be sure, the occurrence of homosequential speciation in many Hawaiian species groups of Drosophila and in other genera of dipterans demonstrates that speciation is possible without gross chromosomal changes. Nevertheless, the frequency with which closely related species of most groups of organisms differ in chromosome structure testifies to the frequency at which chromosomal reorganization seems to accompany speciation.

The explanatory value of this observation, however, is quite limited. First of all, chromosomal rearrangements have at least two very different effects. The first is that any breaks may affect the neighborhood of genes, hence various kinds of position effects in the broadest sense of the word. The second effect is that such rearrangements often inhibit or prevent crossing over and thus protect coadapted linkage groups against being broken up. Also rearrangements often cause difficulties during meiosis. There is another difficulty: The discovery of transpositional DNA and of the various kinds of repetitive DNA have somewhat blurred the sharp demarcation between genes and gross chromosomal changes. How does this affect the question: Is the genetic basis of speciation genic or chromosomal?

There is one point, however, that needs to be emphasized. Michael White and other cytogeneticists have shown that, broadly speaking, chromosomal rearrangements fall into two classes. In one class heterozygotes are of normal viability or even superior—ie heterotic—and this situation leads to chromosomal polymorphism. In the other class heterozygotes are inferior or even semilethal. This is precisely the class to which so often the chromosomal differences between closely related species belong. As I have emphasized for more than 10 years there is no surer way to get quickly through this deleterious heterozygous condition to a new superior homozygosity than in a highly inbred founder population. Such a rapid chromosomal replacement may accompany or even cause a genetic revolution. The new homozygote may be established within two generations, but the chance of two heterozygotes finding each other in a large population is infinitely smaller.

A third possible major genetic cause of speciation is some as yet unidentified fraction of DNA. I mentioned this possibility more than 15 years ago [11], and I am sure that it will be an important theme of several presentations at this symposium. Numerous cases of incipient or complete speciation, evidently caused by genetic factors other than enzyme genes, have been described in the literature. For instance, in several North American species of Lepidoptera there is no discernible geographic variation of enzyme genes, yet there is a considerable amount of hybrid inviability in crosses between different populations [12]. To what class of DNA these inviability (or in other cases sterility) factors belong is unknown. It could be repetitive DNA, but it could also be a category of single-coding genes, genes outside the class of enzyme genes revealed by electrophoresis.

It has become fashionable to say that regulatory genes are the genetic agents of speciation. Hedrick and McDonald [13] rightly remark that "changes in genetic regulation would be the favored genetic strategy for a population adapting to a sudden and substantial environmental change." For no other kind of population would this be so true as for a founder population undergoing a genetic revolution.

When finally identified, I am quite certain that the genetic speciation factors will consist of several classes. For instance, they might well be different in species in which the premating isolating mechanisms developed first, from species in which postmating mechanisms (eg, sterility) developed first.

That a breaking up of the internal balance of the genotype may be an important component of peripatric speciation is indicated by the frequency by which the resulting neospecies give rise to further new rapidly speciating founder populations. As each neospecies becomes adapted to the new environment beyond the ancestral species range, it becomes especially suited to serve as source of colonists of new environments. The distribution of the neospecies of Spalax and Proechimys, as well as the successive speciational steps in Hawaiian Drosophila, illustrates this process particularly well.

NONGEOGRAPHICAL KINDS OF SPECIATION

Although no one can any longer question the importance of the various categories of allopatric speciation, and although many zoologists think that they are the prevailing modes of speciation among animals, it must not be forgot that other forms of speciation have been proposed, and I would like to make now some comments on them.

A saltational origin of new species was postulated long before there was any theory of evolution and was popular among many of Darwin's contemporaries, such as T. H. Huxley, Galton, and Koelliker. Bateson, de Vries, and most early Mendelians favored it, and so did several later paleontologists, including Schindewolf, and the geneticist Goldschmidt with his "hopeful monsters." The crucial aspect of this postulated type of speciation is the typological concept of the production of a single individual, which then becomes the progenitor of a new species.

It took a long time before the virtual impossibility of this happening in a sexually reproducing species was understood. The demonstration that de Vries's new Oenothera species were not species, greatly helped to clear the air. There are, however, three potential processes by which such instantaneous speciation can indeed occur. Two of these are autopolyploidy in a sexual species, and any kind of macromutations in uniparentally reproducing organisms. There is still much uncertainty as to what role these two types of speciation play. Autopolyploidy presumably is quite unimportant.

There is, however, a third and reasonably common type of instantaneous speciation, and that is the production of stabilized species hybrids. This was first discovered for plants in which the doubling of the chromosome number of more or less sterile species hybrids may produce *allopolyploids,* capable of sexual reproduction. Such polyploidy is very common in plants, but, as White has once more clearly demonstrated in Chapter 8 of his "Modes of Speciation," it is exceedingly rare in animals. What happens, however, not infrequently in animals, is that a product of hybridization shifts to uniparental reproduction and forms a new parthenogenetic or thelytokic species [1, Chapter 9]. There are some special cases of persistent hybridity, typified by Rana esculenta and some American freshwater fishes, in which the paternal chromosome set is invariably lost in meiosis, and restored by mating with the respective paternal species. As successful as some of these stabilized species hybrids may be in the short run, they appear to have little long-range evolutionary success.

SYMPATRIC SPECIATION

The form of nongeographic speciation that is considered most probable and most frequent by many students of speciation is *sympatric speciation* by host specialization. Since this type of speciation will be dealt with by other speakers, I will keep my comments short. Sympatric speciation is favored particularly by authors who have little experience with geographical variation. I must have in my files some 15 or 20 reprints proposing sympatric speciation in which the authors have ignored even the most elementary potential objections to their schemes. I tabulated in 1947 [14] the difficulties encountered by hypotheses of sympatric speciation, but most of these difficulties are not at all taken into consideration by these modern authors. The difficulties encountered by the proposals of sympatric speciation have recently also been emphasized by others [15].

I am sometimes accused of having always adamantly denied any possibility of sympatric speciation. A glance at my previous publications [eg, 16, p 215; 3, pp 449–480; 2] shows that this accusation is not justified. I have always taken the possibility of sympatric speciation seriously, but what I objected to was the superficial manner by which the claims of sympatric speciation were advanced. In nearly all cases listed in the literature, allopatric speciation is as probable an explanation for the particular cited example or more so than sympatric speciation. Let me mention only one case. The existence of sympatric groups of closely related species of beetles on small oceanic islands (St. Helena, Rapa) would seem, at first sight, conclusive proof of sympatric speciation. But when one looks at these cases more carefully, one is struck by three facts: First, most of these species are not at all host-specific; second, they are flightless; and third, these are volcanic islands, crisscrossed by old lava flows. Carson [17] quite rightly points to the parallel situation with speciation of Drosophila on the island of Hawaii. As he says: "Oceanic volcanic shields invariably have a long period of violent growth wherein lava flows from the rifts periodically destroy the forests in haphazard mosaic patterns (Kipuka formation). These continuing cycles of building, destruction, isolation, and recolonization would be expected to impose a pattern of spatial isolation and founder effects even over very short distances. These influences would appear to leave ample opportunity for allopatric speciation." I fully agree with this interpretation. I do not object to the proposals of sympatric speciation, but I deplore it when authors completely ignore alternate explanations.

Among the numerous claims of sympatric speciation none has been promoted with more vigor than the derivation of an "apple species" of Rhagoletis pomonella fruit flies from a hawthorne population [18]. However, not even this case is securely established, and serious questions have been raised about it. Even the nearest relative of pomonella, the Rhagoletis suavis group, clearly exhibits a pattern of allopatric speciation [15]. Furthermore, as I pointed out in 1947, the existence of species-rich genera of host specialists by no means proves sympatric speciation. It merely means that they have more niches available than generalists.

The conclusion we can draw from the literature is that sympatric speciation by a shift of host preference is a possibility but that it has not yet been conclusively substantiated. After analyzing the documentation, Paterson [15] concludes: "I am yet to be convinced that sympatric speciation has ever occurred." I likewise doubt that any of the putative cases of sympatric speciation, reported in the recent literature, will survive a truly critical analysis. This includes the numerous cases of postulated sympatric speciation in freshwater.

Bush et al [19] have suggested that in the Equidae (horse family) the social-reproductive structure of the small herds was inducive to sympatric-stasipatric speciation. This ignores the fact that the young individuals of one sex always join other herds, or establish their own herd, and that this results in very active gene flow. The distribution pattern of related species of horses, asses, and zebras documents allopatric speciation quite graphically. The rather drastic chromosomal repatterning that seems to have occurred during some of the speciation events in this group was surely much more easily achieved in a peripherally isolated founder population than inside the population continuity of the parent species.

STASIPATRIC SPECIATION

The fact of chromosomal speciation poses a problem. If a chromosomal rearrangement produces superior heterozygotes, its origin will lead to polymorphism, but not to speciation. If it produces deleterious heterozygotes, these will be eliminated by natural selection in a large panmictic population before they have any chance to attain the relatively high frequency required before the production of the new type of homozygotes becomes probable. Yet, heterozygously deleterious chromosomal rearrangements must sometimes be able to pass through the bottleneck of heterozygosity: How else could we explain the frequent presence of chromosomal differences between closely related species, which are deleterious in heterozygous condition? I have previously [2, pp 310–319] discussed this problem in detail and have proposed that these chromosomal reconstructions are best achieved peripatrically in small inbred founder populations, where homozygosity of the new chromosomal type can be attained in two generations.

White [20], by contrast, has proposed a different process, which he calls *stasipatric speciation*. According to this model a new gene arrangement may turn up anywhere within the range of a species, gradually become more and more common, and spread out from the place of origin, until the number of heterozygotes has reached such a high frequency that homozygotes will finally occur. White states clearly the two reasons that induced him to reject the model of chromosomal speciation in peripheral isolates. First, White could not conceive of any isolating barriers on continents, and secondly, he believed that genetic revolutions in founder populations are restricted to genetic changes. Both of these assumptions are misconceptions, as I have pointed out previously [2, and above].

What is involved is the question of primacy. White has stated: "More and more it appears as if chromosomal rearrangements . . . have played the primary role in the majority of speciation events" [1, p 336], and he makes it quite clear that by "primary" he means preceding in time any kind of spatial segregation of the speciating individuals. For White, the chromosomal change comes first and the new chromosomal type creates its own deme. For me, the new chromosomal type has no chance to achieve homozygosity, unless it occurs in an isolated deme; hence the isolation of the deme comes first. In other words, peripatric speciation is involved.

Some of White's own followers are confused about the problem of chronological primacy. When Bush et al describe chromosomal speciation as follows: "A karyotypic mutation that has become fixed in a given deme can act as a sterility barrier" [19, p 3945], they imply that the existence of a (an isolated) deme precedes the fixation of the karyotypic mutation, but such a process would clearly be peripatric and not stasipatric speciation.

The causal relation between chromosomal rearrangements and genetic revolutions in peripheral isolates is asymmetrical: The homosequential Drosophila species of Hawaii prove that not every speciation in a peripheral isolate is correlated with a chromosomal rearrangement, on the other hand there is much evidence to suggest that chromosomal restructuring occurs preferentially, if not always, in peripheral isolates. I have been unable to find any evidence to substantiate White's claim of a temporal precedence of chromosomal over geographical factors. The proof he offers, that the distribution maps of currently allopatric species indicates that the most primitive species are peripheral, is inconclusive since both Ken Key and I looking at these very same maps, have concluded exactly the reverse.

The model of stasipatric speciation has one great virtue: It can be tested readily. According to White a new center of stasipatry (a new chromosomal heterozygote) may turn up anywhere in the species range. Consequently, in any widespread species subject to chromosomal speciation, one should find numerous enclaves of new chromosomal species with smaller or more extended ranges, like oases in a desert. Actually, such a distribution pattern is totally unknown. New chromosomal rearrangements that can serve as isolating mechanisms invariably occur peripherally parapatric, as is consistent with the founder model, but not with the stasipatric one. I am sorry to have to say that the stasipatric model has been completely falsified by this test. If we look at cases of chromosomally speciating mammals, like Spalax, Proechimys, or Thomomys, to mention only three genera, the patterns of the new chromosomal species are invariably allopatric. The same is, of course, true even for the species groups of Australian morabine grasshoppers. Their ranges are invariably allopatric or parapatric, as one would predict from the peripatric explanation, and no oases or enclaves are ever found, as the stasipatric model would predict. As a consequence, White's own co-worker in the grasshopper work, Ken Key, rejects the stasipatric hypothesis, and has accepted allopatric speciation for the origin of species in these animals [8]. The adherents of stasipatric speciation may object to my claim that no cases of stasipatric enclaves are known anywhere and may cite the rodent Ellobius talpinus. This species has indeed 54 chromosomes from the Crimea to Mongolia but also a 31 chromosome population seemingly in the middle of that area. However, this aberrant population occurs in a single isolated valley in the Pamir Mountains, which is peripheral in the total species range.

Key [21] has rightly pointed out that in species with low dispersal facilities (and/or rather specific habitat requirements), one can also find "internal" founder populations—that is, populations colonizing previously unoccupied spots within the species range. These new founder populations may be quite isolated for some time and undergo peripatric speciation during this isolation exactly like peripherally isolated populations. The chromosomal races of Mus musculus in the Alpine valleys may be due to such a process. Mouse populations in these valleys are furthermore confined (as commensals) to isolated human habitations. In these small founder populations the occurrence of chromosomal fusion homozygotes is made possible by stochastic processes, and if the new karyotype is adaptively superior, the newly formed population will be able to spread even against other karyotypes. The same interpretation is applicable to the origin of new mouse karyotypes at other localities.

SPECIATION AND MACROEVOLUTION

I am delighted over the inclusion of three papers in this program that deal with the findings of the paleontologists in relation to speciation. Needless to say it flatters me that both Eldredge and Gould [22] in their theory of *punctuated equilibria* and Stanley [23] in his of *rectangular evolution* base their models on my 1954 theory of peripatric speciation. There are numerous reasons why founder populations are, so to speak, preadapted for a drastic ecological reorientation and for the incorporation of evolutionary novelties. These have been so well discussed by these and other recent writers that I need add nothing further to what I have said above.

Some of these paleontologists, unfortunately, have telescoped the rapid process of peripatric speciation into a single moment. For a paleontologist "50,000 years are like a moment," as Gould has once said. That the genetic revolution is a gradual process, no matter how rapid it is, is concealed by the paleontologist's terms "rectangular evolution" or "punctuated equilibria." If one applies a higher magnification than is available to the paleontologist, one perceives that one is *not* dealing with a "hopeful monster"—like typological phenomenon, but with gradually changing normal diploid, sexually reproducing populations, no matter how small. Indeed, 50,000 years would be time enough for a whole series of sequential gradual speciations. Bock's [24] analysis of the processes by which the remarkable adaptive radiation of the Hawaiian honeycreepers (Drepanididae) has occurred, and the even more remarkable and detailed analysis of speciation in the Hawaiian drosophilids by Carson and co-workers have confirmed the gradual nature of these shifts beyond any doubt.

This recognition also proves that the usual alternative, phyletic gradualism vs punctuated equilibria, is somewhat misleading. The polarity is not between continuity and a mutational discontinuity, but between evolutions at different rates. Even the replacement of one chromosomal type by another, one that has to pass through a heterozygous condition, is a gradual process. Furthermore, as I pointed out before, there are intermediate conditions between founder populations and widespread, populous species. I think, for instance, of the populations in Pleistocene refuges and other cases of a drastic reduction of population size. In view of this evidence one can establish the very general rule that the rate and the scope of speciation are inversely proportional to population size. From a purely genetical point of view this was already realized by Haldane [25], who showed how evolutionarily inert large, widespread, populous species are.

The problem of species selection, articulated first by Charles Lyell and perhaps other pre-Darwin authors, appears now in a new light. Every successful neospecies is a potential competitor for one or several already existing species. This competition, if relatively mild, will lead to a finer partitioning of the ecological space and to a correlated character divergence, but if such competition is severe it may lead eventually to the extinction of one of the competitors. Such a process has been largely responsible for generating what is usually called evolutionary progress. The acorn barnacles are a well-studied case of species selection [26].

THE ECOLOGY OF SPECIATION

Time does not permit me to say much on this subject, except to point out that since speciation is as much a population phenomenon as it is one of genetic mechanisms, any aspect of the structure of populations is also of potential relevance to speciation. The botanists were the first to understand this, perhaps because plants have a so much richer repertory of modes of reproduction. Population size, as I mentioned already, is inversely related to rate of speciation. When J. Diamond and I (ms in preparation) assigned the species of Northern Melanesian birds to five classes, based on their proclivity for dispersal, we discovered that each of these five classes had a class-specific speciation pattern. High rates of dispersal mean high rates of gene flow, and such gene flow leads to the formation of widespread, relatively uniform populations and species.

As stated by Patton and Young [27, p 714], "Gene flow is a key element in the population biology of T(homomys) bottae and perhaps other gophers as well." The claims of the unimportance of gene flow, based on highly arbitrary and unrealistic assumptions made by some recent authors, have been rejected with sound reasons by Jackson and Pounds [28]. The uniformity of species is, of course, not caused solely by gene flow, but also by the regulatory system of species [2, p 300], but gene flow is an important factor in determining species structure and pattern of speciation. A comparison of the speciation patterns of marine species with high and of such with low larval dispersal confirms this most impressively. Great care must be exercised in all species of animals with a social population structure in which, for instance, all the young of only one sex may be expelled from the family group. Gene flow in such species is largely controlled by the expelled individuals. Factors such as this have, unfortunately, been entirely ignored in some recent speculations.

CONCLUSIONS

My report is largely a critical summary of a vast amount of speciation literature. It would be absurd to try to summarize it even further. However, I would like to emphasize once more that we will never achieve a balanced understanding of the process of speciation unless we accept two basic truths about speciation:

1. The principle of simultaneity. During speciation several processes, for instance populational and genetic ones, always proceed simultaneously. To present them as "either-or" choices reflects a failure to understand what really goes on in nature.

2. The principle of pluralism. There are multiple possible answers to every aspect of speciation. Only by comparing the modes of speciation of many, preferably all, groups of organisms can we discover the extent of variation in speciation patterns found in organic nature. In all cases, however, the attainment of reproductive isolation is the key feature.

ACKNOWLEDGMENTS

Walter Bock, Douglas Futuyma, and Robert Selander have read an earlier draft of this paper and have made numerous helpful suggestions. I am deeply indebted to their kindness.

REFERENCES

1. White MJD: Modes of Speciation. San Francisco: Freeman, 1978.
2. Mayr E: Populations, Species, and Evolution. Cambridge: Harvard University Press, 1970.
3. Mayr E: Animal Species and Evolution. Cambridge: Harvard University Press, 1963, pp 483–486.
4. Mayr E: Change of genetic environment and evolution. In Huxley J, Hardy, Ford EB (eds): Evolution as a Process. London: Allen and Unwin, 1954, pp 157–180. Also in Evolution and the Diversity of Life; Cambridge: Harvard University Press, 1976, pp 188–210.
5. Templeton has recently suggested to revive Galton's term *transilience* (1894) for the process I have referred to as *genetic revolution*. Galton, however, believed in blending inheritance and, therefore, postulated that evolutionary change depended on "sports" or "transiliences"—that is, on single deviant individuals that could not "regress backwards towards the typical centre" (p 368). The process suggested by Galton is so completely different from the rapid reorganization of populations during genetic revolutions that it would be altogether misleading to apply Galton's term to this gradual populational phenomenon.

6. Lewis H: Catastrophic selection as a factor in speciation. Evolution 16:257–271, 1962.

7. Haffer J: Speciation in Amazonian forestbirds. Science 165:131–137, 1969.

8. Key K: Species, parapatry and the morabine grasshoppers. Syst Zool 30, 1981, p 425–458.

9. Parapatric speciation is particularly favored by Clarke (1968) and Murray (1972): Clark B: Balanced polymorphism and regional differentiation in land snails. In Drake ET (ed): Evolution and Environment. New Haven: Yale University Press, 1968, pp 351–368. Murray JJ: Genetic Diversity and Natural Selection. Edinburgh: Oliver and Boyd, 1972. Mosaic distributions where rather different populations meet discontinuously (as in cases of area effects), are frequent in land snails. This does not lead to speciation, as shown by Cepaea memoralis, and numerous other examples.

10. Meise W: Rassenkreuzungen an den Arealgrenzen. Verh Dtsch Zool Ges 1928:96–105. Many of the suture lines go for hundreds or thousands of kilometers through regions that had been uninhabitable during the preceding isolation thus documenting their secondary origin.

11. Mayr E: Selektion und die gerichtete Evolution. Naturwissenschaften 52:173–180, 1965.

12. Oliver CG: Genetic differentiation and hybrid viability within and between lepidopterous species. Am Nat 114:681–694, 1979.

13. Hedrick PW, McDonald JF: Regulatory gene adaptation: An evolutionary model. Heredity 45(1):83–97, 1980.

14. Mayr E: Ecological factors in speciation. Evolution 1:263–288, 1947. (Reprinted in Evolution and the Diversity of Life; Cambridge: Harvard University Press, 1976, pp 144–175.)

15. Futuyma DJ, Mayer GC: Non-allopatric speciation in animals. Syst Zool 29:254–271, 1980. Jaenike J: Criteria for ascertaining the existence of host races. Am Nat 117:830–834, 1981. Paterson HEH: The continuing search for the unknown and unknowable: A critique of contemporary ideas on speciation. South Afr J Sci 77:113–119, 1981.

16. Mayr E: Systematics and the Origin of Species. New York: Columbia University Press, 1942.

17. Carson HL: Chromosomes and species formation. Evolution 32:925–927, 1979.

18. Bush GL: Modes of animal speciation. Ann Rev Ecol Syst 6:339–364, 1975.

19. Bush GL, Case SM, Wilson AC, Patton JF: Rapid speciation and chromosomal evolution in mammals. Proc Natl Acad Sci USA 74:3942–3946, 1977.

20. White MJD: Models of speciation. Science 159:1065–1070, 1968.

21. Key KHL: The concept of stasipatric speciation. Syst Zool 17:14–22, 1968.

22. Eldredge N, Gould SJ: Punctuated equilibria: An alternative to phyletic gradualism. In Schopf TJM: Models in Paleobiology. San Francisco: Freeman, Cooper and Co, 1972, pp 82–115.

23. Stanley SM: Macroevolution. Pattern and Process. San Francisco: WH Freeman, 1979.

24. Bock WJ: Microevolutionary sequences as a fundamental concept in macroevolutionary models. Evolution 24:704–722, 1970.

25. Haldane JBS: The cost of natural selection. J Genet 55:511–522, 1957.
26. Stanley SM, Newman WA: Competitive exclusion in evolutionary time: The case of the acorn barnacles. Paleobiology 6:173–183, 1980.
27. Patton JL, Yang SY: Genetic variation in Thomomys bottae pocket gophers: Macrogeographic patterns. Evolution 31:697–720, 1977.
28. Jackson JF, Pounds JA: Comments on assessing the dedifferentiating effect of gene flow. Syst Zool 28:78–85, 1979. Also Pounds JA, Jackson JF: Riverine barriers to gene flow and the differentiation of fence lizard populations. Evolution 35:516–528, 1981.

DISCUSSION

Comment and question by F. Ehrendorfer: The populations of a species are in a "front-line" position not only at the periphery of its distribution area but wherever they encounter unsuitable habitats within this area. If your "peripatric" model of speciation is applied to such internal situation, an obvious link to "parapatric" or even more or less "sympatric" models of speciation can be seen.

Mayr: As Key [21, p 20] has quite rightly remarked, one can recognize an "internal" allopatry in species that have vacant areas inside the overall species range, owing to vegetational or physiographic barriers. Species with low dispersal power may be able to establish founder populations in these vacant areas if they find suitable spots. This is consistent with all requirements of peripatric speciation. The origin of new karyotypic "species" within the range of Mus musculus in isolated human habitations in remote peripheral Alpine valleys may be an illustration of "internal" peripatric speciation. It is not parapatric speciation which, according to its defenders, is a rupture of population continuity through selection. Nor is it sympatric, because in the isolated area of the founder population there is no other population of the species. In questions of speciation one must think in terms of populations, and not in terms of gross geography.

Mechanisms of Speciation, pages 21–39

Plant Speciation

G. Ledyard Stebbins

INTRODUCTION

Flowering plants (angiosperms) are particularly favorable organisms for analyzing processes of speciation for the following reasons. First, many different genera, belonging to a diversity of families and orders and native to a great variety of natural habitats, can be and have been cultivated. Their species can be analyzed by quantitative examination of morphological, cytological, and biochemical differences, and similar analyses performed upon interspecific hybrids and their progeny. Second, their sedentary condition makes possible direct estimates in natural sites of population size, breeding structure, and within population differences in response to microhabitat differences. Third, many closely related species of plants are separated from each other by barriers of partial hybrid sterility, sufficient to permit their populations in nature to exist side by side and still pass the test of sympatry without destructive gene exchange, but able to yield segregating F_2 and later-generation progeny under cultivation. All stages can be found from populations that exchange genes freely, through those having varying restrictions to gene flow, and finally to fully isolated but closely related species [1,2].

During the past 60 years, plant geneticists have taken full advantage of the opportunities that these organisms afford. The pioneer research studies of Turesson; Clausen, Keck and Hiesey; Babcock; Müntzing; Sinskaja; Gregor; and others have contributed much to the development of a biological species concept, and are reviewed in several books: [1,3,4,].

Building upon this foundation, a second generation of research workers, including Ehrendorfer [5–7], Lewis [8–11], Verne Grant [2], De Wet and Harlan [12], and Zohary [13] have added information that led to preliminary analyses of the dynamics of the speciation process itself. During the past fifteen years, a third generation, taking advantage of newer techniques such as electrophoresis, measurement of nuclear DNA content, fine structure of chromosomes as revealed

TABLE I. Distinctive Characters of Plants That Affect Speciation

Character	Effects
Longevity	Sterile hybrids persist, "bottlenecks" of partial sterility easily overcome, rare fertile progeny of hybrids can succeed.
Simpler developmental patterns	Wide hybrids develop more easily, hybrid inviability and weakness less likely, polyploidy and other chromosomal changes more likely.
Lack of sense organs	Ethological isolation impossible, speciation can be based upon complex flower-pollinator interactions.
Hermaphroditism and autogamy	When present, facilitates "budding" of new species via uniparental origins of populations and "founder effect."

by banding techniques, and computer-based analyses of population dynamics, have carried even further basic research on the species problem. They include RW Allard and his associates [14,15], MD Bennet [16], GD Carr [17,18], RB Flavell and his associates [19], LD Gottlieb [20–22], J Greilhuber [23,24], JL Harper [25], SK Jain [26,27], DA Levin [28–31], GE Marks and G Schweizer [32], W Nagl [33–36], HJ Price and K Bachmann [37], H Rees and M Hazarika [38], and OT Solbrig [39]. The ultimate significance of this most recent research cannot yet be evaluated. I believe, however, that satisfying answers to the questions: "How can barriers of reproductive isolation and consequent biological species evolve?" and "What are the dynamics and time spans of the speciation process?", are not far off.

DISTINCTIVE BIOLOGICAL PROPERTIES OF PLANTS

The purpose of this review is to show how research on plant species can contribute to a better understanding of the speciation process in organisms generally. First, therefore, a summary of the most distinctive characteristics of plants is necessary. They are summarized in Table I. Most of them contribute toward blurring the boundaries between biological species. The open system of growth and the absence in plants of carryover via messenger RNA that is coded and transcribed by maternal genes but is translated into protein during embryonic development of the offspring, a phenomenon that is frequent in animals, produces in plants a more direct interaction between environment and organism during embryogeny. It also reduces the probability of a clash between gene products of maternal origin and those coded by genes of the offspring. In animals, the clash between products coded by an F_1 hybrid and those coded by its segregating F_2 offspring may be an important cause of hybrid breakdown, which is often a highly significant barrier to gene exchange [40]. Weaker barriers of reproductive isolation cause plant hybrids to exist that encompass an entire spectrum of gametic failure from zero to 100%. The plant geneticist must often make arbitrary decisions as to what degree of gametic sterility is to be regarded as an effective barrier to gene exchange. Comparative rarity of highly effective barriers of prezygotic reproductive isolation, which in animals is often reinforced by direct selection against the production of weak or sterile hybrid offspring, also contributes to a blurring of boundaries between species. Finally facultative autogamy has two significant effects. In many groups it enables a single individual, derived from a hybrid progeny, which has acquired a rare combination of genes and chromosome segments that render it both vigorous and fertile, to give rise to a new evolutionary line that can qualify as a new diploid species of hybrid origin [2,3]. Even more often, segregating progeny from fertile intraspecific hybrids between closely related biotypes can form constant, fertile subpopulations that temporarily mimic sibling species. The excessive splitting that some plant taxonomists have practiced on groups like Panicum sect. Dichotoma and Draba subg Erophila is due largely to this phenomenon.

All of these properties have contributed to the skepticism with which many highly competent plant taxonomists view the biological species concept as recognized by zoologists [41–44]. Their skepticism is well founded, particularly when one considers the fact that reproductive isolating mechanisms cannot be studied directly in the great majority of the species that form the richest communities in the world, those of tropical rain forests; and that in them even indirect evidence about these barriers is very difficult to obtain. Nevertheless, whatever may be the most convenient way of delimiting species for purposes of classification and floristics, exploration of reproductive isolating barriers must remain an essential part of research on species and speciation in plants just as in other organisms. This is because unless effective barriers have evolved between species that are sympatric over a wide area, evolution of different ways of exploiting similar environments cannot continue. In plants as in animals, a high proportion of diversity is generated by such differential exploitation.

The reader will note that no reference has been made to a characteristic that many evolutionists regard as the most distinctive of plants: polyploidy. This omission is intentional. The present discussion aims to show how knowledge about plant speciation can generate guidelines for understanding the origin of species in all kinds of organisms. Consideration of polyploidy is irrelevant to this aim. For recent discussions of polyploidy and plant speciation, see the papers edited by Lewis [43].

PLANT SPECIATION PATTERNS THAT RESEMBLE THOSE OF ANIMALS

How effectively can one extrapolate from information about plant speciation to formulate hypotheses about speciation in general? This question can be answered best by considering certain groups of plants that in some of their biological characteristics resemble animals more nearly than do the majority of flowering plants.

The most numerous genera of this kind consist of annual species that are self incompatible, and therefore obligately outcrossed. In them, there is no overlap between generations; fecundity based upon seed production has a high adaptive value (as contrasted with long-lived, vegetatively reproducing perennials, in which low fecundity can be balanced by great vegetative vigor); and in which population structure is determined by spatial distribution of individuals more than by internal restrictions that are prevalent in autogamous annuals. Two analyses of genera of this kind are outstanding: those of Clausen and his associates [1,44] followed by Carr [17,18,45] on the tarweeds (subtribe Madiinae) of the sunflower family (Asteraceae, tribe Heliantheae) and of Grant [2] on the leafy-stemmed species of Gilia. In genera of tarweeds like Layia there are species having rich gene pools that have generated distinctive ecotypes or subspecies, as in L platyglossa, as well as others, such as L carnosa, that are restricted to specialized habitats and have correspondingly restricted gene pools. Species are for the most part easily recognized morphologically, but sibling species occur, such as L platyglossa and L septentrionalis. The distinctness of species appears to be well correlated with number and kinds of chromosomal differences that separate them. In another genus, Holocarpha, most of the barriers of reproductive isolation, consisting chiefly of chromosomal differences, separate a cluster of sibling species, imperfectly explored, but probably large. In Calycadenia, Carr [17,18] has found that one species complex, that of C pauciflora-ciliosa, is subdivided into six partly sympatric races or semispecies that differ in chromosome number and segmental arrangement but are morphologically very similar, while two closely related species, C multiglandulosa and C hispida, are subdivided into morphologically different allopatric races that are similar in chromosome number and structure. After considering several hypotheses, Carr was unable to suggest an explanation for this differential rate of chromosomal as compared to morphological differences. His results are, however, highly important in three respects. First, they support other research on plant species indicating that complete reproductive isolation is rarely achieved by occurrence and establishment of one or a few chromosomal rearrangements. Second, they show that even within a group of closely related, outcrossing and similarly adapted species morphological divergence can in one evolutionary line proceed more rapidly than chromosomal change, while in another parallel line the reverse is the case. Finally, Carr [18] found individuals heterozygous for a chromosomal translocation in 12.5% of the populations of the multiglandulosa-hispida complex that are not differentiated from each other as populations, but only 6.9% of chromosomally heterozygous individuals in the differentiated and partly isolated populations of the C pauciflora-ciliosa complex. In other words, the spontaneous occurrence of chromosomal changes in individuals of a population is not correlated and is perhaps inversely correlated with the chromosomal differentiation of populations from each other. Carr concludes from his data that in Calycadenia, natural selection rather than chance plays the dominant role in the diversification of races or species at the level of populations.

Another group of genera resembles animals in a different way. These have specific, delicately balanced interactions with other organisms so that feedback interactions favor coevolution. The most numerous of them have flowers of complex structure, adapted in highly specific ways to pollination by specialized insects, particularly Hymenoptera. Coevolution between flowers and pollen vectors is in many ways comparable to that between predators and prey or between parasites and their hosts. With respect to complexity and specificity it is also comparable to interaction between males and females in courtship. In all of these examples, the more specific and complex is the interaction, the more it promotes the evolution of numerous, diverse variants upon a theme that become stabilized as new species in both of the kinds of organisms that are coevolving with each other. The most outstanding example of this kind involves the fig genus, Ficus, and wasps of the genus Blastophaga that pollinate it [46,47].

More than 600 species of Ficus are known, and almost every one of them is pollinated by a different species of Blastophaga. This is one of the most remarkable examples in the plant kingdom of extensive speciation. It is still more remarkable when we consider the fact that 40 other genera of the family Moraceae, to which Ficus belongs, have only from one to ten species, and the second largest genus in the family, Artocarpus, has only 50 species.

Another example is the family Orchidaceae, which contains more than 30,000 species, the largest of any plant family. Its largest genera, Epidendrum and Pleurothallis, both contain about 400 species. Among tropical epiphytic orchids, as many as 30 species belonging to a single genus (Pleurothallis) have been found on a single tree [LB Smith, personal communication]. Speciation in Orchidaceae is comparable in activity to that in a genus like Drosophila. I have been told by Dr. Herman Spieth [personal communication] that genera of Diptera that lack the elaborate courtship behavior found in Drosophila have many fewer species. Accurate data need to be obtained, but at present a correlation between highly complex and specific biological interactions and a tendency for extensive speciation appears to be highly probable.

VARIATION AMONG GROUPS OF ANGIOSPERMS WITH RESPECT TO THE NUMBER, DIVERSITY, AND DISTINCTNESS OF SPECIES

Plant cytogeneticists and taxonomists have long been aware of the fact that the number, diversity, and particularly the distinctness of species one from the other differ greatly from one group of flowering plants to another [3]. This topic has been discussed in depth by Grant [2]. The most obvious correlation is between growth habit and speciation. Woody plants (Table II) tend to have fewer species per genus, and reproductive isolating mechanisms appear to be poorly developed as compared to annual herbs (Table III). Perennial herbs are more diverse but in general intermediate (Table IV). Are differences in growth habit themselves responsible for differences in number and distinctness of species, or must we look for other differences that are correlated with growth habit?

With respect to numbers of species and development of prezygotic isolating mechanisms, the examples of Ficus and the Orchidaceae, just cited, as well as many others indicate that with respect to prezygotic barriers, not growth habit by itself but correlated factors are involved. Is the same explanation correct for the differential development of postzygotic barriers, particularly hybrid sterility? Unfortunately, no data exist by which woody groups might be identified in which hybrids between recognized species are usually or always completely sterile, even if such groups exist. Because of the long generation times involved, relatively few artificial hybridizations have been made between woody species, except for fruit trees, ornamentals, and a few of the most important commercial trees. Nevertheless, in many woody genera indirect evidence, in the form of aneuploid series of chromosome numbers, extensive polyploidy or both of these features, suggests the existence of well developed barriers. Examples are Astroloma and Leucopogon in the Epacridaceae [48] and several genera of Rosaceae, such as Cotoneaster, Crataegus, and Rubus. None of these genera contain trees of climax forests. They are best adapted to pioneer habitats, an adaptation that favors frequent colonization of new areas, and occasional or frequent drastic fluctuations in population size. Among annual species, exceptional examples exist of species that have undergone little diversification during tens of thousands or millions of years. Examples are Plantago ovata, Polygonum arifolium, and P sagittatum, Microsteris gracilis (Polemoniaceae) and Alchemilla occidentalis. The two species of Polygonum live in mesic habitats in which great fluctuations in population size would not be expected. With respect to the others, either direct casual observation suggests that their populations fluctuate less in size than do those of other annuals; or their occurrence in relatively stable mesic habitats points in the same direction. Consequently, a reasonable hypothesis is

TABLE II. Modal Speciation Patterns in Flowering Plants, Trees, and Shrubs

Modal population structure	Morphological divergence between species	Reproductive isolation
Large populations	Relatively great	Prezygotic barriers rare
Constant size	Sibling species rare	All barriers often imperfectly developed
Outcrossing predominant	Vegetative parts, fruits more than flowers	Semispecies common
		Polyploid species per genus usually less than 30%

TABLE III. Modal Speciation Patterns in Annual Flowering Plants

Modal population structure	Morphological divergence between species	Reproductive isolation
Small or large populations Often drastic fluctuation in size Partial or complete autogamy frequent	Great or small, sibling species frequent Emphasis upon flowers, fruits, and seeds or both	Prezygotic barriers occasional Hybrid inviability and sterility common, semispecies rare Polyploid species per genus modally 10–30%

TABLE IV. Modal Speciation Patterns in Flowering Plants, Perennial Herbs

Modal population structure	Morphological divergence between species	Reproductive isolation
Large or small populations Usually constant size Outcrossing obligate, occasional or autogamy predominant	Usually great, but sibling species occasional All characters affected, but sometime emphasis upon flowers, fruits or both	Prezygotic barriers slightly developed in families Semispecies occasional Polyploid species per genus modally 20–60%

that the distinctness of species and strong development of hybrid sterility among annual flowering plants is due not primarily to their annual growth habit but to frequent colonization of new habitats accompanied by drastic fluctuations in population size. Another important contributing factor is autogamy, as was first pointed out by Baker [49]. This hypothesis is well supported by Grant's [2] comparison between the moderate amount of speciation found in the outcrossing annual species of Gilia with the much greater proliferation of species in the facultatively autogamous species belonging to the G inconspicua group [2]. I conclude from these comparisons that if annual species are not subjected to frequent fluctuations in population size, accompanied by colonization of new, ecologically distinctive sites, their rate and frequency of speciation will be little if any greater than it is in outcrossing herbaceous or woody perennials. Regardless of growth habit, one might expect facultative autogamy to stimulate speciation, because of the opportunity it affords for reducing or eliminating gene flow between sympatric populations.

If these deductions are correct, then we might expect that one or more of these factors would stimulate speciation in woody groups as well as in herbs. With respect to morphological and ecological diversification, this is clearly evident in some woody genera of California, even though their species have not become differentiated chromosomally. A particularly good example is Ceanothus subg. Cerastes, in which taxonomists recognize 20 species, even though all of them can form vigorous, fertile hybrids in both the F_1 and F_2 generations [50]. In nature, several species pairs (C crassifolius-megacarpus; C cuneatus-prostratus; C cuneatus-jepsonii) exist sympatrically in large populations but still remain distinct, so that according to the biological species concept at least six distinct species can be recognized. This subgenus has been carefully analyzed by Nobs [50] from morphological, anatomical, and cytogenetic data. Its fossil record, reviewed by Nobs from the data of Axelrod, is scanty, but nevertheless provides a yardstick to estimate periods of speciation and of species stability.

The species can be divided into three principal categories. Three of them, C crassifolius, megacarpus (of which C insularis may be only a subspecies), and C verrucosus are confined to coastal Southern California, are easily distinguished from each other, and two of them, verrucosus and megacarpus share characteristics with the much larger subgenus Euceanothus. Three species, C Greggii, cuneatus, and prostratus form a second category. They are widespread in interior California, or (C Greggii) Nevada, Utah, Arizona, New Mexico, and Mexico. The remainder that form the third category, are all more or less highly localized in the Sierra Nevada and the coast ranges of central and northern California. The concentration of highly localized, endemic populations is particularly great in the coast ranges north of San Francisco Bay. Many of them are restricted to rock formations that yield distinctive soil types, particularly serpentine outcrops that were first exposed during the Pleistocene epoch and volcanic flows of upper

Pliocene age. The complex pattern of distinctive rock and soil formations is highly correlated with the equally complex pattern of localized populations in Ceanothus subgenus Cerastes. Moreover, during the Pleistocene and Recent epochs the advance and retreat of ice sheets far to the north caused the climate of central California to fluctuate greatly from mesic to xeric and back again. The recent speciation of Ceanothus subgenus Cerastes in this region, giving rise to the third group, probably reflects the stimuli produced by edaphic diversity, fluctuating climatic change, and adaptation of populations to sites that were affected most strongly by variation in physical factors. The phylogenetic chart (Fig. 1), compiled from the data of Nobs and Axelrod, supports the concept of species stability in response to environments that remain constant through time, and diversification in response to environmental diversity in space and time, plus the added stimulus of crossing between differently adapted populations. This same region in the North Coast Ranges is the scene of extensive speciation in Calycadenia [17], Streptanthus [51], *Hesperolinon* [52], and probably other groups.

SEVEN QUESTIONS AND TENTATIVE ANSWERS TO THEM

Hopefully, the examples just reviewed demonstrate that speciation in plants is sufficiently similar to that in animals so that conclusions of general significance can be drawn from the plant evidence. In this section tentative answers are suggested to seven questions that evolutionists are now asking frequently.

1) How important is sympatric speciation in diploid cross fertilizating organisms? In a recent review of the subject, Bush [53] concluded that wide allopatric separation, which has been emphasized by proponents of the geographic theory of speciation, is not necessary for the origin of species. On the other hand, populations within and between which there is continuous gene flow are highly unlikely to generate new species. Some degree of spatial separation is very important if not absolutely necessary, except for speciation by parasites that are becoming adapted to new hosts. The plant evidence is fully compatible with this intermediate point of view. In all of the examples cited in this paper, except for those that involve autogamy and uniparental reproduction, species that appear to have evolved recently are either still separated from each other by at least a

---

Fig. 1. Diagram showing the probable phylogeny of shrubs belonging to the genus Ceanothus, subg. Cerastes (Rhamnaceae). Data from Nobs [50]. The time scale is that of the upper Tertiary and Quaternary Periods. Each millimeter (on the 5 × 7 print) represents about 200,000 years. The vertical lines represent species that have remained essentially constant during the time period represented by the length of the line. Solid circles represent the approximate age of fossil leaves of these species. The length of the nearly horizontal lines does not express exactly the morphological distance between

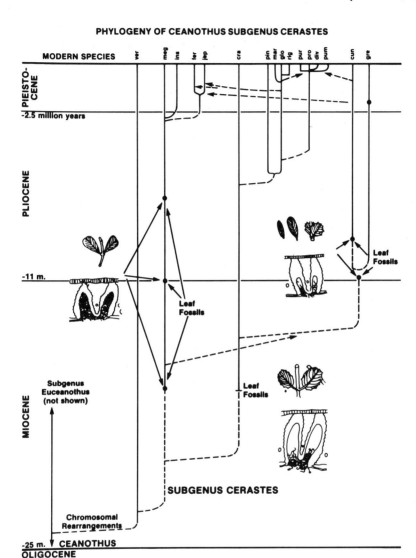

PHYLOGENY OF CEANOTHUS SUBGENUS CERASTES

species; for that purpose a 3-dimensional or 4-dimensional diagram would be required. Note that the slight upward tilt in these lines represents periods of about 100,000 years during which speciation could have taken place. Three sets of drawings and cross sectional diagrams are included: These show gross leaf morphology and cross sectional leaf anatomy of species having each one of the major types of stomatal crypts. The broken lines that terminate in arrows show probable influences of hybridization. Abbreviations of the modern species are as follows: ver,verrucosa; jep,Jepsonii; glo,gloriosus; div,divergens; meg,megacarpa; cra,crassifolia; rig,rigidus; pum,pumilus; ins,insularis; pin,pinetorum; pur,purpureus; cun,cuneatus; fer,Ferrisae; mar,maritimus; pro,prostratus; gre,Greggii.

short distance, or their present contacts appear to be secondary. Nevertheless, the great majority of these presumably recent species live within a few kilometers of supposed parental species. In outcrossing populations of plants, whether or not sympatric speciation can be recognized depends largely upon how the term sympatry is defined. Even if not absolutely necessary, spatial separation for a few generations strongly promotes speciation, and has usually occured at some time during the process.

2) What is the relative frequency of speciation by fission as compared to budding? Fission is defined as allopatric separation, preceded and followed by approximately equal divergence, so that two species result that are about equidistant from an extinct common ancestor. Budding is the process emphasized by Eldredge and Gould [54] and Stanley [55] in promoting the concept of punctuated equilibria. According to this concept each new species arises as a "bud" that has become separated via migration from the parental species and differentiates during and after the migration process. The parental species persists unchanged for a varying length of time, and may even become sympatric with its offshoot, daughter species.

The phylogeny of Ceanothus, subgenus Cerastes suggests that at least the majority of its species have originated by budding. Most of the Pleistocene or recent species of north central California are best explained either as buds from the widespread C prostratus or as stabilized segregates from hybridization between C prostratus and either C gloriosus or C cuneatus. Every fossil leaf that can be assigned definitely to subgenus Cerastes can be referred to a modern species. Even those from the Miocene Tehachapi flora, 17 million years old, are clearly within the morphological range of either one of two modern species: C megacarpus or C crassifolius. Because these two old, stabilized species form fertile hybrids with even the most strongly differentiated and localized species, such as C purpureus, the hypothesis of hidden genetic changes in untenable.

In another genus in which recent speciation has been recognized, Clarkia, the localized C lingulata came from an isolated "bud" of the more widespread C biloba [56]. Localized species of Clarkia found in the foothills surrounding the southern end of the San Joaquin Valley are offshoots by budding from the widespread C unguiculata [57–59]. I do not know of any examples that are best explained according to the concept of fission as defined above.

3) What are the rates of speciation? How many generations are necessary for a population that has become established in a new site to evolve effective barriers of reproductive isolation from its parental species?

There is no single answer to this question. Some examples from higher plants show that the fixation in a newly founded population of two or three structural rearrangements of chromosomes provides a sufficiently strong barrier of reproductive isolation to maintain the integrity against gene flow in the newly founded population. An example of this kind, now classic, is *Clarkia lingulata* [56,60].

This species became effectively isolated from its parental species, C biloba, by virtue of the establishment of new chromosomal rearrangements. This process could have been completed in ten to 20 generations, and therefore during the same number of years. Speciation at this almost instantaneous rate could easily be the model for annual plant species that are facultatively autogamous. On the other hand, indirect evidence suggests that for outcrossed, self-incompatible species longer time spans are required. Two examples from the Asteraceae, tribe Madiinae are suggestive. The first one consists of three closely related species, Layia Jonesii, L Munzii, and L leucopappa [1]. Their populations are allopatric, being separated from each other by distances of 60–65 km. Cytogenetically, they must be considered as semispecies rather than full species, since they produce vigorous partly fertile hybrids. The same degree of postzygotic reproductive isolation has evolved between the "races" of the Calycadenia ciliosa-pauciflora group [17,45]. These races are to a large extent allopatric, but sympatric along border zones. For both of these complexes to reach their present state (clusters of weakly isolated semispecies) must have required hundreds of generations. Subpopulations derived from any one of these species, colonizing a new locality, could achieve full species status by the fixation of two or three more structural arrangements that could be established in them after ten or twenty generations. Consequently a reasonable deduction from these examples is that among facultatively autogamous annuals, new species can evolve and usually have evolved in twenty years or less, while for outcrossing annuals the time required would be more like 100 to 500 years. The latter time span is reasonable for many perennial herbs. For woody plants that have much longer generation times and much less commonly establish new allopatric populations, time spans of 1,000 to 5,000 years appear to me to be more plausible.

Since even 5,000 or 10,000 years is too short a period to be recognized or measured by paleontologists or geologists, my conclusion is that in flowering plants, normal rates of speciation may be so slow that the natural process cannot be observed and followed by contemporary evolutionists, but that it has always been so rapid as to be instantaneous according to the geological time scale for any period older than the Quaternary Period (Pleistocene and Recent epochs).

4) What role have chromosomal rearrangements played in the origin of species? In his review of this topic, Grant [2] cites many examples from both perennial and annual herbs in which meiosis of F_1 interspecific hybrids reveals heterozygosity for chromosomal rearrangements. Furthermore, many F_1 interspecific hybrids have nearly normal meiosis but, nevertheless, high pollen sterility. Since in some instances polyploids derived from such hybrids prove to be fertile, I concluded [3] that speciation can often occur via fixation of a large number of chromosomal differences too small to be detected by the usual observations of meiosis in hybrids. Recent analysis of gene structure and of repetitive, non-coding DNA that is interspersed between or even within genes

suggests that transposition of relatively short sequences of nucleotides from one chromosome to another occurs commonly [61,62], and that such sequences diverge rapidly from each other [63]. These observations strengthen greatly the case for cryptic structural differences as contributors to reproductive isolation. The term "cryptic structural differences" may now be discarded and replaced by a more specific term, *submicroscopic structural differences*. The hypothesis I presented in 1950, that sterility of hybrids depends to a much greater degree upon the number of chromosomal arrangements for which they are heterozygous rather than the size of individual arrangements, is still valid. Hence, when all of these facts are considered, the conclusion may be reached that the origin of the majority of plant species is associated with chromosomal repatterning.

5) How easily can reproductive isolation arise between populations that are separated from each other by long distances, but have a common ancestor and have retained similar adaptive complexes? Can spatial or geographic isolation lead eventually to the origin of new species, unaided by any change in population structure or new adaptation?

For plants, the answer to the second question is, "No." Numerous examples of species having widely disjunct distributions are well known to plant geographers [3]. The examples of Polygonum arifolium and P sagittatum which have disjunct areas of distribution in eastern North American and Eastern Asia, are similar to a whole series of temperate-to-warm-temperate species that occur in these two widely disjunct areas [3]. Other well known examples are skunk cabbage (Symplocarpus foetidus) and tulip tree (Liriodendron). The latest paleobotanical evidence [64] indicates that migration across a temperate Beringia could have taken place late in the Miocene epoch, 12 to 16 million years ago, but not more recently. In the Old World, disjunct distributions between different parts of Eurasia have also been recognized. A striking example is that of two very similar races of the grass genus Dactylis, D glomerata ss. aschersoniana and subspecies himalaica. These two subspecies are almost indistinguishable morphologically and are completely interfertile [65]. They both live in mesic deciduous forests. At present, they are separated from each other by the distance between southeastern Europe and the Himalaya, about 5,500 km. Continuous forests that would have permitted migration across the now arid intervening areas ceased to exist during the Pliocene epoch, between five and ten million years ago. The most reasonable conclusion from these examples is that in all cross-fertilizing animals and plants spatial separation of populations is a necessary but not sufficient factor to bring about speciation.

6) Does migration into a new habitat on the part of one or a few individuals or propagules (giving rise to a "founder principle" [66]) by itself bring about the origin of a new species?

With respect to the origin of visibly different populations that can be recognized as subspecies or semispecies, the answer is probably, "Yes." The morphologically recognizable species of Ceanothus subgenus Cerastes, mentioned in a previous section, support this affirmation. Similar examples in many other genera of the California flora, such as Arctostaphylos, Mimulus, sect. Diplacus, Prunus, Quercus, and several herbaceous genera contain subspecies or semispecies that may well have originated in this fashion. On the other hand, the very existence of a large number of spatially isolated, morphologically differentiated entities that have not become separated by barriers of postzygotic or internal reproductive isolation supports my belief that the status of a fully evolved species, separated from other species by effective barriers of postzygotic isolation, rarely if ever evolves following a single founder event.

7) Does the occurrence of great fluctuations in populations size that might be called "flush-crash" cycles necessarily cause a population to evolve into a new species? In arid and semi-arid regions throughout the world such cycles are commonplace among species of annual plants. [25]. If fluctuations by themselves inevitably bring about speciation, the number of species in genera of desert and semi-desert annuals would be far greater than it is. The answer to this question, therefore, is also, "No."

Nevertheless, combined with other changes, flush-crash cycles may often have helped to establish enough chromosomal or other differences to bring about strong reproductive isolation. Annual plant species are much more likely to pass through these cycles than are perennials or woody species. The relatively large number of distinctive, strongly isolated species among annuals may be due to repeated flush-crash cycles, acting in combination with other factors, such as colonization and founder effect, as well as continuous trends of climatic change.

SUMMARY AND CONCLUSIONS

Species cannot be established by virtue of a single mutation, having multiple effects upon morphology, physiological adaptation, and reproductive isolation. Evolution of a new species following spatial or geographic isolation is not inevitable, no matter how long the isolation persists. Moreover, no single factor can by itself cause the origin of a new species. Most new reproductively isolated species arise during a period of time, such as ten to a hundred or a few thousand generations, that may be long relative to the duration of controlled experiments, but is so short relative to the geological time scale as to be almost instantaneous from the viewpoint of most macroevolutionists.

The factors that promote speciation are: 1) Migration into and colonization of a new area on the part of one or a few individuals of an existing species; 2) repeated flush-crash cycles that drastically alter population size; 3) highly specialized interactions between mutually dependent organisms, such as flower-pollinator, predator-prey, and parasite-host; 4) establishment of newly balanced gene combinations that drastically alter rates of development; 5) origin and establishment of few-to-many differences in chromosome structure, including submicroscopic rearrangements of nucleotides; 6) facultative autogamy that permits rapid fixation of isolating factors via uniparental reproduction.

None of these factors is usually effective by itself. Any one of several different combinations of them may in a particular genus be responsible for the origin and diversification of its species. The "species problem" is not a single problem, but rather a collection of allied problems. Many more examples must be analyzed in depth, using biologically diverse kinds of organisms, before generalizations will emerge that are more precise than the ones just mentioned. Evolutionists will be studying and debating the species problem for a long time to come. I believe that open mindedness and readiness to accept different but related factors as causes of speciation will be more profitable than attempts to fit the problems into an all-inclusive theoretical mold.

REFERENCES

1. Clausen J: Stages in the Evolution of Plant Species. Ithaca: Cornell University Press, 1951.
2. Grant V: Plant Speciation. New York: Columbia University Press, 1971.
3. Stebbins GL: Variation and Evolution in Plants. New York: Columbia University Press, 1950.
4. Grant V: The Origin of Adaptations. New York: Columbia University Press, 1963.
5. Ehrendorfer F: Differentiation hybridization cycles and polyploidy in Achillea. Cold Spring Harbor Symp Quant Biol 24:141–152, 1959.
6. Ehrendorfer F: Geographical and ecological aspects of intraspecific differentiation. In Heywood VH (ed): Modern Methods in Plant Taxonomy. New York: Academic Press, 1968, pp 261–296.
7. Ehrendorfer F: Evolution: Pathways and patterns. Evolutionary patterns and strategies in seed plants. Taxon 19:185–195. 1970.
8. Lewis H: Speciation in flowering plants. Science 152:167–172, 1966.
9. Lewis H: Evolutionary processes in the ecosystem. Bioscience 19:223–227, 1969.
10. Lewis H: The origin of diploid neospecies in Clarkia. Am Nat 107:161–170, 1973.
11. Lewis H, Raven PH: Rapid evolution in Clarkia. Evolution 12:319–336, 1958.
12. De Wet JMJ, Harlan JR: Apomixis, polyploidy and speciation in Dichanthium. Evolution 24:270–277, 1970.
13. Zohary D: Colonizer species in the wheat group. In Baker HG, Stebbins GL (eds): The Genetics of Colonizing Species. New York: Academic Press, 1965.
14. Allard RW, Kahler AL: Multilocus genetic organization and morphogenesis. Brookhaven Symp Biol 23(Basic Mechanisms in Plant Morphogenesis): 329–343, 1974.
15. Clegg MT, Allard RW: Patterns of genetic differentiation in the slender wild oat species Avena barbata. Proc Nat Acad Sci USA 69:1820–1824, 1972.
16. Bennett MD: Nuclear DNA content and minimum generation time in herbaceous plants. Proc R Soc 181:109–135, 1972.
17. Carr GD: Chromosome evolution and aneuploid reduction in Calycadenia pauciflora (Asteraceae). Evolution 29:681–699, 1976.

18. Carr GD: A cytological conspectus of the genus Calycadenia (Asteraceae); an example of contrasting modes of evolution. Am J Bot 64:694–703, 1977.
19. Flavell RB, Bennett MD, Smith JB, Smith DB: Genome size and the proportion of repeated nucleotide sequence DNA in plants. Biochem Genet 12:257–269, 1974.
20. Gottlieb LD: Genetic differentiation, sympatric speciation and the origin of a diploid species of Stephanomeria. Am J Bot 60:545–553, 1973.
21. Gottlieb LD: Biochemical consequences of speciation in plants. In Ayala FJ (ed): Molecular Evolution. Sunderland: Sinauer Assoc, 1976, p 123.
22. Gottlieb LD: Electrophoretic evidence and plant systematics. Ann Missouri Bot Gard St Louis 64:161–180, 1977.
23. Greilhuber J: Why plant chromosomes do not show G-bands. Theor Appl Genet 50:121–124. 1977.
24. Greilhuber J, Speta F: Quantitative analysis of C-banded karyotypes, and systematics in the cultivated species of the Scilla siberica group (Liliaceae). Plant Syst Evol 129:63–109, 1978.
25. Harper JL: Population Biology of Plants. New York, London: Academic Press, 1977.
26. Jain SK: Patterns of survival and microevolution in plant populations. In Karlin S, Nevo E (eds): Population Genetics and Ecology. New York: Academic Press, 1976, p 49.
27. Holland RF, Jain SK: Insular biogeography of vernal pools in the central valley of California. Am Nat 117:24–37, 1981.
28. Levin DA: The challenge from a related species: a stimulus to saltational speciation. Am Nat 103:316–322, 1969.
29. Levin DA: Developmental instability and evolution in peripheral isolates. Am Nat 104:343–353, 1970.
30. Levin DA: The origin of reproductive isolating mechanisms in flowering plants. Taxon 20:91–113, 1971.
31. Levin DA, Wilson AC: Rates of evolution in seed plants: Net increase in diversity of chromosome number and species numbers through time. Proc Nat Acad Sci USA 63:2086–2090, 1976.
32. Marks GE, Schweizer D: Giemsa banding: Karyotypic differences in some species of Anemone and in Hepatica nobilis. Chromosoma 44:405–416, 1974.
33. Nagl W: Role of heterochromatin in the control of cell cycle duration. Nature 249:53–54, 1974.
34. Nagl W: Mitotic cycle time in perennial and annual plants with various amounts of DNA and heterochromatin. Dev Biol 39:342–346, 1974.
35. Nagl W: Endopolyploidy and Polyteny in Differentiation and Evolution. Amsterdam, New York, Oxford: North Holland Publishing, 1978.
36. Nagl W, Ehrendorfer F: DNA content, heterochromatin, mitotic index, and growth in perennial and triannual Anthemideae (Asteraceae). Plant Syst Evol 123:35–54, 1974.
37. Price HJ, Bachmann K: DNA content and evolution in the Microseridinae. Am J Bot 62:262–267, 1975.
38. Rees H, Hazarika MH: Chromosome evolution in Lathyrus. In Darlington CD, Lewis KR (eds): Chromosomes Today. Edinburgh: Oliver and Boyd, 1969, Vol 2, p 158.

39. Solbrig OT: Plant population biology: an overview. Syst Bot 1:202–208, 1976.
40. Dobzhansky T: The Genetics of the Evolutionary Process. New York: Columbia University Press, 1970.
41. Ehrlich P, Raven PH: Differentiation of populations. Science 165:1228–1232, 1969.
42. Cronquist A: Once again, what is a species? Biosystematics Agricult Res 2:3–20, 1978.
43. Lewis WH (ed): Polyploidy: Biological Relevance. New York: Plenum Press, 1979.
44. Clausen J, Keck DD, Hiesey WM: Experimental Studies on the Nature of Species. II. Plant Evolution Through Amphiploidy and Autoploidy, with Examples from the Madiinae. Carnegie Inst. Wash. Publ. 564, 1945.
45. Carr GD: Experimental evidence for saltational chromosome evolution in Calycadenia pauciflora Gray. Heredity 45:109–115, 1980.
46. Baker HG: Ficus and Blastophaga. Evolution 15:378–379, 1961.
47. Ramirez BW: Host specificity of fig wasps (Agaonidae). Evolution 24:680–691, 1970.
48. Smith-White S: Cytological evolution in the Australian Flora. Cold Spring Harbor Symp Quant Biol 24:273–289, 1959.
49. Baker HG: Pollination mechanisms and inbreeders. Rapid speciation in relation to changes in the breeding systems of plants. In Bailey DL (ed): Recent Advances in Botany. Toronto: University Toronto Press, 1961, p 881.
50. Nobs MA: Experimental Studies on Species Relationships in Ceanothus. Carnegie Inst. Wash. Publ. 623, 1963.
51. Kruckeberg AR: Variation in fertility of hybrids between isolated populations of the serpentine species Strepthanthus glandulosus Hook. Evolution 11:185–211, 1957.
52. Sharsmith HK: The genus Hesperolinon. Univ Calif Publ Bot 32:235–314, 1961.
53. Bush GL: Modes of animal speciation. Ann Rev Ecol Syst 6:339–361, 1975.
54. Eldredge N, Gould SJ: Punctuated equilibria as alternative to pyletic gradualism. In Schopf TJM (ed): Models in Paleontology. San Francisco: Freeman and Cooper, 1972, p 82.
55. Stanley SM: Macroevolution: Pattern and Process. San Francisco: W.H. Freeman Co, 1979.
56. Lewis H, Roberts MR: The origin of Clarkia lingulata. Evolution 10:126–138, 1956.
57. Vasek FC: A cytogenetic study of Clarkia exilis. Evolution 14:88–97, 1960.
58. Vasek FC: The evolution of Clarkia unguiculata derivatives adapted to relatively xeric environments. Evolution 18:26–42, 1964.
59. Vasek FC: The relationships of two ecologically marginal, sympatric Clarkia populations. Am Nat 102:25–40, 1968.
60. Gottlieb LD: Genetic confirmation of the origin of Clarkia lingulata. Evolution 28:244–250, 1974.
61. Murray MG, Peters DL, Thompson, WF: Ancient repeated sequence in the pea and mung bean genomes. Carnegie Inst Wash Year Book 79:112–114, 1980.
62. Tashima M, Calabretta B, Torelli G, Scofield M, Maizel A, Saunders GF: Presence of a highly repetitive and widely dispersed DNA sequence in the human genome. Proc Natl Acad Sci USA 78:1508–1512, 1981.

63. Denison RA, Van Arsdell SW, Bernstein LB, Weiner AM: Abundant pseudogenes for small nuclear RNA's are dispersed in the human genome. Proc Natl Acad Sci USA 78:810–814, 1981.

64. Wolfe JA: An interpretation of Alaskan Tertiary floras. In Graham A (ed): Floristics and Paleofloristics of Asia and eastern North America. 1972, pp 201–233.

65. Stebbins GL, Zohary D: Cytogenetic and evolutionary studies in the genus Dactylis. I. Morphology, distribution and interrelationships of the diploid species. Univ Calif Publ Bot 31:1–40, 1959.

66. Mayr E: Systematics and the Origin of Species. New York: Columbia University Press, 1942.

Mechanisms of Speciation, pages 41–49
© 1982 Alan R. Liss, Inc., 150 Fifth Avenue, New York, NY 10011

Speciation and the Fossil Record

Steven M. Stanley

It seems evident that the modern synthesis of evolution did not focus upon speciation as the site of most evolutionary change. Instead, it fostered a number of ideas that implied a dominant role for phyletic evolution (transformation of established species) [1]. Among these ideas are (a) sexual reproduction prevails among eukaryotic organisms because it is essential for more-or-less continuous evolutionary response to environmental change; (b) coevolution is a widespread phenomenon of reciprocal phyletic evolution within ecologically associated species; (c) Homo sapiens has arisen by the gradual, progressive humanization of an ape-like australopithecene; and (c) living fossils—taxa that have evolved very little for long intervals of time—must be considered highly problematical.

SPECIATION AS A LOCUS OF CHANGE

The fossil record contradicts the idea that phyletic evolution achieves major evolutionary transitions. Fossils, instead, testify that speciation is the site of most change. Their testimony for the most part takes the form of negative evidence: Major evolutionary transformations have occurred during particular intervals of geologic time but, meanwhile, species within the phylogenies embracing the transformations have persisted as highly stable entities. The remarkable geologic longevity of species has only recently come to light [1–3]. It is demonstrated most effectively for numerous taxa by species of the Cenozoic Era, where the fossil record is of especially high quality and where fossil species can be compared to living ones. We find, for example, that for fossil faunas of marine bivalve mollusks seven Myr (million years) old, approximately half of the species are so similar to living forms as to be judged conspecific. We can conclude that an average bivalve species survives for much longer than seven Myr (extant species have a future). Furthermore, many (and probably most) of the seven Myr old species not assigned to living species underwent true extinction, not "pseudoextinction" (artificial extinction that simply reflects a taxonomic decision that a lineage has evolved enough to deserve a new name). Clearly an average bivalve species undergoes little phyletic transformation during five or ten Myr. For other well-studied animal taxa, approximate ages of fossil faunas comprising 50% extant species are as follows [1,4]: marine gastropods, three point five Myr; benthic foraminifera, > fifteen Myr; planktonic foraminifera, > ten Myr; freshwater fish, terrestrial mammals, zero point seven Myr. For beatles (taxonomy based on faithfully preserved genitalia) nearly all species younger than two Myr are extant [5].

Data revealing extraordinary longevity for species of various kinds of fossil plants are also now available. The approximate ages of floras in which 50% of the species are extant are [6]: seed-bearing vascular plants (taxonomy based on seeds), four Myr; marine diatoms, twelve Myr; and bryophytes, > ten Myr (nearly all known Miocene and Pliocene species are extant).

Observations on the great durations of species have led me to make the following generalization [6]: Barring extinction, a typical established species— whether a species of land plants, insects, mammals, or marine invertebrates— will undergo little measurable change in form during 10^5–10^6 generations.

Species stability is now evident in human evolution as well (Fig. 1). Particularly striking here is the great geologic longevity of Homo erectus, an especially well-known species that survived for more than a million years without undergoing enough evolutionary change in traditionally measured cranial characters for this change to be statistically significant on the basis of the data presently available [7].

Observing that nearly all living species of European mammals have survived for substantial intervals of late Pleistocene time, I have argued that the dramatically new genera which have appeared in Europe during the Pleistocene must have developed by way of rapidly divergent speciation events. A similar, but even stronger, argument can be made for the origin of not only generic, but also suprageneric, taxa during Early Eocene time, when major evolutionary transitions were taking place as the modern Mammalia underwent their initial adaptive radiation. For a large sample of North American species that lived during this critical interval, evolution has been assessed by the study of a nearly continuous stratigraphic section in the Bighorn Basin of Wyoming [8].

As shown in Figure 2, sixty nine lineages here (a sample with no obvious bias) exhibit virtually no evolutionary change within an interval representing three or four Myr. During this interval, at least 20 new families of mammals appeared in the world, along with several hundred new genera. Dozens of new genera obviously evolved in North America, yet not a single phyletic transition to a new genus is recorded in the 69 lineages depicted in Figure 2. In fact, within this set of well-documented lineages, which must represent a large percentage of all Early Eocene mammal species of North America, there are recognized only two possible phyletic transitions between species. Furthermore, although the stratigraphic ranges plotted in Figure 2 show many species lasting more than two Myr, these are only partial ranges.

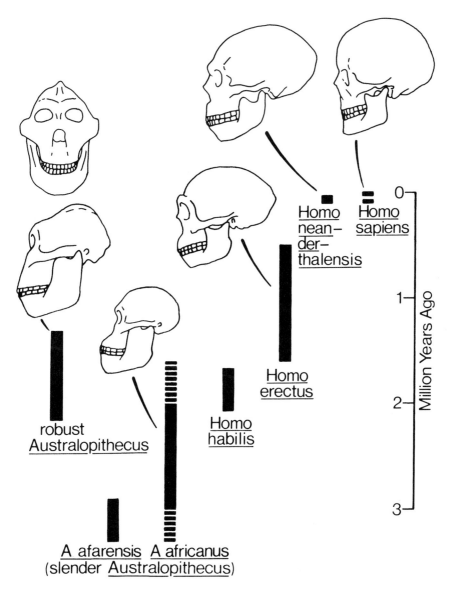

Fig. 1. Geologic ranges of species of the Hominidae [16].

From the kinds of evidence described above, it can be concluded that, although the concept of a genus is not consistent from taxon to taxon, the large majority of the entities that we recognize as genera must have developed their distinctive

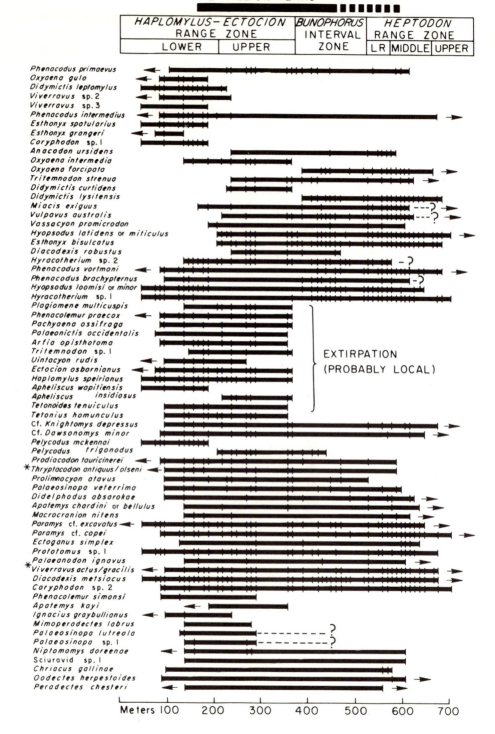

characters rapidly, in small populations, by way of what Grant [9] has termed "quantum speciation." In other words, phyletic evolution has accounted for little of the change that has produced new genera. If localized speciation has been responsible for most of this change, by extrapolation we can reason that localized speciation has also been primarily responsible for the origins of families and orders.

Another test that employs fossil data to evaluate the efficacy of phyletic evolution is the examination of clades (monophyletic segments of phylogeny) of a certain kind. These are clades of narrow breadth—clades never including more than a very small number of species—that have survived for long intervals. The traditional view of evolution, laying great importance to phyletic evolution, would predict that phyletic evolution within these narrow clades should have proceeded at varying rates, often producing substantial morphological restructuring, like that observed within other clades. The punctuational view of evolution, which states that speciation is the site of most morphological change [10], would predict that long slender clades would display very little evolution because they experience very little speciation [1,2]. This second prediction is, in fact, borne out by the fossil record. Every well-documented clade that has survived for a long geologic interval without expanding significantly exhibits evolutionary stagnation. Extant representatives of such clades are invariably what we call "living fossils." Among numerous examples are horseshoe crabs (Limulacea), bowfin fish (Amiidae), sturgeon fish (Acipenseridae), snapping turtles (Chelydridae), alligators (Alligatorinae), and aardvarks (Tubulidentata). The apparently universal presence of living fossils as the recent representatives of long, narrow clades emphatically supports the idea that without speciation there is little evolution.

It is important to clarify two points relating to the inferences described in the preceding paragraphs. One relates to taxonomy. It is true that paleontological data are sometimes too incomplete to permit taxonomists to distinguish between nearly identical species. The particular groups for which I have assessed data on species longevity are, however, groups for which the morphological features preserved as fossils have been shown to be reliable for species identification. Our failure to recognize sibling species has no bearing on the issue at hand anyway. The total evolutionary change encompassed by the development of

---

Fig. 2. Partial geologic ranges (horizontal bars) of species of mammals, as documented by the remarkably complete fossil record of the central part of the Bighorn Basin, Wyoming. The total time interval depicted here approximately spans the Early Eocene. Broken portion of the two Myr time scale indicates degree of uncertainty. An arrow beside a range bar indicates that the range plotted for this locality is known to be incomplete for the species as a whole: The species also has a younger or older fossil record elsewhere. An asterisk indicates the possibility of phyletic transition between one species and another.

groups of sibling species is too trivial to play a significant role in the origins of higher taxa.

The second point is that when asserting that most large-scale evolution is associated with speciation, one is not claiming that all speciation events are rapidly divergent. Such a claim would be ridiculous, given the existence of sibling species. One is simply claiming that some speciation events are rapidly divergent and that these account for most large-scale evolution.

RATES OF SPECIATION

As first explored in some detail by Simpson [11] who employed higher taxa as units, the fossil record is a unique repository of data documenting rates of evolution. The data of the fossil record show that rates of speciation in adaptive radiation, which are approximately exponential, vary enormously from taxon to taxon [1]. The assessment of such rates is especially important because adaptive radiation is the site of most large-scale evolution. Here, as in assessing the geologic longevity of species, one can employ techniques that circumvent most of the inadequacies of the fossil record [1,2,12]. If one has a reasonable estimate of the geologic antiquity of an extant higher taxon in the midst of adaptive radiation and the number of living species belonging to the taxon, then it is possible to calculate a net exponential rate of diversification (percentage increase per unit time). This rate (R) amounts to the difference between rate of speciation (S) and rate of extinction (E). The latter is approximately the reciprocal of mean species longevity.

As it turns out, values of R and E vary enormously among taxa, but tend to be correlated with each other [1]. Because $S = R + E$, S is also correlated with E. Thus, groups characterized by low rates of extinction (many groups of marine invertebrates) are also saddled with low rates of speciation during adaptive radiation. This constraint may, in large part, reflect the fact that rate of extinction and rate of speciation are governed in similar ways by inherent biological properties of taxa: Both E and S appear to decrease with dispersal ability and increase with level of behavioral complexity [1]. Thus, it is understandable that both rate of extinction and rate of adaptive radiation are lower for most marine invertebrates than for birds and terrestrial mammals.

THE ROLE OF QUANTUM SPECIATION

The idea that speciation runs rampant in adaptive radiation is deceiving. It seems to suggest that rate of large-scale change is high early in adaptive radiation because rate of speciation is high. What is misleading is that while early in adaptive radiation the exponential rate (the increase per species per unit time) is high, the number of species is low. This means that the overall rate of speciation (number of new species per unit time) is relatively low [1]. This observation has important implications because the fossil record shows that enormous evolutionary changes tend to be concentrated in the early stages of adaptive radiation. Because total number of speciation events per unit time is relatively low here, we can conclude that the degree of divergence per speciation event is unusually great: The incidence of quantum speciation must be unusually high within episodes like the great mammalian radiation that took place during the interval depicted in Figure 2.

THE DIRECTION OF LARGE-SCALE CHANGE

The fossil evidence that the normal fate of established lineages is to evolve very little before being terminated by extinction creates a fundamental problem: How are we to explain the presence of large-scale trends in evolution?

Here we can consider the macroevolutionary fate of a clade (in which species form units) to be analogous to the macroevolutionary fate of a population (in which individuals form units). This analogy brings to light three potential processes of macroevolutionary change [1].

(a) A process analogous to genetic drift would be what can be termed "phylogenetic drift." Here the morphological character of phylogeny would change by random fluctuations in the frequencies of various kinds of species. To accept the idea that phylogenetic drift is a dominant form of macroevolutionary change (given the assumption that most evolution is associated with speciation) would be to allege that macroevolutionary trends are accidental. By analogy with genetic drift, we can assume that phylogenetic drift seldom plays a dominant role except in small clades.

(b) A process rather crudely analogous to mutation pressure on a microevolutionary scale would be what I have called "directed speciation," or a tendency for speciation to move in a certain direction. There is no question that within particular taxa there are biases in the direction of speciation. Probably many examples of evolutionary convergence and parallelism reflect such biases. On the other hand, as Wright [13] has stressed, there is a strong random element in the directionality of speciation. This is not to say that all possible directions of speciation are equally probable any more than all possible directions of mutation are equally probable. The point is that there is a high degree of unpredictability associated with the direction of speciation. The course of a particular speciation event may be somewhat predictable if we know where, when, and in what subpopulation it will occur, but if we stood back and awaited the next speciation event within a sizeable segment of phylogeny, we would have no knowledge of these influential circumstances. Where and when the next speciation event would occur and what external conditions would obtain would depend to a large degree on historical accidents—the vagaries of geographic, geologic, and climatic change, for example. The result is that we cannot at any time predict with a high degree of certainty the direction of the next speciation event.

(c) The strong random element in the directionality of speciation largely decouples macroevolution from microevolution [1,2]. This forces us to lay greater importance than previously granted to what I have called "species selection." Darwin was the first worker to fully recognize that selection operates at the species level, by virtue of differential rates of survival and differential rates of speciation among the species of a higher taxon. Darwin failed to emphasize this process, however, because of his belief in the dominance of phyletic evolution. Not recognizing the discrete nature of most animal species [14], he also blurred the distinction between selection at the level of the species and selection at the level of the individual.

Species selection may guide a phylogenetic trend in a particular direction through the dominance of a gradient in rates of speciation, through the dominance of a gradient in rates of extinction, or through both kinds of gradient acting in concert [1,15]. In many cases, the agents of species selection will be ecological limiting factors, which also represent agents of extinction: competitive interactions, susceptibility of predation, or vulnerability to changes in the physical environment. In other cases, a phylogenetic trend may develop when certain kinds of taxa simply speciate at an unusually high rate because of inherent traits relating to such things as reproductive behavior or dispersal ability.

In effect, the idea that species selection prevails in the guidance of large-scale trends implies that the dominant long-term macroevolutionary role of speciation is one of generating raw material for selection among species. Where directed speciation prevails—presumably most commonly in small segments of phylogeny—the direction of speciation governs the direction of phylogenetic trends.

Certainly in any large segment of phylogeny, phylogenetic drift, directed speciation, and species selection all have some effect, and superimposed on them are whatever minor trends are wrought by phyletic evolution.

REFERENCES

1. Stanley SM: Macroevolution: Pattern and Process. San Francisco: Freeman, 1979.
2. Stanley SM: A theory of evolution above the species level. Proc Natl Acad Sci (USA) 72:646, 1975.
3. Stanley SM: Chronospecies' longevities, the origin of genera, and the punctuational model of evolution. Paleobiol 4:26, 1978.
4. Stanley SM, Addicott WO, Chinzei K: Lyellian curves in paleontology: possibilities and limitations. Geology 8:422, 1980.
5. Coope GR: Interpretations of Quaternary insect fossils. Annu Rev Entomol 15:97, 1970.

6. Stanley SM: Macroevolution and the fossil record. Evolution (in press).

7. Rightmire GP: Patterns in the evolution of Homo erectus. Paleobiology 7:241, 1981.

8. Schankler D: Faunal zonation of the Willwood Formation in the central Bighorn Basin. Univ Michigan Pap Paleont 24:99, 1980.

9. Grant V: The Origin of Adaptations. New York: Columbia University Press, 1963.

10. Eldredge N, Gould SJ: Punctuated equilibria: an alternative to phyletic gradualism. In Schopf TJM (ed): Models in Paleobiology. San Francisco: Freeman, Cooper, 1972, p 82.

11. Simpson GG: Tempo and Mode in Evolution. New York: Columbia University Press, 1944.

12. Yule GU: A mathematical theory of evolution, based on the conclusions of Dr. J.C. Willis, FRS. R Soc London Proc 213:21, 1924.

13. Wright S: Modes of selection. Am Nat 90:5, 1956.

14. Mayr E: Isolation as an evolutionary factor. Proc Am Phil Soc 103:221, 1959.

15. Vrba E: Evolution, species and fossils: how does life evolve? S Afr J Sci 76:61, 1980.

16. Stanley SM: The New Evolutionary Timetable: Fossils, Genes and the Origin of Species. New York: Basic Books, 1981.

Mechanisms of Speciation, pages 51–66

Gradualism Versus Punctualism in Speciation: Reproductive Isolation, Morphology, Genetics

Francisco J. Ayala

In sexually reproducing organisms, species may be defined as "groups of interbreeding natural populations that are reproductively isolated from other such groups" [1]. A species is a natural unit or system defined by the possibility of interbreeding between its members. The ability to interbreed is of considerable evolutionary import because it establishes species as discrete and independent evolutionary units. Consider an adaptive mutation or some other genetic change originating in a single individual. Over the generations, the adaptive mutation may spread by natural selection to all members of the species, but not to individuals of other species. This can be stated differently: Individuals of a species share in a common gene pool which is, however, not shared by individuals of other species. Owing to reproductive isolation, different species have independently evolving gene pools.

Interest in the process of speciation has recently been renewed due, in part, to arguments advanced by proponents of the "punctuated equilibria" model of macroevolution [2–4]. These have argued that, according to paleontological evidence, "species have tended to last for such long intervals of geological time that, once formed, they must have evolved very slowly. . . This condition, when compared to the rapid pace of large-scale evolution, implies that most sizable evolutionary steps in the history of life must have occurred cryptically from a paleontological vantage point, during the rapid origination of certain species from small, localized populations of pre-existing species" [3, p 3].

Three claims are made in the statement just quoted; (a) that species originate rapidly from small, localized populations of pre-existing species; (b) that the sizable morphological changes that characterize macroevolution occur during speciation; and (c) that species change morphologically little, if at all, after they are formed. Thus, punctualists argue that speciation is not a slow, gradual process, but a bursting one. Species arise rapidly and considerable morphological change occurs during the short time spans involved.

Whether macroevolution is punctuated or gradual does not concern us now. But it should be noted from the beginning that the alleged relevance of punctuational evolution to speciation is based, at least in part, on two misunderstandings. The first is a definitional artifact: Paleontologists recognize species by their different morphologies as preserved in the fossil record. Thus, speciation events yielding little or no morphologically different products go totally unrecognized [5]. Speciation as seen by the paleontologist always involves substantial morphological change because only when such change has occurred is the paleontologist able to recognize the presence of a new species. The second misunderstanding concerns the time scale. When punctualists argue that paleontological evidence indicates speciation is a rapid process [3,4,6], they are using a geological time scale. "Instantaneous" events in the paleontological scale, as in the transition between different geological strata, may involve at times many thousands of years [7]. In the microevolutionary scale of the population biologist, a thousand years is a long time, not an instant.

In order to evaluate whether speciation is a gradual or a punctuated process, one needs to deal separately with three different kinds of change. First is the development of reproductive isolation. This is the essential constituent of the process according to the definition advanced above: Species are reproductively isolated populations; speciation requires that such reproductive isolation takes place. Second is change in morphology and other aspects of the phenotype, such as behavior and ecology. The issue here is whether morphological (phenotypic) change occurs in bursts, concomitantly with the development of reproductive isolation, rather than in a more-or-less gradual manner independent, at least in a good part, of the speciation process. Third is the genetic change underlying both the development of reproductive isolation and morphological change. The question now is whether or not the rate of genetic change is greatly accelerated during speciation, so that speciation is invariably accompanied by major changes in the genotype.

THE ORIGIN OF SPECIES

Species are reproductively isolated groups of populations. The question of how species come about is, therefore, equivalent to the question of how reproductive isolation arises between groups of populations. Two theories have been advanced to explain the origin and development of reproductive isolation between populations. One theory considers isolation as an accidental by-product of genetic divergence. Populations that become genetically more and more different (as a consequence, for example, of adaptation to different environments) may eventually be unable to interbreed because their gene pools are disharmonious. The other theory regards isolation as a product of natural selection. Whenever hybrids are less fit than nonhybrid individuals, natural selection will directly promote the development of reproductive isolation because gene variants interfering with hybridization have greater fitness than those favoring hybridization since the latter will often be present in poorly fit hybrids.

The two theories of the origin of reproductive isolation are not mutually exclusive. Reproductive isolation may indeed come about as an incidental by-product of genetic divergence between separated populations. Consider, for example, the evolution of many endemic species of plants and animals in the Hawaiian archipelago. The ancestors of such species arrived in the Hawaiian islands long ago, perhaps several millions of years before the present. There they evolved and became adapted to the local conditions. Although natural selection did not directly promote reproductive isolation between the species evolving in Hawaii and the continental populations from which the colonizers came, reproductive isolation has nevertheless become complete in many cases.

In fact, the process of speciation commonly involves the two kinds of phenomena postulated by the two theories of species. Reproductive isolation may start as an incidental by-product of genetic divergence, but is completed when it becomes directly promoted by natural selection. Speciation may occur in a variety of ways, but two stages may often be recognized (Fig. 1).

Stage I

The onset of the speciation process requires, first, that gene flow be somehow interrupted (completely, or nearly so) between two populations of the same species. Absence of gene flow makes it possible for the two populations to become genetically differentiated as a consequence of their adaptation to different local conditions or to different ways of life (and also as a consequence of genetic drift, which may play a lesser or greater role depending on the circumstances). The interruption of gene flow is necessary because otherwise the two populations would, in fact, share in a common gene pool and fail to become genetically different. As populations become more and more genetically different, reproductive isolating mechanisms may appear. This occurs because different gene pools are not mutually coadapted; hybrid individuals will have disharmonious genetic constitutions, and have reduced fitness in the form of reduced viability or fertility.

Stage II

This stage encompasses the completion of reproductive isolation. Assume that external conditions disappear which interfered with gene exchange between two populations in the first stage of speciation; this might occur, for example, if two previously geographically separated populations would expand and come to exist together in the same territory. Two outcomes are possible: (a) a single

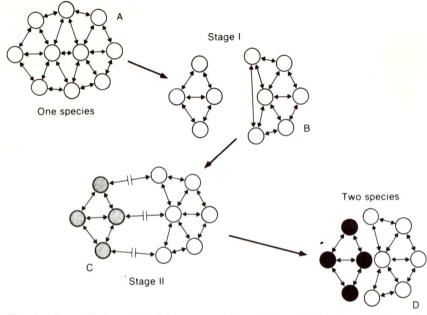

Fig. 1.   Generalized model of the process of speciation. (A) Local populations of a single species are represented by circles; arrows indicate that gene flow occurs between populations. (B) The populations have become separated into two groups between which there is no gene flow. This may be due to splitting of the pre-existing populations or to the budding-off of one or a few new local populations. The two groups gradually become genetically different. As a consequence of genetic differentiation, reproductive isolating mechanisms (usually postzygotic) arise between the two groups. This is the first stage of speciation. (C) Individuals from different population groups are now able to intermate. However, owing to the pre-existing reproductive isolating mechanisms, little if any gene flow takes place, which is represented by the small lines crossing the arrows that connect the two population groups. Natural selection favors the development of additional reproductive mechanisms, particularly prezygotic ones which avoid matings between individuals from different population groups. This is the second stage of speciation. (D) Speciation has been completed because the two groups of populations are fully reproductively isolated. There are now two species which may coexist in the same territory without gene exchange.

gene pool comes about, because the loss of fitness in the hybrids is not very great (or because one population is eliminated by the other through ecological competition); and (b) two species ultimately arise, because natural selection favors the development of reproductive isolation.

The first stage of species is reversible: If it has not gone far enough it is possible for two previously differentiated populations to fuse into a single gene pool. However, if matings between individuals from different populations leave progenies with reduced fertility or viability, natural selection would favor genetic variants promoting matings between individuals of the same population.

Nevertheless, speciation may take place without the occurrence of Stage II. Populations in the absence of gene exchange may develop complete reproductive isolation if the process of genetic differentiation continues long enough; eg when they remain indefinitely separated in two islands. However, whenever the opportunity for gene exchange arises after previous genetic differentiation, natural selection will accelerate the development of reproductive isolating mechanisms; Stage II, then, speeds up the speciation process. Most often, speciation involves both stages.

GEOGRAPHIC SPECIATION

Speciation is a multifarious process. A number of classifications of modes of speciation have been proposed [8–12]. From the present point of view, we may reduce these various modes to two: geographic speciation and quantum speciation [13]. In geographic speciation, Stage I begins as a result of geographic separation between populations. As a result of natural selection, geographically separate populations become adapted to local conditions, and thus become genetically differentiated. Random genetic drift may also contribute to genetic differentiation particularly when populations are small, or are derived from only a few individuals. If geographic separation continues for some time, incipient reproductive isolation may appear, particularly in the form of postzygotic mechanisms; the populations will then be in the first stage of speciation.

The second stage of speciation begins when previously separated populations come into geographic contact, at least over part of their distributions. This may happen, for example, by topographic changes on the earth's surface or by ecological changes in the intermediate territory that make it habitable for the organisms, or by migration of members of one population into the territory of the other. Matings between individuals from different populations may then take place. Depending on the strength of the pre-existing isolating mechanisms and of the extent of hybridization, the two populations may fuse into a single gene pool or may develop additional isolating mechanisms and become separate species.

The two stages of the process of geographic speciation may be illustrated with a group of closely related species of Drosophila that live in the American tropics [14]. This group encompasses six sibling species (i e , morphologically virtually indistinguishable). Some of the siblings (such as D willistoni and D equinoxialis) consist of two or more subspecies, between which there is incipient reproductive isolation in the form of hybrid sterility, but no ethological (sexual) isolation or other prezygotic mechanisms. Thus, the subspecies are groups of populations in the first stage of speciation.

Stage II of the speciation process can also be found within the group: D. paulistorum is a species consisting of six semispecies or incipient species. The semispecies exhibit hybrid sterility similar to that found between the subspecies just mentioned. Crosses between males and females of two different semispecies yield fertile females but sterile males. But two or three semispecies have come into geographic contact in many places, and there the second stage of speciation has advanced to the point that ethological isolation has developed to a lesser or greater extent. When females and males from two different semispecies are placed together in the laboratory, the results depend on the geographic origin of the flies [15]. The proportion of homogamic matings is greater when both semispecies are from the same locality than when they are from different localities, indicating that ethological isolation is well advanced among the former but not among the latter. The semispecies of D paulistorum thus provide a remarkable example of the action of natural selection during the second stage of speciation; reproductive isolation has been nearly completed where the semispecies are sympatric but not elsewhere, because the genes involved have not yet fully spread throughout each semispecies.

QUANTUM SPECIATION

In the case of geographic speciation, Stage I involves gradual genetic divergence of geographically separated populations. The development of postzygotic isolating mechanisms as by-products of genetic divergence may require a long period of time—thousands, perhaps millions, of generations. However, there are other modes of speciation where the first stage of speciation may require only relatively short periods of time. Quantum speciation (which may also be called "rapid speciation" or "saltational speciation") refers to these modes of speciation that involve an acceleration of the process, particularly of the first stage.

One form of quantum speciation is polyploidy, the multiplication of entire chromosome complements. Polyploid individuals may arise in one or a few generations. Polyploid populations are reproductively isolated from their ancestral species, and thus are new species. In polyploidy, the suppression of gene flow that is required for the onset of the first stage of speciation is due not to geographic separation, but to cytological irregularities. Reproductive isolation in the form of hybrid sterility does not require many generations but follows immediately due to chromosomal imbalance. Polyploidy is a rare mode of speciation in animals, but it is common in plants. Nearly half (47%) of the species of flowering plants and a majority of ferns (95%) are polyploids.

Other modes of quantum speciation besides polyploidy occur in plants as well as in animals. One plant example, studied by H. Lewis [16], involves the two diploid species, Clarkia biloba and C lingulata. Both species are native to California, but C lingulata has a narrow distribution, being found only at two sites in central Sierra Nevada at the southern periphery of the distribution of C biloba. The two species reproduce by outcrossing although they are capable of self-fertilization; they are similar in external morphology although differences exist in flower shape. The narrowly distributed species, C lingulata, has arisen from C biloba by a rapid series of events involving extensive chromosomal reorganization. Chromosomal rearrangements, such as translocations, reduce the fertility of individuals heterozygous for the arrangements. The first stage of speciation may thus be accomplished through chromosomal rearrangements without extensive allelic differentiation.

Rapid speciation initiated through chromosomal rearrangements has also occurred in animals; for example, in some flightless Australian grasshoppers such as Moraba scurra and M viatica studied by White [17,18]. Incipient species differing by chromosomal translocations are found in adjacent territories. A translocation establishes itself at first in a small colony by genetic drift. If members of this colony possess high fitness, they may subsequently spread and displace the ancestral form from a certain area. The ancestral and the derived population may then coexist contigually, their individuality maintained by the low fitness of the hybrids formed in the contact zones, since the hybrids are translocation heterozygotes. The first stage of speciation is thus rapidly accomplished, and natural selection favors the development of additional isolating mechanisms (Stage II). This mode of speciation seems to be common in several animal groups, particularly in rodents living underground and having little mobility, such as mole rats of the group Spalax ehrenbergi in Israel [19], pocket gophers of the group Thomomys talpoides in the southern Rocky Mountains of the United States [20], and spiny rats of the genus Proechimys in South America [21].

Quantum speciation may be a common phenomenon in certain groups of insects [9]. Carson [22,23] proposed a flush-and-crash model to account for the enormous diversification of Drosophila species in Hawaiian archipelago. New species may have derived from small propagules colonizing a new island, or a new valley, where absence of competitors leads to a population explosion ("flush") during which natural selection is largely relaxed. The expanded population may, then, be subject to strong natural selection ("crash") and to a substantial reorganization of the gene pool, eventually yielding a new species. Incipient speciation has been ascertained by Bullini and Coluzzi [24] between Culex pipiens pipiens, a rural form, and C pipiens autogenicus, an urban form, in a process that must have involved a very short period of evolutionary time. The process of "sympatric host race formation" in phytophagous parasitic insects represents another mode of quantum speciation [9].

The preceding discussion of modes of speciation makes it clear that reproductive isolation may come about in just a few generations, and thus in intervals of time so brief as to appear as "instants" in the geological time scale. But in the mode of geographic speciation, the development of incipient reproductive isolation during the first stage is a gradual process, which may require many thousands of generations. Thus, reproductive isolation may be either rapid or slow, depending on the mode of speciation involved. The question can then be raised as to whether geographic speciation or quantum speciation is the prevailing mode. The geographic mode is thought by some authors [25,26] to occur in a greater diversity of organisms. Bush [9] claims that rapid modes of speciation are preponderant in some groups of insects; because the number of species of arthropods is so large, it may be the case that more instances of speciation occur by the quantum than by the geographic mode, even if the latter were to prevail in most phyla. Be that as it may, rapid as well as slow modes of speciation are both common in life. The evolution of reproductive isolation is punctuated in some cases, gradual in others.

GENETIC CHANGE DURING SPECIATION

Is the process of speciation accompanied by a burst of genetic change as punctualism would imply? The discovery that genes code for proteins together with the techniques of gel electrophoresis have made it possible to estimate the amount of genetic change during the speciation process. Before these methods became available, there already was evidence suggesting that a fair number of allelic substitutions might be involved in speciation, since it was known that even closely related species are genetically quite different. Baur [27], for example, had crossed two species of snapdragons, Antirrhinum majus and A molle, which produce fertile hybrids. Considerable phenotypic variability appeared in the F_2. Most individuals showed various combinations of the parental traits, but some had characteristics not present in either parent but found in other species of Antirrhinum or related genera. Baur estimated that more than one hundred genetic differences exist between A majus and A molle. It was not possible, though, to estimate what proportion of the genes are different—invariant genes cannot be detected by Mendelian methods.

Estimates of genetic differentiation between two populations can be obtained by studying a sample of proteins in both populations, chosen without knowing whether or not they are different in the populations. The genes coding for the proteins represent, then, a random sample of all the structural genes with respect to the differentiation between the populations. Therefore, the results obtained from the study of a moderate number of gene loci can be extrapolated to the whole genome [8].

Studies of protein variation by gel electrophoresis provide estimates of genotypic and allelic frequencies in populations. The degree of genetic differentiation can be quantified using the statistic D, genetic distance, which estimates the number of allelic substitutions per locus that have occurred in the separate evolution of two populations. There is an allelic substitution when one allele is replaced by a different allele, or when a set of alleles is replaced by a different set [13,28]. Genetic distance, D, may range in value from zero (no allelic changes at all) to infinity; D can be greater than one because each locus may experience complete allelic substitutions more than once as evolution goes on for long periods of time.

First, we shall consider geographic speciation. The Drosophila willistoni group of species was earlier described as a model of geographic speciation, because both stages of the process can therein be identified. This group of species has been extensively studied using electrophoretic techniques [14]. The results are summarized in Table I. Five levels of evolutionary divergence are represented. The first involves comparisons between populations living in different localities but without any reproductive isolation between them; the genetic distance is 0.03, indicating a very low degree of genetic differentiation.

The second involves comparisons between different subspecies, populations that are in the first stage of speciation and exhibit reproductive isolation in the form of hybrid sterility. These exhibit a fair amount of genetic differentiation, D = 0.230; 23 complete allelic substitutions have occurred, on the average, for every 100 gene loci.

The third level of evolutionary divergence involves comparisons between the incipient species of the D paulistorum complex. These are populations in the second stage of speciation exhibiting some prezygotic, as well as postzygotic, reproductive isolation. These populations apparently are not genetically more differentiated than those in the first stage of speciation. This means that the second stage of speciation has not required much genetic change, which is perhaps not surprising. During the first stage of speciation, reproductive isolation comes about as a by-product of genetic change: A fair amount of genetic change needs to take place over the whole genome before incipient reproductive isolation appears. However, during the second stage of speciation, natural selection directly favors the development of prezygotic isolation; only a few genes—affecting

TABLE I. Genetic Differentiation (Mean and Standard Error) Between Populations of the Drosophila willistoni Group at Various Levels of Evolutionary Divergence [14].

Level of comparison	D
1. Local populations	0.031 ± 0.007
2. Subspecies	0.230 ± 0.016
3. Incipient species	0.226 ± 0.033
4. Sibling species	0.581 ± 0.039
5. Morphologically different species	1.056 ± 0.068

courtship and mating behavior, for example—need to be changed to accomplish it.

The fourth level involves comparisons between sibling species. In spite of their morphological similarity, these species are genetically quite different; about 58 allelic substitutions have occurred, on the average, for every 100 loci. Species are independently evolving groups of populations. Once the process of speciation is completed, species continue to diverge genetically. The results of this gradual process of divergence are also apparent in the comparisons between morphologically different species of the D willistoni group (fifth level in Table I). On the average, somewhat more than one allelic substitution per gene locus has occurred in the evolution of these nonsibling species.

Using the techniques of gel electrophoresis, comparisons between populations at various levels of evolutionary divergence have been carried out during the past few years in many kinds of organisms. Evolution is a complex process determined by the environmental conditions as well as by the nature of the organisms, and thus the amount of genetic change corresponding to a given level of evolutionary divergence is likely to vary from organism to organism, from place to place, and from time to time. The results of electrophoretic studies confirm this variation, but also show some general patterns (Table II). With few exceptions, the genetic distance between populations in either the first or the second stage of speciation is about 0.20 (most comparisons fall in the range between 0.16 and 0.30) for organisms as diverse as insects, fish, amphibians, reptiles, and mammals. These results are consistent with the conclusions derived from the study of the Drosophila willistoni group. The first stage of the geographic speciation process requires a fair amount of genetic change (of the order of 20 allelic substitutions for every 100 gene loci), whereas little additional genetic change is required during the second stage.

How much genetic change takes place in the quantum mode of speciation? It is clear that when a new species arises by polyploidy no genetic changes other than the chromosome duplications are required; the new species has the alleles present in the parental species and no others. However, because polyploid species usually start from only one individual of each progenitor species, they possess at the beginning less genetic variation than the parental species.

Other modes of quantum speciation start with chromosomal rearrangements that cause either partial or total hybrid sterility. As in the case of polyploidy such rearrangements do not necessarily involve changes in allelic constitution, although there is often a reduction of genetic variation because the derivative population starts from only one or a few individuals. The first stage of speciation is, therefore, accomplished with little or no genetic change at the level of the individual genes.

What about genetic change in the second stage of quantum speciation? The second stage of speciation is similar in geographic and in quantum speciation.

TABLE II. Genetic Differentiation at Various Stages of Evolutionary Divergence in Several Groups of Organisms [29]*

Organisms	Genetic distance			
	Local populations	Subspecies	Semispecies	Species and closely related genera
Drosophila	0.013	0.163	0.239	1.066
Other invertebrates	0.016	—	—	0.878
Fish	0.020	0.163	—	0.760
Salamanders	0.017	0.181	—	0.742
Reptiles	0.053	0.306	—	0.988
Mammals	0.058	0.232	0.263	0.559
Plants	0.035	—	—	0.808

*The values given are for genetic distance, D.

TABLE III. Genetic Differentiation (D) in Quantum Speciation*

Populations compared	D
Plants:	
Clarkia biloba vs. C. lingulata[a]	0.128
Stephanomeria exigua vs. S. malheurensis[b]	0.057
Rodents:	
Spalax ehrenbergi[b]	0.022
Thomomys talpoides[b]	0.078
Proechimys guairae complex[b]	0.032
Insects:	
Drosophila sylvestris vs. D. heteroneura[a]	0.063
Culex pipiens pipiens vs. C. p. autogenicus[b]	< 0.050

*From various sources.
[a]Comparison between two recently arisen species.
[b]Comparisons between incipient species; i.e., populations completing the second stage of speciation.

In both cases, the populations already exhibit some (usually postzygotic) isolation and are developing prezygotic reproductive isolation due to natural selection. If in the case of geographic speciation the second stage requires genetic changes in only a small fraction of the genes, the same should be true in quantum speciation. Experimental results confirm such prediction (Table III). The first comparison involves two annual plant species, Clarkia biloba and C lingulata, previously discussed as examples of quantum speciation. These species remain genetically similar: D = 0.128, indicating that only about 13 allelic substitutions for every 100 gene loci have occurred in their separate evolution [30].

The second comparison is also between two annual plants, Stephanomeria exigua and S malheurensis, the latter having derived from exigua only very recently. Gottlieb [31] has shown that the original and the derivative populations differ by one chromosomal translocation and by their mode of reproduction—the original species reproduces by outcrossing while the derivative species reproduces by selfing. As expected, the two species are genetically very similar (about six allelic substitutions for every 100 loci).

The third, fourth, and fifth comparisons in Table III involve rodents. Spalax ehrenbergi is a species of mole rats consisting of four groups of populations differing in the number of chromosomes (52, 54, 58, and 60). The populations are largely allopatric although they enter in contact with one another in narrow zones at the edge of their distributions and some hybridization takes place there. The differences in chromosome number due to chromosomal fusions or fissions provide effective postzygotic isolation; moreover, some ethological isolation has developed. Laboratory tests show greater preference for matings between individuals of the same chromosomal type, although they appear morphologically indistinguishable. These four populations in the second stage of quantum speciation are, on the average, genetically similar. Only about two allelic substitutions for every 100 gene loci have taken place in their separate evolution [19].

Thomomys talpoides is a species of pocket gopher, consisting of more than eight populations differing in their chromosomal arrangements. They live in the north central and northwest United States and neighboring areas of southern Canada. As in the case of Spalax, the populations of Thomomys are mostly allopatric but are in geographic contact at the margins of their distributions. The chromosomal rearrangements keep the populations from interbreeding in the zones of contact. Nevertheless, the average genetic distance between these populations is quite small (about eight allelic substitutions for every 100 loci) [20]. The data for the Proechimys guairae complex involve comparisons between populations of spiny rats with number of chromosomes 2n = 46, 48, 50, and 62. The degree of genetic differentiation between these allopatric or parapatric populations is D = 0.032, even smaller than between the Thomomys populations [21].

Drosophila sylvestris and D heteroneura are endemic in the island of Hawaii. These species are morphologically easily differentiated and largely sympatric, indicating that the second stage of speciation has been completed, even though this must have occurred in recent geological time. Yet the degree of genetic differentiation (D = 0.063) is very small, only slightly greater than that found between local populations in various kinds of organisms [32].

The final comparison in Table III is between two subspecies of Culex pipiens that exhibit prezygotic isolation due to substantially differentiated nuptial flights.

The genetic differentiation between these two forms, which may also be considered to be in the second stage of speciation, is very small, with D ranging from 0.01 to 0.05 [Bullini, personal communication].

In conclusion, quantum speciation can occur with little change at the level of the genes; that is, neither Stage I nor Stage II require substantial allelic evolution in this mode of speciation. This result, in turn, confirms the conclusion reached with respect to geographic speciation, namely that Stage II—when natural selection directly promotes prezygotic isolation—does not require major genetic changes. It is apparent that speciation per se does not require allelic substitutions in a large fraction of the structural genes. The substantial amount of genetic change observed during the first stage of speciation reflects the long time involved. The present evidence, as well as other molecular studies of evolutionary differentiation, indicate that structural genes evolve, by and large, in a gradual manner and that no substantial acceleration of the rate of genetic change occurs during speciation [33]. Whether or not speciation may require major changes in other parts of the genome, such as regulatory genes, is at present a moot question, because relevant evidence does not exist.

PHENOTYPIC CHANGE DURING SPECIATION

Proponents of the punctualist model argue that most of the morphological change that appears in the fossil record takes place in association with the process of speciation. They allege that paleontological evidence shows that bursts of morphological change occur in "geological instants" as new species bud-off from their ancestors. Thereafter, species retain their newly acquired morphologies without significant change for the rest of their existence, up to many millions of years.

The alleged evidence, however, is flawed by a definitional artifact. New species are identified in the paleontological record by the appearance of new morphologies. Any new species that does not differ morphologically from its ancestral lineage will not be detected as a new species. And if morphological differentiation occurs without speciation, this could nevertheless be recorded by the paleontologist as an instance of speciation.

Evidence from living species demonstrates, in any case, that speciation needs not be accompanied by detectable morphological change. The most obvious evidence comes from the common occurrence of sibling species, which are morphologically indistinguishable but are nevertheless fully reproductively isolated. One such example is the D willistoni group, where six sibling species exist. In spite of their morphological similarity, these sibling species are genetically quite different (D = 0.58, on the average; see Table I). In the well-studied genus Drosophila, many other sets of sibling species are known, such as D persimilis, D pseudoobscura, and D miranda; D melanogaster, D simulans,

D yakuba, D teissieri, D erecta, D mauritiana, and D orena [34]. The siblings of these sets are, like the siblings of the D willistoni group, genetically quite different from one another.

Sibling species may, however, exhibit little allelic differentiation from each other. Spiny rats of the Proechimys trinitatis and P guairae complex are one example: The average genetic distance among the guairae siblings is only 0.03 (see Table III). These species are reproductively isolated by their considerable chromosomal differentiation; the diploid chromosome number ranges from 24 to 62.

Sibling species are common among rodents and other mammals and, indeed, they are known in every major group of animals in which the required studies for their detection have been conducted. Sibling species are not restricted to the animal kingdom. Plant examples include tarweeds such as Layia platyglossa and L septentrionalis, and a large cluster in the genus Holocarpha [see Stebbins, this volume].

The occurrence of sibling species demonstrates that speciation does not necessarily involve morphological differentiation. Moreover, allopatric populations of the same species sometimes do exhibit clear-cut morphological differences in many animals, such as lady beetles, snails, salamanders, birds, as well as in many kinds of plants [35,36]. Indeed, the available evidence indicates that morphology and genotype largely evolve independently of the development of reproductive isolation (and of each other), although some acceleration of the rates of morphological and of genetic evolution may take place during speciation. The punctualists' proposal that morphological (and genetic) change occurs for the most part during speciation finds little support in the evidence obtained from the study of living organisms. This does not represent, however, conclusive evidence against the punctuational model of macroevolution [7]. Because of the different scales involved, slow morphological evolution in living organisms may appear as sudden change in the paleontological scale.

REFERENCES

1. Mayr E: Populations, Species and Evolution. Harvard University Press, 1970.
2. Eldredge N: The allopatric model and phylogeny in paleozoic invertebrates. Evolution 25:156, 1971.
3. Stanley SM: Macroevolution, Pattern and Process. San Francisco: W. H. Freeman, 1979.
4. Gould SJ: Is a new and general theory of evolution emerging? Paleobiology 6:119, 1980.
5. Levinton JS, Simon CM: A critique of the punctuated equilibria model and implications for the direction of speciation in the fossil record. Syst Zool 29:130, 1980.
6. Vrba ES: Evolution, species and fossils: how does life evolve? S Afr J Sci 76:61, 1980.

7. Stebbins GL, Ayala FJ: Is a new evolutionary synthesis necessary? Science 213:967, 1981.
8. Ayala FJ: Genetic differentiation during the speciation process. Evol Biol 8:1, 1975.
9. Bush GL: Modes of animal speciation. Annu Rev Ecol Syst 6:339, 1975.
10. Endler JA: Geographic Variation, Speciation and Clines. Princeton University Press, 1977.
11. White MJD: Modes of Speciation. San Francisco: W. H. Freeman, 1978.
12. Templeton AR: Modes of speciation and inferences based on genetic distances. Evolution 34:719, 1980.
13. Ayala FJ, Valentine JW: Evolving: The Theory and Processes of Organic Evolution. Menlo Park: Benjamin/Cummings, 1979.
14. Ayala FJ, Tracey ML, Hedgecock D, Richmond RC: Genetic differentiation during the speciation process in Drosophila. Evolution 28:576, 1974.
15. Ehrman L: Direct observation of sexual isolation between allopatric and between sympatric strains of the different Drosophila paulistorum races. Evolution 19:459, 1965.
16. Lewis H: Speciation in flowering plants. Science 152:167, 1966.
17. White MJD: Animal Cytology and Evolution. Cambridge University Press, 1973.
18. White MJD: Modes of Speciation. San Francisco: W. H. Freeman, 1978.
19. Nevo E, Shaw CR: Genetic variation in a subterranean mammal, Spalax ehrenbergi. Biochem Genet 7:235, 1972.
20. Nevo E, Kim YJ, Shaw CR, Thaeler CS, Jr.: Genetic variation, selection, and speciation in Thomomys talpoides pocket gophers. Evolution 28:1, 1974.
21. Benado M, Aguilera M, Reig OA, Ayala FJ: Biochemical genetics of Venezuelan spiny rats of the Proechimys guairae and Proechimys trinitatis superspecies. Genetica 50:89, 1979.
22. Carson HL: The population flush and its consequences. In Lewontin RC (ed): Population Biology and Evolution. Syracuse University 1968.
23. Carson HL: The genetics of speciation at the diploid level. Am Nat 109:83, 1975.
24. Bullini J, Coluzzi M: Ethological mechanisms of reproductive isolation in Culex pipiens and Aedes mariae complexes (Diptera Culicidae). Monit Zool Ital 14:99, 1980.
25. Mayr E: Animal Species and Evolution. Harvard University Press, 1963.
26. Dobzhansky T: Genetics of the Evolutionary Process. Columbia University Press, 1970.
27. Baur E.: Einführung in die Verenbungslehre. Berlin: Borntraeger, 1930.
28. Nei M: Genetic distance between populations. Am Nat 106:283, 1972.
29. Ayala FJ, Kiger JA: Modern Genetics. Menlo Park: Benjamin/Cummings, 1980.
30. Gottlieb LD: Enzyme differentiation in Clarkia franciscana, C. rubicunda and C. amoena. Evolution 27:205, 1973.
31. Gottlieb LD: Allelic diversity in the outcrossing annual plant Stephanomeria exigua ssp. carotifera (Compositae). Evolution 29:213, 1975.
32. Sene FM, Carson HL: Genetic variation in Hawaiian Drosophila. IV. Allozymic similarity between D. sylvestris and D. heteroneura from the island of Hawaii. Genetics 86:187, 1977.

33. Avise JC and Ayala FJ: Genetic change and rates of cladogenesis. Genetics 81:757, 1975.
34. Tsacas L, David J: Une septième espèce appartenant au sous-groupe Drosophila melanogaster Meigen: Drosophila orena spec. nov. du Cameroun. (Diptera: Drosophilidae). Beitr Ent 28:179, 1978.
35. Carr GC: Chromosome evolution and aneuploid reduction in Calycadenia pauciflora (Asteraceae). Evolution 29:681, 1976.
36. Carr GD: A cytological conspectus of the genus Calycadenia (Asteraceae): An example of contrasting modes of evolution. Am J Bot 64:694, 1977.

Mechanisms of Speciation, pages 67-74

Genetic and Nongenetic Mechanisms of Speciation

Jeffrey R. Powell

A CLASSIFICATION OF SPECIATION MECHANISMS

In discussing mechanisms of speciation, it is useful to define exactly what we mean by the term "mechanism of speciation." Of the many mechanisms or theories proposed, some have features in common and features distinctly contradictory. In order to understand which theories agree and disagree on what points, a scheme of classification will be presented which should enable a clear understanding of the issues involved. Next this paper will explore how previously proposed theories fit into this scheme and suggest ways the scheme can be used in interpreting data. Finally a somewhat novel, or at least ignored, mechanism of species formation will be discussed.

Any fully developed theory of speciation must deal with four, possibly five, different aspects of the process: ecological context, genetic change, cause of the genetic change, and the result of the change. The easiest way to understand what is meant by these four phrases is to examine Figure 1. This is not meant to be a complete listing of all the possibilities in each heading. Even with this partial listing there are 840 different ways of combining each aspect; that is, choosing one item in each column leads to 840 possible "pathways" through this chart. A fifth aspect, purposely left out, is temporal. Different theories of speciation predict different rates of speciation. This would add another column to Figure 1 and would increase the number of possible choices. However, for the sake of clarity, the issue of rates will not be explicitly dealt with in this paper.

GENETIC MECHANISMS OF SPECIATION

Of the myriad possible theories of speciation (Fig. 1), relatively few have been clearly articulated and supported by data. In fact there are only five. The

Ecological Context	Genetic Change	Cause	Result
Allopatric (Large)	Polygenic	Selection	Hybrid Sterility/ Inviability
Founder Events (Allopatric Small)	Coadapted Complexes	Recombination	Mate Recognition Change
	Chromosomal	Genetic Drift	Habitat Preferences
Parapatric	Polyploidy	Mutation	Temporal Isolation
Clinal	Single (Few) Major Gene	Migration	:
Sympatric	None (Little)	Combination Of Above	:
:	:	Symbiosis	:
Etc.	Etc.	Etc.	Etc.

Fig. 1. Scheme of classification of speciation mechanisms. The listing is not intended to imply a temporal sequence nor a complete listing.

first completely articulated and thorough model of speciation was delineated by Dobzhansky around 1940 [1,2]. This model can be outlined as follows:

Ecological context	Genetic change	Cause	Result
Allopatric, large	Polygenic	Selection (combination)	Hybrid sterility/ inviability

In outlining these models one can generally identify one phrase to describe the model in each category, although there is sometimes ambiguity. While Dobzhansky stressed the importance of natural selection in causing genetic divergence, for example, he explicitly included a role for mutation and recombination.

The first significant modification of the Dobzhansky model was presented by Mayr in 1954 [3]:

Ecological context	Genetic change	Cause	Result
Founder populations allopatric, small	Polygenic	Combination	Hybrid sterility/ inviability

Mayr stressed the importance of small peripherally isolated populations and the rapid "genetic revolutions" which may occur in such isolates. This author believes Mayr's use of the term "isolate" can be interpreted as allopatric (or "peripatric," see Mayr, this volume); the term genetic revolution can be interpreted as "polygenic." Mayr stresses a combination of causes of genetic differentiation: decrease in genetic variation (drift), relaxed selection, migration, etc. While not ruling out other forms of isolation, Mayr places emphasis on hybrid sterility or inviability.

Carson [4,5, and this volume] has elaborated on the Mayr model arriving at a significantly modified speciation mechanism:

Ecological context	Genetic change	Cause	Result
Founder events	Coadapted complexes	Recombination (drift) (migration)	Mate recognition

Two main aspects distinguish Carson's model from previously proposed models. First, he places great emphasis on the role of recombination breaking up coadapted gene complexes, especially those "closed" complexes. He also includes a role for drift in the founding event as well as a possible role for migration causing the initial population flush [6]. The second unique aspect of Carson's model is that reproductive isolation results as a by-product of the founder-flush-crash cycle(s); adaptive divergence is to a large degree independent of and follows reproductive isolation. While the form of isolation may be variable, Carson does stress premating barriers and my own work on this model [7] supports the notion that mate recognition changes may be the first isolating mechanisms to evolve.

White [8,9] deviates from these truly allopatric models in proposing his stasipatric model:

Ecological context	Genetic change	Cause	Result
Stasipatric	Chromosomal	Drift (selection)	Hybrid sterility/ inviability

White envisions the fixation, due to drift, of a chromosomal change in a small deme within the range of a species. In order to maintain and spread the new chromosomal type a selective advantage must be associated with it. The chromosomal change must be of a sort, e.g., translocation, such that hybrids (heterozygotes) are selected against.

The final model is that of Bush [10,11] who champions truly sympatric speciation via host race formation especially in phytophagous insects:

Ecological context	Genetic change	Cause	Result
Sympatric	Single (few) major gene	Mutation	Habitat preference

While sympatric speciation has not been widely accepted, Bush has made the strongest case for its existance; it is far from proven [12].

Figure 2 summarizes these five modes of speciation. With this scheme it is explicit just where the various models converge and where they differ. It is also clear that speciation is a multifaceted problem. Two interesting questions can be asked of a scheme such as illustrated in Figures 1 and 2: Of the large number of possible modes of speciation, why have only five been given serious consideration? Do different groups of organisms tend to follow different pathways through Figure 1? (It should be emphasized that by following pathways through this scheme, there is no implication to any temporal sequence of events.)

A NONGENETIC MECHANISM

While most mechanisms of speciation involve genetic divergence, there is a way in which reproductive isolation can arise with no (or little) genetic change. This mode of speciation has been ignored to a large extent although it was proposed, at least in one form, many years ago [13]. The mechanism involves infection by a microorganism which induces sterility between previously compatible populations. Two examples illustrate this phenomenon.

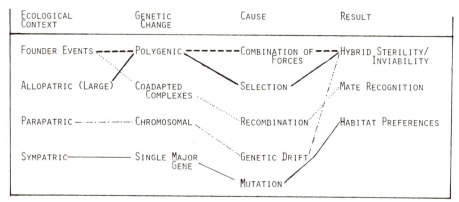

Fig. 2. An attempt to classify existing speciation mechanisms. (Solid dark line) Dobzhansky; (dashed line) Mayr; (dotted line) Carson; (dash-dot line) White; (solid light line) Bush. This chart indicates the primary factors in each theory; many authors include the possibly of two or more factors in a column.

TABLE I. Results of Crosses of Culex pipens Treated or Nontreated with Tetracycline Which Renders them Free of the Wolbachia Symbionts [15,16]

Female		Male	Result	Offspring
Infected	x	Infected	100% Compatible	Infected
Aposymbiotic	x	Aposymbiotic	80–90% Compatible	Uninfected
Infected	x	Aposymbiotic	80–90% Compatible	Infected
Aposymbiotic	x	Infected	Incompatible	None

The first case is that of the common house mosquito, Culex pipiens. Strains derived from different areas often exhibit incompatibility which may be either reciprocal or nonreciprocal. The cause of the incompatibility was shown to be maternally inherited, i.e., resides in the egg cytoplasm [14]. More recently Barr and his associates [15] have demonstrated that the causitive agent is a rickettsia-like bacteria called Wolbachia. Rearing strains with an antibiotic such as tetracycline renders them aposymbiotic and changes their reproductive compatibilities. Table I summarizes the situation. Presumably the mechanism involved here is some kind of "proto-immune" response. If sperm produced in a male infected with a particular strain of Wolbachia enter the cytoplasm of an egg infected with the same strain of Wolbachia, fertilization proceeds normally. If the sperm enter the cytoplasm of an uninfected egg or an egg infected with a different strain of Wolbachia, the egg cytoplasm responds by killing the sperm before they affect fertilization. While this phenomenon of cytoplasmic incompatibility has long been known for strains of different geographic origin, more recently Barr [17] has observed incompatibility among sympatric strains. The population dynamics of such symbiote-induced incompatibility have been treated mathematically and conditions for its maintenance and spread are known [18].

We have been interested in determining whether a symbiont might be responsible for the sterility observed in crosses between certain populations of Drosophila pseudoobscura. Prakash [19] and Dobzhansky [20] report that crosses between females derived from Bogotá, Colombia and males from anywhere in North America, yield completely sterile F_1 males; F_1 females are fertile as are both sexes in reciprocal crosses. It should be noted that the Bogotá strains used by both of these workers had been in laboratory culture for ten or more years. Some authors [21] have cited this Bogotá isolate as a paradigm of the first stages of speciation.

In January 1979, I obtained several freshly collected strains of D pseudoobscura from the vicinity of Bogotá. A series of crosses with N. American strains were made; the strains had been in culture about two generations at the start. Some of the crosses were done on standard medium and others on the same medium treated with tetracycline. F_1 males were tested for fertility with their

TABLE II. Results of Crosses Between Bogotá Females and North American Males of D. pseudoobscura*

Bogotá strains in culture		Number of crosses	Mean number of F_2 adults per cross
Two generations	Control	4	43.50
	Treated	12	254.25
Four to five generations	Control	32	2.35
	Treated	48	11.05
5 Years	Control	16	0
	Treated	24	0

*In each cross about ten flies of each sex were the parents; approximately 100 F_1 males per cross were tested for fertility.

sibling females in mass culture. Table II summarizes the results. In the first experiment when the strains had been in culture two generations, tetracycline-reared males were six times as fertile as the control males. However, contrary to previous reports [19,20], the controls (untreated medium) exhibited a low level of F_1 male fertility. A second set of crosses were done after the Bogotá strains had been in culture four or five generations. The F_1 male fertility was again greater for individuals reared on antibiotic-treated medium as compared to untreated controls, although the controls still exhibited a low level of fertility. However for both controls and treated crosses the level of fertility was much lower than in the first experiment.

Because of this evident decrease in fertility after the strains were in culture a few generations, several crosses were made with Bogotá strains which had been in laboratory culture about five years. Both untreated and tetracycline-treated medium was used. Of the total of 40 different crosses made, none of the F_1 males exhibited any fertility.

Two conclusions can be drawn from these experiments. First, rearing F_1 males on medium treated with an antibiotic increases their fertility. This is not "proof" that a microorganism is the cause of the infertility. However, the results certainly hint at a role for a symbiont in this "paradigm" of the first stages of speciation.

Second, and equally important, is the evident change in reproductive isolation associated with laboratory culturing. This is not an unique finding. It has been noted in another group of Drosophila, the D paulistorum complex [22,23]. Significantly, microorganism have been implicated as a cause of sterility in this group [24]. An explanation for these findings is that laboratory culturing some-how alters the host-symbiont relationship such that crossing compatibilities are changed. This should be taken as a note of caution, especially to Drosophilists.

That endosymbionts may play an important role in the evolution of their hosts, has been noted by others [25]. It may be a more widespread phenomenon than has been appreciated heretofore. Describing a role for symbionts in speciation with respect to the scheme previously outlined, we arrive at:

Ecological context	Genetic change	Cause	Result
?	Little or none	Symbiosis	Hybrid sterility/ inviability

While most often, symbiotic relationships of the sort described here probably evolve in allopatry, the finding of three crossing types of Culex pipiens in sympatry [17] leaves open the possibility such phenomena could occur without geographical isolation. Whether any genetic change occurs in the host to "coadapt" to the symbiont is also unknown. The fact that "cured" strains may be nearly 100% interfertile [16] indicates that the type of genetic change required to live with symbionts may not affect reproductive isolation in the absence of the symbiont. This type of speciation mechanism may therefore be called nongenetic. While it is unknown how important and widespread such phenomena may be, in keeping with the theme of this volume, they should be considered in our view of the plurality of speciation mechanisms.

REFERENCES

1. Dobzhansky T: Speciation as a stage of evolutionary divergence. Am Nat 74:312, 1940.
2. Dohzhansky T: Genetics and the Origin of Species. 2nd Ed. New York: Columbia University Press, 1941.
3. Mayr E: Change of genetic environment and evolution. In Huxley J (ed): Evolution as a Process. New York: Macmillan, 1954, pp 157–180.
4. Carson HL: Speciation and the founder principle. Stadler Symp 3:51, 1971.
5. Carson HL: Genetics of speciation. Am Nat 109:83, 1975.
6. Carson HL: Increase in fitness in experimental populations resulting from heterosis. Proc Natl Acad Sci (USA) 44:1136, 1958.
7. Powell JR: The founder-flush speciation theory: an experimental approach. Evolution 32:465, 1978.
8. White MJD: Models of speciation. Science 159:1065, 1968.
9. White MJD: Modes of Speciation. San Francisco: W.H. Freeman, 1978.
10. Bush GL: Modes of animal speciation. Annu Rev Ecol Syst 6:339, 1975.
11. Bush GL: Sympatric speciation in phytophagous parasitic insects. In Price PW (ed): Evolutionary Strategies of Parasitic Insects and Mites. New York: Plenum, 1975, pp 187–206.
12. Futuyma DJ, Mayer GC: Non-allopatric speciation in animals. Syst Zool 29:254, 1980.

13. Kitzmiller JB, Laven H: Current concepts of evolutionary mechanisms in mosquitoes. Cold Spring Harbor Symp Quant Biol 24:173, 1958.

14. Laven H: Speciation by cytoplasmic isolation in the Culex pipiens complex. Cold Spring Harbor Symp Quanta Biol 24:166, 1958.

15. Yen JH, Barr AR: The etiological agent of cytoplasmic incompatibility in Culex pipiens. J Invert Pathol 22:242, 1973.

16. Barr AR: Personal communication or Symbiont control of reproduction in Culex pipiens. In Advances in Invertebrate Reproduction. eds W.H. Clark and T.S. Adams, p. 382. Elsevier/North Holland Press.

17. Barr AR: Cytoplasmic incompatibility in natural populations of a mosquito, Culex pipiens L. Nature 283:71, 1980.

18. Fine PEM: On the dynamics of symbiote-dependent cytoplasmic incompatibility in culicine mosquitoes. J Invert Pathol 30:10, 1978.

19. Prakash S: Origin of reproductive isolation in the absence of apparent genetic differentiation in a geographic isolate of Drosophila pseudoobscura. Genetics 72:143, 1972.

20. Dobzhansky T: Genetic analysis of hybrid sterility within the species Drosophila pseudoobscura. Hereditas 77:81, 1974.

21. Lewontin RC: The Genetic Basis of Evolutionary Change. New York: Columbia University Press, 1974.

22. Dobzhansky T, Pavlovsky O: Experimentally created incipient species of Drosophila. Nature 230:289, 1975.

23. Dobzhansky T, Pavlosky O: Unstable intermediates between Orinocan and Interior semispecies of Drosophia paulistorum. Evolution 29:242, 1975.

24. Ehrman L, Williamson DL: On the etiology of the sterility of hybrids between certain strains of Drosophila paulistorum. Genetics 62:193, 1969.

25. L'Héritier P: Drosophila viruses and their role as evolutionary factors. Evol Biol 4:185, 1970.

Mechanisms of Speciation, pages 75–103
© 1982 Alan R. Liss, Inc., 150 Fifth Avenue, New York, NY 10011

Rectangularity, Speciation, and Chromosome Architecture

M.J.D. White

For some time now there have been signs that the consensus that existed among biologists in the period 1940–1965 with regard to the fundamental nature of evolutionary processes is being subjected to critical re-examination. While it would be premature to say that a new consensus has been reached, it is probable that a major scientific revolution is under way. Advances in biological knowledge, coming from the application of new cytogenetic, molecular and biochemical techniques to a wide variety of organisms have accumulated to such an extent that the much vaunted neo-Darwinian "modern synthesis" of a quarter of a century ago is now seen as the beginning rather than the culmination of a truly profound interpretation of evolution, a preliminary body of theory that must now be modified and superseded in many respects. At the same time the gradual accumulation of palaeontological data on a large number of groups of organisms has forced a major re-evaluation of ideas on macroevolution [1]. Gould [2] has actually asked the question: "Is a new and general theory of evolution emerging?"—and has answered it decisively in the affirmative. Elsewhere [3], the same author has expressed the view that the 'modern synthetic' version of neo-Darwinism, although "still a reigning orthodoxy among evolutionists" is "on the verge of crumbling." Similar views have been expressed by Stanley.

These are strong statements which come from palaeontologists who have found the fossil record simply incompatible with the accepted orthodoxy. However, traditional views are still expressed by some palaeontologists [4,5].

Speciation or cladogenesis, the splitting of evolutionary lineages, is a complex phenomenon that has many aspects. On the one hand, we can consider the populational and adaptive changes that accompany speciation, as they are revealed by morphologic, biometric, ethologic, ecological, and biogeographic studies. This is what the majority of evolutionists do. But, equally, we need to understand the whole range of complex molecular changes that take place within the chromosomes in the course of speciation and evolution. Molecular biology as well as palaeontology calls out for a new interpretation of evolution.

For a complete understanding of any single case of speciation, studies in molecular and developmental biology, cytogenetics, ethology, ecology, biometry, and biogeography are all necessary. Because there are many elements or components in the speciation process, it is natural that they should interact differently in organisms having diverse forms of morphological organizations, genetic systems, population dynamics and life styles. It is the patterns formed by the interaction of these components that generate the different modes of speciation.

For some time past a controversy has continued regarding the geographic component of speciation. The issue here has been as to whether complete geographic separation of the speciating populations is an essential precondition of speciation. Such separation was the basis of the original allopatric model of Mayr [6] and the modified one of the same author [7], invoking the founder principle. The alternative view, namely that speciation can also occur sympatrically or from contiguous populations that remain in contact (so-called parapatric populations) has been put forward by a number of authors [8–17] and is now widely accepted. On the latter interpretation we have several different modes of speciation that differ in the geographic component and we face the task of determining their relative roles in particular groups of organisms. However, some evolutionists continue to support Mayr's original position. Thus Futuyma and Mayer [18] find the evidence for non-allopatric speciation unconvincing.

These geographic questions are undoubtedly important, so much so that for some they have seemed to be the main issue in speciation studies. However, recent discoveries in cytogenetics and particularly advances in our understanding of the molecular architecture of eukaryote chromosomes have now opened up an entirely new area of inquiry in speciation theory. We must now ask the question: to what extent is speciation due to the accumulation of ordinary allelic substitutions (the classical view), to major chromosome rearrangements such as translocations, fusions and inversions, or to changes in the repetitive DNA sequences in the chromosomes that do not seem to code for specific proteins?

The classical neo-Darwinian view of evolution, of which the allopatric models of speciation were an integral part, relied (in the words of Gould [3]) "on the belief that genetic variation is copious, small in extent and available in all directions." It neglected chromosomal rearrangements and assumed tacitly that the number of genes (ie structural genetic loci) was very large, of the order of 10^5 or 10^6. Within neo-Darwinism there was a good deal of argument about population structure, the importance or unimportance of random drift and peak shifts, and the role of genic pleiotropism [19]. But in the absence of molecular biology the viewpoints were necessarily schematic and abstract. The advent of gel electrophoresis to study allelic variation did not fundamentally alter this situation.

A relative lack of interest in the precise genetic causes of speciation was natural so long as it was generally supposed that the process was invariably allopatric, because in that case the "cause" of speciation would be the extrinsic geographic separation of populations that would simply diverge, in isolation, by ordinary phyletic evolution—special genetic causes of cladogenesis would not exist. There can, of course, be no doubt that allopatric types of speciation do occur, and that in some groups of organisms they are probably the predominant method. However, evolutionists have recently come more and more to accept the idea that a great many speciation events do not involve complete geographic separation. In these cases of non-allopatric speciation (semi-geographic and sympatric in Mayr's terminology) we must look to specific genetic events as the primary causes of the original divergence. In non-allopatric speciation it is fairly clear that the initial dichotomy-producing events precede phenotypic change of a type that would be detected by taxonomic study. Thus taxonomists will almost always assign to infra-specific categories populations that have only recently become irreversibly committed to divergence; the essential first stage of non-allopatric speciation, whatever the precise mechanism, is invisible to them unless they employ genetic or cytogenetic techniques of analysis.

The search for special primary causes of speciation should hence be directed on the one hand towards unusual ecological opportunities (new unoccupied niches) and on the other to special types of genetic changes, perhaps especially those that are above the level of single allelic substitutions. The negative and pessimistic conclusions of Lewontin [20] regarding mechanisms of speciation were the outcome of attempts to interpret it as the result of mutations of structural genes of the type revealed by gel electrophoresis and, more generally, as generated by the "normal" processes of population genetics as they operate in allelically polymorphic populations. The entirely different approach adopted here, even if somewhat speculative, seems likely to prove more fruitful. It is clear that in creating a general theory of speciation we shall have to distinguish very clearly between different stages of the process.

The modern picture of the architecture of the eukaryote genome is an increasingly complex one. Ever since Britten and Kohne [21] showed that a large fraction of the nuclear DNA consisted of short sequences of bases tandemly repeated tens of thousands or even millions of times, it has gradually become apparent that only a relatively small fraction of the total DNA consists of 'unique' sequences organised into structural genes (of which there may be no more than 10,000 in most eukaryotes), and that the rest of the DNA consists of "moderately repetitive", "highly repetitive", and "foldback" nucleotide sequences. Associated with the single copy DNA of the structural genes are some adjacent spacer sequences and the so-called introns, a special type of spacer occupying interstitial positions in "split" genes. The role of movable DNA segments like the *copia*

sequences of Drosophila melanogaster in eukaryote genomes is not yet understood [22]. In fact, the whole complicated picture of the molecular architecture
of chromosomes has not really been brought into a meaningful relationship with
evolutionary theory as yet.

Before attempting to discuss the possible roles of the different DNA fractions
in speciation, we must clarify and define what we mean by genetic changes (of
whatever kind) "causing" speciation. There can be no doubt that a wide variety
of genetic and cytogenetic changes regularly accompany speciation, so that
species that have diverged in evolution and maintained their distinctness for
some time will show allelic differences and differences in karyotype, the latter
including both those due to major structural rearrangements and those resulting
from reorganization of the detailed molecular architecture. At least some of these
differences will have accumulated after the completion of the initial dichotomy
and will have had nothing to do with its causation. All kinds of changes associated
with speciation should be investigated and are clearly relevant to a full understanding of the evolutionary process as a whole. But the thing we especially need
to know in each case is what type of genetic event has played the primary role
in initiating divergence, to the point where it became irreversible.

Identifying the primary genetic change in any particular speciation event is
obviously difficult, for several reasons. Speciation can only be detected post
factum, when subsequent genetic changes that have had nothing to do with the
original dichotomy may have accumulated. Moreover, to a considerable extent
we do not know what we are looking for. We may be seeking to identify a single
genetic event that has played a key role. But there is a possibility that in many
instances speciation is due to the occurrence in rapid succession of several genetic
changes of the same kind, having a cumulative effect in a population. The
chromosomal fusions in certain populations of mice (Mus musculus) in Italy and
other Mediterranean countries is discussed by Capanna elsewhere in this volume.
That these rearrangements are very recent is suggested by the fact that although
they act as powerful isolating mechanisms on account of the reduced fecundity
of the natural hybrids [23, 24] there are no significant allozyme differences
between 22- and 24-chromosome populations and neighboring 40-chromosome
populations [25]. In spite of the absence of allozymic differentiation between
these mouse populations they are clearly potential incipient species. Equally
striking examples of multiple repeated chromosomal changes have led, no doubt
after longer periods of time, to new species of cotton rats, Sigmodon and musk
deer, Muntiacus [26, 27]. A situation somewhat similar to the Mus musculus
one exists in the rodent Ellobius talpinus which has a uniform karyotype of 2n
= 54 acrocentric chromosomes from the Crimea to Mongolia but has produced
a population with chromosome numbers ranging down to 2n = 31 in one valley
of the Pamir-Alay mountains [28].

It is certainly not only major chromosomal rearrangements such as fusions that have established themselves in high frequency in certain populations and evolutionary lineages. Many cases are known where there have been sequential amplifications of numerous heterochromatic chromosome segments in particular races and species.

Such an acceleration of the rate of occurrence, or incorporation into the population, of particular types of chromosomal change, may be a genetic equivalent of the punctuational or rectangular model of evolution [1, 29–33] according to which a major part of the phenotypic changes that occur in evolution are concentrated in speciation events. This model has not been accepted by all palaeontologists, but it is now supported by a considerable body of evidence. Clearly, however, "rectangularity" is not a universal concomitant of speciation, or we should not have the existence of "sibling species," with minimal morphological gaps between them, which are frequent in many groups of invertebrates. What the palaeontologists have been postulating are periodic bursts of morphological change, without regard to the underlying genetic causes. But if the model corresponds to reality, at least in some groups, it is clear that the causative genetic events, whatever their precise nature (allelic substitutions, structural rearrangements of chromosomes or amplifications or de-amplifications of repetitive nucleotide sequences) must also exhibit rectangularity. However, even if rectangularity is a genuine phenomenon, the question is still open as to how far "rectangular" events cause speciation or merely follow it.

The concept of rectangularity in speciation appears to have developed from Mayr's [7] idea of a "genetic revolution" occurring when a small localized population of a species is isolated allopatrically. Thus Eldredge and Gould [28] derive "punctuated equilibria" from 'allopatric theory' and the views of Stanley [1] do not seem essentially different. However, it will be argued below that rectangularity is far more applicable to cases of non-allopatric speciation. A highly fruitful scientific concept appears to have had a dubious intellectual ancestry.

The genetic processes that occur in allopatric speciation by large geographically isolated populations should be essentially similar to those that go on all the time in phyletic evolution. The separated populations will have similar modes of life and even if their habitats are somewhat different there seems little reason to expect any marked rectangularity to occur as a result of natural selection. Mayr's concept of a "genetic revolution" following on speciation by the "founder principle" remains of doubtful general importance. Speciation by a very small number of founding individuals migrating to an unoccupied territory must be a rare mode except in archipelagos, like the Hawaiian one that has figured so largely in recent evolutionary discussions, but which is really a very exceptional geographic situation.

If it is true that little or no morphological rectangularity is to be expected in strictly allopatric speciation, and if the limited geological time available for great morphological transitions (eg primitive unspecialized mammals to bats and whales, following the Cretaceous-Cenozoic catastrophic mass extinction*, as discussed by Stanley forces us to the view that rectangularity has indeed occurred, it inevitably follows that it must have been associated with non-allopatric speciation events. Direct evidence to this effect, based on living species, is not lacking. Probably sympatric speciation occurs mainly or exclusively by trophic specialization, a switch to a new food source or other vital requisite. A change in basic life style will impose strong selective pressures for rapid change in a whole range of morphological characters. Cases of so-called explosive speciation in freshwater fishes exemplify this situation. A striking instance of trophic specialization under conditions of sympatry has been described in the Mexican fish genus Ilyodon [36]. The trophic types of Ilyodon, which are probably incipient species, differ greatly in morphology, especially as regards head shape and dentition. They may represent a very early stage in rectangular-sympatric speciation. A much later stage would be the 18 species of cyprinid fishes (very diverse morphologically) that have probably arisen from a single ancestral form in Lake Lanao, in the Philippines [37], possibly in no more than 10,000 years. It is likely that the species of Cyprinodon from Death Valley and adjacent river basins [38] fall into the same general picture, although here the evidence for the speciation processes having been sympatric is much less clear and some, at least, may have been allopatric.

Rectangular-sympatric speciation is also known in other families of fishes. Kirkpatrick and Selander [39] investigated two types of Whitefishes (Coregonus clupeaformis) in the Allegash River Basin of northern Maine. These differed

---

*It is probable that the mass extinction of all large animals (approximately those >25 kg) and most surface-living marine organisms at the end of the Cretaceous was caused by a global catastrophe (an encounter of the earth with cometary or intergalactic material in dispersed form or with an asteroid about 10 km in diameter). The postulated collision, whatever its precise nature, deposited greatly increased levels of iridium and osmium in a layer of rock of precisely this age at a number of widely separated localities [34, 35]. Such an astronomical catastrophe would have caused great clouds of dust that would have cut off solar radiation for a number of years and may have also given rise to extensive chemical changes of a global character. If this catastrophe actually occurred (and the evidence for it is now very strong), conditions for speciation in many of the surviving lineages may have been extremely abnormal for some time afterwards; the main adaptive radiation of the mammalian orders may have been facilitated by the creation of numerous vacant niches that were made available for occupation by 'relaying' groups. It is only fair to point out that not all palaeontologists accept an astronomical catastrophe at the end of the Cretaceous as the cause of mass extinctions, which the dissenters claim were not precisely synchronous.

greatly in size ("normals" and "dwarfs"). They are partially isolated by a difference in spawning time and appear to have separate gene pools. The overall genetic distance between them, based on allozymes, is no greater than between populations of a single morphotype from different lakes. The authors postulate that speciation has involved regulatory systems governing rate of maturation and time of spawning.

The now-classic example of speciation by trophic specialization (host-plant change) in the Tephritid fruit flies of the genus Rhagoletis [10, 11, 40, 41] has involved changes in relative body size, number of post orbital bristles and ovipositer length as well as in the seasonal cycle of the insects. These differences have arisen in the course of the last 116 years in sympatric populations that now live on different species of host plants. On the geological time scale such changes are indeed rapid and indicate strong rectangularity. However, in this case as in Coregonus and Cyprinodon there has only been minimal change in the allozyme loci. We shall discuss this apparent paradox later (p 95).

To summarise this discussion we may say that, as concluded by Stanley [1] there is now strong evidence that morphological evolution during anagenesis (phyletic evolution) is slow and that most morphological change occurs during cladogenetic episodes. But, contrary to the view of the original protagonists of the concept, it now seems that there is little reason to associate rectangularity with allopatric speciation and that it is most conspicuous in the various types of non-allopatric speciation, where the new incipient species must rapidly adapt to an entirely new mode of life. Rectangularity is then produced by natural selection, not by any sudden outburst of new macromutations. For present purposes it is convenient to distinguish three modes of non-allopatric speciation:

1) invasive chromosomal speciation, 2) stasipatric speciation, and 3) sympatric speciation.

The fact that the great majority of animals species differ in karyotype from even their closest relatives [14] is prima-facie evidence that chromosomal rearrangements such as fushions and fissions, inversions and translocations have frequently played a part in speciation. It is true that in Drosophila and some other genera of Diptera with giant polytene chromosomes there exist some complexes of species that have identical karyotypes, including the finest details of the banding pattern as seen in the polytene chromosomes. However, such homosequential species [42–46] are rare, in those genera (Drosophila, Anopheles, Chironomus, Simulium) in which they are known to exist. Moreover, it is not certain that the heterochromatic chromosome segments, which contain the highly-repetitive DNA's, are identical in size, molecular organization, and location in homosequential species pairs, since these regions are not easily studied in the polytene chromosomes where they are relatively underreplicated. In fact, one pair of Mexican homosequential *Drosophila* species do differ in respect to heterochromatic segments [47]. How far homosequential species occur in organisms not having polytene chromosomes is uncertain, but so many karyotypic differences between closely related species are now known, in almost all groups, that they must be generally rare.

Although homosequential species prove that speciation can occur without major chromosomal rearrangements, it is now clear that most speciation events do involve such rearrangements. However, it is less certain that chromosomal rearrangements usually or frequently play a direct causative role in speciation; the alternative being that they follow on speciation. But there is plenty of evidence for chromosome rearrangements frequently occurring very early in the speciation process, in the origin of "chromosomal races."

Major chromosomal rearrangements exhibit a spectrum of effects on fecundity in the heterozygotes. At one end of the range are those, such as paracentric inversions in Drosophila, many pericentric inversions in grasshoppers and centric fusions and dissociations in many (but not all) cases; these when heterozygous do not reduce fecundity at all, or only by a minimal amount. At the other end of the scale are those, like pericentric inversions in Drosophila, tandem fusions, and translocations of many types, that invariably lead to a considerable degree of sterility (up to 50%) in the heterozygote. Obviously, the former kind (which frequently exist in natural populations as floating polymorphisms) are unlikely to act as primary isolating mechanisms in speciation. The latter, on the other hand, can constitute effective isolating mechanism, even if they do not lead to complete hybrid sterility, and must be suspected of being important in speciation, at least in those groups where a cytotaxonomic survey demonstrates precisely these types of karyotypic differences between species.

Those authors who minimize or doubt the role of chromosomal rearrangements in speciation, tend to stress those cases (eg the grasshoppers Podisma pedestris and Caledia captiva) where their direct effect on fecundity in the heterozygote appears to be small or non-existent. But it is equally legitimate to point to cases where the fecundity of the heterozygotes is much reduced. The XY and XO "races" of the grasshopper Vandiemenella "P45b" (probably good biological species, even by Key's [48] stringent criterion) differ by a tandem fusion of the X-chromosome to an autosome, that would automatically reduce the fecundity of the heterozygous females by 50% [49]. Tandem fusions between autosomes, such as have undoubtedly occurred in the phylogeny of the grasshopper Dichroplus silveiraguidoi [50, 51] and probably in some other grasshoppers [52] as well as in such mammalian genera as Muntiacus [27] and Sigmodon [26] will have an equally drastic effect on heterozygote fecundity, in this case manifested in both sexes. Of course, tandem fusions are uncommon in phylogeny. But the above instances, as well as several in the Chironomidae, are enough to dispose of arguments according to which tandem fusions aught never to establish themselves in natural populations. Clearly, they could never exist for more than a few generations in a polymorphic state in a population; they must either achieve homozygosity, thereby initiating a cladogenesis, or else be eliminated by natural selection. Many translocations are in essentially the same situation. Some centric fusions between large acrocentric chromosomes lead to full fertility in the heterozygote because trivalent formation is quite regular. But centric fusions between short acrocentrics, of which at least twelve have occurred in the phylogeny of the Australian morabine grasshoppers, are very liable to diminish the fecundity of heterozygotes, because trivalent formation frequently fails. Hence they are most unlikely to establish themselves as floating polymorphisms but liable to act as isolating mechanisms between speciating populations.

Many evolutionists still feel, however, that the establishment in natural populations of chromosomal rearrangements that are deleterious in the heterozygote is so difficult to conceive that such rearrangements cannot play a significant role in speciation. This is essentially the standpoint of Futuyma and Mayer [18] who, however, present no new evidence on the matter. But recent mathematical studies have put the matter in a somewhat different light. Thus Lande [53] has shown [1] that the initial fixation of a rearrangement leading to heterozygote disadvantage can only happen by random genetic drift in a small deme (a few tens to a few hundreds of individuals), but that in such populations it is plausible, and [2] that "once established in a deme, a negatively heterotic gene arrangement can spread in homozygous form through a subdivided population by random local extinction and colonization," with a probability approximately equal to its rate of establishment in a single deme. Clearly, most terrestrial invertebrates and many species of reptiles, rodents, etc. do have populations that are subdivided in such a way as to render this process plausible. It is also possible that in many instances meiotic drive, especially in the heteropolar female meiotic divisions, may favor the success of a newly arisen rearrangement [14].

The mathematical analysis has been taken a stage further by Hedrick [54] who concludes that there are four situations that, theoretically, can lead to the fixation of a chromosomal rearrangement of the type we are discussing: 1) meiotic drive alone, 2) meiotic drive plus genetic drift, 3) inbreeding plus a selective advantage of the new homozygous type, and 4) inbreeding and genetic drift. Bengtsson and Bodmer [55] may be quite correct in their conclusion that "fixation of chromosome mutations by drift only occurs under special, and presumably very rare, circumstances. But, from the population genetics standpoint speciation events are "special and very rare" episodes (a point that too many population geneticists have ignored).

If then, chromosomal speciation based on heterozygote inferiority has been a major mechanism of speciation in many groups of organisms, the evidence suggests two modes which may be designated *invasive* and *stasipatric*. The first occurs when a chromosomal rearrangement establishes itself in a peripheral population of a species that is then able to invade an area previously unoccupied by the species. A particularly good example seems to be the repeated speciation of the Venezuelan spiny rats of the genus Proechimys [56, 57]. The rodents of the genus Spalax in Israel are almost certainly another [58]. Probably the chromosomal speciation of the Australian geckos of the Phyllodactylus marmoratus complex followed this model [59], although that of the Diplodactylus vittatus complex seems to have been of the stasipatric type [60].

Stasipatric speciation [13, 14, 49] results from chromosomal rearrangements establishing themselves within the range of a species and spreading out from the point of origin. It is an "internal" mode of speciation, as contrasted with the "external" invasive model, and results in a derived species having a central location with the ancestral species peripheral to it. The original example of this mode of speciation was in certain wingless Australian grasshoppers of the genus Vandiemenella (formerly referred to as the "viatica group") [49, 61]. Other particularly clear examples (although undoubtedly representing different stages of speciation) are the forms of the Australian stick insect Didymuria violacea complex [62, 63], although there is room for argument as to just what stage in speciation has been attained in each of these cases.

The operation of these types of chromosome speciation is especially impressive when they have occurred repeatedly in the same manner. In the Proechimys case six incipient species seem to have been produced *seriatim* by chromosomal fusions or dissociations. In Didymuria the original $2n\male = 39$ population gave rise to another 39-chromosome form (by an inversion in the X-chromosome) and by successive fusions to incipient species having $2n\male = 37, 35$ and 33 (the latter apparently now extinct), 32 (XY), 31 32 (XY), 30(XY), 28 (XY, two different races) and 26 (XY). The distributions of all these races are apparently contiguous (parapatric) and some natural hybrids have been found where they meet. The case of the chromosomal fusions in populations of *Mus musculus* is dealt with by Capanna elsewhere in this volume. Any complete interpretation of this interesing case of chromosomal speciation in progress must take into account the peculiar sociosexual behavior of the mouse, which involves strong heterogamy [64–66]. It was suggested [67] that the chromosomal fusions, which clearly generate powerful postmating isolating mechanisms, might protect local populations with coadapted genotypes from introgression that would destroy their "area effects" (complexes of genes adapted to a local environment). Such an effect might be important in a species like the mouse, with strongly xenophilic females that accept alien males with especial readiness and even abort already fertilized ova and go back into estrus if they meet a male from another population.

In both invasive and stasipatric speciation it must be supposed that the new homozygous type has a biological advantage. In the latter model this must be a selectional advantage over both the heterozygote and the old homozygous type; in the invasive model the advantage must be an adaptational one that enables the new genotype to survive in an environment where the original one could not. Some evolutionists have argued against a role for chromosomal rearrangements in speciation on the basis of cases where two parapatric races differ karyotypically in respect of rearrangements for which one or both are polymorphic. There may be some legitimate arguments against the stasipatric speciation model. But to attack it on the basis of cases that are clearly or probably *not* instances of stasipatric speciation does not seem to be a justified procedure. Furthermore, to argue that heterokaryotype fitness is not reduced simply because some backcross of F_2 hybrids have been found in nature is clearly wrong.

The zone of hybridization in south-eastern Queensland between the 'Torresian' and 'Moreton' incipient species of the Caledia captiva grasshopper complex has been studied in depth by Shaw and colleagues [68–75] and is probably not a case of chromosomal speciation in the strict sense. Unlike the Spalax and Mus musculus cases the genetic distance between the parapatric taxa, based on allozyme data, is quite high ($\overline{D} = 0.18$, compared with 0.022 in Spalax and 0.008 in Mus). In general, populations speciating according to the invasive or stasipatric models are characterised by very low levels of allozymic divergence, either because they only develop allozymic differences very slowly, or because the cases that have been studied are of very recent origin. The much greater divergence of the Moreton and Torresian taxa of Caledia suggests either that they are more ancient, or that the contact between them is a secondary one (ie that parapatry was preceded by allopatry in this case).

The karyotype of the Torresian taxon consists of acrocentric or telocentric chromosomes which have only small blocks of heterochromatin around the centromeres. The Moreton taxon, on the other hand, is extensively polymorphic for chromosomal rearrangements, with metacentric and submetacentric chromosomes frequent (the "pure Moreton" populations are homozygous for seven metacentric autosomes). Furthermore the Moreton chromosomes carry numerous interstitial and terminal segments of constitutive heterochromatin [71].

Mating appears to be random in the hybrid zone which is less than 1 km wide. F_1 hybrids are quite normal in viability and fertility, but F_2 individuals are inviable and in backcross generations viability ranges from zero to 50%. The embryonic mortality appears to be due to the production of new types of lethal or sublethal recombinant chromosomes by the F_1 parent. It is quite uncertain whether recombination in this case is between structural genes or involves the C-bands present in the Moreton taxon but absent from the Torresian one.

Sympatric speciation occurs where there is no geographic separation of the two taxa, but only an ecological one, or a trophic one (eg feeding on different plant species). The best studied cases are undoubtedly in phytophagous insects. Species that have differentiated sympatrically seem, in general, to show little difference in respect of allozyme loci. Thus Berlocher and Bush [76] state that "some species of Rhagoletis have essentially identical gene pools" and "Differentiation of enzyme genes is minimal between R pomonella, R mendax and zephyria, as would be expected if these genes are not involved in host selection, host survival, etcetera." Bush's model of sympatric speciation in Rhagoletis involves an allelic substitution at a *host selection* locus (H_1H_1 individuals recognize only the original food plant, H_1H_2 flies recognize both hosts, H_2H_2 ones only the new host). This is then followed by mutation at a *survival* locus (S_1S_1 larvae survive on the old host plant fruits but not on the new species of fruit, S_1S_2 survive on either and S_2S_2 ones survive only on the new host. Later mutations affect diapause and emergence times.

The speciation of Diprionid sawflies on North American conifers seems to have followed essentially the same course as in Rhagoletis, with two significant differences [12]: males are haploid and arise from unfertilized eggs, so that the population-genetics aspect is different, and the conifers contain various toxic chemicals, so that successful speciation necessarily involves the acquisition of novel detoxification mechanisms.

Sympatric speciation has probably played a major role in all those groups of monophagous phytophagous insects. It has probably been much less important in polyphagous, scavenging or saprophytic arthropods. A case where it seems almost impossible to avoid a sympatric interpretation is that of the flower-inhabiting species of Drosophilella described by Carson and Okada [77]. Both in New Guinea and on Taiwan-Okinawa there exist pairs of species one of which breeds in the male flowers, the other in the female flowers on the hermaphroditic inflorescences of various species of Aroids. The obvious interpretation would seem to be that in each region there has been a case of sympatric speciation involving a switch from male to female flowers or vice-versa. But Carson and Okada prefer an interpretation according to which the cladogenesis between male flower inhabiting and female flower inhabiting lineages took place a considerable time ago, perhaps on the mainland of Asia. They may well be correct, but that only pushes the sympatric speciation event further back in time. An allopatric interpretation would have to involve a geographic migration of an ancestral species from female flowers in one region to male flowers in another (or vice-versa), a process that seems highly improbable. Numerous analogous cases where closely related species of insects live on different parts of the same plant species are known in the gall-forming Cecidomyidae.

Clearly, the "primary" genetic event or events in any case of speciation must create a genetic isolating mechanism of some kind. For a long while it was believed by many evolutionists that the first type of isolating mechanism to arise (and hence the initiator of divergence) was usually or always an ethological one concerned with sexual or trophic behavior. Mayr [78] stated unequivocally that "Ethological barriers to random mating constitute the largest and most important class of isolating mechanisms in animals," and Brncic Nair and Wheeler [79], whose ideas were based on Drosophila speciation, wrote: "Speciation phenomena depend primarily on behavioral traits such as those that determine reproductive isolation, or a particular manner of niche exploitation and co-existence."

Now it is undeniable that this is sometimes the case. Bush's [10, 11] model of speciation in monophagous trypetid flies depends on mutation at a host-recognition genetic locus and Tauber and Tauber's [16, 17] model is similar in principle. However, there are many instances where the evolution of post-mating isolation seems to precede any pre-mating isolating mechanisms; this is so in the case of Drosophila pseudoobscura with its subspecies bogotana and D willistoni with the subspecies quechua [14, 80, 81].

Genetic models of sympatric speciation by trophic specialization have assumed the primary events to be a small number of allelic substitutions at critical gene loci. No morphological rectangularity is assumed, there is no general "genetic revolution" and no chromosome rearrangements, either of a major or a minor kind are postulated. We must, however, ask the question: once the critical dichotomy has occurred and a population has become established on a new food source, what happens next? Unfortunately, there is a lack of karyotypic evidence on cases where sympatric speciation may reasonably be supposed to have occurred in the not-too-distant past. We simply do not know whether such species generally differ in karyotype or not, neither do we know whether they differ in the amount, distribution and organization of the repetitive DNA components of the chromosomes. Clearly, however, it must be expected that once the dichotomy has occurred and there are two sympatric populations living on different food resources or other essential requisites, there will be strong selection for divergence and that this will rapidly produce marked rectangularity as far as morphological features are concerned.

Rectangularity may also be expected in invasive and stasipatric chromosomal speciation. In the former mode the new taxon must develop a genotype capable of adapting it to a territory that the original species population had been unable to occupy. In stasipatric speciation only those incipient species that are genetically sufficiently different from the original population will be capable of displacing it from previously occupied territory. Divergent selection will probably be maximal in the case of truly sympatric speciation. In this case the two taxa will be unable to coexist unless they diverge fairly completely so that they become adapted quite rapidly to different life styles.

These considerations suggest that there is a strong *a priori* case for expecting disruptive selection to produce rapid and marked rectangularity following the initial dichotomy in all types of non-allopatric speciation. It should be noted that the rectangularity is not absolutely synchronous with the initial dichotomy, but on the geological time scale they would always appear to be so.

The available evidence suggests, moreover, that the type of morphological rectangularity we have been considering is due, in the main, to genetic regulatory mechanisms rather than to mutations at structural genetic loci. In the first place, there is now much evidence that populations that have diverged in karyotype according to the stasipatric or invasive models differ very little in their proteins, so that the genetic distances between them based on electrophoretic studies are generally much smaller than those of populations that are speciating allopatrically [25, 81].

The direct evidence for regulatory changes in chromosomal speciation has been presented by Wilson and his colleagues [82–85]. These authors have also demonstrated a correlation between the number of major chromosomal rearrangements that have occurred in a group of animals and the amount of morphological change, interpreted as due to regulatory evolution. While their claim appears well-founded, at least in vertebrates, it does not follow that a direct causal connection exists. It seems unlikely that the chromosomal rearrangements that lead to changes in chromosome number or in the number of chromosome arms would themselves directly produce regulatory genetic changes. At least in Mus musculus centric fusions can take place with a completely negligible loss of satellite DNA sequences. But if such chromosomal changes are important primary causes of non-allopatric speciation, as suggested above, and if this is usually followed by morphological rectangularity, a causal connection would exist, but it would be less direct than supposed by Wilson and his collaborators.

Before discussing the possible nature of the genetic changes underlying morphological rectangularity, it is necessary to consider the architecture of eukaryote chromosomes as revealed by modern molecular biology.

Broadly speaking, the chromosomal DNA of the higher eukaryotes is of four types. There are some short segments that renature in solution immediately, after thermal denaturation, being probably in all instances reversed repeats or palindromes. This kind of DNA is only a small fraction of the total amount. A larger fraction, in most organisms, is constituted by the highly repetitive DNA's. These frequently consist of a simple sequence of bases repeated in tandem thousands or millions of times. For example, in Drosophila melanogaster there is an ATAAT sequence, with a variant form ATATAAT (A indicates an adenine nucleotide, T a thymidine one) that together constitute about 4% of the total nuclear DNA. Most species of eukaryotes that have been investigated possess several different satellite DNA's in their genomes [86–90]; these form the major part of the heterochromatin of classical cytology. The moderately repetitive DNA's are less highly repeated than the satellite DNA's, but nevertheless exist as hundred or thousands of copies in the genome. They include sequences that code for the histones and ribosomal and transfer RNA's, but most of them are still of unknown function. Finally, there are the unique or single-copy DNA's; they include most of the structural genes, together with some adjacent spacer sequences and the so-called introns—sequences occupying interstitial positions in "split" genes.

This is obviously a very complex picture. It is one that has not yet been brought into a meaningful relationship with evolutionary theory. We need to find an answer to the question: what part do genetic changes in these various DNA fractions play in generating cladogenetic dichotomies, in producing rectangularity and in phyletic evolution? The issue is complicated by the fact that the molecular biologists themselves are in considerable disagreement as to the role of repetitive DNA's in the cell, and hence in evolution. Thus it has been claimed by some [91–93] that these nucleotide sequences are functionless "junk DNA" with little or no effect on the phenotype and owing their persistence solely to molecular mechanisms whereby the sequence undergoes amplification through forming additional copies of itself within the genome (the "selfish DNA" concept).

Evolutionists familiar with the whole range of cytogenetic information regarding the distribution of these components in the genomes of a vast number of eukaryotes will be unlikely to accept the "junk DNA" viewpoint. There are far too many precise regularities in the distribution of the highly repeated sequences for this interpretation to be accepted in its entirety. In some species they are concentrated in pericentromeric segments, in others they may be interstitial or telomeric, but their distribution, in general, is similar in the chromosomes of a given karyotype. In the grasshopper Caledia captiva (a complex of incipient species) the Daintree species has its C-banding material forming 24% of the karyotype, while in the Moreton "race" it constitutes 35.5% and in the Torresian "race" it forms only 14% of the total chromosome length [71, 94]. The important point here is that these evolutionary increases and decreases of C-banding material have not been capricious or chaotic but, on the contrary, are distributed very regularly over the members of the karyotype.

In another grasshopper species, Atractomorpha similis, there are similar regularities in the distribution of the heterochromatin [94]. In individuals from the Cape York Peninsula of Australia C-banding material constitutes about 39% of the total chromosome length, while in ones from near Sydney it forms 52% of the karyotype. The evolutionary increases or decreases in heterochromatic material are distributed very uniformly over the members of the karyotype. Numerous other examples of the same type could be cited. The regularities in distribution of the C-banding material are just as pronounced in a parthenogenetically reproducing grasshopper species as they are in many sexual species; that is to say most of the bands are individually recognizable from one clone to another, and have been handed down, over a period of at least half a million years, from two ancestral species that, by hybridization, produced the parthenogenetic one [95, 96]; moreover in this case it has been shown that the sequence of DNA replication along the length of the chromosomes has been strongly conserved from the ancestral species to their hybrid derivative [97]. Where the precise satellite sequences forming the heterochromatin have been isolated and sequenced, regularities of distribution are found, on a smaller scale [86, 98–100].

In other cases we have what are clearly balanced polymorphisms for extra blocks of heterochromatin existing in natural populations. Numerous instances are now known and it is quite certain that they are incompatible with the view that heterochromatin is without significant effects on the phenotype.

Workers who reject the view that repetitive DNA's are parasitic "junk" have suggested a wide variety of functions for them, some concerned with chromosome organization and cell metabolism, others with speciation and evolution. A determined attempt to ascribe an adaptive significance to variation in the size and distribution of heterochromatic segments has been made by John and Miklos [101, 102] who emphasize the effects of the distribution of heterochromatic segments on the distribution of the chiasmata at meiosis, although they wisely do not exclude other possible cellular effects.

Now it is certain that a number of studies have demonstrated effects of naturally occurring heterochromatic segments (in grasshoppers) or experimentally induced ones (in Drosophila) on the distribution of genetic recombination, whether observed cytologically, as chiasmata, or genetically, in the form of crossover products. It is equally clear, however, that in some cases no such effect can be demonstrated, and in many species the distribution of chiasmata is quite different in the two sexes, although that of the heterochromatic segments is the same.

The question is, however, whether this effect is one that is sufficiently important, from the adaptive standpoint, to be acted upon by natural selection or whether it may be regarded as a relatively unimportant side-effect, a mere epiphenomenon resulting from a system of variation that is controlled by natural selection in virtue of some much more important property or properties.

It is clear that any explanation of the existence or distribution of heterochromatic segments based on changes they produce in the distribution of recombination along the length of the chromosome relies on a second-order effect. That is to say, individuals are selected, not because of any trait that enhances their own survival or fecundity, but on the basis of some benefit to their descendants. Ordinarily, we might dismiss a selective interpretation of this kind by saying that the effect would be so weak as to be insignificant compared to direct first-order selective forces. Clearly it relies on a form of kin selection.

Now kin selection and group selection in general should probably not be entirely dismissed as far as chromosome evolution is concerned. The present author [67] has invoked this kind of explanation in the case of the multiple chromosome fusions that have established barriers to gene flow in some Italian mouse populations. But certainly we should regard claims based on kin selection with a good deal of critical suspicion. In the case of geographic races of grasshoppers that differ in the size or distribution of their heterochromatic blocks it is conceivable that an effect on the positioning of chiasmata in a particular chromosome might be selected for, if we assume that there is a chromosome segment in which high recombination is advantageous in one geographic region or habitat while tight linkage is advantageous in another area. But such an explanation stretches one's credulity very far indeed. Where it seems to break down completely is on two counts. In the first place, there are cases [71, 94,

103] where several or many chromosomes in the karyotype have all got added blocks of heterochromatin in one geographic race but not in another (karyotypic orthoselection). It is surely inconceivable that there could exist a number of similarly positioned segments in different chromosomes such that free genetic recombination in them would be adaptive in one geographic area or habitat but not in another.

The other situation where the adaptive interpretation of heterochromatic segments based on their effects on chiasma localization seems to break down is in the very common case (in grasshoppers at any rate, but examples in other organisms are not wanting) where we have what is obviously a balanced population polymorphism for the presence or absence of a heterochromatic segment. Group selection is a possible type of explanation for cases of interpopulation variation, but is implausible in that of intrapopulation polymorphism. It is simply not credible that there should simultaneously be selection for high and low levels of recombination in a population.

One argument that has been used in support of the view that an important role of the satellite DNA's is the control of meiotic recombination is that they are sometimes eliminated from the soma (eg in Ascarid worms and Cecidomyid midges). This argument [104] is almost certainly fallacious. In all cases they are present at the beginning of development when fundamental morphogenetic processes are determined.

An entirely different interpretation of the evolutionary role of the satellite DNA's is based on the assumption that they assist in the meiotic pairing process, providing a means of recognition between homologous chromosomes. If this is so it is argued [105, 106] that changes in the distribution of satellite sequences might be important in speciation, since they would lead to imperfect synapsis and hence impair fertility in hybrids between incipient species.

Clearly, facilitating synapsis is not the sole function of the satellite DNA's, since the same satellite species occur on different, non-homologous chromosomes in Drosophila [98]. Satellite sequences may, however, be important in the maintenance of specific chromosomal spatial patterns in interphase nuclei which may be of great importance in cell biology [107].

As far as directly furnishing the basis for hybrid sterility by disturbing meiotic synapsis, however, the evidence regarding the satellite DNA's concerned is almost entirely negative. Thus in Drosophila, individuals heterozygous for large deficiencies of heterochromatin in the autosomes appear to have fairly normal synapsis and segregation [108, 109]. Mus molossinus has about 40% less of a particular satellite DNA than M musculus [110, 111] but synapsis appears to be normal in hybrids. In the case of M musculus and M spretus there are considerable differences in the amount of satellite DNA and some differences in organization and sequence composition, yet hybrids between the two species are fully fertile [100]. It seems, therefore, that on present evidence we should exclude changes in satellite DNA's as primary causes of speciation. This does not mean that they play no part in speciation, only that they do not lead directly to cladogenesis.

The third group of suggestions regarding the functions of repetitive DNA's assigns to them a rather miscellaneous range of roles in regulatory processes and in cell metabolism. At least some satellite DNA's are transcribed, although they may not be translated [112, 113, 114]. Some satellites or regions of complex satellites may be important for binding or proteins [115]. Thus an embryonic protein in Drosophila melanogaster that binds specifically to the 1.688 g/cm^3 satellite has been isolated [116].

The general view of speciation that is increasingly gaining acceptance is that mutation of structural gene loci is less important than changes in regulatory systems [82, 83, 117]. Many of these are certainly concerned with the timing of gene action leading to protein synthesis and hence with the whole architecture of morphogenetic processes in early development. It would be entirely natural if variation in these developmental patterns were adaptive to different environments at both the intra-population and interpopulation levels. In a species whose range extends over a large geographic area particular rates and types of development will be adaptive to different climates and habitats. Thus in the case of grasshopper species with eggs developing in the soil there are likely to be different embryonic optima for many processes, in relation to absorbtion of water from the soil, dormancy or diapause under conditions of cold or drought and in soils of different types, porous or impervious. Neither will the importance of specific developmental patterns cease at the end of the embryonic period. There are still the crises of hatching from the eggshell, emergence through the soil, the first moult and the beginning of feeding. It is not surprising that there are high mortalities at these critical stages. The whole question of timing in embryological development has been discussed recently [118]; clocks controlling the various developmental processes are postulated, but their precise nature and genetic determination are unknown. It has been pointed out [119] that in development the initial activation of the maternal and paternal alleles of a particular gene is almost always synchronous but that in some species hybrids the maternal allele become activated at the appropriate time while the activation of the paternal allele is delayed. Clearly this is likely to be due to the genomes of the two species having different regulatory systems. That closely related species may, in fact, have different regulatory systems (we hesitate to speak of regulatory *genes,* because of lack of evidence as to their precise nature) is clear from a recent study [120] on the expression of the ADH gene in different tissues of Drosophila grimshawi and D orthofascia. In this case there is evidence that the regulatory elements are *cis*-acting, ie that they only control genic activity in the same chromosome and not in the homologous one. Hedrick and McDonald have given reasons for supposing that "changes in genetic regulation would be the favored genetic strategy for a population adapting to a sudden and substantial environmental change" [122]. According to the viewpoint presented here such environmental changes would be especially pronounced in all types of non-allopatric speciation.

It has been claimed that negative correlation exists between the amounts of repetitive DNA in species genomes and extent of polymorphism for structural gene loci [121]. That species with much repetitive DNA can dispense with adaptive polymorphism for structural enzyme loci is possible. But whether it is the "highly" or the "moderately" repetitive DNA fraction that is responsible is not yet known, and the implications for speciation theory are still obscure. The main role of regulatory systems may be "fine tuning" of the activity of structural gene loci [123]. But some regulatory systems affecting major developmental patterns could have much more general and fundamental effects.

In the genus Drosophila, the extreme difference in head shape between the Hawaiian sibling species D silvestris (normal head) and D heteroneura (head laterally extended) is a striking example of rectangularity. Since the species can be crossed (they even hybridize naturally at one locality) and F_2 and backcross progenies can be obtained, it was possible to carry out a genetic analysis [124, 125]. The results were interpreted as indicating that the difference in head shape was due to mutation at a minimum of eight loci. The evidence, however, is not adequate to discriminate between allelic changes at structural gene loci and the effects of repetitive DNA segments distributed throughout genome. These species are homosequential and show insignificant allozymic differences [126].

One measure of the extent to which the regulatory mechanisms of related species have diverged is the ability to produce viable F_1 hybrids, whether in nature or in the laboratory. However, failure to produce hybrids may be due to causes that have nothing to do with regulatory systems. It has been shown that birds have lost the potential for interspecific hybridization much more slowly than mammals [127]. Based on albumin immunological distances, which are known to evolve at a rather steady rate, the average hybridizable bird species pair diverged about 22 million years ago, while the corresponding period for placental mammals is only about a tenth of this. The data were interpreted as indicating that evolutionary changes in regulatory systems have been much slower in birds, which (on the view advocated above) is consistent with their having speciated predominantly according to the allopatric model. No evidence exists as to the nature and extent of satellite DNA divergence between related species of birds; if the viewpoint presented here is correct the differences may be minor compared with those in mammals. Both karyotype evolution and rate of speciation have been exceptionally rapid in most groups of placental mammals [128], which may be due to the prevailing modes of social structuring of mammalian populations that lead to significant levels of inbreeding [129].

If differences in regulatory systems do not cause hybrid inviability, they may nevertheless produce sterility of the F_1 due to abnormalities of gonadal development. In insects at any rate such abnormal development is generally or always determined within the germ line and the hybrid soma has no effect.

Obviously any assessment of the possible role of repetitive DNA sequences in speciation must rest on a detailed comparison of satellite DNA sequences and their chromosomal location in closely related species of organisms. The situation has been studied in the virilis group of Drosophila [130] in man, chimpanzee, gorilla and orang-utan [131], in Mus musculus and M spretus [100], in the African Green Monkey, Cercopithecus aethiops and the Baboon, Papio cynocephalus [132], in Drosophila melanogaster and D simulans [90, 98], in species of the melanogaster subgroup [133] and in ten species and subspecies of tsetse flies [134]. A preliminary study on the willistoni group of Drosophila [135] has demonstrated some differences between the satellite DNA's of D willistoni and D paulistorum and also between the well-known "semi-species" of the latter.

The kinds of evolutionary transformations that satellite DNA segments may undergo have been described by several authors [115, 136]. Complex satellites are formed from shorter repeats by amplification events in which units differentiated by mutational changes are multiplied to give tandem arrays of identical repeats. Thus in the 1.708 g/cm³ satellite of the red-necked wallaby it was possible to identify eight mutation-amplification events whose temporal sequence, starting from a uniform ancestral repeated unit could be identified [137]. However, in some satellites the units of amplification are not necessarily unit repeats.

In general, closely related species possess similar or identical satellite DNA's, but these may be amplified to very different extents. Thus Drosophila melanogaster and D simulans have the same satellite DNA's but their copy numbers are very different in the two species. But in the same melanogaster species group, D teisseiri and D orena share a 1.684 g/cm³ satellite that is not known to be present in the five other members of the subgroup. The seven species of the melanogaster subgroup contain altogether fifteen major satellite DNA's [133, 134].

One thing that is clear about these amplification and sequence modification events is that they are very numerous. It is thus improbable that each one, by itself, will have a major effect in causing speciation. But a series of such events may be imagined to have a cumulative effect on regulatory systems that would cause a major rectangularity to occur. If, however, the sequence changes and amplification events that occur in the satellite DNA's are not coincident in time with cladogenetic events but occur subsequently (as suggested here), there is the problem of how they spread through a species population [135]. In principle there seem to be three possibilities: meiotic drive, founder effects, and selection; but some purely molecular mechanisms have been suggested [136].

Some satellite DNA's have been conserved for extremely long periods of time. Thus the HSα satellite of the kangaroo rat Dipodomys ordii is also present in the guinea pig and the antelope ground squirrel, species belonging to three rodent suborders that diverged 40 to 50 million years ago [107, 138]. It is unknown whether this particular satellite subserves some vital function which would be subject to strong stabilizing selection; possibly it may play some mechanical "housekeeping" role in the chromosome architecture. The highly repetitive DNA sequences of the sea urchins have also been highly conserved in evolution, the rate of sequence divergence being a half to a third of that of the unique sequence DNA [139]. Considerable evolutionary conservation in respect of the sequences of some highly repetitive DNA's may be imposed by the structure of nucleosome subunits of chromatin [140].

When the several satellite DNA's of a species are compared, it is sometimes found that they are very similar and clearly have a common origin. This is the case with the three satellites of Drosophila virilis, heptamers that differ by single nucleotide substitutions [130]. More frequently, however, the main satellites of a species appear to be unrelated; for example the 5'TTAGGG3' HSα satellite of Dipodomys ordii (which exists in at least a dozen sequence variants) is not related in any way to the other two satellites that occur in this and other species of Dipodomys [106].

Comparative studies of speciation in different groups of organisms may be susceptible to several alternative interpretations. An attempt was made to test for rectangularity by comparing the means and variances of genetic distances between species in the fish families Cyprinidae (minnows) and Centrarchidae (sunfishes) [33]. The former have speciated extensively, the latter much less [50]. No evidence of accelerated rates of protein evolution in the Cyprinidae was found. However, if the present interpretation is correct, Avise's failure to obtain evidence of rectangularity in the Cyprinidae might be due to these fishes having speciated allopatrically rather than according to one of the non-allopatric modes. It might also have been expected if the phenomenon of rectangularity is due mainly to changes in regulatory systems rather than to mutations of structural loci detectable by gel electrophoresis. Similarly, the demonstration [141] of extensive interspecific genetic compatibility between fishes of the genus Cyprinodon would indicate absence of any large scale evolution of regulatory systems which (on the hypothesis propounded here) would likewise suggest that speciation had been predominantly allopatric in the Cyprinodontidae. The karyotypes of Cyprinodon species are extremely similar or even indistinguishable, so it is fairly certain that chromosome modes of speciation have not prevailed in this genus. The mean levels of genic heterozygosity were not significantly different in the species of Cyprinidae and Centrarchidae examined, [142] thus failing entirely to support earlier hypotheses according to which the mean level of heterozygosity should be correlated with rate of speciation (intra-species genic variation being considered as "raw material" for speciation).

Modern evolutionary studies have led to a far greater appreciation of the diversity of the mechanisms involved than was possible a few decades ago. This is so at all levels of analysis: the morphological, the adaptational, and the molecular. It is now much more difficult to formulate universal laws or principles with regard to any evolutionary process. The evolutionary picture becomes ever more complex. Perhaps a great simplification is coming, but its outlines are hardly visible at present.

REFERENCES

1. Stanley SM: Macroevolution: Pattern and Process. San Francisco: Freeman, 1979.
2. Gould SJ: Is a new and general theory of evolution emerging? Paleobiology 6: 119,1980.
3. Gould SJ: The evolutionary biology of constraint. Daedalus 1980:39,1980.
4. Boucot AJ: Community evolution and rates of cladogenesis. Evol Biol 11: 545,1978.
5. Chaline J, Mein P: Les Rongeurs et l'Evolution. Paris: Doin, 1979.
6. Mayr E: Systematics and the Origin of Species. New York: Columbia University Press, 1942.
7. Mayr E: Change of genetic environment and evolution. In Huxley J, Hardy AC, Ford EB (eds): Evolution as a Process. London: Allen and Unwin, 1954, p 157.
8. Thoday JM, Gibson JB: Isolation by disruptive selection. Nature 193: 1164,1962.
9. Maynard Smith J: Sympatric speciation. Amer Nat 100: 637,1966.
10. Bush GL: The mechanism of sympatric host race formation in the true fruit flies (Tephritidae). In White MJD (ed): Genetic Mechanisms of Speciation in Insects. Sydney: Australia and New Zealand Book Co, 1974, p 3.
11. Bush GL: Modes of animal speciation. Ann Rev Ecol Systemat 6: 339,1975.
12. Knerer G, Atwood CE: Diprionid sawflies: polymorphism and speciation. Science 179: 1090,1973.
13. White MJD: Models of speciation. Science 159: 1065,1968.
14. White MJD: Modes of Speciation. San Francisco: Freeman, 1978.
15. Rosenzweig ML: Competitive speciation. Biol J Linn Soc London 10: 275,1978.
16. Tauber CA, Tauber MJ: Sympatric speciation based on allelic changes at three loci: evidence from natural populations in two habitats. Science 197: 1298,1977.
17. Tauber CA, Tauber MJ: A genetic model for sympatric speciation through habitat diversification and seasonal isolation. Nature 268: 702,1977.
18. Futuyma DJ, Mayer GC: Non allopatric speciation in animals. Syst Zool 29: 254,1980.
19. Wright S: Genic and organismic selection. Evolution 34: 825, 1980.
20. Lewontin RC: The Genetic Basis of Evolutionary Change. New York: Columbia University Press, 1974.

21. Britten RJ, Kohne DE: Repated sequences in DNA. Science 161: 529,1968.
22. Ashburner M: Drosophila at Kolymbari. Nature 288: 538,1980.
23. Tettenborn U, Gropp A: Meiotic nondisjunction in mice and mouse hybrids. Cytogenet 9: 272,1970.
24. Ford CE, Evans EP: Robertsonian translocations in mice: segregational irregularities in male heterozygotes and zygotic unbalance. Chromosomes Today 4: 387,1973.
25. Britton-Davidian J, Bonhomme F, Croset H, Capanna E, Thaler L: Variabilité génétique chez les populations de souris (genre Mus) à nombre chromosomique réduit. C.R. Acad Sci Paris 290D:195, 1980.
26. Elder FFB: Tandem fusion, centric fusion and chromosomal evolution in the cotton rats, genus Sigmodon. Cytogenet Cell Genet 26: 199,1980.
27. Liming S, Yingying Y, Xingshen D: Comparative cytogenetic studies on the red muntjac, Chinese muntjac and their F_1 hybrids. Cytogenet Cell Genet 26: 22,1980.
28. Lyapunova EA, Vorontsov NN, Korobitsyna KV, Ivanitskaya EYu, Borisov YuM, Yakimenko LV, Dovgal VYe: A Robertsonian fan in Ellobius talpinus. Genetica 52/53: 239,1980.
29. Eldredge N, Gould SJ: Punctuated equilibria: an alternative to phyletic gradualism. In Schopf TJM (ed): Models in Paleobiology. San Francisco: Freeman and Cooper, 1972, p 82.
30. Gould SJ, Eldredge N: Punctuated equilibria: the tempo and mode of evolution reconsidered. Paleobiol 3: 115,1977.
31. Stanley SM: Stability of species in geologic time. Science 192: 267,1976.
32. Stanley SM: Macroevolution: Pattern and Process, San Francisco: Freeman, 1979.
33. Avise JC: Is evolution gradual or rectangular? Evidence from living fishes. Proc Nat Acad Sci USA 74: 5083,1978.
34. Birkelund T, Bromley RG, Christensen WK (eds): Proc Cretaceous Tertiary Boundary Events Symp I, II, University Copenhagen, 1979.
35. Alvarez LW, Alvarez W, Asaro F, Michel HV: Extraterrestrial cause for the Cretaceous-Tertiary extinction. Science 208: 1095,1980.
36. Turner, BJ, Grosse DJ: Trophic differentiation in Ilyodon, a genus of stream-dwelling goodeid fishes: speciation versus ecological polymorphism. Evolution 34: 259,1980.
37. Myers GS: The endemic fish fauna of Lake Lanao, and the evolution of higher taxonomic categories. Evolution 14: 323,1960.
38. Turner BJ: Genetic divergence of Death Valley pupfish species: biochemical versus morphological evidence. Evolution 28: 281,1974.
39. Kirkpatrick M, Selander RK: Genetics of speciation in lake whitefishes in the Allegash basin. Evolution 33: 478,1979.
40. Bush GL: Taxonomy, cytology and evolution of the genus Rhagoletis in North America (Diptera:Tephritidae). Bull Mus Comp Zool Harvard Univ 134: 431,1966.
41. Bush GL: Sympatric host race formation and speciation in frugivorous flies of the genus Rhagoletis. Evolution 23: 237,1969.
42. Wasserman M: Cytological studies of the repleta group of the genus Drosophila. V. The mulleri subgroup. Studies in Genetics II. Univ Texas Publ 6205: 119,1962.
43. Wasserman M: Cytology and phylogeny of Drosophila. Amer Nat 97: 333,1963.

44. Carson HL, Clayton FE, Stalker HD: Karyotypic stability and speciation in Hawaiian Drosophila. Proc Nat Acad Sci USA 57: 1280,1967.

45. Craddock EM, Johnson WE: Genetic variation in Hawaiian Drosophila. V. Chromosomal and allozymic diversity in Drosophila silvestris and its homosequential species. Evolution 33: 137,1979.

46. Bedo D: Cytogenetics and evolution of Simulium ornatipes Skuse (Diptera:Simuliidae). II. Temporal variation in chromosomal polymorphisms and homosequential sibling species. Evolution 33: 296,1979.

47. Ward BL, Heed WB: Chromosome phylogeny of Drosophila pachea and related species. J Hered 61: 248,1970.

48. Key KHL: Species, parapatry, and the morabine grasshoppers. Syst Zool, 30:425, 1981.

49. White MJD, Blackith RE, Blackith RM, Cheney J: Cytogenetics of the viatica group of morabine grasshoppers. I. The "coastal" species. Austral J Zool 15: 263,1967.

50. Saez FA: An extreme karyotype in an Orthopteran insect. Amer Nat 91: 259,1957.

51. Cardoso H, Saez FA, Brum-Zorilla N: Location, structure, and behavior of C-heterochromatin during meiosis in Dichroplus silveiraguidoi (Acrididae:Orthoptera). Chromosoma 48: 51,1974.

52. Hewitt GM: Orthoptera. In John B (ed): Animal Cytogenetics 3 Insecta 1:1. Berlin and Stuttgart: Borntraeger, 1979.

53. Lande R: Effective deme size during long-term evolution estimated from rates of chromosomal rearrangement. Evolution 33: 234,1979.

54. Hedrick PW: The establishment of chromosomal variants. Evolution 35: 322,1981.

55. Bengtsson BO, Bodmer WF: On the increase of chromosome mutations under random mating. Theor Pop Biol 9: 260,1976.

56. Reig O, Aguilera M, Barros MA, Useche M: Chromosomal speciation in a Rassenkreis of Venezuelan spiny rats (genus Proechimys, Rodentia, Echimyidae). Genetica 52/53: 291,1980.

57. Benado M, Aguilera M, Reig OA, Ayala F: Biochemical genetics of chromosome forms of Venezuelan spiny rats of the Proechimys guairae and Proechimys trinitatis superspecies. Genetica 50: 89,1979.

58. Nevo E: this volume.

59. King M, King D: An additional chromosome race of Phyllodactylus marmoratus (Gray) (Reptilia: Gekkonidae) and its phylogenetic implications. Austral J Zool 25: 667,1977.

60. King M: Chromosomal and morphometric variation in the Gekko Diplodactylus vittatus (Gray). Austral J Zool 25: 43,1977.

61. Mrongovius MJ: Cytogenetics of the hybrids of three members of the grasshopper genus Vandiemenella (Orthoptera:Eumastacidae:Morabinae). Chromosoma 71: 81,1979.

62. Craddock EM: Chromosomal evolution and speciation in Didymuria. In White MJD (ed): Genetic Mechanisms of Speciation in Insects. Sydney: Australian and New Zealand Book Co, 1974, p 24.

63. Craddock EM: Interspecific karyotypic differentiation in the Australian phasmatid Didymuria violescens (Leach). I. The chromosome races and their structural and evolutionary relationships. Chromosoma 53: 1,1975.

64. Mainardi D: La Scelta Sessuale. Torino: Boringhieri, 1975.
65. Yanai J, McClearn GE: Assortative mating in mice. I. Female mating preference. Behavior Genetics 2: 173,1972.
66. Bruce HM, Parrot DMV: Further observations on pregnancy block in mice caused by the proximity of strange males. J Reprod Fertil 1: 311,1960.
67. White MJD: Chain processes in chromosomal speciation. Syst Zool 27: 285,1978.
68. Shaw DD: Population cytogenetics of the genus Caledia (Orthoptera: Acridinae). I. Inter- and intra-specific karyotype diversity. Chromosoma 54: 221,1976.
69. Shaw DD, Moran C, Wilkinson P: Chromosomal reorganization, geographic differentiation and the mechanism of speciation in the genus Caledia. In Blackman RL, Hewitt GM, Ashburner M (eds): Insect Cytogenetics, Symp X Royal Entomological Society of London. Oxford: Blackwell, 1980, p 171.
70. Shaw DD, Wilkinson P: Chromosomal differentiation, hybrid breakdown, and the maintenance of a narrow hybrid zone in Caledia. Chromosoma 80: 1,1980.
71. Shaw DD, Webb GC, Wilkinson P: Population cytogenetics of the genus Caledia (Orthoptera: Acridinae). II. Variation in the pattern of C-banding. Chromosoma 56: 169,1976.
72. Shaw DD, Wilkinson P, Moran C: A comparison of chromosomal and allozymic variation across a narrow hybrid zone in the grasshopper Caledia captiva. Chromosoma 75: 333,1979.
73. Moran C: The structure of a narrow hybrid zone in Caledia captiva. Heredity 42: 13,1979.
74. Moran C, Wilkinson P, Shaw DD: Allozyme variation across a narrow hybrid zone in the grasshopper Caledia captiva. Heredity 44: 69,1980.
75. Daly, JC, Wilkinson P, Shaw DD: Reproductive isolation in relation to allozymic and chromosomal differentiation in the grasshopper Caledia captiva. Evolution, 35:1164, 1981.
76. Berlocher SH, Bush GL: Modes of speciation and genetic divergence in frugivorous flies of the genus Rhagoletis (Diptera:Tephritidae). Manuscript.
77. Carson HL, Okada T: The ecology and evolution of some flower-breeding Drosophilidae of New Guinea. Abstracts XVI Internat Congr Entomol, Kyoto, p 7, 1980.
78. Mayr E: Animal Species and Evolution. Cambridge: Harvard University Press, 1963.
79. Brncic D, Nair PS, Wheeler MR: Cytogenetic relationships within the mesophragmatica species group of Drosophila. Univ Texas Publ 7103: 1,1971.
80. Prakash S: Origin of reproductive isolation in the absence of apparent genic differentiation in a geographic isolate of Drosophila pseudoobscura. Genetics 72: 143,1972.
81. Dobzhansky Th: Analysis of incipient reproductive isolation within a species of Drosphila. Proc Nat Acad Sci USA 72: 3638,1975.
82. Wilson AC: Evolutionary importance of gene regulation. Stadler Symp, Univ Missouri 7: 117,1975.
83. Wilson AC: Gene regulation in evolution. In Ayala FJ (ed): "Molecular Evolution," Sunderland, Mass: Sinauer Association 1976, p 225.

84. Wilson AC, Carlson SS, White TJ: Biochemical evolution. Ann Rev Biochem 46: 573,1977.
85. Wilson AC, Sarich VM, Maxson LR: The importance of gene arrangement in evolution: evidence from studies on rates of chromosomal, protein, and anatomical evolution. Proc Nat Acad Sci USA 71: 3028,1974.
86. Appels R, Peacock WJ: The arrangement and evolution of highly repeated (satellite) DNA sequences with special reference to Drosophila. Int Rev Cytol Suppl 8: 69,1978.
87. Barnes SR, Webb DA, Dover GA: The distribution of satellite and main-band DNA components in the melanogaster species subgroup of Drosophila. Chromosoma 67: 341,1978.
88. Brown SDM, Dover GA: Conservation of sequences in related genomes of Apodemus: constraints on the maintenance of satellite DNA sequences. Nucl Acids Res 6: 2423,1979.
89. Mizuno S, Andrews C, Macgregor HC: Intrespecific "common" repetitive DNA sequences in salamanders of the genus Plethodon. Chromosoma 58: 1,1976.
90. Peacock WJ, Appels R, Dunsmuir P, Lohe AR, Gerlach WL: Highly repeated DNA sequences: chromosomal localization and evolutionary conservatism. In Brinkley BR, Porter KR (eds): International Cell Biology 1976-1977. New York: Rockefeller University Press, 1977, p 494.
91. Ohno S: So much "junk" DNA in our genome. In Evolution of Genetic Systems Brookhaven Symposia in Biology. Brookhaven Natl. Laboratory, Upton, N.Y. 23:366,1972.
92. Doolittle WF, Sapienza C: Selfish genes, the phenotype paradigm, and genome evolution. Nature 284: 601,1980.
93. Orgel LE, Crick FHC: Selfish DNA: the ultimate parasite. Nature 284: 604,1980.
94. King M, John B: Regularities and restrictions governing C-band variation in Acridoid grasshoppers. Chromosoma 76: 123,1980.
95. Webb GC, White MJD, Contreras N, Cheney J: Cytogenetics of the parthenogenetic grasshopper Warramaba (formerly Moraba) virgo and its bisexual relatives. IV. Chromosome banding studies. Chromosoma 67: 309,1978.
96. White MJD: The genetic system of parthenogenetic grasshopper Warramaba virgo. In Blackman RL, Hewitt GM, Ashburner M (eds): Insect Cytogenetics, Symp X Royal Entomological Society of London. Oxford: Blackwell, 1980, p 119.
97. White MJD, Webb GC, Contreras N: Cytogenetics of the parthenogenetic grasshopper Warramaba (formerly Moraba) virgo and its bisexual relatives. VI DNA replication patterns of the chromosomes. Chromosoma 81: 213, 1980.
98. Brutlag DL, Appels R, Dennis ES, Peacock WJ: A highly repeated DNA in Drosophila melanogaster. J Mol Biol 112: 31,1977.
99. Peacock WJ, Lohe AR, Gerlach WL, Dunsmuir P., Dennis ES, Appels R: Fine structure and evolution of DNA in heterochromatin. Cold Spr Harb Symp Quant Biol 42: 1121,1977.
100. Brown SD, Dover GA: Conservation of segmental variants of satellite DNA of Mus musculus in a related species: Mus spretus. Nature 285: 47,1980.
101. John B, Miklos GLG: Functional aspects of satellite DNA and heterochromatin. Internat Rev Cytol 58: 1,1979.

102. Miklos GLG, John B: Heterochromatin and satellite DNA in man: properties and prospects. Amer J Hum Genet 31: 264,1979.
103. John B, King M: Heterochromatin variation in Cryptobothrus chrysophorus. II. Patterns of C-banding. Chromosoma 65: 59,1977.
104. Jones KW: Speculations on the functions of satellite DNA in evolution. Z Morph Anthropol 62: 143,1978.
105. Corneo G: Do satellite DNA's function as sterility barriers in eukaryotes? Evol Theor 1: 261,1976.
106. Fry K, Salser W: Nucleotide sequences of HS-α satellite DNA from Kangaroo rat Dipodomys ordii and characterisation of similar sequences in other rodents. Cell 12: 1069,1977.
107. White MJD: Chromosoimal repatterning-regularities and restrictions. Genetics 79:63,1975.
108. Yamamoto M, Miklos GLG: Genetic studies on heterochromatin in Drosophila melanogaster and their implications for the functions of satellite DNA. Chromosoma 66: 71,1978.
109. Yamamoto, M: Cytological studies of heterochromatin function in the Drosphila male: autosomal meiotic pairing. Chromosoma 72: 293,1979.
110. Rice NR, Strauss NA: Relatedness of mouse satellite deoxyribonucleic acid to deoxyribonucleic acid of various Mus species. Proc Nat Acad Sci USA 70: 3546,1973.
111. Dev VG, Miller DA, Tantravati R, Schreck RR, Roderick T, Erlanger BF, Miller OJ: Chromosome markers in Mus musculus: differences in C-banding between the subspecies M m musculus and M m molossinus. Chromosoma 53: 335,1975.
112. Harel J, Hanania N, Tapiero H, Harel L: RNA replications by nuclear satellite DNA in different mouse cells. Nucl Acids Res 4: 4425,1977.
113. Varley JM, Macgregor HC, Erba HP: Satellite DNA is transcribed on lampbrush chromosomes. Nature 283: 686,1980.
114. Varley JM, Macgregor HC, Nardi I, Andrews C, Erba HP: Cytological evidence of transcription of highly repeated DNA during the lampbrush stage in Triturus cristatus carnifex. Chromosoma 80: 289,1980.
115. Peacock WJ, Dennis ES, Gerlach WL: Satellite DNA-change and stability. Chromosomes Today 7: 30,1981.
116. Hseih T, Brutlag D: A protein which preferentially binds Drosophila satellite DNA. Proc Nat Acad Sci USA 76: 726,1979.
117. Wilson AC, Maxson LR, Sarich VM: Two types of molecular evolution. Evidence from studies of interspecific hybridization. Proc Nat Acad Sci USA 71: 2843,1974.
118. Snow MHL, Tam PPL: Timing in embryological development. Nature 286: 107,1980.
119. Ohno S: The preferential activation of maternally derived alleles in development of interspecific hybrids. In Defendi V (ed): Heterospecific Genome Interaction. Philadelphia: Wistar Inst Press, 1969 p 137.
120. Dickinson WJ, Carson HL: Regulation of the tissue specificity of an enzyme by a cis-acting genetic element: evidence from interspecific Drosophila hybrids. Proc Nat Acad Sci USA 76: 4559,1979.

121. Pierce BA, Mitton JB: The relationship between genome size and genetic variation. Amer Nat 116: 850,1980.
122. Hedrick PW, McDonald JF: Regulatory gene adaptation: an evolutionary model. Heredity 45: 85,1980.
123. Ayala FA, McDonald JF: Continuous variation: possible role of regulatory genes. Genetica 52/53: 1,1980.
124. Val FC: Genetic analysis of the morphological differences between two interfertile species of Hawaiian Drosophila. Evolution 32: 611,1977.
125. Templeton AR: Analysis of head shape differences between two infertile species of Hawaiian Drosophila. Evolution 31: 630,1977.
126. Sene FM, Carson HL: Genetic variation in Hawaiian Drosophila. IV. Allozymic similarity between D silvestris and D heteroneura from the island of Hawaii. Genetics 86: 187,1977.
127. Prager EM, Wilson AC: Slow evolutionary loss of the potential for interspecific hybridization in birds: a manifestation of slow regulatory evolution. Proc Nat Acad Sci USA 72: 200,1975.
128. Bush GL, Case SM, King MC: Social structuring of mammalian populations and rate of chromosomal evolution. Proc Nat Acad Sci USA 72: 5061,1975.
129. Wilson AC, Bush GL, Case SM, King MC: Social structuring of mammalian populations and rate of chromosomal evolution. Proc Nat Acad Sci USA 72: 5061,1975.
130. Gall JG, Atherton DD: Satellite sequences in Drosophila virilis. J Mol Biol 85: 633,1974.
131. Gosden JF, Mitchell AR, Seuanez HN, Gosden IM: The distribution of sequences complementary to human satellite DNA's I, II and IV in the chromosomes of the chimpanzee (Pan troglodytes), gorilla (Gorilla gorilla) and orang-utan (Pongo pygmaeus). Chromosoma 63: 253,1977.
132. Singer D, Donehower L: Highly repeated DNA of the baboon: organization and sequences homologous to the highly repeated DNA of the African Green Monkey. J Mol Biol 134: 835,1979.
133. Dover GA: The evolution of "common" sequences in closely related insect genomes. In Blackman RL, Hewitt GM, Ashburner M (eds): Insect Cytogenetics Symp X Royal Entomological Society London. Oxford: Blackwell, 1980 p 13.
134. Dover GA: Problems in the use of DNA for the study of species relationships and the evolutionary significance of genomic differences. In Bisby FA, Vaughan JG, Wright CA (eds): Chemosystematics: Principles and Practice. New York: Academic Press, 1980.
135. Corneo G, Ceccherini Nelli L, Meazza D, Ayala FJ: Satellite DNA sequences and reproductive isolation in the Drosophila willistoni group. Experientia 36: 837,1980.
136. Dover G, Strachan T, Brown S: The evolution of genomes in closely-related species. In Scudder GGE, Reveal JL (eds): Evolution Today. Pittsburgh: Hunt Institute for Botanical Documentation, 1981, p 337.
137. Dennis ES, Dunsmuir P, Peacock WJ: Segmental amplification in a satellite DNA: restriction enzyme analysis of the major satellite of Macropus rufogriseus. Chromosoma 79: 179,1980.

138. Dover G: DNA conservation and speciation: adaptive or accidental? Nature 272: 123,1978.

139. Harpold MM, Craig SP: The evolution of repetitive DNA sequences in sea urchins. Nucl Acids Res 4: 4425,1977.

140. Musich PR, Maio JJ, Brown FL: Subunit structure of chromatin and the organization of eukaryotic highly repetitive DNA: indications of a phase relation between restriction sites and chromatin subunits in African Green Monkey and calf nuclei. J Mol Biol 117: 657,1977.

141. Turner BJ, Liu RK: Extensive interspecific genetic compatibility in the new world killifish genus Cyprinodon. Copeia 1977: 259,1977.

142. Avise JC: Genic heterozygosity and rate of speciation. Paleobiology 3: 422,1977.

Mechanisms of Speciation, pages 105–121
© 1982 Alan R. Liss, Inc., 150 Fifth Avenue, New York, NY 10011

Genetic Architectures of Speciation

Alan R. Templeton

INTRODUCTION

In his review of MJD White's book, Modes of Speciation, Ernst Mayr [1] stated, "One of White's rather startling, but I think quite legitimate findings is how little population genetics has contributed to our understanding of speciation." Mayr is expressing a widespread, but erroneous, perception. For example, many models of speciation stress the importance of small population size. In virtually all of these models, small size interacts with some sort of selection to produce evolutionary transitions that are unlikely or even impossible to happen by selection alone. This concept of interaction between selection and population size stems directly from the population genetic studies of Sewall Wright, and, moreover, Wright [2,3] was the first to argue that such interactions could play an important role in speciation. This population-genetic linkage of small population size, selection, and speciation obviously has greatly contributed to our understanding of speciation, as is patent from the many papers presented at this symposium invoking this linkage. Population genetics has not only contributed to our understanding of speciation in the past, but its relevancy to speciation continues and is expanding. I intend to illustrate this relevancy by briefly discussing how descriptive, experimental, and theoretical population genetic studies can be applied to the problem of speciation. This, of course, is a very broad topic, so after some general comments upon speciation I intend to focus upon just one mechanism of speciation.

THE DESCRIPTIVE POPULATION GENETICS OF SPECIATION

Interspecific differences may be studied by examining homologous genetic elements between species such as karyotypes, enzyme coding loci, DNA sequences, etcetera, as well as traits that are presumably genetically based, such as morphology, ecological niche, etcetera. These studies have revealed that speciation can occur in the absence of, or is uncorrelated in some groups with karyotypic change, significant DNA sequence divergence, significant isozyme differentiation, morphological change, and shifts in niche or habitat [4]. This body of evidence does not mean that all these changes are irrelevant to speciation, only that there is no single universal marker of speciation. These data do indicate that speciation is a set of processes having very diverse genetic consequences. However, all of the above-mentioned measures of species differences are plagued by a common indeterminancy: It is virtually impossible to sort out which differences are actually associated with the process of speciation and which are consequences of evolution subsequent to the speciation process. Hybridization experiments have shown this to be a real problem: Many species differences— morphological, karyotypic, isozyme, etcetera—contribute little or nothing to reproductive isolation [4]. Hence, we must always keep in mind the distinction between the genetics of speciation versus the genetics of species differences. Only the genetics of speciation is relevant to this discussion.

In view of this confounding factor, hybridization experiments, although of limited applicability, provide the best current tool for discriminating between the genetics of speciation versus the genetics of species differences. For this reason, I have recently reviewed the extensive literature on plant and animal hybridization [4]. Three major conclusions came out of this review. First, all the standard types of pre- and post-mating isolating barriers [5] have been analyzed by hybridization experiments, including total F_1 inviability and sterility (eg see [6–9] for studies on the genetic basis of F_1 sterility and lethality in hybrids between Drosophila melanogaster and D simulans). Second, reproductive isolation is due to a breakdown or incompatibility at a specific part of growth or life history rather than due to some diffuse incompatibility between the species. Third, the genetic basis of the isolating barriers falls into one of three categories: 1) Many segregating units, each of small effect (the phrase "segregating units" is used rather than "loci" because mapping studies are rarely performed, and sometimes the barrier is associated with a chromosomal mutation); 2) one or a few major segregating units, commonly with many epistatic modifiers; and 3) complementary or duplicate pairs of loci. It is difficult to state just how common these three categories are relative to one another since categories 2 and 3 are more amenable to genetic analyses; hence, the possibility of an experimental or reporting bias cannot be excluded. However, the idea that speciation must always be due to large numbers of genes each of small effect is clearly not tenable.

The above three conclusions indicate that the genetics of speciation is not a hopelessly complex problem. The small number of genetic architectures underlying the isolating mechanisms is particularly important for construction of population-genetic models. Moreover, as I will soon argue, interactions exist between the mode of speciation and the genetic architecture of the isolating mechanisms. Hence, information on the genetic architectures is needed both for constructing plausible genetic models of speciation and in testing predictions generated from those models.

A POPULATION GENETIC CLASSIFICATION OF SPECIATION MECHANISMS

A barrier to the integration of speciation and population genetics is the traditional geographical classification of modes of speciation. Much of the population genetic literature relevant to speciation appears at first glance to be muddled and self contradictory primarily because it is forced into compartments that are ill-defined for population genetic purposes. For this reason, I have proposed a population genetic classification of modes of speciation [10], reproduced here in Table I. There are two basic categories: divergence and transilience. Under divergence, the isolating barriers evolve in a continuous fashion (but not necessarily slowly or uniformly over long periods of time), with some form of natural selection, either directly or indirectly, being the driving force leading to reproductive isolation. Transilience modes involve a discontinuity in which some sort of selective barrier is overcome by other evolutionary forces. These definitions are consistent with the original distinction between "transilience" and

TABLE I. Modes of Speciation

	Type of speciation	Basic mechanism
1	Divergence	
	a adaptive	erection of extrinsic isolating barrier followed by independent microevolution
	b clinal	selection on a cline with isolation by distance
	c habitat	selection over multiple habits with no isolation by distance
2	Transilience	
	a genetic	founder event causing rapid shift in previously stable genetic system
	b chromosomal	inbreeding and drift causing fixation of strongly underdominant chromosomal mutation
	c hybrid maintenance	hybridization of incompatible parental species followed by selection for maintenance of hybrid state
	d hybrid recombination	hybridization of incompatible parental species followed by inbreeding and selection for stabilized recombinant

"divergence" made by Francis Galton [11], who described transilience as "a leap from one position of organic stability to another." In a somewhat oversimplified sense, divergence occurs because of selection, transilience in spite of selection. (Nevertheless, it should be kept in mind that selection is often the primary force defining the "leap" made during a transilience as well as the force responsible for the "organic stability.")

I will now show how theoretical and experimental population genetics can be applied to this classification scheme, using adaptive divergence as my example.

ADAPTIVE DIVERGENCE

Adaptive divergence occurs when a population is split into two subpopulations by some extrinsic barrier (not necessarily geographical) to gene flow such that the two gene pools have effectively independent evolution. Isolating barriers arise as a pleiotropic consequence of microevolutionary divergence between the two independent gene pools.

Because all that is required is time and isolation, population-genetic considerations may seem irrelevant other than in the trivial sense of being important in microevolution. On the other hand, explicit population genetic considerations are built into the very definitions of the other modes of speciation given in Table I. As a consequence, much population genetic theory is immediately applicable to those modes (eg, see [12] for a population genetic theory of genetic transilience). Hence, adaptive divergence is the worst possible case for illustrating the applicability of population-genetic considerations to speciation.

Population-genetic considerations become important in adaptive divergence for two principle reasons. First, the chances for adaptive divergence are a function of how frequently extrinsic barriers arise, and this in turn depends upon geographical, ecological, and population-genetic structural attributes. However, for lack of space, I will not discuss this aspect of the problem.

The second reason stems from the fact that many extrinsic barriers are temporary: glaciers advance and retreat, seas rise and fall, lava flows subdivide a forest into "kipukas" (islands of vegetation) which gradually reunite through ecological succession, etcetera. These phenomena often occur rapidly on a geological or even ecological time scale. Because of the temporary nature of many extrinsic barriers, the chances of speciation depend upon how rapidly adaptive divergence occurs.

The speed of adaptive divergence apparently can vary greatly. Incipient isolation has occurred in the laboratory in just a few generations in some cases [4]; yet other populations have been separated thousands of generations and up to 30 million years with no evidence for substantial divergence [13,14]. I will now examine the various factors that can influence the speed of divergence.

The ancestral population structure is an important determinant of the rate at which isolating barriers arise. If the ancestral population is widespread (in an ecological sense) and gene flow limited, locally adapted or geographical races may arise within the species even before an extrinsic barrier splits the population. These local races and populations can show considerable pre- and/or post-mating isolation from one another, sometimes to the extent that certain local populations display total reproductive isolation from some others even though they are interconnected by gene flow through intermediate populations [4]. If such a species were split so as to put similarly adapted demes together (as most geographical and ecological barriers will do), the resulting sub-populations could display considerable initial isolation, thereby increasing the odds of ultimate speciation given a temporary extrinsic barrier. Even if the ancestral population occupies a uniform habitat, subdivision into small, inbred demes can sometimes lead to incipient isolation between demes due to occupation of different adaptive peaks as described by Sewall Wright [15] in his shifting balance theory. An interesting experiment relevant to this prediction was performed by Santibanez and Waddington [16]. They established two large subpopulations from a large base population of Drosophila melanogaster and subjected them to different selective regimes. No detectable reproductive isolation arose between them. In another experiment, they isolated several small inbred lines from the base population and raised all of them in a homogeneous environment. Interestingly, a significant degree of reproductive isolation arose between a majority of these lines. In this case, reproductive isolation arose between small, inbred demes in a homogeneous environment much more rapidly than isolation between large subpopulations subjected to divergent selection. In general, from Wright's shifting balance theory, populations subdivided into demes with variance effective sizes of less than 100 to 150 should display initial conditions optimal for rapid adaptive divergence.

Another important determinant of the chances for adaptive divergence is the type of "mate recognition system" [17] found in the ancestral population. The mate recognition system refers to the totality of phenotypic attributes (even non-genetic) that contribute to species identification for purposes of mating. Fisher [18] and Lande [19] have generated models of sexual selection that can lead to bursts of rapid divergence in both mate preference and mating cues. Such outbursts become more likely as the complexity of the mate recognition system increases [20], and even the non-genetic factors can greatly contribute to such instability [12,21]. Hence, ancestral populations characterized by complex and elaborate mate recognition systems would be particularly liable to rapid adaptive divergence through pre-mating isolating barriers. Birds might be a good candidate for a group displaying this type of speciation because they show indications of rapid speciation primarily through pre-mating barriers [22].

Events that occur after the extrinsic split are also important determinants of the rate of divergence. Stebbins [23] noted that divergence occurs slowly in plants occupying stable habitats but rapidly in the presence of changed environments. Hence, if one or both of the two subpopulations are subjected to very different selective pressures, adaptive divergence should be more rapid. A large number of experiments confirm this prediction, although they also indicate that isolated subpopulations can evolve incipient isolation even under identical environments [4]. In these experiments, reproductive isolation was not selected per se, thereby indicating that isolation does indeed arise as a pleiotropic consequence of selection in different environments and/or selection operating upon new mutants and genotypic combinations arising in independent gene pools.

In most of these experimental studies pre-mating isolation arose, often in the absence of post-mating isolation. This observation is not surprising in light of the sexual selection theories of Fisher [18] and Lande [19] and because adaptive divergence can lead to changes in morphology, behavior, niche, etcetera—all attributes which can contribute to a species' mate recognition system. Consequently, pleiotropic effects are not limited to post-mating attributes. These facts contradict the idea that adaptive divergence yields primarily post-mating barriers, with pre-mating barriers evolving after secondary contact due to selection against individuals that hybridize—the hypothesis of "reinforcement" [17]. The problem of reinforcement is very amenable to population genetic analysis, and indeed a considerable population genetic literature already exists upon this subject [4]. However, for lack of space all I shall say here concerning reinforcement is that theoretical and experimental population genetic studies imply that reinforcement is possible but extremely unlikely. Hence, reinforcement cannot be regarded as a major contributor to speciation via adaptive divergence.

Population structure is another important determinant of rate of adaptive divergence after the split has occurred. Wright [15] predicted with his shifting balance theory that the most rapid adaptive divergence occurs when the population is subdivided into small demes with restricted gene flow between demes. Moreover, the course of adaptation in this case is strongly influenced by stochastic forces making divergence likely even if the subpopulations inhabit identical environments. Nevertheless, different selective regimes would greatly contribute to speed of divergence in this case as well. Thus, population subdivision both before and after the split should allow rapid adaptive divergence and increase the likelihood of speciation in the face of temporary barriers.

There is another aspect of Wright's theories that is extremely important here and, unfortunately, not widely appreciated. The phrase "shifting balance" can be used not only to describe adaptive peak changes, but also the manner in which basic evolutionary forces (natural selection, genetic drift, and gene flow) interact with one another upon an adaptive landscape. To illustrate this, let me draw an analogy that is in some ways the inverse (quite literally) of Wright's original analogy of the adaptive landscape. Wright's knowledge of developmental genetics convinced him there are generally many ways of adapting to an environment, although pleiotropy insures that they are rarely equivalent in a fitness sense. Wright illustrated this conclusion with the adaptive landscape concept with its adaptive peaks of unequal height separated by adaptive valleys. However, in my analogy, consider turing this adaptive landscape upside down, thereby transforming the peaks into pits and valleys into ridges. Now let a ball correspond to a deme. When this ball is placed upon the inverted adaptive landscape and released, gravity will cause it to roll down to the bottom of a pit, corresponding to natural selection causing the demes to "climb" to the top of peaks. Note also that gravity causes the balls to roll to the nearest pit, not the deepest, just as natural selection causes the demes to climb the nearest adaptive peak, not the highest. Now let us put some lateral motion into the balls by randomly shaking the inverted landscape. As we shake the inverted landscape, the balls begin to roll around, even rolling up the sides of the pits against the force of gravity, just as random genetic drift causes demes to move around the adaptive landscape, even in directions opposed by natural selection. The intensity of the shaking corresponds to the strength of genetic drift; that is, the more the inverted landscape is shaken, the smaller the deme sizes in Wright's model. During this shaking process, some balls will actually roll up the side of a pit and over a ridge, at which point gravity once again causes the balls to roll to the bottom of a new pit. This corresponds to an adaptive peak change. If you place several balls in this inverted landscape either at random or in a shallow pit and then begin shaking, you would find that the balls would preferentially come to be located in the deeper pits. The reason for this is very straightforward—it is harder to roll out of a deep pit than a shallow pit; hence, as the balls roll around the inverted landscape, the ones in shallow pits are very likely to continue to roll into different pits, but the ones landing in deep pits are unlikely to make any further transitions. The result of this process over time is for most of the balls to be found in deep pits and none or only a few in shallow pits. Another way of saying this is that there is a shift in the balance between the relative importance of gravity and random shaking in determining the movement of the balls depending upon whether the balls are in a shallow or deep pit. Similarly, there is a shift in the balance between natural selection and genetic drift as a deme makes transitions from peaks of unequal height and steepness. The results of this shifting balance is that populations go preferentially from lower to higher peaks through time. Many people portray Wright's theory as if genetic drift induces peak transitions at random. However, what Wright realized is that genetic drift plus natural selection would consistently cause evolution from low to high adaptive peaks. Hence, adaptive evolution is far more efficient when natural selection is not in sole control even though, paradoxically, natural selection is the only force actually causing adaptation. Similarly, gravity plus random shaking is a far more effective procedure than just gravity alone in getting all the balls at the bottoms of deep pits in my inverted landscape; yet gravity is the only force that actually causes the balls to roll to the bottom. It took the genius of a mind like Wright's to come up with such a simple yet subtle insight into the nature of adaptive evolution.

There is yet one more property of the shifting balance theory which must be pointed out, and this is the property most critical to understanding the relationship of the shifting balance theory to speciation via adaptive divergence. Note that in shaking such an inverted landscape pit changes (adaptive peak transitions) are very common at the beginning of the process, but as more and more of the balls come to be in deep pits such transitions become less likely, often to the point that there is no movement at all, even though the amount of random shaking may be constant throughout the entire process. This of course is also a result of the progressive shift in the favor of gravity in the balance between gravity and shaking. Similarly, in Wright's theory, peak shifts become less likely as the shifting balance process operates through time even if the deme sizes remain constant. Moreover, there are further attributes of the shifting balance model that accentuate this tendency towards stasis that are not readily modeled by balls rolling in an inverted landscape. Demes are on many peaks during the initial phases of the process; hence, gene flow between them frequently acts as a random perturbing factor aiding genetic drift. But as the shifting balance process proceeds and more and more demes end up on the highest peak or a small set of high peaks, gene flow becomes more and more of a deterministic force attracting demes to the highest peaks. As more demes are brought to a single peak by the action of this type of gene flow (Wright's "interdemic selection"), the more gene flow between demes acts as a factor to maintain all the demes on that single peak. In other words, gene flow becomes more and more of a static force reinforcing natural selection and less and less of a perturbing force reinforcing genetic drift as the shifting balance process operates through time. This of course accentuates the shift in the balance between selection and drift, and thereby accentuates the tendency to go from dynamism to stasis. Wright [15] has also argued that as the demes move onto the higher adaptive peaks their population sizes might also tend to increase, thereby decreasing the importance of genetic drift. Thus, the shifting balance theory predicts that periods of evolutionary transition will be intense but brief and lead directly to a very static adaptive situation. This stasis will only be broken if the environment (and thus the landscape) is altered, if the system of mating is altered (which also determines the topology of the adaptive landscape), if the population structure is altered to shift the balance back towards genetic drift, or if new genetic variability occurs that alters the adaptive landscape. In summary, just as a thermos bottle can keep things either hot or cold through the very same underlying mechanism, the shifting balance theory explains why adaptive evolution is rapid and why it is static; both of which are caused by the same underlying mechanism. There has been a tendency to emphasize the dynamic aspect in most accounts of the theory, including Wright's, but the progressive shift towards stasis inherent in the shifting balance theory is equally important in making sense out of speciation and macro-evolutionary patterns.

This tendency to have rapid adaptive change followed by stasis is particularly important in understanding speciation via adaptive divergence. The extrinsic split itself is a factor that will inevitably alter population structure and perhaps mating systems and therefore potentially open up an ancestral population in the static phase of shifting balance to the more dynamic phase. This transfer to the more dynamic phase is much more likely if one or both of the subpopulations created by the split must adapt to a new environment, (ie a new adaptive landscape), and often the factors causing an extrinsic split are associated with environmental changes. With altered landscapes and/or population structure the shifting balance theory then predicts a phase of rapid adaptive transition shortly after the split that quickly evolves into stasis. By concentrating most of the adaptive changes into a short period following the extrinsic split, the shifting balance mode of phyletic evolution greatly increases the chances for speciation given a temporary extrinsic barrier.

Another important factor with respect to the shifting balance theory is that population structure is often not uniform throughout the entire range of the ancestral species. Ancestral populations inhabiting ecologically and/or geograph-ically marginal areas will often tend to be more subdivided and have smaller deme size than populations in the ecological and/or geographical centers of the species range. Thus, the central populations should be very static under the shifting balance theory, whereas the marginal populations have more potential to enter the dynamic phase of shifting balance. Nevertheless, this potential is often not realized because gene flow from the more numerous central populations represents a strong conservative force that can maintain the static phase of shifting balance even in the marginal areas and that can interfere with local adaptation. However, this static situation can be radically altered if gene flow from the central populations is severed, and it would also seem reasonable to assume that most geographical and ecological barriers that split a species would preferentially occur between marginal and central populations. Once gene flow from central to marginal areas has been severed, the marginal (but not the central) population could enter into the dynamic phase of shifting balance, particularly since the severing of gene flow also allows more effective local adaptation and thus the exploration of a somewhat novel adaptive landscape. Hence, ecologically and/or geographically marginal populations will often play a critical role in the speciation process resulting in a peripatric pattern of species distributions.

Earlier in this symposium, Dr. Mayr proposed that such peripatric patterns are due to speciation via "genetic revolution;" that is, founder events in the marginal areas lead to considerable losses of genetic variation and increased levels of inbreeding and homozygosity. This altered genetic environment then interacts with natural selection to effect drastic evolutionary transitions. Unfortunately, there are several flaws in this argument which make "genetic revolution" an unlikely explanation for peripatric patterns. First, the population-genetic implications of founder events are distorted in this model. As Lewontin [24] pointed out, founder events per se do not lead to considerable losses of genetic variation, nor do they result in greatly enhanced levels of homozygosity. This fact is illustrated very well by studies on Hawaiian Drosophila, the group of organisms with the strongest evidence for founder-event associated speciation. In the Hawaiian Drosophila, even the species derived from recent founder events show no tendency whatsoever to display lowered levels of genetic variation or increased homozygosity [25]. Hence, both population-genetic theory and population-genetic observations indicate the "genetic revolution" hypothesis is founded upon a false premise (also see [26]). Second, founder populations will only display considerable losses of genetic variation if, after the founder event, the population remains very small for many generations [12]. In this case, the genetic variation is lost primarily due to genetic drift in the generations following the founder event. However, "genetic revolution" by necessity requires intense selection, and if the founder population remains small and loses much genetic variation, the opportunity for responding to selection is correspondingly diminished. Consequently, "genetic revolutions" will not occur under these conditions [12]. Finally, the marginal demes often will be somewhat inbred themselves, so founder events stemming from marginal areas will not greatly alter the genetic environment and hence are not likely to induce "genetic revolution" [12]. I am not stating that founder events cannot induce speciation: they can [12], but only under conditions and with implications extremely different than those portrayed by Mayr [5]. In particular, the above population-genetic arguments imply that "genetic revolution" is a far less likely explanation for peripatric speciation than the induction of dynamic shifting balance in marginal areas as described earlier.

THE GENETIC ARCHITECTURES OF ADAPTIVE DIVERGENCE

Adaptive divergence can lead to isolating mechanisms with any of the three genetic architectures discussed earlier. However, genetic systems with many segregating units of small effect are more immune to the effects of genetic drift than systems characterized by one or a few major loci with epistatic modifiers [12]. The reason is that there is no direction to drift; hence in a system with many segregating units of small effect, drift on one element is often phenotypically counterbalanced by drift at the others in a manner analogous to the central limit theorem in statistics. However in a system with only a few major elements this central limit theorem argument is inapplicable. Thus, the shifting balance mode of phyletic evolution should be effective on traits with type II genetic architectures but rather ineffective on traits with type I architectures. Experimental studies on Drosophila by Rathie and Nicholas [26] support this prediction that type I architectures are not affected in any major fashion by shifting balance processes, and my own experimental work [27] indicates type II architectures are sensitive to factors inducing adaptive peak changes. Moreover, Parsons [28] has amassed considerable data indicating that intense selection under stress environments favors tolerant phenotypes that depend upon only a few major loci. Finally, Whitten et al [29] have shown that most cases of evolution of insecticide resistance in natural populations involve type II architectures, whereas many laboratory instances of insecticide resistance have type I architecture [30]. In the field, the intensity of the selective forces can be very large, but in the laboratory the selective regimes are usually more uniform and less intense. These alternative selective regimes apparently are differentially effective upon the various genetic architectures present in the gene pool, an observation consistent with Mather's [31] contention that the genetic architecture for a trait reflects the type of selective regime that had operated upon that trait. These results together imply that rapid adaptive divergence—whether caused by a radical change in environment, a shifting balance mode of phyletic evolution, or both—should be biased towards type II architectures, but that type I architectures should predominate under slow adaptive divergence [26,32]. Similar predictions about genetic architecture can be made for the other modes of speciation as shown in Table II (see [4]). The role of genetic architectures as a constraint on selection and speciation is an important, but much neglected, aspect of evolutionary biology. It is an aspect which we can no longer afford to ignore if we wish to truly integrate population genetics and speciation.

DISCUSSION

I hope the above account has convinced you that population genetics is indeed relevant to our understanding of speciation even in this worst possible case. I would now like to discuss one reason why the particular perspective offered by population genetics is of critical importance in evolutionary biology.

One of the more hotly debated issues in recent evolutionary biology is the relationship between microevolution and macroevolution [33]. The process of speciation lies directly at the interface of these two areas. Population genetics has traditionally dealt with microevolutionary phenomena, and its extension into the field of speciation represents the first step in generating a unified theory of

TABLE II. The Genetic Architectures of Isolating Barriers Most Commonly Associated With the Various Modes of Speciation

Mode of speciation	Basis of isolation[a]
Divergence	
adaptive	
slow	I̲, II, III
rapid	I, I̲I̲, III
clinal	I, I̲I̲, III
habitat	I, I̲I̲, I̲I̲I̲
Transilience	
genetic	II, III
chromosomal	II
hybrid maintenance	polyploidy̲, hybridogenesis, parthenogenesis
hybrid recombination	I, I̲I̲, I̲I̲I̲

[a]The most likely architecture is underlined.

micro- and macroevolution. For lack of space, I cannot discuss the avenues the population genetic approach is opening up, but I do wish to give warning of some of the potential cul-de-sacs this approach has already revealed.

The first danger concerns making broad generalizations from just one group of organisms. As mentioned earlier, descriptive population genetics indicates that speciation is not a process but a set of processes with very diverse evolutionary implications. Even within a single mechanism of speciation (Table I), population genetic considerations indicate that very diverse outcomes are possible, as I have illustrated with my discussion of slow versus rapid adaptive divergence (obviously, these are but two points upon a continuum). However, the really critical point is that groups of organisms subjected to particular types of genetic, population-structural, and ecological constraints will preferentially display certain modes or sub-modes of speciation (and all the implications of that mode) and not others. For example, in developing my theory of speciation via genetic transilience (founder-effect induced speciation) [12], I have emphasized that most organisms do not have the requisite attributes to allow founder events to directly induce isolating barriers. Consequently, this mode of speciation is a very rare one on a global basis. Nevertheless, the Hawaiian Drosophila have the requisite contraints on system of mating, karyotype, ecology, life history, etcetera, which make this mode probable in this group, even though it is very unlikely in the rest of the genus. Hence, generalizations from the Hawaiian Drosophila even to continental Drosophila would be erroneous, or vice versa. Another example is provided by paleontological data. Such data are subject to a large number of sampling biases [34,35], but certain groups have extremely good fossil records [36]. However, making broad macroevolutionary generalizations from the fossil record of just a few groups cannot be justified.

An even more fundamental bias plagues the fossil evidence—it cannot distinquish between morphological transition and speciation, rather it equates the two. It is therefore not exactly surprising that some authors (eg [36]) have concluded the fossil data indicate that morphological changes are concentrated into "speciation" events. However, the area of descriptive population genetics of speciation indicates that morphological transition cannot be equated to speciation. First, there is the common occurrence of sibling species; that is, perfectly good biological species that are morphologically identical. One must also keep in mind that fossil inferences are based upon only a small subset of morphology, and hence the occurrence of sibling species will be even more extensive in the fossil record than among living organisms. On the other hand, there are polytypic species; that is, one species can exist in several morphological forms. For example, by morphological criteria, there were 35 species of otter that were subsequently reduced to four biological species [37]. In cases such as this, the fossil record could lead to an incorrect inference of speciation when no speciation has in fact occurred. These difficulties and others are well known to workers in this area, but what is not so widely appreciated is that these biases exist in correlated sets that can differ consistently from one group to another. As mentioned above, different groups of organisms are subjected consistently to certain speciation modes because of their genetic, population-structural, and ecological attributes. These same attributes can influence the incidence of polytypic and sibling species. For example, Wilson et al [38] and Bush et al [39] have argued that there are very consistent differences in population structure between frogs and mammals, such that mammals are much more likely to display higher degrees of population subdivision. In terms of the models I have discussed in this paper, this implies that mammals are far more subject to rapid adaptive divergence than frogs. Another implication of this population structure is that mammals are more prone to form polytypic species than frogs, but on the other hand frogs are more prone to form sibling species complexes [5]. The result is that the number of fossil frog species will be consistently underestimated, whereas the number of fossil mammalian species could actually be overestimated. Obviously, these biases will confound any conclusions about how rates of morphological evolution are related to rates of "speciation." Moreover, times of speciation in frogs will tend to be underestimated because morphological divergence often occurs only long after speciation in sibling species complexes, but times will be overestimated in mammals because with polytypic species morphological divergence precedes speciation. In addition, genetic distances will overestimate time of speciation under conditions of rapid adaptive divergence (under the usual assumption that initial genetic distances are zero), whereas this bias disappears under conditions

TABLE III. Relative Importance of Speciation Mechanisms*

Divergence modes
adaptive $>$ clinal $>>$ habitat
Transilience modes
hybrid maintenance and recombination $>>$ genetic $>>>$ chromosomal

*Global importance only. What is otherwise a rare mode of speciation may predominate in certain groups or situations.

favoring slow adaptive divergence [10]. Note that this bias in genetic distance is in the same direction as the bias in inferring times of speciation from morphological data. Hence, the observation that both types of data yield a consistent picture does not necessarily imply that both are accurate—only that both are biased in a consistent direction. This is the primary danger of correlated suites of biases—they can yield internally consistent patterns—and population genetics warns us that such correlated suites are exactly what we can expect to find. Thus, my first warning was that patterns revealed from one group cannot be generalized to other groups; my second warning is that the patterns themselves may not be real, particularly if they are inferred by contrasting groups with known differences in genetic systems, population structure, or ecological constraints.

Another danger is the tendency to make inferences about evolutionary mechanisms from fossil data. Fossil data simply do not have the appropriate level of resolution of time scale to justify these inferences. This point was made very well by Schindel [35] who, in discussing fossil records so complete and exhaustive that microstratiographic sampling was possible, concluded, "Owing to discontinuous or low rates of sedimentation, it is either impossible or impractical to recover a continuous series of discrete life assemblages of fossil populations by collecting microstratigraphic samples from continuously fossiliferous intervals." He also stated, "Paleonotologists should be aware of processes operating over short time spans, but not for the purpose of testing or seeking the results of these processes in the fossil record." Nevertheless, several paleontologists have attempted to do just this. For example, Stanley [36] has concluded that phyletic evolutionary mechanisms are unimportant in macroevolution because many fossil lineages display stasis interspersed with short periods of rapid change. Instead, he argues that evolutionary mechanisms concentrated in the process of speciation itself and absent in phyletic evolution are of paramount importance. In the terminology of my model, this means that transilience modes of speciation are more important than divergence modes in affecting macroevolutionary patterns (although note that the factors inducing transiliences are still found in phyletic evolution). Once again, population genetics is very relevant to this question. Table III presents the relative likelihood of the speciation modes given in Table I as inferred from population genetic arguments [4]. Of the transilience modes, only the hybrid transilience modes are at all common in the organic world. Moreover, Anderson and Stebbins [40] have predicted that groups susceptible to speciation via these modes will be characterized by bursts of speciation followed by long periods of stasis, which is quite consistent with Stanley's argument. However, due to genetic and other biological constraints, these modes of speciation are common only in plants and are extremely rare in animals [4], but most fossil inferences are based on animals, not plants. The only other likely transilience mode is the genetic transilience, but as I have pointed out elsewhere [41] the population genetic conditions that make this mode possible are not found in many of the groups Stanley [36] analyzed. However, these same groups have population-genetic conditions ideal for rapid adaptive divergence, and as I have pointed out this mode of speciation results in a pattern of rapid evolutionary transition followed by periods of stasis. However, speciation is a pleiotropic byproduct of phyletic processes under adaptive divergence; hence Stanley's inference about the unimportance of phyletic mechanisms is unwarranted. I emphasize that I am not doubting Stanley's description of the observations, only his inference about mechanisms from those observations.

The primary strength of the population-genetic approach is that it deals directly with the genetic mechanisms causing the evolution of reproductive isolation, and hence it is complementary to most biogeographical, systematic, and paleonotological studies that deal most directly with the results of these mechanistic processes. Thus, population genetics offers a unique type of understanding into that most fundamental evolutionary problem—the origin of species.

ACKNOWLEDGMENTS

Supported by NSF Grant DEB-79 08860 and NIH Grant 1 RO1 GM27021-01.

REFERENCES

1. Mayr E: Review of modes of speciation by MJD White. Syst Zool 27: 478, 1978.
2. Wright S: Breeding structure of populations in relation to speciation. Am Nat 74: 232, 1940.
3. Wright S: On the probability of fixation of reciprocal translocations. Am Nat 75: 513, 1941.
4. Templeton AR: Mechanisms of speciation—a population genetic approach. Ann Rev Ecol 12: 23, 1981.
5. Mayr E: Populations, Species and Evolution. Cambridge: Belknap Press of Harvard University Press, 1970.
6. Muller HG, Pontecorvo G: Recombinants between Drosophila species the F_1 hybrids of which are sterile. Nature 146: 199, 1940.

7. Pontecorvo G: Viability interactions between chromosomes of Drosophila melanogaster and Drosophila simulans. Genetics 45: 51, 1943.
8. Pontecorvo G: Hybrid sterility in artifically produced recombinants between Drosophila melanogaster and Drosophila simulans. Proc R Soc Edin B 61: 385, 1943.
9. Watanabe TK, Lee WH, Inoue Y, Kawanishi M: Genetic variation of the hybrid crossability between Drosophila melanogaster and D simulans. Jpn J Genet 52: 1, 1977.
10. Templeton AR: Modes of speciation and inferences based on genetic distances. Evolution 34: 719, 1980.
11. Galton G: Discontinuity in evolution. Mind NS 3: 362, 1894.
12. Templeton AR: The theory of speciation via the founder principle. Genetics 94: 1011, 1980.
13. Stebbins GL: Variation and Evolution in Plants. New York: Columbia University Press, 1950.
14. Stebbins GL, Day A: Cytogenetic evidence for long continued stability in the genus Plantago. Evolution 21: 409, 1967.
15. Wright S: Evolution and the Genetics of Populations, Vol 3, Experimental Results and Evolutionary Deductions. Chicago: University of Chicago Press, 1977.
16. Santibanez Koreb S, Waddington CH: The origin of sexual isolation between different lines within a species. Evolution 12: 485, 1958.
17. Paterson HEH: More evidence against speciation by reinforcement. S Afric J Sci 74: 369, 1978.
18. Fisher RA: The Genetical Theory of Natural Selection. New York: Dover, 2nd ed, 1958.
19. Lande R: Models of speciation by sexual selection on polygenic traits. Proc Natl Acad Sci USA 78: 3721, 1981.
20. Templeton AR: Once again, why 300 species of Hawaiian Drosophila? Evol 33: 513, 1979.
21. Kalmus H, Smith SM: Some evolutionary consequences of pegmatypic mating systems (imprinting). Am Nat 100: 619, 1966.
22. Avise JC, Patton JC, Aquadro CF: Evolutionary genetics of birds: Comparative molecular evolution in New World warblers and rodents. J Hered 71: 303, 1980.
23. Stebbins GL Jr: Evidence on rates of evolution from the distribution of existing and fossil plant species. Evol Mono 17: 149, 1947.
24. Lewontin RC: Comment. In Baker HG, Stebbins GL (eds): The Genetics of Colonizing Species. New York: Academic Press, 1969, p 481.
25. Carson HL, Kaneshiro KY: Drosophila of Hawaii: Systematics and ecological genetics. Ann Rev Ecol Syst 7: 311, 1976.
26. Lande R: Genetic variation and phenotypc evolution during allopatric speciation. Am Nat 116: 463, 1980.
27. Templeton AR: The unit of selection in Drosophila mercatorum. II. Genetic revolutions and the origin of coadapted genomes in parthenogenetic strains. Genetics 92: 1283, 1979.
28. Parsons PA: Isofemale strains and evolutionary strategies in natural populations. Evol Biol 13: 175, 1980.

29. Whitten MJ, Dearn JM, McKenzie JA: Field studies on insecticide resistance in the Australian sheep blowfly, Lucilia caprina. Aust J Biol Sci 33: 725, 1980.
30. Crow JF: Genetics of DDT resistance in Drosophila. Proc Int Genet Symp 1956: 408, 1957.
31. Mather K: Variability and selection. Proc R Soc London B 164: 328, 1966.
32. Lande R: Review of "Macroevolution: Pattern and Process," by S Stanley. Paleobiology 6: 233, 1980.
33. Lewin R: Evolutionary theory under fire. Science 210: 883, 1980.
34. Raup DM: Species diversity in the Phanerozoic: an interpretation. Paleobiology 2: 289, 1976.
35. Schindel DE: Microstratigraphic sampling and the limits of paleontologic resolution. Paleobiology 6: 408, 1980.
36. Stanley SM: Macroevolution: Pattern and Process. San Francisco: Freeman, 1979.
37. Rensch B: Das Prinzup geographische Rassenkreise und das Promleme der Artibildung. Berlin: Borntraeger, 1929.
38. Wilson AC, Bush GL, Case SM, King MC: Social structuring of mammalian populations and the rate of chromosomal evolution. Proc Natl Acad Sci USA 72: 5061, 1975.
39. Bush GL, Case SM, Wilson AC, Patton JL: Rapid speciation and chromosomal evolution in mammals. Proc Natl Acad Sci USA 74: 3942, 1977.
40. Anderson E, Stebbins GL: Hybridization as an evolutionary mechanism. Evolution 8: 378, 1954.
41. Templeton AR: Review of "Macroevolution: Pattern and Process," by S.M. Stanley." Evolution 34: 1224, 1980.

Mechanisms of Speciation, pages 123–142

DNA Sequence Changes and Speciation

W.J. Peacock, E.S. Dennis, and W.L. Gerlach

Speciation is a process of separation of organisms into noninterbreeding units. Both phenotypic and genotypic variation are associated with the differentiation of species, and the question arises as to whether any class of variation can be cited as causative rather than consequential or associative. Since there must be many paths of speciation a universally applicable answer is unlikely, but we might still meaningfully ask whether there are any particular aspects of DNA sequence variability which are associated with the events of speciation.

It is possible that DNA sequence changes accumulate until, at a certain level, they are compatible with speciation occurring (Fig. 1a). On this hypothesis individuals of two recently diverged species would have more DNA sequence differentiation than individuals within these species, and individuals of two species diverged for a longer period of time would show an even greater level of sequence differentiation. Alternatively, we could suppose that speciation occurs during a period in which there is an effective acceleration, at a population level, of DNA sequence divergence. This would also imply significantly greater levels of sequence differentiation at the interspecific than at the intraspecific level (Fig. 1b). Finally, it is possible that a particular type of DNA sequence change may induce speciation (Fig. 1c). A comparison between species would show more sequence changes than are evident in an intraspecific comparison, but would also show sequence changes of a type not occurring between individuals within a species.

In order to understand evolution and speciation we will need to know the principal sources and kinds of changes in the genome and the mechanisms by which these changes alter gene expression.

DNA SEQUENCE COMPONENTS IN THE GENOME

DNA sequence changes have been measured by quantifying certain parameters of bimolecular reassociation of DNA strands. The rate and extent of reassociation

DNA SEQUENCE CHANGES IN SPECIATION

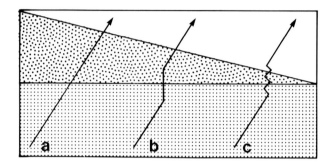

Fig. 1. DNA sequence changes in evolution: a, b, and c represent different possibilities in terms of DNA sequence divergence that might occur during the process of speciation. Species divergence has occurred in the upper part of the diagram.

and the thermal stability of the reannealed duplex have provided measures of sequence differentiation between taxa on the basis of correlation between the extent and fidelity of base pairing in the heteroduplex and their phyletic relationship [1]. However, the technique has had limitations because not all DNA sequences have the same rates of change, and secondly, because the annealing technique does not adequately take into account the existence in the genome of sequences with vastly different reiteration frequencies.

The DNA of eukaryote genomes can be classified in terms of sequence reiteration frequency (Fig. 2). There are unique or single-copy sequences, and moderately and highly repeated sequence components [2]. Another way of looking at the genome is to classify sequences as transcribed or nontranscribed. In general, nontranscribed sequences correspond to the most highly repeated sequences of the genome. Within the transcribed class of sequence there are moderately repeated sequences as well as single-copy sequences [3]. The genome organization is even more complicated since often a transcribed region in eukaryotes includes segments of DNA coding for amino acid sequences in proteins (exons) interspersed with other segments (introns) which are initially transcribed but not included in the final mRNA molecule which is translated in the cytoplasm [4].

Recently another criterion of sequence differentiation has been recognized, some sequences varying in both number and position within the genome. In Drosophila melanogaster these mobile sequence elements may constitute between 10 and 15% of the genome [5].

TYPES OF DNA SEQUENCE

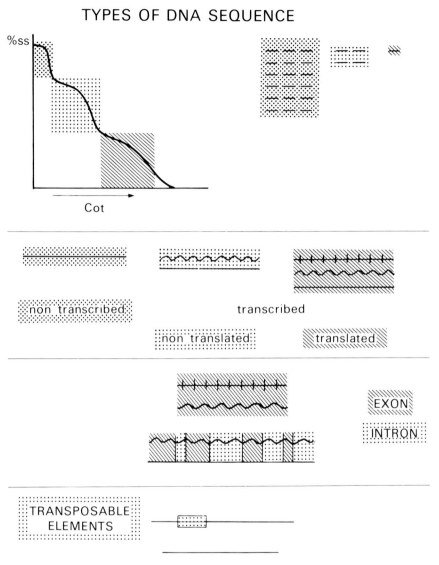

Fig. 2.   DNA sequences in eukaryote genomes. This figure shows in diagrammatic form four ways of classifying DNA sequences in the genomes of higher organisms. Sequences may be classified by reiteration frequency, whether they are transcribed, translated, or if they are stable or nomadic sequences in the genome.

TABLE I. Nucleotide Substitutions in the Coding Region of Three Gene Systems [5]

	Rabbit and mouse α-globin genes	Histone genes two species of sea urchins	Human growth hormone and chorionic somatomammotropin
Silent substitutions	39/100	37/100	7/100
Amino acid replacement substitutions	18/100	3/100	15/100
	18% amino acid/100 million years 1% nucleotide/2 million years		

Because the different sequence components are likely to have different rates and mechanisms of change, they must be examined separately. This has been made possible by recombinant DNA technology. Prior to the development of these techniques it had only been possible to examine in detail highly repeated DNA sequences because these sequences were easily purified.

RATES AND PATTERNS OF SEQUENCE CHANGE

Unique Sequences

Most genes or coding sequences occur as a few or single copies in the genome. Most repetitive sequences, with the exception of rDNA sequences, cannot be considered to be genes. But not all unique or single-copy DNA is accounted for by hybridization experiments with total polyA mRNA populations. In most organisms only about 15% of unique sequence DNA can be readily attributed to mRNA-related sequences [6]. It is likely, as will be discussed later, that at least as much unique sequence DNA corresponds to intervening sequences or introns within coding genes, and spacer sequences between genes.

Nucleotide sequence divergence for total single-copy DNA has been measured between primate species to be of the order of one-half of a percent per million years [7]. This estimate corresponds to those obtained by the examination of specific single-copy sequences but the nucleotide substitution evolutionary clock varies over a wide range. In protein coding regions the proportions of silent changes to changes which result in amino acid substitutions can vary markedly (Table I). Amino acid data from protein sequences suggests a mean rate of change of 18% residue replacement per 100 million years, a figure in accord

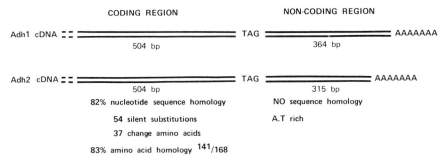

Fig. 3. The alcohol dehydrogenase genes of maize. Nucleotide sequencing of cDNA clones derived from the Adh1 and Adh2 genes in maize has revealed a distinctive pattern of sequence differentiation between the two loci. In the coding regions nucleotide substitutions which are silent and give rise to amino acid substitutions occur but to a lesser extent than the sequence divergence occurring in the 3' nontranslated region of the genes.

with data obtained from sequencing the coding regions of mRNA where there is approximately 1% nucleotide substitution per two million years [7].

But we need only compare two related gene sequences to be reminded that these average figures, even within a reiteration class of DNA, have limited value. For example, in the two alcohol dehydrogenase (Adh) genes in maize, which probably have a common origin, the coding region shows about 18% nucleotide substitution with about 60% of the substitutions being silent changes not causing amino acid substitutions (Fig. 3). Most of these silent changes are in the third base of codons. In the noncoding region at the 3' end of the genes there is no detectable sequence homology other than the one in four correspondence expected by chance. In contrast, in another example of plant genes, comparisons between different copies of the leghemoglobin gene in soybean show a high degree of sequence conservation in the 3' noncoding region, whereas there is considerable sequence divergence in the 5' noncoding region [8].

Obviously base substitution changes are a common feature of all parts of gene sequences but they are under selection pressures which differ substantially between genes and for different regions of the same gene.

Introns

A majority of genes in both plant and animal genomes contain intervening sequences or introns which interrupt coding segments of the gene. The number of introns per gene may vary from 0 up to the current record of 52. The sequences of introns in different species have not been compared structurally but several points have become clear. There is considerable variability in intron length and in nucleotide sequence, but there is conservation of sequence at intron/exon junctions and to a lesser extent in nearby regions of the introns [9]. The position of introns within a gene is also constant. A striking example is the conservation of the position of introns 1 and 3 in the leghemoglobin gene of soybean which are inserted in the coding sequence exactly where two introns are inserted in globin genes of animals [8]. The three introns of soybean leghemoglobin genes also provide an example of the action of selection pressures on sequence change. The 5' intron is far more conserved than either of the remaining two introns, the 3' intron being the most variable.

A number of hypotheses have been presented for functional roles of introns. Introns may have been important in the evolution of certain genes by associating different sequence elements together into a new functional genetic unit. This suggestion [10] certainly seems to apply to immunoglobulin proteins. In mouse immunoglobulin heavy chain constant regions there are three distinct domains which are important for light-chain attachment, complement fixation, and cell surface interactions, respectively. The domains have similar numbers of amino acids and similar molecular configurations suggesting they were all derived from a common precursor gene. The DNA coding sequences corresponding to the functional units of the protein are separated by introns [11]. Thus, the present gene appears to be a patch-work of pre-existing coding regions. If introns generally function to create new genes, this is a form of genome restructuring which is not likely to be of significance at a speciation level but could be important for larger phyletic units.

Another suggestion is that introns have effects on gene expression. This could be important in speciation. An example of an effect of intervening sequences on gene expression is seen in the 28S ribosomal RNA gene in Drosophila melanogaster. Approximately half of the ribosomal gene repeats on the X chromosome are interrupted by an insertion at a specific location in the 28S gene. The repeat units which contain this insertion are transcribed three orders of magnitude less than are the other repeats in the tandem array of ribosomal RNA genes [12]. If introns can affect the level and timing of gene expression, then their presence and absence at specific gene locations could be of significance in genetic changes underlying speciation.

Moderately Repeated Sequences

A substantial proportion of most eukaryotic genomes is in the form of interspersed, moderately repeated sequence DNA [13]. These sequences are generally a few hundred nucleotides in length and have reiteration frequencies ranging from one hundred to some tens of thousands per genome. Moderately repeated sequences are interspersed with single-copy sequences and two general types of interspersion can be recognized. In some species, such as sea urchins, the moderately repeated sequences of about 300 base pairs are positioned between single-copy segments of approximately 1000–2000 base pairs. In Drosophila melanogaster different moderately repeated sequence units tend to be grouped together in long arrays, these in turn being interspersed between arrays of single-copy sequences some several thousand base pairs in length. These two categories do intergrade and in any one genome some representatives of each may occur. In sea urchin species analyses have shown that most families of moderately repeated sequences are transcribed and at least some of them are known to be represented in mRNAs in the cytoplasm. There are generally transcripts of both strands of the repeated sequence, implying that the sequence occurs in opposite orientation with respect to the nearest promoter, in different locations in the genome [14].

In some families of moderate repeats there is considerable divergence of sequence between members but in others the sequence is highly conserved. These analyses have been facilitated by cloning technology which has also established that individual sequence families may be conserved over quite long evolutionary periods, periods in which there could have been expected to have been considerable sequence decay. The use of individual moderately repeated sequences as specific probes has revealed two surprising features. One is that the number of copies of a particular repeat family can vary dramatically between related species even though the nucleotide sequence itself may be conserved. The second point is that a particular sequence family may be present as a homogeneous sequence population in one species with a slightly different sequence variant constituting the population in a related species. These two features imply that there are mechanisms in the genome for rapidly changing repeat frequencies of these dispersed sequences and that there are mechanisms which can bring about replacement or correction of repeat units.

For any given repeat the frequency of its representation in nuclear and mRNA populations in different tissues can differ dramatically. The degree of representation of related moderately repeated sequences in mRNA transcripts in different organs seems to be conserved between species, although the nucleotide sequence per se may change between species.

All of these properties suggest that a variety of selection pressures operate on this dispersed sequence class.

Mobile DNA Sequences

A new class of moderately repeated DNA has been demonstrated in Drosophila melanogaster [5] and yeast [15] and almost certainly exists in maize [16]. This class of sequence is similar to the transposons or insertion sequences which have been described in bacteria. The sequences are mobile in the genome, with a high probability of varying in both number and location even between geographical races in Drosophila melanogaster. In Drosophila they show marked responses in location and number when cells are put into culture. This may be a response to stress conditions, but could signal that movement of the sequences could be a regular feature of development. These nomadic elements may participate in

MOBILE REPEATED DNA SEQUENCES

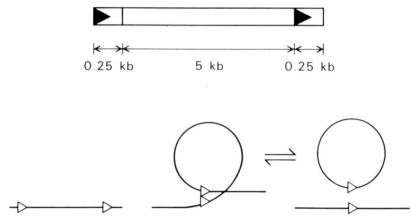

Fig. 4. Generalized structure of mobile sequence elements. In both Drosophila and yeast the structure of some mobile sequence elements shows terminal direct repeat regions bounding a unique region which is well conserved among the many copies occurring in the genome. The occurrence of the direct repeat indicates that excision and reinsertion of these sequences could occur by recombination events.

processes of regular restructuring of the genome. An example of changes in genome architecture is found in antibody genes where plasticity of genome organization is an essential attribute for the production of antibody diversity [17].

In Drosophila the 80 or so families of these transposable elements could account for as much as 10–15% of the genome. In other organisms it is not known how much of the moderately repeated DNA is of this type. In Drosophila they are abundantly transcribed and at least one of the sequence families (Copia) can be translated in an in vitro system. Copia sequences have now been identified in certain mutations in Drosophila, and there are indications that many naturally occurring mutations are a result of insertions of this class of sequence. When elements are inserted into regions other than the coding regions it is highly probable that they affect levels of gene expression. This is best documented in the genetic analyses of mobile controlling elements in maize which can abolish gene activity or alter gene expression in amount or tissue specificity.

In both yeast and Drosophila the sequence families of transposable elements show a common molecular organization. They have an internal region which varies from 1 to 5 kb in different families and show a high degree of sequence conservation in the many copies dispersed around the genome. In every case the internal sequence is bounded at both ends by a sequence of approximately 500 bases arranged in the same orientation. This direct terminal repeat structure suggests mechanisms by which the elements may excise and reinsert into the genome (Fig. 4).

These sequences may be a component of the genome which is of importance in speciation events. For example, different races of Drosophila melanogaster show hybrid dysgenesis when crossed. In these cases the F1 hybrid is frequently sterile or near sterile, there is usually a high frequency of mutation, and gene recombination characteristics can be markedly altered. Recent data have shown that hybrid dysgenesis is associated with the presence and destabilization of a mobile sequence [18]. Races displaying hybrid dysgenesis show marked differences in the frequency and locations of certain transposable sequences, which appear to be the cause of the mutations observed.

Since these transposable elements cause mutations, can be associated with structural rearrangements of the chromosome, and can change the extent and timing of gene expression, they may have a considerable role in evolution. The pleiotropic genetic effects associated with hybrid dysgenesis could contribute to isolation barriers between populations which could lead to speciation. They may provide the mechanism for the spread and replacement of other sequences in the genome. The condensed timescale of genomic changes associated with this class of sequence contrasts sharply with the timescale previously cited for nucleotide substitution in gene sequences.

Highly Repeated Sequences

The sequences in the class of highly repeated (satellite) DNA often have very short repeat units with distinctive base composition enabling them to be isolated by buoyant density procedures and to be detected by restriction enzyme analyses. Highly repeated sequences are usually organized in long tracts of tandemly arrayed repeats and located in heterochromatin. They commonly account for 10–40% of a genome [19]. Most highly repeated DNAs have several chromosomal locations (Fig. 5) implying the existence of mechanisms for transposing the sequences to several sites in the genome [20].

In a number of analyses it has been shown that the total population of repeat units of a given sequence is built up by a series of successive amplification events. In the major satellite DNA of the red-necked wallaby, which accounts for 20% of the total DNA, restriction enzyme digests showed the presence of subpopulations within the satellite [21]. The related but different sequences are arranged in clusters of tandem arrays making up the total of 500,000 repeat units. Successive digests with different enzymes showed that each tandem array of similar repeats must have arisen by an amplification of a single variant repeat

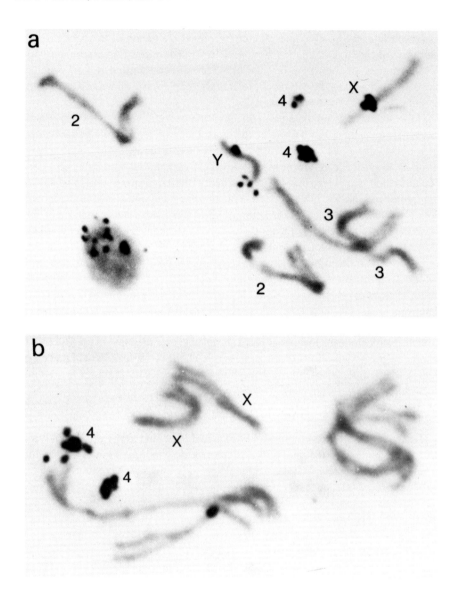

Fig. 5. In situ hybridizations of highly repeated DNAs in Drosophila. (a) Mitotic chromosomes of D simulans hybridized with cRNA to the 1.695 g/cc satellite isolated from D simulans DNA. The asymmetry of labeling on the fourth chromosomes was not reproducible in other cells. (b) The same probe hybridized to a D. melanogaster stock which carries fourth chromosomes derived from D simulans. There are no sites on the X or Y and only one minor site on an autosome. (Photographs supplied by Dr. Alan Lohe).

MECHANISMS OF AMPLIFICATION

UNEQUAL
SISTER
CHROMATID
EXCHANGE

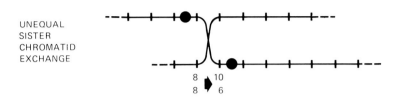

REPLICATION

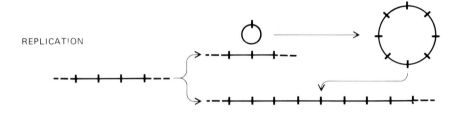

Fig. 6.   Possible mechanisms of sequence amplification. There are two general classes of mechanism proposed for sequence amplification: Unequal exchange events between the members of tandem arrays of repeat units which could result in an increase or decrease of repeat numbers (upper), or replication external to the chromosome coupled with mechanisms for excision and reinsertion into chromosomal DNA (lower).

in a previously existing array. In this satellite it was possible to deduce the temporal pattern of amplification events since the existing subpopulations of related sequences could be arranged in a unique hierarchy.

The mechanism of amplification has not been established but may be either a consequence of unequal exchange events between a series of tandemly arrayed units or replication processes coupled with excision and reinsertion events (Fig. 6). In either case it is possible to model both the contraction and expansion of related sequences with the ultimate result of one variant replacing another. Although unequal exchange is attractive because sister chromatid exchange is a well-documented chromosomal phenomenon and also because unequal exchange has been demonstrated to occur in recombination between duplicated gene elements, it does not explain the transposition of sequences from one chromosomal location to another.

Exchange between homologous chromatids or chromosomes does not lead to any genetic problems but there are limitations in the number of exchanges occurring between nonhomologous chromosomes. A single or odd number of ex-

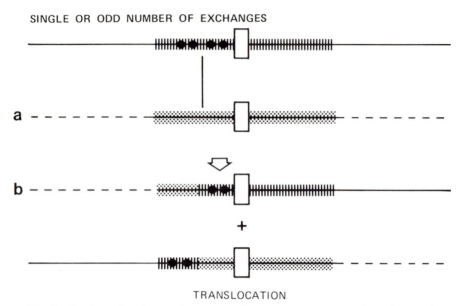

EXCHANGE BETWEEN NON HOMOLOGOUS CHROMOSOMES

SINGLE OR ODD NUMBER OF EXCHANGES

a

b

+

TRANSLOCATION

Fig. 7. Reciprocal exchanges between tandem repeat arrays on nonhomologous chromosomes. The figure indicates the consequences of odd and even numbers of exchanges.

changes will lead to reciprocal translocations and an even number of exchanges is required to bring about a transposition (Figs. 7 a, b). An excision and reinsertion mechanism has an additional attraction because it can account for the replacement and spread of moderately repeated sequences which are dispersed, as well as of repeats arranged in tandem arrays. Whatever the mechanism of change in reiteration frequency of highly repeated DNA, it appears to be under cellular controls since although polymorphisms for frequency of repeats have been described [22, 23], for the most part there is stability in the size and locations of arrays of particular highly repeated sequences within a species.

Interspecific comparison of these highly repeated sequences discloses that they are not species specific, although initially buoyant density analyses indicated this to be the case. The availability of purified sequences has established that many related species share overlapping complements of highly repeated DNA sequence families in their genomes. However, there are often large differences in reiteration frequency of any given sequence. Detailed comparisons between the sibling species Drosophila melanogaster and Drosophila simulans show that chromosomal locations of particular satellite DNAs appear to remain constant

TABLE II. Relative Amounts (kb) of Specific Highly Repeated Sequences in the Chromosomes of Drosophila melanogaster (M) and Drosophila simulans (S)

Chromosomal region labelled	1.672 g/cc(M)		1.705 g/cc(M)		1.695 g/cc(S)	
	M	S	M	S	M	S
X proximal	680	60	330	5	—	50
X distal	—	180	—	280	30	1600
YS + YL proximal	1580	530	1830	160	420	600
YL mid	2560	1820	1720	120	—	—
YL distal	4060	1740	2720	270	140	590
2	150	30	4660	—	240	—
3	980	140	390	5	—	<50
4	2780	1520	280	5	<50	1870

but that reiteration frequencies at certain chromosomal sites can be drastically changed [24] (Table II). In some of the satellites there appears to be coordinate change in all chromosomal sites but this is not always the case.

An example of noncoordinate change among chromosomal sites of a sequence is seen in kangaroo species which have been examined for the presence of the major highly repeated sequence DNA isolated from the wallaroo [23] (Fig. 8). In the wallaroo this satellite is present in the pericentromeric heterochromatin of all chromosomes of the complement and on the X chromosome it occurs at a centromeric location, near the nucleolus organizer region and on the long arm of the X chromosome. In the red-necked wallaby the same sequence is detectable and relative to the wallaroo is amplified in the X chromosome. However, it is not detectable on any of the autosomes of the complement. In yet another species, the red kangaroo, the sequence is again present on the X chromosome, once more associated with the nucleolus organizer region, and it is present on four telocentric autosomes in the complement. These chromosomes differentiate the chromosome complement of the red kangaroo from most other species in this group. These nonrandom changes in reiteration frequency among chromosomal sites imply that selection pressures are operating on what may well be a randomly occurring underlying mechanism of change.

Although a particular highly repeated sequence may be represented in the genomes of several species, the changes in reiteration frequency result in a species-specific composition of the total complement of repeated DNA species. Despite the magnitude and short-time scale of these changes, consideration of several highly repeated sequence families over a range of taxa can provide information on evolutionary relationships. For example, in the kangaroo and wallaby species in the macropod group of marsupials, monitoring of several

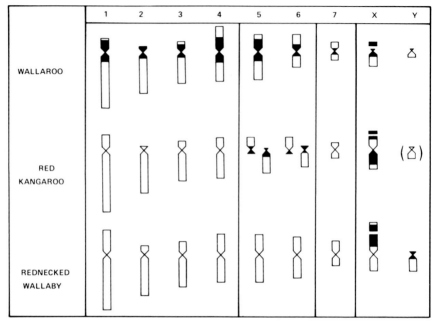

Fig. 8. The chromosomal locations in three kangaroo species of sequences homologous to the major highly repeated sequence of the wallaroo genome.

highly repeated DNA sequences points to a close relationship of the red-necked wallaby, red kangaroo, and euro-wallaroo complex [25]. This conclusion was based on satellite DNAs isolated initially from the wallaroo, the red-necked wallaby, and several from the red kangaroo [21,23,26]. The relationship of the red-necked wallaby and red kangaroo to the wallaroo complex was largely unsuspected but other data are consistent with the conclusion.

Examination of individual highly repeated DNA sequences between species shows that there is relative conservation of some of these sequences. This has been observed in most groups of related species that have been studied. In the sibling species Drosophila melanogaster and Drosophila simulans all repeated sequences checked occur in both species [24], and Fry and Salser [27] proposed the "library" hypothesis of satellite DNA as a consequence of finding common sequences in widely separated rodent species. Identical highly repeated sequences are also known among species of plants [28] and in many other organisms.

Interspecies comparisons also show changes in the long-order periodicity of repeats as well as the changes in reiteration frequency of particular sequences. When a satellite DNA sequence is present in related species the length of the repeat, defined by restriction enzyme digests, is often different in each species [31]. These different long-order periodicities may be the result of amplification events where a particular length of segment is amplified, either by sister chromatid exchange or by replication mechanisms as previously mentioned. In some cases an amplification cycle has as a base unit, a multimer length of a repeat series, rather than a variant monomer. In a situation where long-order periodicity is changed, the amplification may involve an unrelated, adjacent sequence as well as in the repeat sequence [29]. However, detailed sequencing has established that in some cases of repeat length alteration, intrarepeat sequence changes are involved (e.g., calf satellites [30] and maize knob heterochromatin satellite [Dennis and Peacock, unpublished]).

A total replacement of one repeat periodicity by another might imply selection pressure for the new repeat length; however, genetic drift could account for the establishment of a variant repeat dependent on the frequency of events and time of divergence between species.

If selection pressures operate there must be some specific function of a sequence. Although most satellite sequences are not transcribed there are many examples of effects of heterochromatin (and presumably of blocks of highly repeated sequence) on gene functions. The persistence of one repeated sequence on sex chromosomes through a number of groups of reptiles and birds is indicative of a selection pressure [32]. Another example of heterochromatin having cellular effects is in the X chromosome in Drosophila melanogaster where heterochromatin surrounds the nucleolus organizer, the site of the ribosomal cistrons. Provided other ribosomal cistrons are contained within the genome, large amounts of this heterochromatin can be deleted without loss of viability. The sc^4 sc^8 deletion, which removes at least three quarters of the highly repeated DNA of the proximal heterochromatin of the X chromosome, results in marked effects on the course of meiosis, sperm development, time of development of male competence, and other developmental characters [37].

Perhaps the best correlation between cytogenetic effects and a highly repeated DNA sequence is that described in maize [36]. Maize has blocks of heterochromatin in a number of different chromosomal locations—centromeric, at the nucleolar organizer, in B chromosomes, and knobs. Each class of heterochromatin is distinguished by its cytological appearance and cytogenetic effects. The presence of the large knob (K10) causes the preferential recovery of knobbed chromosomes from knobbed/knobless heterozygotes and the induction of neocentromeres at knobs in both the first and second meiotic divisions (Fig. 9).

We have isolated and sequenced a highly repeated DNA which is a major constituent of knob heterochromatin. This sequence is present in all knobs including K10 and absent from centromeric, nucleolar, and B chromosome heterochromatin. We have also isolated and sequenced the knob-specific repeated DNA from the wild relative of maize, teosinte. The genomic organization of the

The K10 KNOB induces all knobs to act as centromeres and causes unequal genetic recoveries.
Other heterochromatin does not respond.

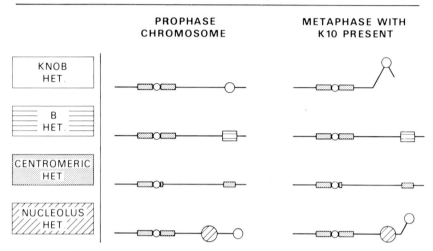

Fig. 9.   Knob heterochromatin behavior in maize meiosis.

sequence is identical to that in maize. The sequence of individual repeats is within the range of variation seen in the maize sequence. Sequence variants present in maize are also present in teosinte.

The conservation of the association of a particular repeated DNA sequence with a class of heterochromatin having a distinctive cytological appearance and cytogenetic effects may reflect a functional role for the sequence.

In a number of animals heterochromatic regions of the genome (in some cases the identity with highly repeated sequences has been firmly established) are eliminated from the soma [33,34]. The somatic elimination and the germline retention suggests specific functions in meiosis for these chromosomal regions. A parallel situation is the underreplication of highly repeated sequence DNA in the polytene chromosomes of Drosophila species [35]; the nuclei of these cells have no mitotic future.

Another reason why highly repeated DNA may be important in evolution is that some heterochromatin contains essential loci. In the pericentromeric heterochromatin of chromosome 2 in Drosophila melanogaster there are 13 essential functions [38]. It is probable that these functions apply to unique sequence DNA interspersed among the highly repeated DNAs which are predominant components of this heterochromatic region. In this case there is clearly a functional role of the heterochromatin but a case is not established for the highly repeated sequences per se, even though they constitute the great proportion of this chromosomal region. There may be an interdependent association of highly repeated sequences and particular gene loci; this could well be true for ribosomal RNA genes and highly repeated sequences in a number of species such as Drosophila [20] and macropods [23].

CONCLUSIONS

We have briefly reviewed the types and frequencies of sequence changes which can occur in the genomes of higher organisms. The recent demonstrations of the complexity of the structure of these genomes necessarily reshapes our concepts of DNA sequence change through evolution. Previously it was recognized that DNA sequences did change over time by nucleotide substitution. There was a correlation between the extent of sequence change and the phyletic distance between taxa. These changes were largely determined by single-copy DNA where nucleotide substitution in protein coding regions could be detected by amino acid replacement, isozyme analysis, or immunological tests on the protein, as well as by direct DNA sequencing. These sequences may only be about 10% of the genome and although studied most they may be of minor importance in speciation.

The demonstration of transposable, or nomadic, sequence elements, of large changes in reiteration frequency of both highly repeated and moderately repeated sequences, and of sequence variant replacement introduce new possibilities for the role of genome changes in evolution. The timescale of change of reiteration frequency and of long-order organization of repeated sequences is in marked contrast to the timescale of nucleotide substitution in sequences coding for amino acid complements of proteins.

Sequence movement and amplification, which may involve up to 80% of sequences in a genome, may account for the rapid establishment of genetic discontinuities which are not likely to be generated by nucleotide substitutions within genes. Movement of sequences around the genome such as has been shown for Copia-like transposable elements in Drosophila can result in changes in the level and timing of gene expression during development, and the inactivation of genes. Hybrid dysgenesis, mentioned earlier, could for example provide a mechanism by which populations become isolated in a condition of incipient speciation.

Amplification of particular repeated sequences could result in alterations of chromosomal properties, such as 3-dimensional positioning in the nucleus. If some gene interactions are dependent on spatial orientation, alterations in chromosome positioning could have major effects on gene expression. Repeated DNAs have effects on recombination and on other cytogenetic properties, so that increases or decreases in particular sequences could have relatively abrupt and large-scale effects on genes. In only rare cases have we recognized relationships and principles that apply to genomic organization. We are only just beginning to have glimpses of functions which apply directly or have been acquired by genomic elements such as spacer sequences, introns, transposons, dispersed moderately repeated sequences, and tandemly arrayed highly repeated sequences.

Study of DNA sequences in different species shows differentiation at the DNA level; species differ in nucleotide sequences in coding regions, they differ in the position and number of moderately repeated DNAs, and they differ in reiteration frequency of particular highly repeated sequences. The divergence of nucleotide sequences in genes provides a measure of evolutionary divergence but these changes occur much less frequently than the other genome changes mentioned, which frequently can be detected at intraspecific levels. Furthermore, a sequence change in a single-copy sequence will only affect a single gene, whereas transposition and amplification of sequence repeats can be pleiotropic. Although none of these genome changes are restricted to interspecific evolution, the relatively high frequency of occurrence and the wide-ranging effects of amplification, transposition, and replacement provide a high potential for their involvement in speciation. This potential could be enhanced if there were concomitant reductions in effective population size, enabling fixation of a variant sequence or genome arrangement which could result in significant genetic isolation between populations of organisms.

REFERENCES

1. Laird CD, McCarthy BJ: Magnitude of interspecific nucleotide sequence variability in Drosophila. Genetics 60:303, 1968.
2. Britten RJ, Kohne, DE: Repeated sequences in DNA. Science 161:529, 1968.
3. Constantini FD, Britten RJ, Davidson EH: Message sequences and short repetitive sequences are interspersed in sea urchin egg poly(A)$^+$ RNAs. Nature 287:111, 1980.
4. Garapin AC, Le Pennec JP, Roskam W, Perrin F, Carri B, Krust A, Breathnach R, Chambon P, Kourilsky P: Isolation by molecular cloning of a fragment of the split ovalbumin gene. Nature 273:349, 1978.
5. Rubin GM, Brorein (Jr) WJ, Dunsmuir P, Flavell AJ, Levis R, Strobel E, Toole JJ, Young E: "Copia-like" transposable elements in the Drosophila genome. Cold Spring Harbor Symp Quant Biol 45 (in press).
6. Rosbach M, Campo MS, Gummerson KS: Conservation of polyA containing RNA in mouse and rat. Nature 258:682, 1975.
7. Jukes TH: Silent nucleotide substitutions and the molecular evolutionary clock. Science 210:973, 1980.
8. Jensen EO, Paludan K, Hyldig-Nielsen JJ, Jorgensen P, Marcker KA: The structure of a chromosomal leghaemoglobin gene from soybean. Nature 291:677, 1981.

9. van den Berg J, van Ooyen A, Mantei N, Schambock A, Grosveld G, Flavell RA, Weissman C: Comparison of cloned rabbit and mouse β-globin genes showing strong evolutionary divergence of two homologous pairs of introns. Nature 276:37, 1978.
10. Gilbert W: Why genes in pieces? Nature 281:501, 1978.
11. Sakano H, Rogers JH, Huppi K, Brack C, Traunecker A, Maki R, Wall R, Tonegawa S: Domains and the hinge region of an immunoglobulin heavy chain are encoded in separate DNA segments. Nature 277:627, 1979.
12. Long EO, Dawid IB: Expression of ribosomal DNA insertions in Drosophila melanogaster. Cell 18:1185, 1979.
13. Davidson EH, Britten RJ: Regulation of gene expression: Possible role of repetitive sequences. Science 204:1052, 1979.
14. Scheller RH, Constantini FD, Kozlowski MR, Britten RJ, Davidson EH: Representation of cloned interspersed repetitive sequences in sea urchin RNAs. Cell 15:189, 1978.
15. Cameron JR, Loy EY, Davis RW: Evidence for transposition of dispersed repetitive DNA families in yeast. Cell 16:739, 1979.
16. McClintock B: Mechanisms that rapidly reorganize genomes. Stadler Symp, University of Missouri, Columbia 10:25, 1978.
17. Seidman JG, Max EE, Leder P: A K-immunoglobulin gene is formed by site-specific recombination without further somatic mutation. Nature 280:370, 1979.
18. Rubin GM, Bingham P, Kidwell M: Personal communication.
19. Brutlag DL, Appels R, Dennis ES, Peacock WJ: Highly repeated DNA in Drosophila melanogaster. J Mol Biol 112:31, 1977.
20. Peacock WJ, Appels R, Dunsmuir P, Lohe A, Gerlach WL: Highly repeated sequences: chromosomal location and evolutionary conservatism. In Brinkley BR, Porter KR (eds): Int Cell Biol, New York: Rockefeller University Press, 1977, p 494.
21. Dennis ES, Dunsmuir P, Peacock WJ: Segmental amplification of a satellite DNA: Restriction enzyme analysis of the major satellite of Macropus rufogriseus. Chromosoma 79:179, 1980.
22. Jones KW, Corneo G: Location of satellite and homogeneous DNA sequences on human chromosomes. Nature [New Biol] 233:268, 1971.
23. Venolia L, Peacock WJ: A highly repeated DNA from the genome of the wallaroo (Macropus robustus). Aust J Biol Sci 34:97, 1981.
24. Peacock WJ, Lohe A, Gerlach WL, Dunsmuir P, Dennis ES, Appels R: Fine structure and evolution of DNA in heterochromatin. Cold Spring Harbor Symp Quant Biol 42:1121, 1977.
25. Peacock WJ, Dennis ES, Elizur A, Calaby JH: Repeated DNA sequences and kangaroo phylogeny. Aust J Biol Sci 34:325, 1981.
26. Elizur A, Dennis ES, Peacock WJ: Satellite DNA sequences in the red kangaroo. Aust J Biol Sci (in press).
27. Fry K, Salser W: Nucleotide sequence of HS α-satellite from kangaroo rat Dipodomys ordii and characterization of similar sequences from other organisms. Cell 12:1069, 1977.

28. Dennis ES, Gerlach WL, Peacock WJ: Identical polypyrimidine-polypurine satellite DNAs in wheat and barley. Heredity 44:349, 1980.
29. Bedbrook J, Jones J, Flavell R: Evidence for the involvement of recombination and amplification events in evolution of Secale chromosomes. Cold Spring Harbor Symp Quant Biol 45 (in press).
30. Pech M, Streeck RE, Zachau HG: Patchwork structure of a bovine satellite DNA. Cell 18:883, 1979.
31. Peacock WJ, Dennis ES, Gerlach WL: Satellite DNA—change and stability. Chromsomes Today 7:30, 1981.
32. Singh L, Purdom IF, Jones KW: Satellite DNA and evolution of sex chromosomes. Chromosoma 59:43, 1976.
33. Meyer GF, Lipps HJ: Chromatin Elimination in the Hypotrichous Ciliate Stylonychia mytilus. Chromosoma 77:285, 1980.
34. Beermann S: The diminution of heterochromatic chromosomal segments in Cyclops (Crustacea copepoda). Chromosoma 60:297, 1977.
35. Gall JG, Cohen EH, Polan ML: Repetitive DNA sequences in Drosophila. Chromosoma 33:319, 1977.
36. Peacock WJ, Dennis ES, Rhoades MM, Pryor AJ: Highly repeated DNA sequence limited to knob heterochromatin in maize. Proc Natl Acad Sci (USA) 78:4490, 1981.
37. Peacock WJ, Miklos GLG: Meiotic drive in Drosophila: New interpretations of the segregation distorter and sex chromosome systems. Adv Genet 17:361, 1973.
38. Hilliker AJ: Genetic analysis of the centromeric heterochromatin of chromosome 2 of Drosophila melanogaster: Deficiency mapping of EMS induced lethal complementation groups. Genetics 83:765, 1976.

Mechanisms of Speciation, pages 143–153
© 1982 Alan R. Liss, Inc., 150 Fifth Avenue, New York, NY 10011

Spatial Distribution of Chromosomal Inversions and Speciation in Anopheline Mosquitoes

Mario Coluzzi

INTRODUCTION

Reviews of the role of chromosomal rearrangements in speciation generally point to their importance in producing postmating reproductive isolation through inviability or infertility of the heterokaryotypes. In organisms characterized by low vagility and/or relatively small interbreeding groups, chromosomal rearrangements might become fixed by random genetic drift and promote the isolation and differentiation of the gene pool involved. This pattern of chromosomal speciation, well documented in various organisms [1,2], does not seem to apply to paracentric inversions. The inversion heterokaryotypes, at least in the case of naturally occurring balanced polymorphisms, are often not only viable and fertile but also heterotic acting as a cohesive force in the gene pool. The involvement of paracentric inversions in speciation has been regarded as an incidental accompaniment of the evolutionary process [3] and the hypothesis of a causal relationship between these rearrangements and the cladogenetic events has apparently no sound theoretical background.

The available data on paracentric inversions are mainly from groups of Diptera with favorable polytene chromosomes such as Drosophila, Chironomus, Simulium, and Anopheles. The evidences accumulated from these materials are in substantial agreement indicating that paracentric inversions are very common both as intraspecific polymorphisms and as fixed rearrangements between related species [1]. However, any attempt to extrapolate this finding to other organisms should take into account that the insect groups studied constitute an obviously biased sample sharing a number of characteristics which, as shall be discussed later, could be relevant in relation to the evolutionary significance of paracentric inversions. Some of these characteristics are: (a) relatively high vagility with respect to the environmental patches which they can perceive; (b) active dispersion largely prevalent over passive; and (c) low number of chromosomes. Anopheline mosquitoes can even represent a border case in this respect since they are able to disperse, in the female adult stage, for more than one Km in a single night, they show complex behavioral patterns for host finding and biting and for the choice of resting and oviposition sites, and they have only three pairs of chromosomes.

Although less deeply investigated cytogenetically than Drosophila, Anopheline mosquitoes have been the subject of detailed polytene chromosome studies, particularly dealing with groups of sibling species of malariological importance, some of which represent recent, man-influenced, speciation events of particular evolutionary interest [4]. These investigations provided a fairly large body of information on paracentric inversions. Moreover, the availability of favorable polytene chromosomes not only from the larval salivary glands but also from the adult ovarian nurse cells [5] allowed the extension of chromosomal examinations directly on adult female samples favoring the study of microspatial variations of inversion frequencies and of relationships between inversions and adult behavior. This article cites the results obtained thus far from polytene chromosome studies in Anopheline mosquitoes and offers a general theory on the origin and maintenance of inversion polymorphism and on its involvement in speciation.

CHROMOSOMAL DIFFERENTIATION BY PARACENTRIC INVERSIONS

The patterns of chromosomal differentiation observed in Anopheline mosquitoes have been substantially summarized by Coluzzi and Kitzmiller [6] and Kitzmiller [7]. The general picture emerging from these reviews has been confirmed and clarified by various recent contributions but their detailed analysis is out of the scope of this paper.

The mitotic chromosomes of Anopheline mosquitoes consist of two pairs of metacentric or submetacentric autosomes and one pair of sex chromosomes with X-Y dimorphism. This karyotype appears remarkably conservative being constant in the genus Anopheles except for quantitative and qualitative changes in the heterochromatin which constitutes the entire Y chromosome and one half at least of the X chromosome. The polytene complement typically shows five banded elements corresponding to the four autosomal arms and to the euchromatic section of the X chromosome which is unpaired in the male. The polytene chromosome differences, as shown by the comparison of closely related species, involve essentially changes of band sequences due to paracentric inversions.

The same type of rearrangement occurs at the intraspecific level resulting in balanced polymorphisms. Homosequential species are extremely rare, the only well-documented case being that of two pairs of sibling species of the maculi-pennis complex. More than sixty species belonging to eight different groups were investigated in detail and at least one fixed inversion generally differentiates each pair of siblings. In most cases the rearrangement consists of two or more inversion steps and frequently involves overlapping inversions.

The distribution of fixed and polymorphic inversions along the polytene complement or within a single arm appears nonrandom. Interspecific fixed inversions are more frequent than expected on the X chromosome taking into account both the very low frequency of polymorphic inversions recorded on this polytene element and its mean length which is less than 15% of the total polytene complement length. An attempt to calculate the mean ratio of polymorphic to fixed inversions from data on various groups of Anopheline mosquitoes gives figures lower than 1:2 for the X chromosome and higher than 1:0.5 for the autosomes. The pattern of distribution of fixed and polymorphic inversions on the autosomal elements is similar but again difficult to explain by stochastic models since some chromosomal arms appear more subject to rearrangements than others. In the gambiae complex, chromosome arm 2R, representing about 1/3 of the autosomal complement, carries 21 out of the 31 autosomal rearrangements described [4]. Moreover, the inversion break points are not uniformly distributed along the arm: As many as three break points are coincident or nearly so and those occurring in the basal fifth are only 2 out of 42. Various cases of parallel inversion polymorphisms were also observed in different groups of Anopheline mosquitoes.

No relationship seems to exist between chromosomal differentiation by par-acentric inversions and genic differentiation as shown by the study of electro-phoretic variants [8]. Homosequential species of the maculipennis complex were found to have higher genetic distances than species of the gambiae complex differentiated by at least three fixed inversions. Moreover, similar levels of allozyme variability were observed in chromosomally monomorphic and poly-morphic populations of A. gambiae and A. arabiensis.

On the other hand, the parallelism generally existing between genic and morphological differentiation does not extend to chromosomal differentiation which, for instance, is higher between some of the sibling species of the gambiae complex than between some of the morphologically well distinct and presumably older species of the Neocellia group. This suggests that differentiation by fixed inversions is mostly related to cladogenesis while allozymic and morphological differentiation marks, to some extent at least, phyletic evolution.

ADAPTIVE SIGNIFICANCE OF INVERSION POLYMORPHISM

The adaptive significance of the genetic difference between the alternative arrangements of a polymorphic inversion is generally assumed to be expressed by the heterokaryotype advantage (e.g., marginal overdominance) while it is not clear to what extent the homokaryotypes are adaptively divergent and how this contributes to the maintenance of the polymorphism and to the population fitness vs. environmental heterogeneities. Direct indications of advantage of the inversion heterokaryotypes in Anopheline mosquitoes are provided by various observations on polymorphic laboratory colonies [9]. Laboratory breeding is not easily achieved with this material and it necessarily involves the "forcing" of various developmental and behavioral patterns imposing on the insects physioethological stresses which presumably amplify marginal overdominance. In such conditions inversion polymorphisms appear extremely tenacious being maintained even in the oldest laboratory colonies in spite of repeated bottle necks. Moreover, the study of adult cage populations frequently shows the heterokaryotypes in highly significant excess when compared to the Hardy Weinberg expectations.

Studies carried out on polymorphic populations of Anopheline mosquitoes in the field do not provide the same direct indications of advantage of the inversion heterokaryotypes. These were rarely found in significant excess while heterokaryotype deficits were recorded in some cases together with patterns of spatial variations supporting the premise that the homokaryotypes are specifically favored in certain environmental patches [4,10]. Clinal geographic variations in inversion frequencies, closely related to climatic and ecological conditions, were observed in members of the A. gambiae complex in the Afrotropical region [4,11]. In both A. gambiae and A. arabiensis some of the changes in inversion frequencies appear to be involved in the extension of the species range from humid forest areas to arid savanna areas or viceversa. These same inversions show microspatial variations (particularly related to man-made environmental contrasts) which can throw further light on the adaptive significance of the various chromosomal variants.

Significant differences in inversion frequencies have been observed between samples of adult females collected in the same locality and at the same time but in different resting sites or with different sampling procedures. The comparison between outdoor and indoor collections is of particular interest since it shows that mosquitoes carrying certain alternative chromosome-2 arrangements have contrasted propensities to bite indoors and, particularly, to rest indoors (endophily). Moreover, in both A gambiae and A arabiensis the arrangements which are more frequent in indoor collections are those favored in drier environments, while the arrangements more frequent in outdoor collections are those increasing in frequency in relatively more humid environments. Since in the study area the indoor environment is characterized, during the night, by a higher saturation deficit than the outdoor environment, the nonrandom distribution of carriers of different arrangements appears to be an expression of different climatic adaptations which result in different house-resting behavior. In other words, each karyotype would have a tendency to keep itself in the environmental conditions for which it is more adapted. A similar situation involving intraspecific habitat selection or optimal habitat choice [12] can lead to the maintenance of polymorphism under conditions much less restricted than marginal overdominance alone [13]. On the other hand, relationships between inversion polymorphisms and behavioral variations in Anopheline mosquitoes are also indicated by laboratory investigations [14–16].

ORIGIN OF THE INVERSION POLYMORPHISM AND ITS INVOLVEMENT IN SPECIATION

The patterns of chromosomal differentiation by paracentric inversions in Anopheline mosquitoes, as previously cited, strongly support the hypothesis that these rearrangements have a causal relationship with speciation events and that they are the product of selection processes favoring chromosomal mutations which include certain gene sequences and/or break points at certain sites. Moreover, some data also support the fact that different coadapted linkage groups or supergenes included into the alternative arrangements of a polymorphic inversion not only confer a general advantage to the heterokaryotype but express, in the homokaryotypic state, specific adaptations for different marginal environmental conditions.

The initial advantage of an inversion in the heterozygous state could be due to its having, by chance, a favorable position effect, or containing heterotic loci or combinations of genes which taken as a whole are capable of expressing heterotic phenomena. It is also possible, as suggested by Dobzhansky and others [17], that further evolution of such initial advantage may occur by a process of coadaptation through heteroselection of the alternative arrangements present in the population. However, it appears to be quite improbable that processes of this sort could lead to the organization of supergenes showing specific adaptive significance for marginal conditions. It would seem more consistent to consider such supergenes the product of directional selection preceeding the origin of the inversion which would essentially act as a chromosomal mechanism preserving the favorable gene association from the disruptive effect of crossing over. Based primarily upon this assumption, a general hypothesis is proposed here which might account for the origin of the inversion polymorphism and its involvement in speciation (Figs. 1 and 2).

Let us consider a species whose range extends along an ecological cline and is subject to periodic expansions and contractions in geographically and/or ecologically marginal zones. This results in frequent formation of temporary isolates which are subject to extinction (A → A1 → . . .) . Alternatively, it results in preservation of the isolate through alternate phases characterized by transitory increases of the population and successive collapses (flush and crush) with consequent drift effects and/or strong selection pressures (A → B). Many authors [18–22] have stressed the evolutionary interest of such situations which are likely

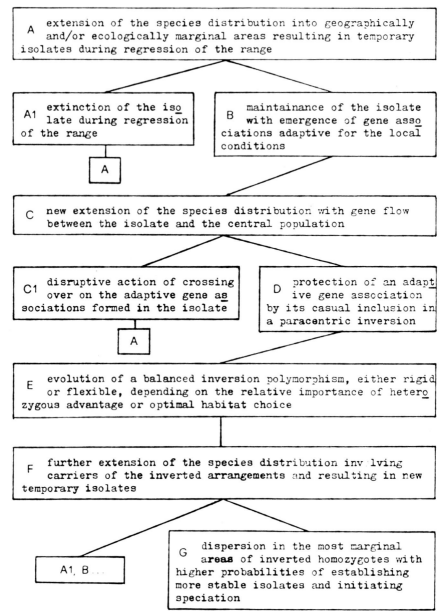

Fig. 1. Hypothetical steps in the origin of inversion polymorphism and its involvement in speciation.

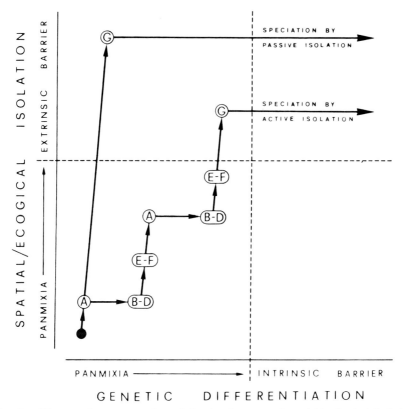

Fig. 2. Diagramatic representation of the development of spatial/ecological isolation and of adaptive genetic differentiation in allopatric speciation processes by passive and active isolation. (A) Temporary isolate; (B–D) evolution of a gene association adaptive for the local conditions and protected by a paracentric inversion; (E–F) extension of the species range involving the inversion polymorphism; (G) more stable isolate. (See Fig. 1 and text for further details.)

to favor a destabilization of the prevalent genetic associations and their reorganization into new adaptive relationships. The process of genetic differentiation is, however, "reabsorbed" when contact with the original population is reestablished (A → B → C → C1 → A) unless intrinsic mechanisms, which are capable of completely or partially preventing gene flow, intervene. The appearance in the isolate of paracentric inversions, therefore, becomes of particular importance. Even when the inversion shows no initial advantage in the isolate, on reunion with the original population a situation occurs which is particularly favorable to the evolution of a balanced polymorphism.

The process of genetic differentiation is "reabsorbed" except for that part of the genome included in the inversion. At this point the inversion heterozygote is not only a chromosomal heterozygote but also a heterozygote for the genetic content of the alternative arrangements. If as hypothesized in Figure 1, the inversion includes a supergene with an adaptive value specific for the marginal condition, the polymorphism has a strong possibility of preserving itself and of spreading not only through the advantage of the heterozygote (eventually increased by coadaptation through heteroselection) but also through optimal habitat choice (A → B → C → D → E).

The relative importance of these two phenomena could constitute the determining factor for the evolution of more rigid or more flexible polymorphisms [23], respectively. In the case of more flexible polymorphisms, the heterozygotes and the inverted homozygotes allow the species to permanently expand its range towards marginal conditions and to form new isolates from which new inversions arise (A → B → C → D → E → F → B → C → . . .). Moreover, the inverted homozygotes, particularly frequent in the new marginal populations, become the best candidates for a directional dispersion along the ecological cline and for overcoming steep "steps" in the cline. Thus, they are also the best candidates for the formation of less temporary isolates and for the initiation of speciation processes characterized by active, nonrandom, isolation (A → B → C → D → E → F → G).

CONCLUSION

In hypothesizing the origin of polymorphic inversions in isolates adapting to marginal conditions, it is possible to outline a coherent picture of the meaning of this chromosomal variability both at the intraspecific and interspecific level. Inversion polymorphism increases the ecological flexibility of the species, permitting on the one hand an expansion of the area occupied in marginal zones and on the other, the more efficient utilization of the spatial and temporal heterogeneities of the environment in the central zones. This greater ecological flexibility is not brought about by a process of despecialization, which may be observed in generalist species, but is achieved through the capture and stabilization, by means of inversions, of blocks of coadapted genes capable of expressing ecotypic divergences.

Inversions are instruments of ecotypic differentiation [24] and inversion heterozygotes resemble F_1 interpopulation hybrids [19]. In terms of perception of the environmental grain [25] the inversion system of a polymorphic species tends to express on the whole an adaptive strategy of a fine-grained type by means of divergent adaptations, each of which in its turn represents a strategy of the coarse-grained type. The polymorphic species would be neither a specialist nor a generalist, but a "multispecialist," ie, an assemblage of different "specialists" evolving as marginal monomorphic populations through the isolation of a section of their genome and tied by the heterokaryotype advantage exploited in the central part of the species range. Incidentally, this would fit well with the apparently puzzling evidence obtained from polytene chromosome studies in Drosophila indicating that chromosomal polymorphism does not play a decisive or leading role in highly specialized species or in domestic, colonizing species which are true generalists [26].

If this formulation is accepted, it can be more easily understood how the inversion might acquire a causal significance in the speciation process. It is possible to consider a role of the inversion in allopatric speciation by introducing the hypothesis of an active and directional dispersion of the carriers of certain inverted arrangements. These would confer on the bearer a partial preadaptation to occupy particular marginal conditions and to cross ecotones with these characteristic conditions. Thus, allopatric speciation would include the following dichotomy: (a) allopatric speciation by active isolation which tends to produce species differentiated chromosomally by fixed inversions; and (b) allopatric speciation by passive isolation, involving either founder effects or the slow genetic divergence of large subpopulations, which is more likely to produce homosequential species (see, for example, the relatively high number of homosequential species in Hawaiian Drosophilids).

It is also possible to hypothesize a role of the inversion within other models of speciation such as the clinal model, the area-effect model, or the sympatric model [2]. In this case the inverted arrangement should have a direct effect on the evolution of reproductive isolation, either by leading to conditions of disruptive selection or by controlling pre- or postmating barriers and thus preserving not a supergene but the entire genome which emerged in the temporary isolate. Perhaps this could especially apply to X chromosome inversions accounting for their frequent involvement in speciation of Anopheline mosquitoes which contrasts with their comparatively rare involvement in balanced polymorphisms. Triads of overlapping inversions could play an important role in determining conditions of disruptive selection as indicated by Wallace [19].

The evolutionary significance of paracentric inversions is not expected to be a general phenomenoon. The assumption of active and directional dispersion of the inversion karyotypes implies a relatively high vagility of the organism involved and its ability to perceive and respond to environmental heterogeneities. On the other hand, the chance for an inversion to capture an adaptively important section of the genome should be primarily related to the number of chromosomes being presumably higher in organisms with low chromosome numbers, such as Anopheline mosquitoes, in which an inversion frequently includes more than 5% of the total complement length.

ACKNOWLEDGMENTS

The data discussed in this paper include various observations on the Anopheles gambiae complex carried out by the author in collaboration with J. Bryan, D.

Di Deco, G. Petrangeli, V. Petrarca, A. Sabatini. Y. Toure, and F. Villani. These observations are part of a research program supported by the Italian Research Council (CNR) grant n. 80.00576.04.

REFERENCES

1. White MJD: Animal Cytology and Evolution. Cambridge: University Press, 1973, p 961.
2. White MJD: Modes of Speciation. San Francisco: Freeman, 1978, p 455.
3. Carson HL, Clayton FE, Stalker HD: Karyotypic stability and speciation in Hawaiian Drosophila. Proc Natl Acad Sci (USA) 57:1280, 1967.
4. Coluzzi M, Sabatini A, Petrarca V, Di Deco MA: Chromosomal differentiation and adaptation to human environments in the Anopheles gambiae complex. Trans R Soc Trop Med Hyg 73:483, 1979.
5. Coluzzi M: Cromosomi politenici delle cellule nutrici ovariche nel complesso gambiae del genere Anopheles. Parassitologia 10:179, 1968.
6. Coluzzi M, Kitzmiller JB: Anopheline mosquitoes. In King RC (ed): Handbook of Genetics. New York: Plenum, 1975, p 285.
7. Kitzmiller JB: Genetics, cytogenetics and evolution of mosquitoes. Adv Genet 18:315, 1976.
8. Bullini L, Coluzzi M: Applied and theoretical significance of electrophoretic studies in mosquitoes (Diptera: Culicidae). Parassitologia 20:7, 1978.
9. Di Deco MA, Petrarca V, Villani F, Coluzzi M: Polimorfismo cromosomico da inversioni paracentriche ed eccesso degli eterocariotipi in ceppi di Anopheles allevati in laboratorio. Parassitologia 22:304, 1980.
10. Bryan JH, Di Deco MA, Petrarca V, Coluzzi M: Chromosomal polymorphism in Anopheles gambiae s.s. in The Gambia, West Africa. Genetica, in press, 1982.
11. Coluzzi M, Sabatini A, Petrarca V, Di Deco MA: Behavioural divergences between mosquitoes with different inversion karyotypes in polymorphic populations of the Anopheles gambiae complex. Nature 266:832, 1977.
12. Taylor CE, Powell JR: Microgeographic differentiation of chromosomal and enzyme polymorphisms in Drosophila persimilis. Genetics 85:681, 1977.
13. Taylor CE: Genetic variation in heterogeneous environments. Genetics 83:887, 1976.
14. Coluzzi M: Inversion polymorphisms and adult emergence in Anopheles stephensi. Science 176:59, 1972.
15. Coluzzi M, Di Deco MA: Cage experiments on homospecific and heterospecific matings with females of Anopheles stephensi carriers of different inversion karyotypes. Rc Accad Naz Lincei 57:683, 1974.
16. Coluzzi M, Di Deco MA, Petrarca V: Propensione al pasto di sangue in condizioni di laboratorio e polimorfismo cromosomico in Anopheles stephensi. Parassitologia 17:137, 1975.
17. Dobzhansky T: Genetics and the Origin of Species. New York: Columbia University Press, 1951.
18. Mayr E: Change of genetic environment and evolution. In Huxley J, Hardy CA, Ford EB (eds): "Evolution as a Process". London: Allen and Unwin, 1954, p 157.

19. Wallace B: The influence of genetic systems on geographical distribution. Cold Spring Harbor Symp Quant Biol 20:16, 1959.
20. Carson HL: The population flush and its genetic consequences. In Lewontin RC (ed): "Population Biology and Evolution." Syracuse: Syracuse University Press, 1968, p 123.
21. Carson HL: The genetics of speciation at the diploid level. Am Nat 109:83, 1975.
22. Powell JR: The founder-flush speciation theory: An experimental approach. Evolution 32:465, 1978.
23. Dobzhansky T: Rigid vs. flexible chromosomal polymorphisms in Drosophila. Am Nat 96:321, 1962.
24. Mayr E: Introduction to symposium on age and distribution patterns of gene arrangements in Drosophila pseudobscura. Lloydia 8:69, 1945.
25. Levins R: Evolution in Changing Environments. New York: Princeton University Press, 1968.
26. Carson HL: Chromosomal morphism in geographically widespread species of Drosophila. In Baker HG, Stebbins GL (eds): Genetics of Colonizing Species. New York: Academic Press, 1965, p 503.

Mechanisms of Speciation, pages 155-177
© 1982 Alan R. Liss, Inc., 150 Fifth Avenue, New York, NY 10011

Robertsonian Numerical Variation in Animal Speciation: Mus musculus, an Emblematic Model

Ernesto Capanna

INTRODUCTION

Ever since Robertson [1] stressed the "significance of the V-shaped chromosomes in shoving taxonomic relationships" some 50 years ago, cytologists have focused special attention in this fusion-fission process which appears to play an important role in animal evolution besides that of Acrididae and Tettigidae.

Rodents also appear to have a tendency towards fusioning of acrocentric chromosomes. Thus any group of species displaying an internal Robertsonian variability would either be incomplete or not up-to-date. This paper explores the Robertsonian karyotypic variability based on my research [2–4] that involves the house mouse.

The case of Robertsonian variability of Mus musculus is, in fact, particularly amusing and worthy of the attention recently devoted to it from a theoretical viewpoint by distinguished evolutionists [5,6]. It displays, in fact, several peculiarities: First, the large number of more or less independent and isolated Robertsonian populations crowded into a limited area; second, the extreme dispersion of the phenomenon. At one time, only a few populations were known in the Alps [7,8] and in the Apennines [9], but with cytogeneticists discovering mice from all parts of the world, more and more populations have been found from northern Scotland [10] to Greece [11], from India [12] to Marion Island—an isolated spot half way between the coast of South Africa and the Antarctic [13].

Furthermore, karyotypic variability in Mus musculus is purely Robertsonian and is not disturbed by the entry of other types of karyotypic transformation. Centromeric heterochromatin polymorphism has been also described in natural

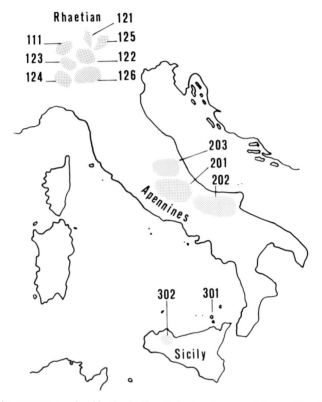

Fig. 1. An attempt to classify the Italian Robertsonian populations. Rhaetian system: 11, Mesolcina subsystem; 111, Mesolcina population 2n 28; 12, Orobian Alps subsystem; 121, Poschiavo valley population 2n 26; 122, Orobian Alps population 2n 22; 123, Milano 1st population 2n 24; 124, Milano 2nd population 2n 24; 125, Upper Valtellina population 2n 24; 126, Cremona population 2n 22. Apennine system: 201, "CD" population (Abruzzi) 2n 22; 202 "CB" population (Molise) 2n 22; 203, "ACR" population (Monti della Laga) 2n 24. Sicily system: 301, Lipari island 2n 26; 302, Palermo 2n 28.

populations of the house mouse [14], but this kind of structural chromosomal rearrangement does not play a relevant role in our speciation models. Consequently, the significance of the Robertsonian mechanism may be isolated and clarified as far as its role in the speciation process. Other interesting peculiarities will be emphasized in this paper.

The summit of this karyological iceberg was discovered by Gropp and his coworkers in 1969 [7] when they detected a population of Mus musculus in Val Poschiavo (Grisons, Switzerland) with a 26-chromosome complement instead of the 40 all-acrocentrics which the standard karyotype of this species possess.

At the time this karyotypic variation of the mouse seemed so rare and extraordinary that the mouse population of Val Poschiavo was assumed to be a different species from Mus musculus i.e., Mus poschiavinus, the older species proposed by Fatio [15], exactly a century earlier for the Val Poschiavo population. Later Gropp [8] detected a new homozygous population on the Swiss side of the Rhaetian Alps, characterized by a 28-chromosome karyotype in Val Mesolcina and other heterozygous Robertsonian conditions in Val Bregaglia and in Val Chiavenna. Gropp then suggested I study the house mouse populations on the Italian side of the Rhaetian Alps and so began a fruitful collaboration between Gropp's group and my own.

This combined research has led to the recognition and precise characterization of the Robertsonian fusion pattern of 11 homozygous populations (Fig. 1) and numerous hybridation areas in the Alps [16–19], in the Apennines [2,9,20], and in the Eolean Islands (Lipari) [21]. Observations by von Leheman and Radbruch [22] led to the twelfth population of the Italian Robertsonian system near Palermo (Sicily).

One peculiar characteristic of the karyological polytypism of Mus musculus is the randomness of the Robertsonian fusion pattern which, in contiguous populations, involves different chromosomes of the all-acrocentric standard mouse karyotype. Consequently, different populations, adjacent as well as placed far apart, with an equal diploid number, $2n = 22$ as in the case of two Apennine populations and one from Alps (Fig. 2) which have completely different karyotypes. One metacentrics is shared by the CD and the CB Apennine karyotypes, i.e., the fusion Rb (9·16), and one is common to the CB Apennine and to the Orobian Alps karyotypes, i.e., the fusion Rb (5·15). Consequently, only two appeared more than once among 25 metacentrics.

The assessment of the pattern of fusion, established on the basis of the G-banding [23–25], was confirmed in the analysis of the meiotic diakinesis of hybrids. During the meiotic prophase, in fact, the Robertsonian metacentrics pair off arm-to-arm thus producing, in a cytological domino effect, amazing patterns of multivalent long chains or rings [3,24,25].

For brevity, this paper deals only with some of the Italian populations, excluding those in Sicily, since they have not yet been extensively studied, and those of Apennines which are previously cited [2,3] and are also exhaustively discussed by White [5]. Consequently, attention is focused on the mosaic of Robertsonian populations distributed in the Alpine system of the Rhaetian Alps and also covering part of the Poplain lying between the Ticino and the Mincio Rivers. This "Rhaetian" system constitutes an enthralling puzzle of six interlocking pieces represented by six populations. Each displays an increasing karyotype diversity, with diploid numbers ranging from 26 to 22. This situation would be extremely useful in discussing a speciation pattern thought to be founded on the progressive accumulation of karyotypic changes.

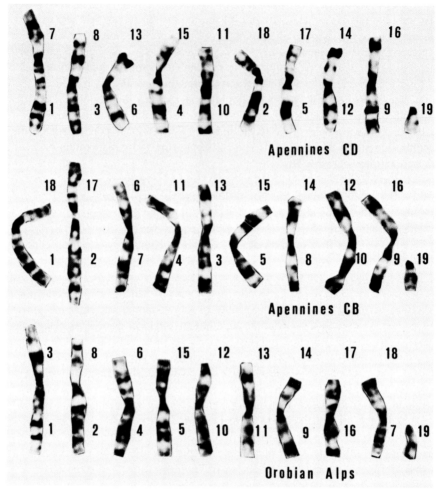

Fig. 2. G-banded autosomal complements of three different 22-chromosome karyotypes, i.e., Apennines "CD," Apennines "CB," and Orobian Alps. The numbers indicate the acrocentric autosomes of the standard karyotype involved in the formation of the metacentrics.

THE GEOGRAPHIC DISTRIBUTION

As previously mentioned, the Rhaetian-Orobian population system includes six populations which are located in an area astride the Italian-Swiss crest line of the Rhaetian Alps, then extending to the entire range of Orobian Alps and to the Po plain between Ticino to the East and Lake Garda to the West; South of the Po we find mouse populations with an all-acrocentric karyotype. Figure 3 shows populations with their diploid number and the denomination which we have introduced to identify them. Hatched areas indicate hybrid zones of Chiavenna, Bregaglia and Western Dolomites as well as the homozygous population of Vel Mesolcina not included in this Orobian-Rhaetian subsystem.

It can immediately be seen from the distribution map that the geographic barrier elements may be recognized in this complex of populations as isolating elements between the individual populations. However, the most important aspect of this group of Robertsonian populations is the complementarity of the centric fusion pattern making up their karyotypes (Fig. 4). Four centric fusions are, in fact, shared by all the populations, whereas a smaller number are shared by fewer populations, until we find that only few metacentrics characterize individual populations. This circumstance, in the light of this randomness of the Robertsonian process, (well documented in other systems of Robertsonian mice [2]) is proof of a chain of succeeding fusions which from a primary ancestral

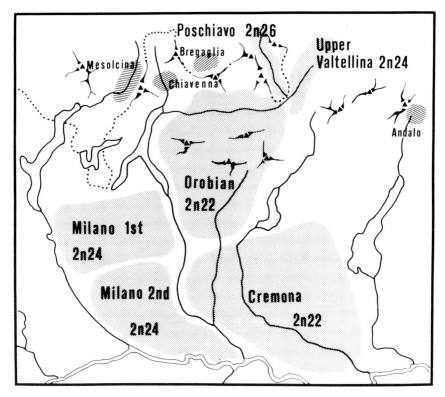

Fig. 3. Distribution map of the Robertsonian populations of the Rhaetian system. Dotted areas indicate six populations of the Orobian Alps subsystem; diagonal lines indicate the population of Val Mesolcina (Mesolcina subsystem) and the main hybrid zones.

Localities	ARM COMPOSITIONS											
Poschiavo	1·3	4·6	5·15	11·13	9·14	16·17					8·12	
Upper Valt.	1·3	4·6	5·15	11·13	9·14	16·17	2·8	10·12				
Orobian	1·3	4·6	5·15	11·13	9·14	16·17	2·8	10·12		7·18		
Cremona			5·15	11·13	9·14	16·17	2·8	10·12	3·4	7·18	1·6	
Milano 1			5·15	11·13	9·14	16·17	2·8	10·12	3·4		6·7	
Milano 2			5·15	11·13	9·14	16·17		10·12	7·8	2 4	3·6	
	North		Paleo	Orobian		South						

Fig. 4. Fusion pattern of Orobian populations. Numbers indicate the acrocentric chromosome of the standard karyotype.

population, "Paleo Orobian," has divided and gradually isolated different populations.

HYBRID FERTILITY AND NATURAL HYBRID ZONES

Also in the Rhaetian system, as in other house mouse Robertsonian population systems, e.g., those of the Apennines [2,9,20], this seems to be a contradictory phenomenon which nevertheless constitutes another peculiarity: On one hand, the experimentally demonstrated hypofertility or sterility of the hybrids [26–33] and, on the other, the presence in nature of narrow hybrid zones between Robertsonian populations and the 40-chromosome populations surrounding them [2,9,20,4,18]. The latter is, in fact, yet another peculiarity and reason of interest in the case of Mus; in mouse, the groups of Robertsonian populations are found to be like islands or archipelagos in a continuous ocean of all-acrocentric mice. The fact that each Robertsonian polytype system is isolated from the others by an all-acrocentric context certainly facilitates the study of the speciation patter, limiting it strictly to the area inhabited by a single group of populations. Nevertheless, as far as the problem of fertility of the hybrids is concerned, this doubles the facets of the problem: That of hybrids between Robertsonian and all-acrocentric mice and that of Robertsonian mice characterized by different fusion patterns. It should be mentioned that narrow hybrid zones are found only in contact areas between Robertsonian and all-acrocentric populations. In fact, no hybrids in nature between carriers of different Robertsonian karyotypes have yet been found, even where it was possible to demonstrate unequivocally a sympatric inhabit of two chromosomically differentiated populations.

However, this second type of hybrid was easily obtained in the laboratory and so it was feasible to experimentally evaluate the relative fertility and/or the sterility of one or other type of hybrid.

The meiotic multivalents, chains, or rings, implicate a disorder in the correct evolution of the diakinesis which would block the meiotic process, and thus spermatogenesis, at the spermatocyte first stage [2–4]. Even in these cases, as in the hybrid CD × CB-appennine, in which rare spermatozoa are present in the epididymis, the percent of eu-haploid gametes present is extremely low, as is clearly demonstrated by microdensitometric evaluation of the DNA content [34].

This is not the case of the double structural heterozygotes. In the first instance, impairment of fertility is very severe [43,44] and in some instances full steriles [44]. From data obtained by Gropp and his co-workers [45,46] regarding double heterozygosity involving eight different chromosomes displaying monobrachial homology, it can be deduced that in the female the production of eu-haploid gametes is 57–74% of the total. Furthermore, in a back-crossing with a normal all-acrocentric male, a quarter of the implanted embryos present severe malformations, such as exencephaly, due to no. 12 trisomy. But what is more important for our theoretical purpose is the incapability of this situation to give rise to a new stable Robertsonian homozygous karyotype. In fact, the only balanced gametes produced by these hybrids will only be able to either reconstruct the double heterozygous situation or rebound to the parental homozygosis (Fig. 6).

In view of these data, the presence of natural narrow hybrid zones would appear less conflicting. It is, in fact, true that heterozygotes are at a disadvantage as far as their fertility is concerned as it heavily influences their reproductive fitness. However, it is also true that within these hybridization bands there are microdemes inhabited only by heterozygotes. Even the limited vagility typical of mouse ethology excludes any comparison between homozygous and heterozygous mice, and any natural verification of the advantage of the former vs. the latter. Only in the immediate peripheral area of the hybridization bands will the homozygotes tend to replace the heterozygotes. They, however, will rapidly reform due to migration and exogamic activity of the young homozygotes which tend to replace the heterozygotes in the demes which have, in part, become available.

GENE FLOW AND REPRODUCTIVE ISOLATION

We now come to a crucial point in our theory: The possibility of a gene flow across these narrow hybrid zones. A possibility of this type would not only allow the passage of genes from the Robertsonian to the all-acrocentric populations, but in a retrograde direction, also to other Robertsonian populations. In other words, the problem is that of whether or not the postmating isolation mechanism, based upon meiotic disorders committed by heterozygosis of Robertsonian metacentrics, may be considered valid. In our opinion this type of Robertsonian isolation works positively, even in spite of natural hybridization areas. In fact, gene clusters may be transferred from a Robertsonian karyotype to an all-acrocentric karyotype merely through chromatid exchange in the meiotic trivalents of the hybrids. But, if we consider that the meiotic trivalents in mouse are paucichiasmatics, and at times even achiasmatics, and that the hybrids, due to the very presence of these trivalents, have imparied fertility, it may be deduced that the gene clusters would therefore have little chance of flowing from a Robertsonian karyotype population to an all-acrocentric one. It is even less likely that the same cluster spread throughout in the large 40-chromosome population would cross a hybridization zone at another equally distant point and ultimately reach a different population with another Robertsonian karyotype.

Thus, the Robertsonian populations may be considered definitely isolated from each other and also, in some measure, from the vast all-acrocentric context surrounding them.

Analysis of the allelic frequency, performed by means of the electrophoretic patterns of allozyme systems, would better demonstrate this reproductive isolation and the presence, or absence, of a gene flow through the hybridization bands. Nevertheless, this approach to the problem should not be proposed with optimism; in fact, Nei's [35] index values for genetic similarity between the Robertsonian populations and between the latter and all-acrocentic populations (Table I) [36] are so close to the identity as to leave little hope as to the usefulness of this approach.

On the other hand, a marked genetic similarity might have been expected also in the mouse as has already been demonstrated in similar cases of rodents with internal Robertsonian variability [37–40].

TABLE I. Genetic Similarity in some Rodent Species Showing Intraspecies Karyotype Variability

Species	Index	Mean	Range	Ecology	Ref.
Spalax ehrenbergi	(I)[a]	0.960	0.990–0.926	Strongly fossorial	[37]
Thomomys talpoides	(S)[b]	0.874	0.967–0.795	Fossorial	[38]
	(I)	0.925	0.996–0.859		
Thomomys bottae	(S)	0.79	0.98 –0.63	Fossorial	[39]
Proechimys guairae	(I)	0.970	0.986–0.956	Forest dwelling	[40]
Mus musculus	(I)	0.992	0.997–0.989	Commensal	[36]

[a]Nei.
[b]Roger.

A MODEL OF SPECIATION

The Robertsonian Phase

At the second Theriological Congress held in Brno in 1978, I presented an interpretation model of this case of parapatric speciation [4] according to the term proposed by Bush [41], or stasipatric, the term proposed by White [42]. This model, despite the numerous and often conflicting findings that have emerged over the last four years, is still largely a valid one (Fig. 5). This speciation pattern comprises two phases: The Robertsonian phase which is the one having taken place so far, and the Darwinian phase of which currently only the first signs can be observed and which will ultimately lead to the genesis of new mouse species.

This antithesis of names indicates that the first phase would figure as a non-Darwinian process, in as much as it does not appear to involve any clear selective phenomena by the natural environment, but is due purely to the interplay of stochastic events.

The first part of the process would be represented by the accumulation of Robertsonian fusions according to the Multiple Succeeding Mutation (MSM) model which I proposed in 1977 [3] and which White [5] adopted and discussed in detail. To my hypothesis of a "concentric" sequence of Robertsonian mutations, he added the possibility of a coalescence of two or more already char-

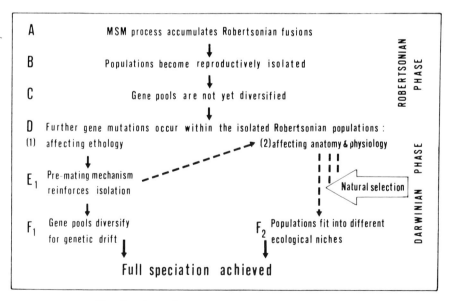

Fig. 5.   A speciation pattern proposed for Mus musculus.

acterized Robertsonian populations, each of which stabilized in homozygosis for a different pair of metacentrics. I shall now add yet another possibility in the accumulation phase of Robertsonian metacentrics, that represented by the introgression of new metacentrics derived from adjacent populations.

One extremely important point, however, should be highlighted here. The free flow of metacentrics across a Robertsonian system of populations during the process of formation due to the accumulation of metacentrics will meet an obstacle when the same acrocentric of the standard karyotype is fused with two other different acrocentrics. In this case metacentrics with monobrachial homology will be formed which will constitute a diverging character of karyotypes of the two subpopulations undergoing differentiation within the same Robertsonian population of origin.

These hybrids between mice with karyotype different for monobrachially homologous metacentrics are referred to as double structural heterozygotes [43,44]. In fact, not only does the degree of gamete aneuploidy affect them differently, but above all the evolutionary prospect of their state of heterozygosity is quite different.

In the case of two completely nonhomologous metacentrics we find fertility impairment which is not particularly severe. Of prime importance to our Robertsonian accumulation model, however, is the fact that of the possible balanced gametes which it produces there are also carriers of both metacentrics which can thus ensure the formation and stabilization of homozygous karyotypes for both metacentrics (Fig. 6).

This is not the case of the double structural heterozygotes. In the first instance, impairment of fertility is very severe [43,44] and in some instances full steriles [44]. From data obtained by Gropp and his co-workers [45,46] regarding double heterozygosity involving eight different chromosomes displaying monobrachial homology, it can be deduced that in the female the production of eu-haploid gametes is 57–74% of the total. Furthermore, in a back-crossing with a normal all-acrocentric male, a quarter of the implanted embryos present severe malformations, such as exencephaly, due to no. 12 trisomy. But what is more important for our theoretical purpose is the incapability of this situation to give rise to a new stable Robertsonian homozygous karyotype. In fact, the only balanced gametes produced by these hybrids will only be able to either reconstruct the double heterozygous situation or rebound to the parental homozygosis (Fig. 6).

The occurrence of metacentrics with monobrachial homology begins to build a barrier within the metacentric flow in a system of Robertsonian populations in formation; but what is more, this overlapping of Robertsonian fusions reduces the number of possible combinations of further fusions which may freely transfer from one population to another within the system, i.e., those involving totally different acrocentrics.

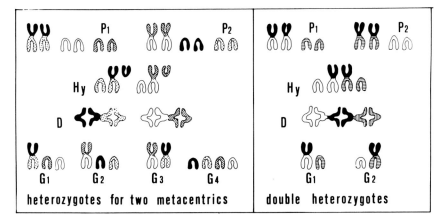

Fig. 6. Differences in the meiotic pattern of hybrid heterozygotes for two nonhomologous metacentris (left column) and hybrid double heterozygotes for monobrachially homologous metacentrics (right column). P_1 and P_2 = parental chromosomal complements; Hy = hybrid karyotypes; D = diakinesis of the hybrids, two trivalents are shown in the first case and one tetravalent in the second; G_1, G_2, etc. = balanced gametes produced by the hybrids.

The fusion patterns of the Rhaetian system population is particularly useful for verifying this hypothesis. The four fusions common to all populations may have been the first to appear either in a typical concentric succession of multiple succeding mutations, or in four different areas of the ancestral all-acrocentric context which gradually came together. In this way a 32-chromosome paleo-Orobian population has been formed (Fig. 7a).

Different fusions appeared in this paleo-Orobian population; 10·12 towards the South and 4·6 in the Northeast. Fusion 1·3 may have penetrated from the limitrophe Val Mesolcina (Fig. 7b); however those instances in which metacentrics implying monobrachial homology appeared in place far apart (such as 2·8 and 12·8 in the North, or 2·4 and 3·4 in the Southwest and Southeast, respectively, and 4·6 in the North) constitute a premise for a subdivision and isolation of subpopulations within the ancestral paleo-Orobian population (Fig. 7c). In any case, the free transit from North to South of fusions 2·8, 10·12, 1·3, and 4·6 would still be possible since they do not involve any monobrachial homology. A new Robertsonian population has thus slowly originated with a 24-chromosome karyotype which we find today relegated to Upper Valtellina. The entry of the paleo-Orobian mouse population into Val Poschiavo must have preceded this stage and it is likely that the 8·12 fusion appearing in this area blocked the introgression of fusion 2·8 and 10·12.

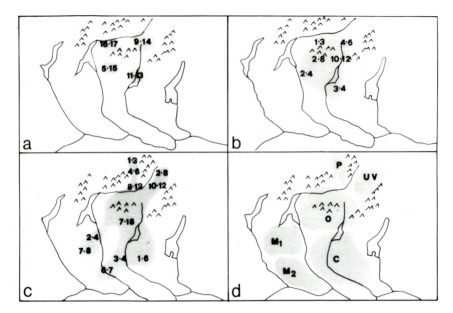

Fig. 7. Hypothesis on the interrelationship and genesis of the Orobian-Valtellina Robertsonian populations.

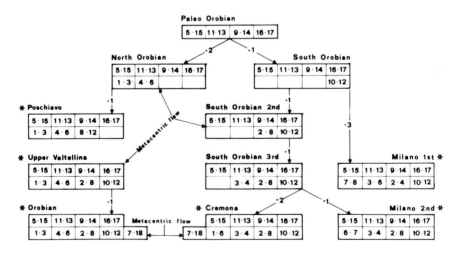

Fig. 8. Block system showing the evolutionary history of the Orobian-Valtellina Robertsonian system by means of the fusion pattern of its populations. Asterisks indicate the actual Robertsonian populations.

The appearance of fusion 7·18 in an area South of the Orobian Alps would have free access in the Orobian-Valtellina karyotype and in the southern Cremona plain, but not in the Milan 1st and 2nd populations where chromosome 7 was already involved in fusions 7·8 and 6·7, respectively. The pattern concerning the genesis of six Rhaetian populations may be summarized using a block system (Fig. 8).

Early History of the Robertsonian Process

An interesting feature of the Robertsonian phase of the Mus musculus model is that concerning the speed of the process. The rapidity of the karyotype transformation process and its independence of the true speciation phase has been emphasized by Bush et al [47,48] as a general character of the speciation process involving karyotype rearrangement; the case of the mouse, therefore, would fit into this general conformation, but what makes this case challenging is evaluating the time of the Robertsonian process.

It is believed [49] that Mus musculus penetrated into Italy later than in the other eastern and central European areas where it already existed in mid-Pleistocene [50,51]. The Alps may have represented an efficient barrier to mouse penetration until neolithic man developed agriculture, an incentive for mouse populations.

It is interesting that the finding of mouse chromosomal polytypism in the Alps coincides with that of the most ancient protohistoric evidence of an alpine people, the Camunes [52]; 5,000 years of their history are represented by over 140,000 rock engravings in Val Camonica and Valtellina in particular, but also in Val Chiavenna and Val Bregaglia, all names of localities also found in the Robertsonian geography of the mouse. Among these figures, yoked oxen (Fig. 9) appear in various configurations dating back to 3,500–3,800 BC; some very large ceramic grain containers belonging to same period have been found along with fossils of domestic dogs. About 3,500 BC the Camunes became farmers and the mouse became commensal. It was perhaps then that, with the transformation of the niche and the decrease in the actual deme size of the mouse as a result of the close commensalism with man, conditions were created which paved the way to the Robertsonian phase of mouse speciation.

About 5,000 years later, 16 different fusions appeared in this alpine area and, furthermore, the number of Robertsonian metacentrics reached 9, the maximum possible, in two populations. This implies a mutation rate of one Robertsonian fusion about every 350 years. The process was even more rapid: On the Marion Island, mice were first introduced by Danish sailors in the early 19th Century [53] and one Robertsonian fusion was already present [13]. Greater advances are seen in the laboratory: Nine years have produced the accumulation of a second fusion, 8·19, in the karyotype of the inbred strain carrying the fusion 6·15.

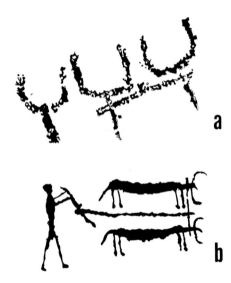

Fig. 9.   Yoked oxen are often depicted among the rock engravings of Valcamonica. The most ancient figure (a) goes back to the 2nd phase of the Valcamonica Neolithic (4,-000–3,500 BC); more explicit are the figures of the Chalcolithic age, e.g., the one engraved on the large stele of Bagnolo II (b), dated about 3,000 BC (redrawn from Anati [52]).

This documented speed of the Robertsonian process makes us realize just how small the genetic distance is between Robertsonian populations and between the latter and the all-acrocentric population. A new problem, however, arises: If the centric fusion process is so easy and so rapid in mice, why then are the Robertsonian populations still an exception with respect to the all-acrocentrics?

Factors Favoring the Robertsonian Accumulation

In my opinion the problem should be tackled from three different viewpoints: The systematic context, the factors favoring accumulation of fusions, and the factors constituting a selective advantage of the Robertsonian homozygotes with respect to the all-acrocentric homozygotes.

With regard to the first point, the data of Britton-Davidian et al [36] have demonstrated that all the Robertsonian populations examined so far belong to the biochemical semispecies Mus 1 according to the terminology of Bonhomme [54] and Thaler [55]. These populations also correspond to the semispecies domesticus-brevirostris referred to by Hunt and Selander [56]. Nevertheless, the large majority of Mus 1 populations remain of the all-acrocentric karyotype. The systematic context alone is thus not sufficient to explain the phenomenon.

We can, therefore, invoke all those conditions of restricted vagility and compartmentalization into microdemes which give rise to area effect [57] and, as such, facilitates the accumulation of Robertsonian fusions. Lengthy discussions have been forthcoming in this regards [2–6,58] and it has been stressed, without doubt, that the mountain ambient, where for the majority of cases Robertsonian populations can be observed, displays the previously mentioned characteristic of extreme compartmentalization. Also the insular ambient provides a well-documented tendency to promote drift phenomena, even in these ambients where Robertsonian mice have been found, as in the Eolian and Orkney Islands.

Other mountainous districts and other small Mediterranean and Northern islands, however, have mouse populations with an all-acrocentric karyotype. Moreover, alpine ambient compartmentalization is not necessary for the mouse to achieve conditions of actual minimal deme size; also in open fields the ethology of mouse, characterized as it is by a restricted vagility [59–62], guarantees this extreme fragmentation of the demes. As an example of a few parameters of mouse vagility, we can cite evaluations of its home range: According to Southern [63], in the domestic ambient, cellar or houses, the home range has been estimated as 6–8 m^2; in open fields Lindicker [64] estimates the home range as 150 m^2; and Quadagno [65] specifies 500 m^2. Again, the very ethological structure of the mouse population [66–69], with dominant males and females choosing mates within the same race, allows rapid attainment of small deme homozygosis and, as far as the problem of the Robertsonian mutations is concerned, a rapid onset of chromosomal homozygosis.

Of course all these conditions leading to an area effect are necessary to prepare the way for the onset of an initial Robertsonian situation, but are not sufficient to explain the phenomenon entirely. In fact, why do the populations fix a condition of Robertsonian homozygosis in certain areas and one of all-acrocentrics in another?

We must, therefore, reconsider the total randomness of the Robertsonian phase and return to a Darwinian view also for this phase of our model. If the onset and attainment of homozygosis may also be a purely stochastic phenomenon within the framework of the this premises, other aspects would have to entail, in order to be rationally explained, a selective advantage in a natural environment of the metacentric karyotype over the acrocentrics. At each stage of the accumulation phase, from the first to the last fusion in a 22-chromosome karyotype, there should be a selective advantage of the homozygotes for metacentrics not only over the heterozygotes but over the all-acrocentric homozygotes or the homozygotes for a lower number of metacentrics.

Furthermore, even the increased tendency of the Robertsonian populations to take away space from the all-acrocentric population suggests a Darwinian role of natural selection in setting up the Robertsonian population. Of particular interest in this regard is the hypothesis of a co-adaptation of some chromosomal rearrangement by Berry and Baker [70] based on data concerning Robertsonian variants of Thomomys bottae and Thomomys umbrinus. For arid environments and high mountains, both species, in fact, display karyotype symmetrization, i.e., an increase in metacentrics at the expense of the acrocentrics. Likewise, Patton [71] describes a karyotype variant distribution following a mountain gradient in Thomomys bottae grahamensis. The same appears to occur in mice. Robertsonian populations are found preferentially in unfavorable mountain ambients such as the Alps or Apennines, in extremely arid environments such as the island of Lipari, or in extreme northern latitudes, e.g., Northern Scotland.

Furthermore, the case of Mus musculus provides clear evidence of natural selection acting against very precise chromosome arrangements [72]. The 171-arm combinations in Robertsonian metacentric autosomes are possible and 56 have been actually detected in natural population while only 15 of them have

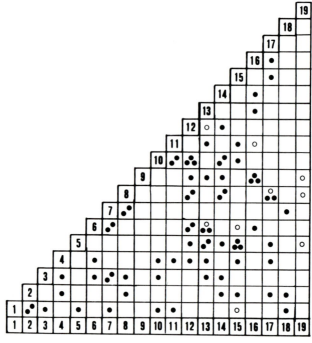

Fig. 10. Random arrangement of the standard acrocentric chromosomes in setting up Robertsonian metacentrics. The autosomes of the all-acrocentrics karyotype are indicated by numbers on the sides of the triangle; each point representing Robertsonian metacentrics identifies, on both triangle sides, the acrocentrics involved in its fusion. Solid circles represent metacentrics from feral populations, open circles identify metacentrics appearing in laboratory strains.

been observed more than once in different independent populations. Out of these 56 centric fusions, no fusion involving chromosome no. 19 has yet been found (Fig. 10). Nevertheless of the 10 fusions appearing in laboratory strains, chromosome no. 19 appeared fusioned three times. Thus, fusion of chromosome no. 19 probably occurs also in nature as in the laboratory, except that it entails a supragenic arrangement which is incompatible with the severe selection of the natural environment, but which is, on the other hand, compatible in a laboratory breeding context. The centric fusion cannot, therefore, be considered in a purely neutralistic light.

It is tempting to suggest, therefore, that beside creating the premises for the isolation between populations, the Robertsonian rearrangement affects the phenotype as to canalize—as proposed by Bickham and Baker [73]—the choice of either chromosome situation in different ways according to different environmental conditions. To suggest that "extreme" or "unfavorable" environmental conditions act in favor of the selection for metacentrics is still a vague statement which cannot be justified until each environmental parameter of the inhabited demes of the Robertsonian mice have been precisely characterized.

Certainly, commensalism does not favor this approach, although laboratory habitat choice tests similar to those successfully performed by Nevo [74,75] for the chromosomal races of Spalax ehrenbergi, might provide interesting information on this respect.

Nevertheless, it will still be difficult to ascertain whether the chromosome rearrangement is in itself co-adapted to that particular habitat or whether is rather a group of genes carried by that particular metacentric. For this reason the main facts of this speciation pattern still need to be elucidated.

An Ethological Approach

Other findings reported by Nevo [76,77] derived from the speciation pattern of Spalax ehrenbergi are of general significance, namely those related to the reinforcement of the isolation [78,79] by means of a premating barrier based on the variation of the breeding and mating behavior. In terms of ethological diversification between the chromosomal races, I speculated on the possibility that ethological barriers reinforced separation between populations [4]; nevertheless, contrasting evidence of natural hybrid narrow zones contradicts the hypothesis of a present day ethological differentiation between chromosomal races.

Nevertheless, preliminary trapping in the Upper Valtellina area provided information on the scattered presence of two chromosomal races, i.e., the 24-chromosome race characteristic of this area and the 26-chromosome one with the Val Poschiavo karyotype [19]. An extensive field search was therefore conducted in order to demonstrate unequivocally whether sympatry actually exists and whether hybridization is possible in a natural environment between these two chromosomal races. The capture were carried out at Migiondo (Upper Valtellina), a village with 500 inhabitants, situated over 1,000 meters above sea level. Traps were placed within a radius of 200 meters in the cellars of two inhabited houses—a cheese factory and a cattleshed. Of the 50 animals trapped, 22 revealed a 26-chromosome poschiavinus karyotype and 28 the 24-chromosome karyotype of Upper Valtellina. It is worth pointing out that no animals with a hybrid karyotype have ever been found, even when numerous animals were trapped in the same room.

Thus a premating barrier would appear to have been documented which acted to reinforce the isolation of two sympatric races. The exact nature of this premating barrier is, at present, under experimental investigation.

Is is worth emphasizing that, in this case, two Mendelian populations, differentiated by their karyotype, behave in sympatry like two bonae species. This finding is relevant as as index of advanced progress in the speciation process.

The Historical Context

This sympatric presence emphasizes another aspect of the speciation pattern of this rodent, which is commensal with man. The cohabitation in Upper Valtellina of these two Robertsonian races should be considered, in fact, as a recent phenomenon with respect to their separation and isolation which occurred in parapatric condition.

Our scheme, in fact, placed the sparation of the Val Poschiavo population from the Orobian-Valtellina context in a somewhat remote age (see Figs. 7 and 8). On the other hand, it can be seen that a primitive isolation between these two mouse populations may have been due to the political separation of the inhabitants of the two valleys. Poschiavo Valley has, in fact, always been part of the Swiss Canton of Grisons, in keeping with the cultural and religious tradition. Valtellina, instead, has been linked to the historical destiny of the Duchy of Milan and thus any traffic involving either goods or people has always been limited to these two adjacent valleys separated rather by enmity and religious bias—Lutherans, on the one hand, and Papists on the other—, than by any geographic barrier. But for the mouse, for whom the only chance of dispersion is passive transport, the blocking of trade implies an insurmountable barrier. Also, the historical trade routes between Switzerland and Lombardy were slightly more to West, across the Bregaglia and Chiavenna Valleys, which in actual fact are inhabited by hybrid populations of mice. The commercial separation between Val Poschiavo and Valtellina ended towards the middle of the last century when the new alpine passes of the Bernina and the Stelvio were opened and led to an increasingly large amount of commerical traffic through Val Poschiavo and Upper Valtellina. Dating back to this period we have the entrance into Italy of the tobacco mouse, already ethologically differentiated in such a way as to condition the present state of sympatric isolation.

In the case, therefore, of animals commensal with man and of episodes of speciation commencing in protohistoric and historic times, even anthropic, political, cultural, and economic factors should be considered important in the isolation of populations in the process of differentiation.

CONCLUSION

Evidence deduced from this case of Robertsonian variability of Mus musculus points to a speciation pattern where the initial fixation of spontaneous chomosomal rearrangement seems to be due purely to the interplay of stochastic events, i.e., a random drift occurring within an isolated deme characterized by an actual minimal size. Commensalism with man, together with any peculiarity of the ethology of the mouse, undoubtedly helps in shattering the mouse populations into microdemes, which is an essential premise to the onset of the initial Robertsonian population. Nevertheless, the mode of speciation of the house mouse does not seem to be totally random; in fact, Robertsonian translocations, other than entailing hybrid fertility impairment, seem to produce phenotypic effects such as canalizing the choice of a symmetrical karyotype situation by means of environmental conditions definable as "extreme" or "unfavorable."

Metacentric accumulation, due to new Robertsonian mutations occurring within the deme as well as derived from an interdeme migration, reinforces the reciprocal isolation between Robertsonian populations, but a definitive isolation will be reached only when metacentrics with monobrachial homology appear. A postmating barrier, however, is an isolating mechanism entailing severe waste of the reproductive potentiality of a natural population, consequently any ethological or physiological mutation fit for activateing a premating isolation will be rapidly and positively selected by the natural environment.

The entire model of mouse speciation, therefore, seems unusual and complex, and even disturbed by the difficult cohabitation with man. But every other speciation model has a dose of originality and peculiarity. We are tempted to paraphrase the aphorism of Linnaeus, "tot sunt speciationis itinera quot sunt speciandae species." But quite apart from the facile Latin, it would be better to propose models each time which are suited to a specific case rather than to propose general models which necessarily have an extremely limited degree of reliability.

ACKNOWLEDGMENTS

Most of the observations reported herein refer to unpublished data and represent the most recent part of the collaborative work carried out with Prof A. Gropp and Dr. H. Winking of the Medical Hochschule of Lubeck and with Dr. C.A. Redi of the University of Pavia. The collaboration of Dr. Vittoria Civitelli and of Dr. Marco Corti of our Istitute is gratefully acknowledged.

REFERENCES

1. Robertson WRB: Chromosomes studies. II. Taxonomic relationships shown in the chromosomes of Tettigidae and Acrididae. J Morphol 27:179,1916.
2. Capanna E, Civitelli MV, Cristaldi M: Chromosomal rearrangement, reproductive isolation and speciation in Mammals. The case of Mus musculus. Boll Zool 44:213,1977.
3. Capanna E, Gropp A, Winking H, Noack G, Civitelli MV: Robertsonian metacentrics in the Mouse. Chromosoma 58:341,1976.
4. Capanna E: Chromosomal rearrangement and speciation in progress in Mus musculus. Folia Zool 29:43,1980.
5. White MJD: Chain process in chromosomal speciation. Syst Zool 27:285,1978.
6. Bickham JW, Baker RJ: Reassessment of the nature of chromosomal evolution in Mus musculus. Syst Zool 29:159,1980.
7. Gropp A, Tettenborn U, von Lehmann E: Chromosomenvariation vom Robertson'schen typus bei der tabakmaus, M. poschiavinus, und ihren hybriden mit der laboratoriumsmaus. Cytogenetics 9:9,1970.
8. Gropp A, Winking H, Zech L, Muller H: Robertsonian chromosomal variation and identification of metacentric chromosomes in the feral mice. Chromosoma 39:265, 1972.
9. Capanna E, Civitelli MV, Cristaldi M: Una popolazione appenninica di Mus musculus L. caratterizzata da un cariotipo a 22 cromosomi. Acc Lincei Rend Sc Mat Fis Nat 54:981,1973.
10. Berry RJ, Brooker PC, Lush IE, Nash HR, Newton MF: Robertsonian translocations in wild mice. Mouse News Letter 64;65,1981.
11. Grohe G, Gropp A, Gropp D, Jüdes U, Kolbus U, Noack G, Winking H: Robertsonian chromosomes in mice from North-Eastern Greece. Mouse News Letter 64:69,1981.
12. Chakrabarta S, Chakrabarta A: Spontaneous Robertsonian fusion leading to karyotype variation in the house mouse. Experientia 33:175,1977.
13. Robinson TJ: Preliminary report of a Robertsonian translocation in an isolated feral Mus musculus population. Mammal Chrom Newsletter 19:84,1978.
14. Forejt J: Centromeric heterochromatin polymorphism in the house mouse: Evidence for inbred strains and natural populations. Chromosoma 43:187,1973.
15. Fatio V: Faune de Vertébrés de la Suisse. Vol I. Mammifères. Genève: H. Georg, 1869, p 410.
16. Capanna E, Civitelli MV, Cristaldi M: Chromosomal polymorphism in an Alpine population of Mus musculus L. Boll Zool 40:379,1973.
17. Capanna E, Valle M: A Robertsonian population of Mus musculus L. in the Orobian Alps. Acc Lincei Rend Sc Mat Fis Nat 62:680,1977.
18. Capanna E, Riscassi E: Robertsonian karyotype variability in natural Mus musculus populations in the Lombardy area of the Po valley. Boll Zool 45:63,1978.
19. Redi CA, Beltrami L, Capanna E, Ferri E, Noack G, Winking H: New data on the karyotype variability of mouse. Boll Zool 46 (suppl):188,1979.
20. Capanna E: Gametic aneuploidy in the mouse hybrids. Chromosomes Today 5:83,1976.

21. Godena G, D'Alonzo F, Cristaldi M: Corrélations entre caryotypes et biotypes chez le Lérot (Eliomys quercinus) et les autres Rongeurs d l'île Lipari. Mammalia 42:382, 1978.
22. von Lehmann E, Radbruch A: Robertsonian translocations in Mus musculus in Sicily. Experientia 33:1025,1977.
23. Zech L, Evans EP, Ford CE, Gropp A: Banding pattern in mitotic chromosomes of tobacco mouse. Exp Cell Res 70:263,1972.
24. Capanna E, Civitelli MV, Cristaldi M, Noack G: New Robertsonian metacentrics in another 22-chromosome mouse population in Central Apennines. Experientia 33:173,1977.
25. Gropp A, Winking H: Robertsonian translocations: Cytology, Meiosis, Segregation pattern, and biological consequences of heterozygosity. Symp Zool Soc (London) 47:141,1981.
26. Tettenborn U, Gropp A: Meiotic nondisjunction in mice and mouse hybrids. Cytogenetics 9:272,1970.
27. Gropp A: Reproductive failure due to fetal aneuploidy in mice. Fertility Sterility, Proc 7th World Congr Tokyo, 326, 1971.
28. Ford CE: Gross gemome unbalance in mouse spermatozoa: Does it influence the capacity to fertilize? In Beatty RA, Gluecksohn-Waelsch S (eds): Edinburg Symposium on the Genetics of the Spermatozoon. Coppenhagen: Bogtrykkeriet Forum, 1972, p 359.
29. Cattanach BM, Moseley H: Nondisjunction and reduced fertility caused by the tobacco mouse metacentric chromosomes. Cytogenet Cell Genet 12:264,1973.
30. Ford CE, Evans EP: Robertsonian translocations in mice: Segregational irregularities in male heterozygotes and zygotic unbalance. Chromosomes Today 4:387,1973.
31. Winking H, Gropp A: Meiotic nondisjunction of metacentric heterozygotes in oocytes versus spermatocytes. Proc Serono Symp 8:47,1976.
32. Ferri E, Capanna: Segregation disorders in multiple heterozygous Robertsonian mice. Acc Lincei Rend Sc Mat Fis Nat 66:598,1979.
33. Beltrami ML, Ferri E, Redi CA: Rate of meiotic malsegregation in hybrid mice: Evaluation by microdensitometric measurement of spermatozoal Feulgen-DNA content. Acc Lincei Rend Sc Mat Fis Nat 68:81,1980.
34. Redi CA, Capanna E: DNA-content variation in mouse spermatozoa arising from irregular meiotic segregation. Boll Zool 45:315,1978.
35. Nei M: Interspecific gene differences and evolutionary time estimated from electrophoresis data on protein identity. Am Natur 105:385,1971.
36. Britton-Davidian J, Bonhomme F, Croset H, Capanna E, Thaler L: Variabilité génétique chez les populations de souris (genre Mus L.) à nombre chromosomique réduit. C R Acad Sci (Paris) 290:195,1980.
37. Nevo E, Shaw CR: Genetic variation in a subterranean mammal Spalax ehrenbergi. Biochem Genet 7:235,1972.
38. Nevo E, Yung JK, Shaw CR, Thaeler CS: Genetic variation, selection and speciation in Thomomys talpoides pocket gophers. Evolution 28:1,1974.
39. Patton JL, Yang SY: Genetic variation in Thomomys bottae pocket gophers: macrogeographic patterns. Evolution 31:697,1977.

40. Benado M, Aguilera M, Reig OA, Ayala FJ: Biochemical genetics of chromosome forms of Venezuelian spiny rats of the Proechimys guairae and Proechimys trinitatis superspecies. Genetica 50:89,1979.
41. Bush GL: Modes of animal speciation. Annu Rev Ecol Syst 6:339,1975.
42. White MJD: Models of speciation. Science 159:1065,1968.
43. Gropp A, Kolbus U, Giers D: Systematic approach of the study of trisomy in the mouse. II. Cytogenet Cell Genet 14:42,1975.
44. Evans EP: Male sterility and double heterozygosity for Robertsonian translocations in mouse. Chromosomes Today 5:75,1975.
45. Gropp A, Kolbus U: Exencephaly in the syndrome of trisomy no. 12 of the foetal mouse. Nature 249:145,1974.
46. Gropp A: Animal model of human disease. Am J Pathol 77:539,1974.
47. Wilson AC, Bush GL, Case SM, King M-C: Social structuring of mammalian populations and rate of chromosomal evolution. Proc Natl Acad Sci (USA) 72:5061,1975.
48. Bush GL, Case SM, Wilson AC, Patton JL: Rapid speciation and chromosomal evolution in Mammals. Proc Natl Acad Sci (USA) 74:3942,1977.
49. Pasa A: Primi risultati dell' indagine paleontologica sui materiali scavati nelle grotte di S. Cassiano (Colli Berici-Vicenza) Ann Univ Ferrara sez 9, 1:169,1953.
50. Jannossy D: Die entwicklung der kleinsäugerfauna Europas in Pleistozän (insectivora, rodentia, lagomorpha). Säugetierkun 26:40,1961.
51. Kretzoi M, Vertes L: Upper Biharian (Intermindel) pebble-industry occupation site in Western Hungary. Curr Anthropol 6:74,1965.
52. Anati E: I Camuni. Milan: Jaca Book, 1979, p 201.
53. Berry RJ, Peters J, Van Aarde RJ: Subantartic house mice colonization, survival and selection. J Zool (London) 184:127,1976.
54. Bonhomme F, Britton-Davidian J, Thaler L, Triantaphyllidis C: Sur l'existence in Europe de quatre groupes de Souris (genre Mus L.) du rang espèce et semi-espèce, démontrée par la génétique biochimique. C R Acad Sci (Paris) 287:631,1978.
55. Thaler L, Bonhomme F, Britton-Davidian J: Processes of speciation and semi-speciation in the house mouse. Symp Zool Soc (London) 47:27,1981.
56. Hunt WG, Selander RK: Biochemical genetics of hybridization in European house mouse. Heredity 31:11,1973.
57. Cain AJ, Currej JD: area effect in Cepea. Phil Trans Roy Soc (London) 246:1,1963.
58. Lande R: Effective deme size during long-term evolution estimated from rates of chromosomal rearrangement. Evolution 33:234,1979.
59. Berry RJ: The natural history of the house mouse. Field Studies 3:219,1970.
60. Berry RJ, Jakobson ME: Vagility in an island population of the house mouse. J Zool (London) 173:341,1974.
61. Berry RJ, Jakobson ME: Life and death in an island population of the house mouse. Exp. Gerontol 6:187,1971.
62. Berry RJ, Jakobson ME, Triggs GS: Survival in wild-living mice. Mammal Rev 3:46,1973.
63. Southern HN: Control of Rats and Mice. Oxford: University Press, 1954, Vol 3, p 372.
64. Lindicker WZ: Ecological observations on a feral house mouse population declining to extinction. Ecol Monogr 36:27,1966.

65. Quadagno DM: Home range size in feral house mouse. J Mammal 49:149,1968.
66. Crowcroft P, Rowe FP: Social organization and territorial behaviour in the wild house mouse (Mus musculus L.) Proc Zool Soc (London) 140:517,1963.
67. Mainardi D: Speciazione nel topo. Fattori etologici determinanti barriere riproduttive tra Mus musculus domesticus e Mus musculus bactrianus. Rend Ist Lombardo (Sc) B 97:135,1963.
68. Mainardi D, Marsan M, Pasquali A: Causation of sexual preferences of the house mouse. Atti Soc Ital Sc Nat, Museo Civ Milano 104:325,1965.
69. De Fries VC, McLearn GE: Behavioral genetics and the fine structure of mouse populations: A study in microevolution. Evol Biol 5:279,1972.
70. Berry DL, Baker RJ: Apparent convergence of karyotypes in two species of pocket gophers of the genus Thomomys (mammalia, rodentia). Cytogenetics 10:1,1971.
71. Patton JL: Karyotypic variation following an elevational gradient in the pocket gopher Thomomys bottae grahamensis Goldman. Chromosoma 31:41,1970.
72. Capanna E, Winking H, Redi CA, Gropp A: Structural genome rearrangement in the mouse: Rb metacentrics in feral populations in Italy. Hereditas 94:8,1981.
73. Bickham JW, Baker RJ: Canalization model of chromosomal evolution. Bull Carnegie Mus Natur Hist 13:70,1979.
74. Nevo E, Shokolnik A: Adaptive metabolic variation of chromosome forms in mole rats Spalax. Experientia 30:724,1974.
75. Nevo E, Guttman R, Haber M, Erez E: Habitat selection in evolving mole rats. Oecologia 43:125,1979.
76. Nevo E: Mole rat Spalax ehrenbergi: Mating behavior and its evolutionary significance. Science 163:484,1969.
77. Nevo E, Naftali G, Guttman R: Aggression patterns and speciation. Proc Natl Acad Sci (USA) 72:3250,1975.
78. Dobzhansky Th: Speciation as a stage in evolutionary divergence. Am Natur 74:312,1940.
79. Dobzhansky Th: Genetics and the Origin of Species, 3rd ed. New York: Columbia University Press, 1951, p 363.

Mechanisms of Speciation, pages 179-190
© 1982 Alan R. Liss, Inc., 150 Fifth Avenue, New York, NY 10011

Does Speciation Facilitate the Evolution of Adaptation?

L. D. Gottlieb

WHAT QUESTIONS ARE RELEVANT?

The analysis of speciation has traditionally been based on the accumulation of information in many data categories. These include geographical distribution, morphology, biochemical polymorphisms, DNA values, karyotypes, reproductive isolating barriers and the results of experimental hybridizations, as well as a description of the habitats and geographical variability [1]. Although this information seems exhaustive, it does not readily identify the characteristics which provide a new species with novel adaptations the ability to live and reproduce in different habitats than those of its progenitor.

The relationship between speciation and adaptation is a central one in evolutionary biology. Indeed, speciation is considered to have the "function" of preventing the disintegration of adaptive gene complexes that might come about by hybridization [2]. The explanation seems reasonable, but whether new adaptations arise before or after speciation is now actively debated [3]. This may be because most studies of speciation have actually not compared progenitor and derivative species, but rather clusters of closely related species in which phylogenetic direction has not been assignable. A second problem has been that most analyses have dealt with phenotypes, not genotypes, with the result that divergence has not been defined in terms of the number and types of genetic differences.

Some of these difficulties can be avoided with annual and other short-lived plants in which progenitor-derivative phylogenies have been identified. The best cases are in Calycadenia [4,5], Chaenactis [6], Clarkia [7–10], Coreopsis [11,12], Crepis [13,14], Gaura [15,16], Haplopappus [17], Holocarpha [18], and Stephanomeria [19,20]. In many of these examples, the derivative species can be experimentally hybridized to its parent to produce F_1 and F_2 progenies, making it possible to probe the genetic basis of their differences. These important advantages permit the following questions: How was the phenotype of the new species assembled? To what extent are its properties a consequence of phylogeny rather than selective and/or stochastic processes since its origin? What specific genetic changes in the genome of the progenitor yield the distinctive properties of the derivative? Does the newly evolved species have distinct adaptations that permit it to live and reproduce in habitats which are different from those of its parent?

In this paper various studies are described that are relevant to these questions for a recently evolved diploid annual plant species in the Compositae family, Stephanomeria malheurensis (MAL) and its progenitor, a population of S exigua ssp coronaria (COR). Since the analyses are still in progress, our present knowledge regarding the species [19–23] will be summarized and the strategy that is being used to assess the origin of the unique attributes of MAL which might provide it with new adaptive capabilities will be illustrated. The experimental research to date, however, has been primarily in the lab and greenhouse whereas ecological analyses in the natural habitat will be required before it can be concluded that a particular complex of characters is adaptive.

STEPHANOMERIA MALHEURENSIS AND ITS PARENT

MAL and COR grow together on an exposed volcanic tuff on a broad hilltop at 5000 ft elevation in the high desert of eastern Oregon. The locality is the only one known for MAL, but COR is widely distributed southwards to southern California. Since the site is isolated by many miles from the next nearest tuff, it can be considered an edaphic island. It represents a "species border" for COR, and is characterized by more extreme and less predictable environmental conditions than those found at the Californian localities. The two Stephanomerias are minor components of the local plant community which is dominated by shrubs such as sagebrush (Artemesia tridentata); other annual species have substantially larger populations.

MAL and COR differ markedly in their population sizes. Observations since 1968 have shown that COR outnumbers MAL by more than 40 to 1. In the best years, defined by high precipitation in the spring months when the seeds of the species germinate, the total number of MAL plants has not exceeded 750 whereas COR numbered about 40,000. MAL plants are not segregated from COR and the two species grow interspersed; indeed their branches are often overlapped. The density of the COR population is low, and even on the few local sites where the majority of individuals grow, an average of only 30.6 plants per square meter has been found.

COR is obligately outcrossed because it has a sporophytic self-incompatibility system that prevents pollen from germinating on stigmas of the plant from which the pollen was shed. In contrast, MAL is self-compatible and highly self-pollinated. The origin of selfing species from outcrossing ones is a common pathway of speciation in annual plants [24].

The two species are morphologically very similar and cannot be distinguished with certainty in the field until they flower and set seed because the most reliable character for separating them is seed size. Those of MAL average one-third longer and two times heavier than the ones of COR, and have longer and more numerous pappus bristles.

TABLE I. Summary of Differences Between MAL and COR

Character	Species differences	
	COR	MAL
Seedlings		
Germination	Cold required	None
Cotyledons	Small	Large
No. days to bolting	Few	Many
No. leaves at bolting	Few	Many
Root dry wt.	Low	High
Root/shoot ratio	Low	High
Mature plants: vegetative		
Stem height	Short	Tall
No. stem nodes	Many	Few
No. leaf nodes/total nodes	Low	High
No. branches	Many	Few
Mature plants: reproductive		
No. heads/cm branch	Many	Few
No. florets/head	4–5	5–6
No. heads/plant	Many	Few
Seed weight	Light	Heavy
Pappus no.	Few	Many

The species are reproductively isolated by three factors: (a) pollen exchange is restricted because of the differences in breeding system; (b) seed set is reduced by about one-half in interspecific cross-pollinations compared to conspecific ones; and (c) fertility of F_1 hybrids, should they be formed, is reduced to about 25% by differences in chromosomal structural arrangement (that includes a reciprocal translocation). Only a few interspecific hybrids have been identified in nature. Thus, the reproductive isolation of the species is highly effective, and they easily pass the "test of sympatry."

THE PHENOTYPE OF THE SPECIES

Important phenotypic differences between MAL and COR have been detected when they were grown under uniform garden conditions (Table I). The larger seeds of MAL germinate to produce seedlings with larger cotyledons, but the MAL rosettes always require significantly more days to initiate stem elongation or bolt. The difference between the species is dramatic, with the average MAL plant bolting about twelve days later than COR (which took forty-one days in a typical experiment). Since leaves are continually added to the rosette prior to bolting, MAL bolts form a much larger rosette than COR.

During the period of rosette growth, MAL also allocates substantially more photosynthate to its roots. When the species were compared at bolting, its root/shoot ratio was 1.6 times that of COR.

TABLE II. Mean Values for Rosette Leaf Number and Number of Stem Nodes for MAL and COR Grown in Three Environments

	Environment 1		Environment 2		Environment 3	
	COR	MAL	COR	MAL	COR	MAL
No. leaves at bolting	15.5	22.5[a]	12.8	22.0[a]	13.0	20.2[a]
No. stem nodes	39.6	27.8[a]	29.2	24.2[a]	32.1	27.0[a]
Total nodes	55.1	50.3	42.0	46.2	45.1	47.2
% Leaf nodes	.28	.45[a]	.30	.48[a]	.29	.43[a]

[a]Species differences were very highly significant ($P < 0.001$). For MAL, n = 14, 42, and 18 in environments 1, 2, and 3, respectively. For COR, n = 12, 114, and 29 in environments 1, 2, and 3, respectively.

The stems of MAL are generally taller than those of COR, but have fewer branches, and the flowering heads, which are borne on the branches, are spaced farther apart. These features reduce the number of heads on the average MAL plant to half that of COR. MAL heads typically have one more floret than those of COR; however, the total number of florets per plant is still very much lower than in its parent.

The reduced branch number in MAL may reflect a change in the timing or expression of a physiological signal which causes bolting. This hypothesis results from the following observation. The plant axis of both species has the same total number of nodes (the sum of the number of nodes or leaves on the rosette plus the number of nodes or lateral buds on the stem) when grown together under the same environmental conditions (Table II). However, the species differ in the proportion on the rosette vs. the stem. In COR, rosette nodes constitute approximately one-fourth of the total, but in MAL the rosette bears about one-half of them (Table II).

Thus delayed bolting in MAL increases the number of nodes in the rosette, thereby reducing the number of branches on the stem and the number of heads. The situation illustrates how a particular physiological change acting early in development and probably brought about by a small number of genes can have a major impact on fitness.

Reduced fecundity in MAL, resulting from the reduction in head number, has also been clearly documented in nature. This was done by weighing shoots (ie the above-ground portion of the plants including stem, branches, and heads) of all individuals of the two species growing within a narrow belt transect (the sample included 104 COR and 83 MAL). The distribution of shoot dry weights was strongly skewed with most plants of both species primarily small and only a few large individuals. The average COR individual was two times heavier than that of MAL, though the median weights were the same. A second important observation was that 13% of the COR plants were heavier than the heaviest MAL. In a separate analysis, using a different set of plants, MAL was found to produce only half as many heads as COR for the same shoot dry weight (Fig. 1). In conjunction with its rarity, these results suggest that MAL responds more poorly than COR to their shared natural habitat.

Two lines of evidence suggest that MAL was extracted from common genetic elements of the richly polymorphic gene pool of COR. The mean expression of most quantitative characters in MAL is well within the COR range, and COR nearly always has larger coefficients of variation [21]. MAL is nearly monomorphic for structural genes coding enzymes detected by electrophoresis, and all but one of its genes are present in moderate to high frequency in COR. The parent, however, possesses many alleles not found in MAL [25].

The absence of unique alleles in MAL, its overall morphological similarity to COR, and its presence in only a single locality where it grows alongside COR, suggests its origin was relatively recent. The origin of MAL probably involved a rapid and abrupt series of events initiated by the occurrence of a mutation at the self-incompatibility locus in a COR plant which permitted self-pollination to take place. Eventually several chromosomal rearrangements occurred, giving MAL reproductive isolation in the midst of its progenitor population. Only a small number of genetic changes appear to have been involved in its reproductive isolation. This is indicated by the substantial increase in mean pollen viability from 25% in F_1 progenies to 62% in the F_2, which also included a significant proportion of fully fertile plants [19].

Thus most if not all of the attributes of MAL can be explained as a consequence of its phylogeny. Their combination produces a distinctive phenotype, yet it is not known if it confers unique adaptations. MAL may be adapted to an infrequent or rare niche in its present locality which is not suitable for COR. Rather than trying to define this niche, and in view of the overwhelmingly greater abundance and vigor of COR in their shared habitat, our studies have concentrated on the genetic processes of MAL's origin.

A TEST OF CHARACTER ASSOCIATIONS

We have begun to analyze the genetic basis of many of the quantitative phenotypic differences between the species. Although some studies are still incomplete, information is available to illustrate several important findings. Figure 2 shows segregations of F_2 progeny obtained by selfing F_1 hybrids between the species. In this experiment, the F_2, the F_1 (grown from seed saved from the original cross), and both parents were grown at the same time in a single uniform environment. The differences between the two species were very highly significant ($P < 0.001$) for number of days to bolting, number of rosette leaves at bolting, stem height, number of rosette leaves/total number of nodes, and seed weight. A difference was not observed for total node number. For each of the different traits, the F_1 was intermediate, and both parental character expressions

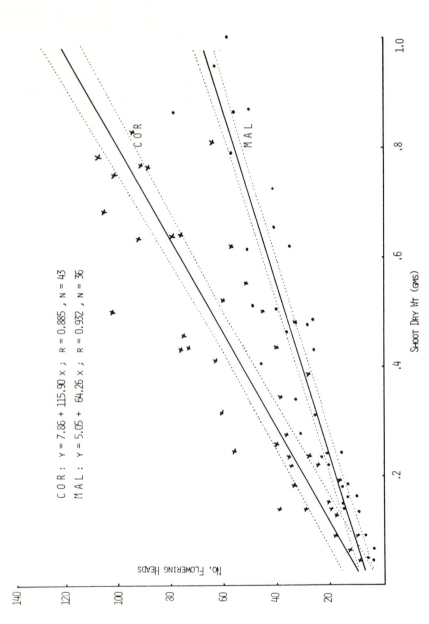

Fig. 1. Regression of shoot dry weight × number of flowering heads on a random sample of MAL and COR plants collected in nature. The dotted lines show the 95% confidence intervals for each species.

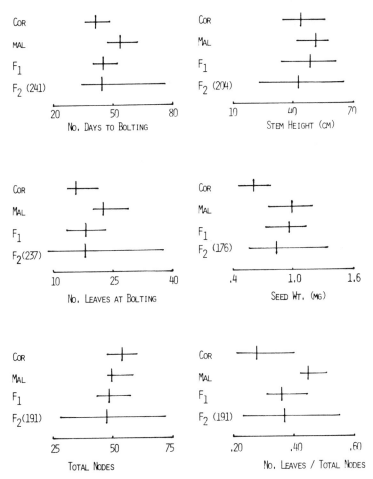

Fig. 2. Means and ranges of variation for six traits in MAL, COR, and their F_1 and F_2 progenies grown at the same time in a uniform environment. The number of individuals in each F_2 progeny is indicated.

were recovered in the F_2 progeny indicating that only a small number of loci were involved in each of them.

Analysis of the correlations between pairs of characters in the F_2 progeny revealed many highly significant r values. Sixty percent of them (48/78) were significant ($P < 0.01$), suggesting the character complexes of the species tended to remain intact. The number of days to bolting, total node number, and number of leaves were significantly correlated with the largest number of other characters.

TABLE III. Correlation Coefficients Between Five Characters and Number of Days to Bolting and to Flowering in F_2 Progeny Between MAL and COR*

Character	Number of days to:	
	Bolting	Flowering
Rosette size	.62	.66
Total branch length	.33	.46
Number rosette leaves/ total number nodes	.64	.61
Number nodes on branches	.38	.44
Seed number	− .29	− .36

*n = 141 for each r value.

Particularly interesting was the similarity of the r values between the number of days to bolting and the number of days to flowering. In this experiment, mean time to bolting in the F_2 was forty-four days and mean time to flowering was eighty-one days. The comparison showed that the correlation at the earlier time accounted for 88% of that for the same set of characters at the later time (Table III). Since bolting was scored when the stems had elongated 5 mm, the actual physiological signal that causes bolting must precede this visible stage by many days. This suggests that much of the final architecture of these plants may be predictable when they are only a few weeks old. The signal is likely to be a specific hormone-mediated event which gives hope that its genetic basis can be elucidated.

Prior to this analysis, the association between large seeds, large roots, and late bolting in MAL was taken to mean that these specific features facilitated its entry into a new niche [21]. Thus, the large seeds were thought to be the basis for niche separation on the hypothesis that they utilized a new safe site for germination and seedling establishment which was not suitable for the smaller seeds of COR. The changes in morphology, branching, and partitioning were considered to be required for the development of the increased seed size. This second hypothesis has now been rejected by the analysis of character correlations in the F_2 progeny because it was observed that early-bolting plants could produce either large or small seeds, and that late-bolting plants could do the same. Thus the particular combination of traits characteristic of MAL was not required for it to produce large seeds. Whether or not these seeds actually germinate in a different safe site than those of COR remains untested.

HOW DO THE SPECIES RESPOND TO EACH OTHER

The significance of the differences between MAL and COR has also been examined by asking if they "perceive" each other as distinct [26]. An experimental design reminiscent of studies of competition was used in order to determine if plants of either species responded differently to plants of the other species than to themselves. In this experiment, plants were grown at several densities (8, 16, and 24 per pot) in pure and mixed (50:50) stands. If the species facilitated each other's growth, increased combined yields were expected in the mixed stands. If they interfered with each other, yields would be reduced. If they did not interact positively or negatively, their responses in mixed stand were expected to be the same as those in pure stand.

Dry weight and stem length are good measures of plant size and vigor; the results for these and other characters are shown in Figure 3. With few exceptions [26], the performance of the two species was the same whether they were grown in pure or mixed stands. The same result was obtained for partitioning of their photosynthate into branches, stems and roots (Fig. 4). Thus to the extent that MAL and COR grow sufficiently close in nature to interact, it is likely that there will be little or no additional competition caused by their being different species.

CONCLUSIONS

This paper presents a brief overview of our ongoing attempt to understand the genetic processes involved in the origin of a self-pollinating annual plant species. The new species is morphologically very similar to its progenitor with which it still grows, and its genome appears to be a limited extraction from the progenitor's gene pool. Much effort has been required to identify differences between the species. Several of these seem significant. The derivative exhibits a new pattern of growth marked by a delay in the time of bolting, a developmental event which initiates the shift from vegetative growth to reproduction. The delay in bolting contributes to a reduction in the number of heads and thereby has an important negative impact on fitness. The new species also allocates a substantially greater amount of photosynthate to its roots. Analysis of character correlations in an interspecific F_2 progeny rejected the hypothesis that the changes in the architecture of plants of the new species were required either because of genetic linkage or developmental constraints for the production of its large seeds. The phenotype of the new species probably was assembled by relatively simple genetic means.

We still do not know if the new species possesses any adaptation(s) that permits it to do something different than its progenitor. The new species exists as a few hundred plants and perhaps only four to five dozen distinguishable types. Since it has a very small population size and its plants are smaller and less fecund than those of its parent in the same habitat, its long-term persistence, at least at its present site, is uncertain.

The presumption that phenotypic divergence between species can be equated with differences in their adaptation is difficult to validate. Evidence from other plant species suggests that speciation is not required for the origin of adaptation. Some perennial species such as Potentilla glandulosa [18] and Achillea millefolium [27] show dramatic ecotypic divergence among populations.

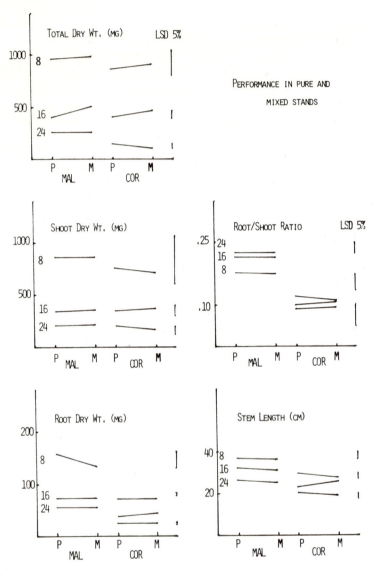

Fig. 3. The expression of five characters in MAL and COR grown in pure (P) and mixed (50:50) (M) stands at three densities (8, 16, and 24 plants per pot). The vertical lines permit evaluation of the differences within and between species for each density.

Many annual species also include populations adapted to diverse habitats [18,28], but species of other annual genera such as Clarkia [8] and those inhabiting vernal pools in California [29] appear adaptively equivalent. In order

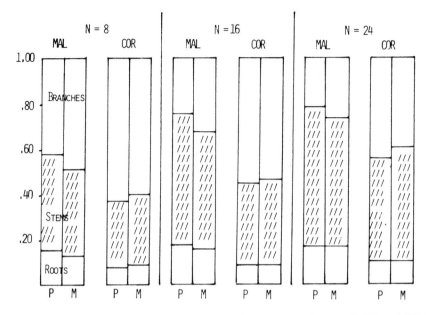

Fig. 4.   Percentage dry matter allocated to branches, stems, and roots of MAL and COR grown in pure (P) and mixed (50:50) (M) stands at three densities (8, 16, and 24 plants per pot). Differences between allocation in the pure and mixed stands were not signficiant ($P > 0.05$).

to determine whether speciation facilitates the origin of adaptation, it will be necessary to identify which particular characters are responsible for adaptation (a problem in ecology) and what changes in the progenitor's genome lead to their expression (a problem in genetics and development). Without such evidence, the assertion that species differences are adaptive is no longer satisfactory.

REFERENCES

1. White MJD: Modes of Speciation. San Francisco: Freeman, 1978.
2. Dobzhansky T: Species of Drosophila. Science 177:664, 1972.
3. Carson HL: Speciation and the founder principle. Stadler Symp 3:51, 1971.
4. Carr GD: Chromosome evolution and aneuploid reduction in Calycadenia pauciflora (Asteraceae). Evolution 29:681, 1975.
5. Carr GD: Experimental evidence for saltational chromosome evolution in Calycadenia pauciflora Gray (Asteraceae). Heredity 45:107, 1980.
6. Kyhos DW: The independent aneuploid origin of two species of Chaenactis (Compositae) from a common ancestor. Evolution 19:26, 1965.
7. Lewis H, Roberts MR: The origin of Clarkia lingulata. Evolution 10:126, 1956.
8. Lewis H: The origin of diploid neospecies in Clarkia. Am Natur 107:161, 1973.

9. Vasek FC: The evolution of Clarkia unguiculata derivatives adapted to relatively xeric environments. Evolution 18:26, 1964.
10. Small E: The evolution of reproductive isolation in Clarkia, section Myxocarpa. Evolution 25:330, 1971.
11. Smith EB: Coreopsis nuecensis (Compsitae) and a related new species form southern Texas. Brittonia 26:161, 1974.
12. Crawford DJ, Smith EB: Allozyme variation in Coreopsis nuecensoides and C. nuecensis (Compositae), a progenitor-derivative species pair. Evolution 36: in press.
13. Tobgy HA: A cytological study of Crepis fuliginosa, C. neglecta and their hybrid, and its bearing on the mechanisms of phylogenetic reduction in chromosome number. J Genet 45:67, 1943.
14. Sherman M: Karyotype evolution: a cytogenetic study of seven species and six interspecific hybrids in Crepis. Univ Calif Publ Bot 18:369, 1946.
15. Gottlieb LD, Pilz G: Genetic similarity between Gaura longiflora and its obligately outcrossing derivative G. demareei. Syst Bot 1:181, 1976.
16. Raven P, Gregory D: A revision of the genus Gaura (Onagraceae). Mem Torrey Bot Club 23:1, 1972.
17. Jackson RC: Interspecific hybridization in Haplopappus and its bearing on chromosome evolution in the Blepharodon section. Am J Bot 49:119, 1962.
18. Clausen J: Stages in the Evolution of Plant Species. Ithaca: Cornell Univ Press, 1951.
19. Gottlieb LD: Genetic differentiation, sympatric speciation and the origin of a diploid species of Stephanomeria. Am J Bot 60:545, 1973.
20. Gottlieb LD: Stephanomeria malheurensis (Compositae), a new species from Oregon. Madrono 25:44, 1978.
21. Gottlieb LD: Phenotypic variation in Stephanomeria exigua ssp. coronaria (Compositae) and its recent derivative species "Malheurensis." Am J Bot 64:873, 1977.
22. Gottlieb LD: Allocation, growth rates and gas exchange in seedlings of Stephanomeria exigua ssp. coronaria and its recent derivative S. malheurensis. Am J Bot 65:970, 1978.
23. Gottlieb LD: The origin of phenotype in a recently evolved species. In Solbrig OT, Jain S, Johnson GB, Raven PH (eds): Topics in Plant Population Biology. New York: Columbia Univ Press, 1979.
24. Grant V: Plant Speciation. New York: Columbia Univ Press, 1971.
25. Gottlieb LD: Biochemical consequences of speciation in plants. In Ayala FJ (ed): Molecular Evolution. Sunderland: Sinauer, 1976.
26. Gottlieb LD, Bennett JP, Taillon P. Am J Bot: in press.
27. Hiesey WM, Nobs MA: Genetic and transplant studies on contrasting species and ecological races of the Achillea millefolium complex. Bot Gaz 131:245, 1970.
28. Clausen J, Keck DD, Hiesey WM: Heredity of geographically and ecologically isolated races. Am Natur 81:114, 1947.
29. Holland RF Jain S: Insular biogeography of vernal pools in the central valley of California. Am Natur 117:24, 1981.

Mechanisms of Speciation, pages 191–218
© 1982 Alan R. Liss, Inc., 150 Fifth Avenue, New York, NY 10011

Speciation in Subterranean Mammals

Eviatar Nevo

INTRODUCTION

Speciation theory has greatly advanced recently by the incorporation of new and improved techniques, multidisciplinary approaches, and theoretical innovations. Reviewed methodologies and theoretical considerations of speciation include the following aspects: molecular [1], cytogenetic [2], ethological [3], mathematical [4–7], and modes of speciation [2,8,9]. The dramatic developments of molecular biology provided powerful tools to assess not only the short-term spatiotemporal changes of the genetic structure of populations, but also the long-term changes of speciation and phyletic evolution. The incorporation of techniques such as electrophoresis and amino-acid sequencing permitted quantification of the number of gene substitutions that occurred during raciation, speciation, and phyletic evolution. Likewise, applications of microcomplement fixation and DNA hybridization techniques contributed substantially to resolution of open evolutionary problems. The most recent and potentially promising aspect of this development relates to genome structure and differentiation through the analysis of sequence DNA organization in animals [10] and plants [11], as well as the emphasis on the relative importance of regulatory rather than structural gene changes in evolution [12]. These new techniques and approaches are expected to answer some basic challenging and unresolved problems of speciation involving the genetics, dynamics, and patterns of the process. Furthermore, attempts are recently being made to develop the predictive rather than the explanatory power of speciation theory: First, by correlating the mode of speciation to the genetic system [8], and second, through the introduction of mechanistic taxonomy of the various forces and genetic mechanisms that underlie the evolution of reproductive isolation [13,14].

Obviously, the incorporation of new techniques and theoretical approaches will substantiate speciation theory. However, the theory cannot be based only on the analysis of detailed and sophisticated genetic, cytogenetic, and ethological differentiation patterns, or mathematical models. It must incorporate into its structure the physical and biotic ecological background in which the process occurs, as advocated by Mayr [15]. Speciation involves the origin and evolution of reproductive isolation between populations, but to be successful the new species must also develop ecological compatibility with its progenitor and other sympatric species [16].

Most importantly, it becomes abundantly clear that species vary in kind in accordance with different adaptive evolutionary strategies and speciation patterns. Reproductive isolation may either follow or precede the genic or chromosomal divergence of gene pools [17]. Likewise, pre- or postmating isolating mechanisms may develop either each alone or successively in this or the reverse order. Speciation may be associated with either minor alterations in structural and/or regulatory genes, as well as with chromosomal and genomic evolution. The extent of genetic differentiation depends on various correlated and interacting factors including the taxon, mode and rate of speciation, population and migration size and structure, breeding and mating system, reproductive and ecological strategies, chromosomal rearrangements and polyploidy, and the ecological selection structure. Finally, speciation rates may vary from very rapid through moderate to very slow ones. To become both predictive and falsifiable, speciation theory must rely on critical multifactorial tests incorporating the genetic system, ecological regime, and historical patterns employing actively speciating populations as its critical tests [18].

My objective is to present a discussion on speciation in subterranean mammals as a multi- and interdisciplinary approach along the lines previously advocated. In particular, I will attempt to substantiate the claim for multidisciplinarity in the study of speciation by exploring the evolutionary patterns of the actively speciating subterranean mole rats of the *Spalax ehrenbergi* superspecies in Israel. This rodent has been intensively studied multidisciplinarily during the last twenty years in terms of speciation and adaptation of evolving species.

SUBTERRANEAN MAMMALS: NATURE, TAXONOMY, AND DISTRIBUTION

Speciation theory should ideally specify the necessary and sufficient determinants of the origin, evolution, and completion of reproductive isolation and ecological compatibility in subterranean mammals, ie mammals that have radiated over space and time into the subterranean ecological zone. These include fossorial species that spend most of their life in sealed burrows and come to the surface only incidentally (Table I). The physical and biotic uniqueness of subterranean ecology provides an excellent evolutionary theater where adaptive convergent and divergent evolution molds populations, species, and communities in similar but geographically distinct environments throughout the world on all levels of organization: genetic, biochemical, physiological, anatomical, behavioral, populational, life historical, and speciational [19]. The objective of this paper is to compare and contrast the evolutionary patterns of speciation of completely subterranean mammals and to suggest the patterns and mechanisms involved in their speciation.

Which mammals radiated underground? Three of the 19 orders of mammals have completely subterranean representatives—the rodents, the insectivores, and the marsupials. Distribution patterns, geographical and paleontological, are listed in the Table. The taxonomic distribution of subterranean mammals is extensive: ten of 132 mammalian families ($\sim$ 7.5%), comprising some 35 of 1004 mammalian genera ($\sim$ 3.5%), and 144 of 4060 mammalian species ($\sim$ 3.5%), based on classical taxonomy [20]. The new discoveries of widespread chromosomal sibling species in mammals may increase severalfold the number of subterranean species. Herbivore speciation underground was often prolific and gave rise to an unusually large number of species, few of which are sympatric. Throughout the world the subterranean environment is divided basically into two trophic modalities—herbivorous [21] and insectivorous—as is suggested by the occurrence of sympatric distributions only between the two, in contrast to the allopatric and parapatric patterns within each. The adaptive convergence and divergence of subterranean mammals in size, structure, and function as well as in patterns and mode of speciation seems to be intimately linked with the physical and biotic structures of the underground environment. Considerations of the origin, geologic history, and environmental dimensions of the subterranean ecological zone are thus essential to an understanding of the adaptive convergent and divergent patterns of speciation of subterranean mammals [19].

EVOLUTIONARY HISTORY AND ORIGINS OF SUBTERRANEAN MAMMALS

Evolutionary convergence and worldwide recurrent radiations of unrelated mammals into the fossorial ecologic zone seem to be related causally to the evolution of open-country biota in the Cenozoic due to mountain formation, extensive sea recessions, and climatic changes [22,23]. The mid- to late Cenozoic cycles of increasing aridity caused the development of nonforest biomes from forests through savannas, to steppes, grasslands, and deserts. New open-country environments provided for rapid evolution and diversification of an open-country fauna. This included large cursorial herbivores (eg horses, camels, and ruminants), small fossorial sedentary herbivores (eg voles, pocket gophers, and mole rats), and insectivores (eg moles). Mammalian adaptive radiation of cursoriality and fossoriality as adaptations to open-country biota emerged as early as the late Paleocene [22,23]. However, impressive open-country adaptations in rodents evolved only in the Miocene. The Plio-Pleistocene evolutionary scenery involved drier climatic conditions followed by a decrease in woodlands and an extension of grasslands, steppes, and deserts. The increasing aridity caused massive extinctions of large ungulates and the survival of small- to medium-sized xeric

TABLE I. Subterranean Mammals: Taxonomical, Geographical, Geological, and Cytogenetic Data

Order	Family and common name		Extant		Karyotype		Geographic distribution	Geological range and major fossil groups
		Genera	Species[a] no.		2n	FN		
Marsupialia	Notoryctidae (Marsupial moles)	Notoryctes	2		20		Australia	Recent
Insectivora	Chrysochloridae (Golden moles)	Amblysomus	5				Africa	Early Miocene to Recent in Africa Prochrysochloris, Kenya
		Chrysochloris	2					
		Cryptochloris	2					
		Eremitalpa	1					
		Chrysospalax	1					
	Talpidae (Moles)	Talpa	2		34–36	62–64	Europe	Late Eocene to Recent in Europe; Late Oligocene- to Recent in N. America Recent in Asia Seven extinct genera
		Mogera	1		32	64	N. America	
		Parascaptor	1		34	64	Asia	
		Scaptochirus	1		34	64		
		Parascalops	1		34	62		
		Scapanus	3		34	64		
		Scalopus	3		34	64		
		Neurotrichus	1		38	72		
Rodentia	Geomyidae (Pocket gophers)	Cratogeomys	10		38–72	68–102	America	Early Miocene to Recent in North America Nine extinct genera
		Geomys	7					
		Heterogeomys	2					
		Macrogeomys	6					
		Orthogeomys	3		36–46	66–86		
		Pappogeomys	2		40–82			
		Thomomys	9					
		Zygogeomys	1					

Family	Genus	No. of species	2n		Distribution	Geological range
Cricetidae (Voles)	Myospalax	5	44–64	80–108	Asia	Oligocene to Recent in Asia
	Ellobius	3				
	Prometeomys	1				
Spalacidae (Mole rats)	Spalax	3	38–62	74–120	Europe (SE) Asia Africa (North)	Upper Pliocene to Recent in Europe Mid. Pleistoc. to Recent in Israel Two extinct genera
Rhizomyidae (Bamboo rats)	Cannomys	1			Asia (SE) Africa (East)	Late Oligocene Europe; Late Miocene to Recent in Asia Pleistoceneto Recent in Africa Five extinct genera
	Rhizomys	3	50[b]			
	Tachyoryctes	14	48[c]			
Octodontidae (Octodonts)	Spalacopus (Coruro)		58	116	America (South)	
Ctenomyidae (Tuco-tucos)	Ctenomys	26[a]	26–68	44–122	America (South)	Pliocene to Recent in S. America Four extinct genera
Bathyergidae (Mole rats)	Bathyergus	2			Africa	Oligocene in Mongolia. Early Miocene and Pleistocene to Recent in Africa
	Cryptomys	15				
	Georychus	1	54			
	Heliophobius	3	60[d]			
	Heterocephalus	1	60			

Total: 35 genera and 144 species

[a] Based on classical systematics.
[b] R sumatrensis [Hsu and Johnson, 1963, in 52].
[c] T splendens and T ruande [Mattey, 1956, 1967, in 52].
[d] H argentocinereus [52].

species. New adaptive radiations of small fossorial rodents followed, including cricetids in the Pliocene and spalacids and ctenomyids primarily in the Pleistocene.

Rodent evolution was determined chiefly by the development of ever-growing gnawing incisors. It seems to have involved three evolutionary levels or grades and at least 11 clades, instead of the three suborders that were formerly recognized [24–27]. Originating from Paleocene rodents, the first grade radiated in the Eocene and involves well-developed gnawing animals, with primitive mammalian jaw musculature. Grade two includes animals that have modified the jaw musculature, and grade three includes animals with very high-crowned, or ever-growing cheek teeth. Burrowing adaptations originated in derivatives of the first grade during the Eocene-Oligocene (eg, Cylindrodontidae, Tsaganomyinae, and bathyergoids). Burrowing trends proceeded independently during the Miocene-Pliocene by parallel and convergent evolution during the second (Mylagulidae and Geomyidae) and third (Spalacidae and Ctenomyidae) evolutionary grades, where extreme hypsodonty was involved not only in grazing in grassland communities, but also in burrowing modes of living. Extremely high crowns, as burrowing adaptations, characterize such unrelated rodents as bathyergoids, spalacids, rhizomyids, and microtine-cricetids, among others.

Burrowing adaptations also evolved independently among the insectivores in the Talpidae (which radiated into subterranean habitats as early as Eocene times) and in Chrysochloridae (which followed the same pattern in early Miocene). Fossorial marsupials display parallel subterranean evolution in the Notoryctidae, which are unknown as fossils—the one genus and two extant species are known only from recent times [20].

THE SUBTERRANEAN ENVIRONMENT

The subterranean ecotope is structurally simple. It is essentially a sealed system, microclimatically relatively stable (ie, permanent and predictable), highly specialized, and presumably low in productivity. According to current theory, the greater buffering and predictability of underground as compared to overground environments, both physically (microclimate) and biotically (food supply, low predation, parapatry), should lead to a greater degree of specialization (ie narrow niches) in the former. The relative constancy and periodicity of physical factors result in high predictability underground [28]. Predictability of temperature, relative humidity (daily or seasonal), darkness and air currents are consistently greater in underground than in overground habitats [29] primarily because of constancy. Other factors (such as oxygen content, range 6–21%, carbon dioxide content, range 0.5–4.8%, and at times soil moisture) display not only a greater range but often a faster shift in values than the respective surface factors [30], ie hypoxia and hypercapnia increase underground primarily after rains or active digging. However, the gaseous environment underground rapidly approaches equilibrium conditions [31].

A VERBAL MODEL OF ADAPTIVE EVOLUTION OF SUBTERRANEAN MAMMALS
The Ecological Background Structure

The subterranean ecotope is relatively simple, stable, specialized, and predictable. It is presumably poor in productivity and carrying capacity, buffered against massive predation, and discontinuous in spatial structure. It is essentially fine grained—a mosaic of unequally distributed sparse resources in both space and time. Consequently, massive convergence ensues in the subterranean ecotope. The major evolutionary determinants of this convergence are specialization, competition, and isolation both within and between species.

The Evolutionary Determinants

Specialization. Subterranean mammals adaptively converge on a variety of levels of organization in specialized patterns. Specialization involves relatively low genetic variation (heterozygosity) [19], stenothermicity and stenohygrobicity, cylindrical body, anatomical reductions (of tail, limbs, eyes, ears, etc.), and hypertrophies (acoustic and tactile sensitivities), food generalism whether herbivorous (rodents) or insectivorous (insectivores and marsupials), ie monomorphic populations of food generalists [32] eating a wide range of foods [33], and 24-h activity patterns. Populations and species are narrow-habitat specialists [19].

Competition. Competition for similar resources may favor convergence [34]. This may be particularly true for the subterranean ecotope, where resource competition (which generally takes the form of both intra- and interspecific aggression) is keen. The results of intraspecific competition are solitariness and high territoriality, generated by aggressive behavior, low-density populations fairly constant and saturated in time near to the carrying capacity of their environments, competitive exclusion as an extreme of MacArthur's broken stick model, keen competition within and between species increasing with genetic relatedness [35], and K-strategy [36]. K-selection in subterranean mammals favors great competitive ability and overall individual fitness reinforced by a relatively slow development, relatively large size, high longevity and low reproductive rate, food generalism, effective predator escape, and low recruitment and mortality rates.

The intense interspecific competition results in largely parapatric distributions, where each species is better adapted to, and more efficient in its preferred microhabitat. The more finely distinct the competing species are genetically and ecologically, the smaller their zone of geographic or habitat overlap and the sharper their abutting ranges. This pattern supports the niche-overlap hypothesis [37]. Consequently, species diversity of subterranean mammals in a given area is low, and the species are distributed in specific microhabitats due to limiting resources. Territoriality within a population and parapatry or allopatry between species increase the harvest of food per individual and species. Moreover, because of the low productivity and low carrying capacity, which reflect the amount of limiting resources available per individual, intraspecific competition is keen. Therefore, populations are relatively small, subdivided, and semi-isolated, particularly in marginal situations such as at the periphery of distribution. The only substantial overlap in subterranean mammals occurs between herbivores and insectivores (Table I), for which food resources are completely separated. This suggests that distribution is dictated primarily by food rather than competition for space.

Isolation. The population structure of subterranean mammals previously discussed displays emigration patterns supporting the principle of isolation by distance [38, 39]. The degree of emigration or dispersal diminishes with distance from a given deme. In its extreme form this is the stepping stone migration pattern [40]. Likewise, gene flow between demes is relatively small due to low vagility exceptions may involve subterranean insectivores [41], colonial Spalacopus [42], and some others. Speciation is often prolific and is greatly facilitated by the population structure and geographic isolation previously discussed. Rapid fixation of chromosomal mutations in relatively small populations often results in the evolution of postmating reproductive isolation, which initiates speciation. Premating isolating mechanisms may be superimposed and increase reproductive isolation gradually and slowly, largely in allopatry, incidental to local adaptive differentiation [43]. Hybrids are mostly inferior to parental types reproductively, and/or in their viabilities and ecological compatibility and are, therefore, selectively eliminated thereby enhancing speciation [44,45].

The homogeneity of the subterranean habitat primarily dictates parapatric or allopatric modes of speciation. In the resulting pattern, vicarious species each adapted to its local microhabitat, replace each other in space due to continuous or discontinuous (eg on mountaintops) climatic variation. Since the evolving species compete for a fine-grained mixture of similar resources, they converge and generalize in their feeding strategies on broad resource curves with extreme interspecific competition leading to competitive exclusion in space. This establishes species identification and divergence [44].

PATTERNS AND MODES OF SPECIATION IN SUBTERRANEAN MAMMALS
Patterns

Chromosomal evolution through major structural rearrangements may lead either to speciation through reproductive isolation caused by postmating barriers, and/or to local adaptive genetic polymorphism through reorganization of new coadapted supergenes [2]. Chromosomal sibling species, based on Robertsonian changes and/or pericentric inversions and reciprocal translocations, are generally widespread in mammals and in unrelated subterranean herbivores (eg Spalax, Thomomys talpoides, Geomys, Ctenomys, etcetera). Notably, classical species of these genera have recently been shown to involve many cryptic, sibling species. Thus the three classical species of Spalax involve over 30 karyotypes (2n = 38–62; FN = 74–120) from Russia, the Balkans, Asia Minor, Israel, and North Africa [46], most of which seem to be distinct species. Similarly, classical Thomomys talpoides may consist of 20 or more chromosomal species (2n = 40–60; FN = 70–82) from the Southern Rockies [47,48]. In the highly polytypic Ctenomys, about eleven karyotypes were described out of the 60 known species in South America (2n = 22–68; FN = 44–122) [49]. For chromosomal numbers in other subterranean mammals see Table I.

Chromosomal speciation characterizing the previously mentioned taxa is not universal in subterranean mammals. In other taxa, chromosomal evolution and speciations seem to be restricted, eg in the subterranean rodents Spalacopus [42], Heterocephalus, and Heliphobius [52], and in subterranean insectivores [41]. Notably, even within the genus Thomomys itself, in North America there are contrasting evolutionary patterns. The valley pocket gopher, T bottae, did not speciate but displays enormous interpopulation non-Robertsonian adaptive karyotypic diversity [50]. In sharp contrast, the largely mountainous northern pocket gopher, T talpoides, speciated profusely and displays enormous interspecific Robertsonian karyotypic diversity that leads to speciation [47].

Isolation, primarily of small, inbred populations due to either geographic, ecologic, or historical bottlenecks, in spatiotemporally changing environments, appear to be substantive for successful speciation in subterranean mammals. Notably, isolation between populations of T bottae during the Pleistocene may have been low, and a considerable amount of gene flow presumably characterizes at present this largely southern species [50]. Conversely, T talpoides complex is known only from the late Pleistocene and became abundant in Wisconsin times around 70,000–10,000 years ago. Its evolution presumably reflects the dynamic changes of physical and biotic conditions during the late Pleistocene in the Rockies, involving vertical movements of ecological zonation from Illinoian to recent times. During that period, changing temperature and humidity regimes resulted in recurrent patterns of isolation and contact among gopher populations. Isolation permitted then, and now still permits among mountaintop ecologically isolated populations, chromosomal experimentation, chiefly by Robertsonian changes and inversions. These may result sometimes, if adaptive superiority is achieved, in successful chromosomal speciation [47,51].

Subterranean rodents may not be a group that necessarily has a different evolutionary potential for chromosome diversity per se than other rodents. The karyotype conservatism described in Heterocephalus and Heliophobius, if further substantiated in more populations and species, may be due to their wide ranging populations in long established ecological conditions [52]. Likewise, wherever gene flow is high, eg in the subterranean rodent Spalacopus [42] and in subterranean insectivores [41], chromosomal evolution and speciation seem to be restricted.

Rapid explosive speciation in fossorial rodents was strongly correlated with a high rate of chromosomal evolution [53], presumably due to a syndrome of frequently subdivided, small, semi-isolated, primarily peripheral populations, patchy distributions, strong territoriality, and historical bottlenecks. These patterns increase inbreeding and interdeme selection, thereby enhancing homozygous chromosome fixation. However, successful speciation will result only if the development of reproductive isolation is complemented by the adaptive and competitive superiority of the descendant over the parental karyotype in the new niche it occupied. Thus, it is the appropriate combination of the potential for chromosomal plasticity, population and migration structures (or isolation), and ecological uniqueness and opportunity that may result in successful speciation. Subterranean rodents often appear to fulfill this syndrome, hence their common, though not universal, successful chromosomal speciation.

Isolating Mechanisms

An important unresolved question in speciation theory is whether the primary factor in initiating speciation is usually a premating isolating mechanism or a postmating one [2]. Obviously, the diversity of organic nature might allow a diversity of solutions to the establishment of reproductive isolation, starting with either pre- or postmating mechanisms or later combining them to substantiate reproductive isolation. Only intensive and extensive studies of evolving species may answer this question which therefore switches from either/or to how much of each.

The nature, origin, and evolution of premating isolating mechanisms have been extensively tested in the four chromosomal species of Spalax ehrenbergi in Israel [43,54,55]. The evidence (see later detailed section on Spalax) suggests that the primary factor in initiating speciation was postmating mechanisms, ie Robertsonian chromosomal changes, which presumably provided partial postmating barriers. These were complemented gradually and slowly by ethological isolation based on olfaction, vocalization, and aggressive behavior. This pattern characterizing Spalax may probably be general in subterranean mammals but this proposition must await substantiation in other species, most of which were not yet studied ethologically.

Stages

Early to late stages of speciation are indicated by the varying extent of gene flow through natural hybridization in both the Spalax ehrenbergi and Thomomys talpoides evolving complexes. In S ehrenbergi, hybrid zone widths decrease progressively with chromosomal differences between parental types [44]. The nature and extent of hybridization reinforces the hypothesis that the four karyotypes are indeed young, closely related sibling species at early stages of evolutionary divergence, as is also indicated by electrophoretic [18] and immunological analyses [56]. Similar patterns have been found in the Thomomys talpoides complex in Colorado where interbreeding is either extensive, limited, or nonexistent [57]. Limited hybridization was also described between Thomomys bottae and T umbrinus [58], as well as between T bottae and T townsendii [59].

Rates

Chromosomal speciation is rapid and datable in the evolving complexes of Spalax and Thomomys. Electrophoretic [18] and immunologic [56] evidence suggest that the S ehrenbergi complex speciated in Israel within the last 250,000 $\pm$ 20,000 years, the last derivative, $2n = 60$, being about 75,000 years old. This is in accordance with the hybrid zone [44] and the fossil [60] evidence. Likewise, electrophoretic evidence suggests that speciation in Thomomys talpoides occurred from late Pleistocene to recent times [61], approximately within the last 250,000 years, and some karyotypes, eg $2n = 44$, may be less than 10,000 years old. While Thomomys fossils appear in early Pleistocene times, the T talpoides complex is known only from late Pleistocene. It became abundant in Wisconsin times (70,000–10,000 years ago) across North America [62].

Parapatric Distribution

Parapatric or allopatric ranges generally characterize the distributions of subterranean mammals. Parapatry has been largely confirmed recently for the different karyotypes of Spalax, Thomomys, and Ctenomys. Exceptions of narrow to broad sympatry have also been described [19, and references therein]. Parapatry, either secondary [16] or primary [2], reflects the combined operation of nearly or totally complete reproductive isolation, as well as competitive exclusion. It prevails in species having similar ecological requirements where barriers in the contact zone are vegetational or climatic rather than geographical. Notably, broad sympatry can exist between subterranean herbivores and insectivores, indicating that food rather than space competition is largely responsible for parapatric distributions in subterranean mammals. Sympatry may also occur between herbivore families (eg Rhizomyidae and Bathyergidae in East Africa, and Cricetidae with both Spalacidae and Rhizomyidae in Asia) as indicated by the distribution maps [19] (Fig. 1). This situation awaits critical clarification.

Competitive Exclusion

Competition theory [63] considers parapatric contacts among rodents [64] and subterranean mammals as due primarily to interspecific competition. Resource competition among subterranean mammals living in structurally simple and indivisible niches is extreme [21,63], and is mediated by aggression resulting in strong territoriality within species and largely parapatric distribution between species [35]. In Colorado gophers, the species with the strictest niche requirements is the superior competitor and is able to displace the other species into less favorable habitats. The order of competitive ability is G bursarius> C castanops> T bottae> T talpoides. The factors involved in competitive interactions are size, territory, aggression, and dispersal [65]. Competitive exclusion has been widely documented and tested [66] in gophers, in which soil depth, texture, and moisture are major critical factors in geographic and habitat distributions. The sympatric pairs of large and dwarf types in gophers [67] and moles [68] may reflect their habitat separation and competitive interaction. Likewise, significant population differences in habitat selection have been found within and between karyotypes of S ehrenbergi in accordance with their respective ecological regimes [69].

Modes

Chromosomal speciation, including allopatric, parapatric, and stasipatric modes, is generally prevalent in mammals [2,8,53], though their relative proportion is unknown. Parapatric or moderately allopatric modes of speciation possibly prevail in subterranean mammals due to the following syndrome, shared at least partly by many: Reproductive strategy: K-strategy (low reproductive rate, equilibrium populations, high competitive ability); Vagility: low; Population size and structure: small, semi-isolated or isolated populations, primarily at the margin of distribution; Niche width and structure: narrow, simple, and highly specialized, speciation accompanied by shift to new niche; Mate selection: positive assortative mating; Breeding system: presumably ranging from outbreeding to inbreeding in central populations, but strongly inbreeding in isolated peripheral populations; Selection: homoselection level high; Chromosome rearrangement: initiate speciation, particularly in rodents by postmating mechanisms and new homozygous fixation; Evolution of reproductive isolation: postmating barriers may be followed by premating barriers after shift to new niche; Genetic changes: speciation may occur through minimal chromosome rearrangements and little changes in structural genes. No information is yet available on sequence DNA organization changes during speciation; Distribution pattern and gene flow: largely parapatric (either primary or secondary) with different niche requirements, high interspecific competition; gene flow largely low or none in the peripherally isolated populations; Speciation rates and hybrid zones: rapid at the initial chromosome stage, but may be gradual and slow at the stage of incidental accumulation of ethological isolation before completion of reproductive isolation and total elimination of hybrid zones.

Only an intensive study of each case of active speciation may unveil its mode of speciation, and no generalization can replace such indispensible studies.

Speciation and Adaptive Radiation

Chromosomal speciation may facilitate adaptive evolution by providing cytogenetic reproductive isolation due to meiotic imbalances, new superior supergenes for ecotypic adaptations, changes in the recombination index and hence the potential for new variants in populations, and/or alteration in gene expression [2,8].

SPECIATION AND ADAPTATION IN THE Spalax ehrenbergi SUPERSPECIES IN ISRAEL: SPECIATION

Subterranean mole rats of the Spalax ehrenbergi complex in Israel involve four, morphologically indistinguishable, homozygous chromosome forms with $2n = 52$, 54, 58, and 60 representing a progressive Robertsonian series (Fig. 1). The four karyotypes inhabit extensive regions which are distributed clinally and parapatrically along a north-south ecological gradient of increasing aridity [70]. Their distribution is correlated with four climatic regimes: $2n = 52$ in cool-humid (Upper Galilee Mts.); $2n = 54$ in cool-dry (Golan Heights); $2n = 58$ in warm-humid (Lower Galilee Mts. and Central Israel); and $2n = 60$ in warm-dry (Samaria, Judea, and Northern Negev) climates. Hence, their distribution is significantly correlated with increasing aridity and temperature both southwards ($2n = 52$, $54 \rightarrow 58 \rightarrow 60$) and eastwards ($2n = 52 \rightarrow 54$) [71]. No extrinsic barriers, except climate and therefore vegetation, separate the ranges of the karyotypes except between $2n = 52$ and 54 (Jordan River and its tributaries), and only partly between $2n = 58$ and 60 (Yarkon River).

Hybrid Zones and Their Evolutionary Significance

The nature and extent of natural hybridization between the four karyotypes is consistent with the hypothesis that the karyotypes are young, closely related, chromosomal sibling species at early stages of evolutionary divergence. Hybrid zones occur along the contacts between all four karyotypes [44]. They decrease progressively in width northward from 2.8 km (between $2n = 58$–60), to 0.725 km (between $2n = 54$–58) to 0.320 km (between $2n = 52$–58). Thus, they regress negatively and linearly on the chromosomal differences between the respective parental types ($r = -0.93$; $P < 0.001$). The larger the chromosomal difference between contiguous karyotypes, the narrower the hybrid zone [44, Fig. 3]. The proportion of hybrids and parental types differs between the three hybrid zones analyzed. Hardy-Weinberg proportions are realized in the $2n = 58$–60 hybrid zone (between the homozygous parental types and the heterozygote $2n = 59$), but not in the $2n = 54$–58 hybrid zone (involving heterozygotes $2n = 55$–57), and in the $2n = 52$–58 hybrid zone (involving heterozygotes ($2n = 53$, 55–57, and homozygote 54). It therefore appears that substantial difference exists be-

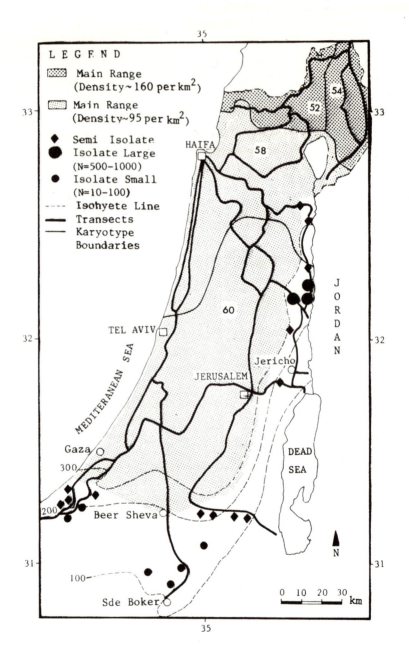

tween the first, southern hybrid zone and the two northern ones. Field evidence based on litter analysis suggests that the 2n = 59 and 2n = 55 heterozygotes are at least partly fertile. Heterozygote 2n = 59 can successfully backcross with both parental types, and the cross 2n = 55 × 55 can produce the homozygote 2n = 54 and the double heterozygote 2n = 56.

The narrowness of the hybrid zones and their progressive decrease in width, as well as the lower than expected proportion of hybrids in the 2n = 54–58 and in the 2n = 52–58 hybrid zones, suggest that selection is currently operating against hybrids. Dispersal of hybrids into parental territories is restricted presumably by a combination of cytogenetic, ethologic, physiologic, and ecologic incompatibilities. Though hybrids are at least partly fertile, their overall fitness appears lower than that of the parental types. Reproductive isolation increases progressively from the southern to the northern hybrid zones, resulting in the perfection of species identification. Therefore, the four karyotypes of S ehrenbergi represent an active process of speciation where a progressive terminalization of species formation is under way. Further evidence presented below indicates that ethological isolation has developed among the presumably older karyotypes.

Ethological Reproductive Isolation in Spalax ehrenbergi: Nature, Origin, and Evolution: Mate Selection

The progressive increase of reproductive isolation evident among the four karyotypes through natural hybridization provides an optimal model for studying the nature, origin, and evolution of premating isolating mechanisms which have been discovered in preliminary laboratory experiments where mate selection was indicated in pooled data involving all four karyotypes [54]. In mating experiments, aggression was significantly higher and copulations lower, in heterogametic compared with homogametic matings $(0.01 > P > 0.001)$. Additional mating experiments revealed positive assortative matings when estrous females of two parapatric karyotypes (2n = 52 and 58) significantly preferred homochromosomal mates $(P < 0.001)$ [55]. Finally, ethological isolation was extensively tested among the four karyotypes through 1095 two-choice mate selection laboratory tests of estral females, involving 680 mole rats from 14 populations [43].

The results suggested that:

(a) Mate selection was similar among females from central and near-hybrid zone populations. This indicates that ethological isolation originated primarily as an incidental by-product of adaptive differentiation, rather than through reinforcement, supporting the allopatric theory of the origin of premating isolating mechanisms [16] in Spalax.

Fig. 1.   Distribution and population structure of the four chromosomal species of the Spalax ehrenbergi complex in Israel.

(b) While mate selection operates among 2n= 52, 54, and 58, the last derivative of speciation, 2n= 60, presumably 75,000 years old [18], displays no or low mate selection suggesting that postmating isolating mechanisms preceded premating ones in Spalax, and incidental accumulation of effective ethological isolation appears to be a very slow process requiring at least tens of thousands of years and can continue after parapatric distributions are established.

(c) Hybrid zones between evolving species of Spalax may be transitory, yet persist over long periods of time, decreasing progressively through the gradual accumulation of ethological isolation. Mate selection in Spalax was shown to be mediated through olfaction, vocalization, and aggression.

Olfaction. Olfactory discrimination was found when females of the parapatric karyotypes 2n= 52 and 58 were tested in the laboratory for male odor discrimination; the source of the odor was either cage litter or urine. Estrous females of both karyotypes significantly preferred homochromosomal odor, whereas diestrous females showed no discrimination [72]. Ongoing gas-chromatograph analysis conducted on urines of the four karyotypes indicates differential chromatogram patterns among the karyotypes, some of which may involve species-specific pheromones which presumably serve as reproductive isolating factors (Nevo et al., unpublished).

Vocalization. Vocal communication plays an important role in the subterranean life of the mole rat, especially during aggressive encounters and in particular during mating [54]. The vocal repertoire of Spalax consists of at least six call types, each occurring in a different behavioral context [73]. One of the call types seems to serve as a "mating call" between the sexes, and is emitted by both. Presumably each karyotype has its species-specific "mating dialect." We are currently testing the effectiveness of the mating dialect as an acoustic isolating mechanism through laboratory female discrimination tests. Our preliminary experiments indicate that estrous females of the two parapatric karyotypes 2n= 52 and 58 significantly discriminate for the mating dialect of their homochromosomal males (Nevo et al., unpublished). Apparently, call differentiation provides effective acoustic isolation among karyotypes.

Aggression. Aggressive behavior is common and adaptive within many animal species chiefly in spacing out individuals, but its evolutionary significance between species is known primarily as an ecological factor rather than as a determinant in speciation. Mole rats are highly territorial and display aggressive behavior both within and between species. We therefore tested their interspecific aggression as a factor in speciation [35]. Preliminary laboratory experiments testing both intra- and interspecific aggression were conducted on 48 adult animals from 10 populations comprising three karyotypes, 2n= 52, 58, and 60. Twelve agonistic, motivational conflict, and territorial behavioral variables were recorded during 72 combats involving homo- and heterogametic encounters between opponents. The results indicated that (a) aggression patterns, involving agonistic, conflict, and territorial variables are higher in heterogametic than in homogametic encounters, and (b) aggression is higher between contiguous karyotypes (2n = 58–60, and 2n = 52–58) than between noncontiguous ones (2n = 52–60). Both (a) and (b) suggest that high interspecific aggression appears to be adaptively selected at final stages of speciation in mole rats as a premating isolating mechanism which substantiates species identification and establishes parapatric distributions between the evolving species. We currently test the validity of these conclusions on a much larger scale.

ADAPTATION

Adaptive Strategies of S ehrenbergi Complex to Subterranean Life and to Climatic Variation

The four chromosomal species of S ehrenbergi complex represent adaptive systems to the subterranean ecotope generally, and to the different climates each occupies, specifically. The evidence supporting this hypothesis involves morphological-anatomical, genetic, physiological, ecological, and behavioral adaptations.

Morphological adaptive strategies. Morphofossorial adaptations involve structural reductions (cylindrical body, short limbs, absence of external tail, ears and eyes, or total blindness) and structural developments (large flat head, cornified snout, ever-growing long incisors, developed pectoral girdle, lip modifications all optimizing teeth burrowing and bulldozing soil with head; olfactory, vocal, and tactile senses developed for efficient spatial orientation in burrows; pineal gland specifically adapted to darkness, etc. [19, pp 277–279 and references therein]. Though the four karyotypes are morphologically indistinguishable, their size decreases clinally and significantly in response to geographic variation in heat load from large populations in the Golan (156 gm in 2n = 54) to dwarf populations in the northern Negev (90 gm in 2n = 60) thus displaying positive correlation of weight with latitude (Bergmann's rule).

Genetic adaptive strategies. *Chromosomal.* Variation in both chromosome number and morphology may be adaptive as previously summarized [2]. In S ehrenbergi in Israel, diploid number increases progressively southwards (2n = 52 → 58 → 60) and eastward (2n = 52 → 54) paralleled by adaptive physiological and behavioral trends. This chromosomal variation appears to be adaptive by presumably providing partial postmating barriers, protecting coadapted supergenes, and increasing the recombination index towards the unpredictable steppic and desert frontiers.

Genic. Living in the relatively stable, simple, and specific underground environment, subterranean mammals are habitat specialists. Since they live in a relatively more predictable environment they are expected, following the niche-width variation hypothesis, to exhibit generally low levels of genetic diversity

than overground species of similar size which live apparently under more fluctuating and complex environments. This prediction for homoselection has been
largely confirmed in fossorial mammals and other subterranean taxa [19]. In the
four karyotypes of S ehrenbergi mean genetic indices based on 25 loci analyzed
in 882 individuals from 21 populations [18,74], for 2n = 52, 54, 58, and 60,
respectively, are: A (mean number of alleles per locus) = mean 1.25 (1.16,
1.28, 1.20, 1.36); P (mean proportion of polymorphic loci per karyotype, criterion 1%) = mean 0.20 (0.12, 0.24, 0.16, 0.28); and H (mean proportion of
loci heterozygous per individual) = mean 0.039 (0.035, 0.016, 0.037, 0.069).
Note that A, P, and H increase southwards towards the Negev desert, and A and
P increase eastwards towards the Golan steppe. This pattern confirms the environmental theory of genetic variation not only generally in Spalax, but also
for each chromosomal species specifically since genic diversity increases (like
2n) towards the ecologically more fluctuating xeric habitats.

Preliminary analysis of the major histocompatibility complex (MHC) in the
four karyotypes of Spalax (Bodmer and Nevo, unpublished) suggest that putative
MHC antigenic diversity may be higher than allozymic diversity, tentative MHC
differentiation among karyotypes appears to be largely quantitative, and antigenic
diversity appears to be higher in 2n = 52, 54 than in 2n = 58 and 60. These
tentative results may suggest that antigenic selection of the MHC recognition
and defense system in Spalax may be related to climatic variation (higher diversity
in more humid regions) in a recently differentiating complex.

It is noteworthy that the Wright Fixation Indices, F, for 2n = 52, 54, 58, and
60, based on eight polymorphic loci in the entire complex, are: 0.038, 0.354,
−0.179, and −0.076, respectively. Thus the only indication for either inbreeding or the Wahlund effect is in 2n = 54.

Genetic identity, I(75) between the four karyotypes is remarkably high, mean
0.966, range 0.931–0.988. Thus, the S ehrenbergi complex represents a notable
case of speciation with relatively very few allelic changes, negating the idea that
genetic revolution in many genes is a necessary corrolary of speciation. Genic
differentiation across karyotypes is quantitative rather than qualitative. Trans-
ferrin, Tf, is the only locus, out of the 25 analyzed, which displays almost
complete alternative fixation, 2n = 54 is nearly fixed for Tf^a, whereas 2n = 52,
58, and 60 are either fixed or nearly so for Tf^b. Thus, at least at the level of
structural genes, the essence of speciation, ie the development of pre- and postmating reproductive isolation per se may occur with little genomic changes,
either genic or chromosomal, pending the analysis of regulatory and/or sequence
DNA organization.

Physiologic adaptive strategies. The unique subterranean ecotope is characterized by poor periodic food supply, atmospheres largely saturated with moisture, and extreme hypoxia and hypercapnia. Consequently, most subterranean
rodents share a linkage of low basic metabolic rate (BMR), high thermal con-

ductances, high ranges of thermoneutrality, low body temperatures (35°–37°C), relatively poor thermoregulation, and respiratory adaptations. This syndrome which characterizes also Spalax, is adaptive since it may reduce heat storage and water exchange, minimize energy expenditure, and meet the severe stresses caused by extreme hypoxia and hypercapnia [19, pp. 289–292, and references therein]. Notably, the BMR of the three analyzed karyotypes of S. ehrenbergi varies in accordance with climatic variation both within and between species [51]. The BMR measured/BMR predicted after Kleiber × 100, for 2n= 52, 54, 58, and 60 are: 91, 82, and 60, respectively, decreasing progressively and significantly towards the desert. This physiological gradient suggests that the BMR is adapted not only generally to the subterranean ecotope, but also specifically to the different microhabitats each karyotype occupies, in accordance with the southward increase in the aridity index. Likewise, adaptive thermoregulatory patterns was recently found in the S ehrenbergi complex [81].

Ecologic adaptive strategies. *Home range, territoriality, and population structure: density and census.* The home range of subterranean mammals, including Spalax, are generally also their exclusive and defended life-long territories (except for short breeding periods), providing adequate energy supply. Territorial size varies with many physical and biotic factors [19]. Average territory size of S. ehrenbergi in Israel is 341 m², range 100–769 m², in accordance with resource availability. Average number of individuals per 1000 m², based on census of 12 dense populations, is three, range 0.8–6.8, correlated (*P* < 0.05) with plant cover. Rough estimates of average population densities per 1000 m², recorded along 1000 km across the ranges of the four karyotypes in Israel decreases southwards, presumably adjusted to productivity, as follows: 2n= 52, 54, 58, and 60: 0.24, 0.29, 0.14, and 0.10, respectively. Rough karyotype census based on their range sizes (770, 970, 3340, and 10,430 km² respectively) is: 181,700; 284,500; 458,500 and 1,061,600, respectively, totaling about 2 million individuals (which is certainly an underestimate) across the 15,828 km² inhabited by Spalax in Israel (Nevo et al., unpublished) (Fig. 1).

Age and sex structure. The equilibrium populations of subterranean mammals contain a relatively high proportion of breeding adults. In S ehrenbergi in Israel, of 386 animals caught in 1976–1978, 19% were juveniles, 40% 1–2 years old, and 41% 2–3 years old. Adult sex ratio in Spalax is unbalanced in favor of females, as is common in subterranean rodents, 39% males of 1091 sexed individuals, possibly due to high territorial male aggression.

Population subdivision. Population size, structure, and distribution affect the dynamics of both phyletic evolution and speciation. Theory predicts faster evolutionary rates in a subdivided population with varying sizes and degrees of isolation of local demes than in a homogeneous panmictic population of comparable total size, owing to local differential selective pressures, gene flows, and random drift [38]. Populations of subterranean herbivores [19, and references therein] are often assumed to conform to the "island model" type of distribution [39]. This pattern, however, seems to be rare in central populations of S ehrenbergi in Israel, but is realized in the periphery of distribution [82]. Across the aforementioned 1000 km studied in the four karyotypes, population continuity prevails in all karyotypes, 2n = 52, 54, 58, and 60 as follows: 89%, 100%, 94%, 97% in their central ranges but only 28% for 2n = 60 in its southern and eastern peripheries facing the Negev and Judean Deserts, respectively, where small, semi-isolated and isolated populations occur, separated by several to tens of kilometers and involving tens to several hundred individuals each (Figure 1).

Life history pattern, population parameters, and dispersal rates. The relatively predictable subterranean ecotope presumably leads to convergent K-selection in subterranean populations resulting in "equilibrium species" in accord with the carrying capacity of the environment. These are achieved by maximizing breeding age and duration of breeding season, and by minimizing offspring number, fraction of breeding population, immigration and emigration rates, and mortality and predation rates [19, and references therein]. Conforming with this pattern, S ehrenbergi starts breeding during its second year, has only one breeding season lasting 4–5 months, and has one litter per year averaging three young, range 1–6 [76]. Notably, the subterranean ecotope is relatively sheltered from predators. Mortality is high primarily during the short subadult dispersal and during parasite and disease plagues and extreme weather conditions. Consequently, the annual population recruitment is low. Finally, vagility in Spalax is low as in other subterranean mammals (eg in pocket gophers migration ranges between tens to several hundred meters [19]. No precise measurements of vagility are available for Israeli Spalax.

Differential survivorship of karyotypes in a standardized laboratory environment. Evidence for differential viability of the four karyotypes was assembled in the Institute of Evolution, University of Haifa, where mole rats caught in the field during 1974–1980 were housed in a standardized environment (20°C; 70% R.H., same vegetable food and caging conditions). From a total of 711 animals only 183 survived in November 1980, belonging to 2n = 54, 52, 58, and 60 as follows: 48.4%, 26.2%, 19.9%, and 17.7%, respectively. Differential survival rate of the order 2n = 54> 52> 58> 60, observed in 1974–1975 was repeated exactly in three out of five successive years (Binomial probability test, $P <$ 0.001, to achieve this order by chance). The mean duration of time (in years) of animals belonging to the different karyotypes which survived in the lab, caught in 1974–1975 was 3.15, 2.20, 1.91, and 1.14 for the 2n = 54, 52, 58, and 60, respectively. This differential survival rate is consistent with the hypothesis that climate plays a decisive differentiating factor. The "climatic niche" of the laboratory resembled more the cool-dry one of 2n = 54 than the cool-humid (2n = 52) or warm-humid (2n = 58), and warm-dry (2n = 60) climates. The latter two karyotypes, but chiefly 2n = 60, experienced consistently greater mortality rates from a variety of unidentified causes. Obviously, the karyotypes vary in their viabilities in the experimental standardized environment largely in accordance with their climatic origins [83].

Behavioral adaptive strategies. *Feeding strategies.* Optimal feeding theory assumes that natural selection maximizes fitness by optimizing net energy gained per unit feeding time [33]. Theory predicts, inter alia, that at low food abundances food generalists are favored over food specialists, increasing food abundance leads to greater food specialization, fitness can be maximized by an optimal foraging space, and optimal feeding period is selected by the environment. These predictions are supported in Spalax and other subterranean mammals [19]. Spalax is a generalized feeder on a wide range and variance of food types, but in cultivated fields food storage reflects specialization related to local crops, eg potatoes and carrots. Spalax is active day and night and its activity is linked with feeding periods. Finally, the exclusive feeding territory of Spalax, its size, structure, patch selection, and foraging path appear to be associated with optimal foraging space [76,77, and Nevo unpublished].

Activity patterns. The hypothesis that endogenous rythmicity is synchronized with, and selected by the environment as an adaptive strategy increasing fitness [78] was tested in the S ehrenbergi complex. Activity patterns were tested in the laboratory in the four karyotypes (2n = 52, 54, 58, 60) which inhabit humid (2n = 52, 58) and dry (2n = 54, 60), as well as cool (2n = 52, 54) and warm (2n = 58, 60) environments, respectively [77]. Experimentals included 98 adult mole rats representing nine populations covering the entire species ranges in Israel. The test apparatus simulated a territorial runway of a solitary occupant and each test lasted 24 hr. The results indicate that the pattern of activity is multiphasic, but activity in all nine populations is remarkably higher during the day than during the night, and differences occur among karyotypes in both levels and pattern of activity.

First, the "humid karyotypes" are more active than the "dry karyotypes," and second, it appears that the activity of the "cool karyotypes" displays a pattern of a smaller number of rest periods as compared with the "warm karyotypes." Lower activity in the "dry karyotypes" may minimize energy and water expenditure in their xeric and low-productive environments. Likewise, the greater number of rest periods characterizing the "warm karyotypes" may minimize overheating and thermal death in their warm regimes. Both results appear to be climatically adaptive and support an optimal activity theory assuming that natural selection maximizes fitness by optimizing net energy gain per unit activity.

Habitat selection. The hypothesis that successful chromosomal speciation involves ecological correlates and is largely adaptive can be also tested by habitat selection of the karyotypes. Differential habitat choice of the karyotypes, kept 1–3 years under standardized laboratory conditions, may suggest that habitat selection is intrinsic. Furthermore, if habitat selection is in accordance with their climatic origins then it suggests that each karyotype is competitively superior in its microclimatic niche space. We have tested the four karyotypes in the laboratory to determine their habitat preference [69]. The testing apparatus simulated four climatic regimes based on temperature and humidity combinations corresponding to the climatic origins of the four karyotypes: cool-humid (2n = 52), cool-dry (2n = 54), warm-humid (2n = 58), and warm-dry (2n = 60). The tests involved 175 adults of all karyotypes representing ten populations. Out of the 139 analyzed animals 88% selected the warm cages and only 12% selected the cool ones. The four karyotypes progressively preferred the warm-dry cage in the following proportions: 53%, 59%, 60%, and 72% for 2n = 58, 52, 54, and 60, respectively, largely in accordance with their increasingly arid climatic origins. Even larger differences were found in populations within karyotypes in accordance with the local climatic variation within a karyotype range.

These results indicate that the chromosomal species and individual populations select the climatic niche in accordance with the climatic conditions of their geographic localities, as was also circumstantially shown earlier in the differential survivorship analysis. The humidity index appears to be the prime differentiator of habitat selection and may have been a substantial factor in the ecological speciation and parapatric distributions of the S ehrenbergi complex in Israel. These results are in accordance with the optimal habitat selection theory [79]. Habitat selection is more likely to be a successful strategy in sedentary and territorial species like Spalax.

Each karyotype may be viewed as an optimal microhabitat specialist which preferentially exploits best the niche space in which it is presumably more fit. If partial reproductive isolation previously developed cytogenetically, each karyotype will be restricted and flourish in its optimal habitat. It is the overall success of the extreme specialist plus the establishment of reproductive isolation that lead to successful speciation.

EVOLUTIONARY OVERVIEW OF Spalax ehrenbergi COMPLEX IN ISRAEL

Ecological speciation of the S ehrenbergi complex in Israel is an actively ongoing process representing progressive stages of late chromosomal speciation. The Plio-Pleistocene climatic scenery of the Eastern Mediterranean is characterized by increasingly drying climatic conditions followed by decreased woodlands and extension of grasslands, steppes, and deserts. The massive extinction in Israel of mesic forest species, such as Apodemus, and the progressive colonization of xeric species, such as Spalax, is highlighted by the Upper Pleistocene fossil record in which Spalax increasingly abounds [60].

Spalacids adaptively radiated into the developing open-country biota through rapid chromosomal evolution with only little allelic differentiation. Major chromosome rearrangements involved whole-arm Robertsonian changes and pericentric inversions. Stepwise Robertsonian dissociations from northern lower (2n = 52, 54) to southern higher (2n = 58, 60) diploid numbers presumably lead to speciation through the initiation of reproductive isolation caused by progressively increasing hybrid inferiority and extensive homozygous fixations. Pericentric inversions presumably contributed to local adaptive genetic polymorphism through reorganization of new coadapted supergenes. The primary postmating mechanisms were complemented progressively by premating ethological isolation that evolved gradually and slowly as an incidental by-product of adaptive differentiation across the ranges of the parental and descendant homozygous karyotypes. Consequently, hybrid zones between evolving species persist over long periods of time but decrease progressively through the gradual accumulation of ethological isolation.

The mode of speciation of the S ehrenbergi complex was presumably parapatric or marginally allopatric in small, semi-isolated, highly inbred populations, mainly at the extreme periphery of the parental species range where marginal conditions prevented its range extension. These unique peripheral populations, isolated several kilometers to tens of kilometers from the main bulk of continuous populations, enabled chromosomal experimentation and budding off and fixation of new homozygous karyotypes. Presumably, severe interdeme selection based on karyotypic variations played a major evolutionary role at this stage. However, successful speciation and colonization of increasingly arid zones depended on whether the new karyotype was competitively superior in overall fitness to its progenitor in the new more xeric microhabitat to which its parents were poorly adapted previously.

The higher chromosome species along the north-south cline appear to be better adapted to the increasingly more arid regions, morphologically, physiologically, ecologically, genetically, and behaviorally representing correlated adaptive systems. The combination of extreme and superior ecological specialization of the new karyotype plus the establishment of reproductive isolation led eventually to successful speciation. The extreme interspecific resource competition between the parental and derivative species, which share similar ecological requirements, resulted in parapatric distributions, mediated by aggressive interactions.

Chromosomal speciation and subsequent adaptive radiation in Spalax appear to be primarily due to ecological causes—the invasion of ever-increasing arid environments during the last 250,000 years of the Upper Pleistocene. The last derivative of speciation, 2n = 60, presumably 75,000 years old, colonized North Africa, but increasing aridity during the early Holocene separated the Egyptian and Israeli populations approximately 10,000 years ago, as is evident by the Sinai disjunction. No chromosomal change has occurred in 2n = 60 since its geographic isolation, reflecting distinct karyotype stability [80].

CONCLUSIONS AND PROSPECTS

The ecological theater of open-country biotas that opened up progressively in the Cenozoic due to expanding terrestrialism and increasing aridity set the stage for a rapid evolutionary play of recurrent adaptive radiation of unrelated mammals on all continents into the subterranean ecotope. The latter is relatively simple, stable, specialized, low in productivity, predictable, and discontinuous. Its major evolutionary determinants are specialization, competition, and isolation. This ecotope involves essentially two nonoverlapping niches—the herbivorous, colonized by rodents, and the insectivorous, colonized by the marsupials and the insectivores. All subterranean mammals share convergent adaptations to their common unique ecotope and divergent adaptations to their separated niches of herbivory and insectivory and to their different phylogenies [19].

Patterns of speciation in subterranean mammals are largely determined by their unique subterranean ecotope and its two niches which determine divergent population structure, vagility, and consequently speciation patterns. Adaptive divergence patterns in subterranean herbivores and insectivores, respectively, involve; large vs. small body size, low vs. higher BMR, high vs. low thermal conductances, small vs. large territories, largely low vs. high gene flow, and finally, largely high vs. low chromosomal speciation and taxonomic diversity [19]. Even within subterranean rodents at least two contrasting evolutionary patterns prevail—prolific speciation vs. prolific and extensive adaptation, largely in accordance with degree of isolation, level of gene flow, and the dynamics of spatiotemporally changing environments.

Modes of speciation in subterranean mammals involve parapatric, allopatric, and stasipatric, possibly in this order. Genetic revolution involving change in many structural genes may not be required. Single major chromosome rearrangements may initiate rapid development of reproductive isolation through postmating barriers in small, isolated, and inbred populations. However, successful speciation depends on the adaptive and competitive superiority of the new homozygous karyotype in the new niche it occupied. Hybrid zones may persist a long time and their progressive decrease and elimination depends on the completion of reproductive isolation which may be slow and gradual through accumulation of incidental ethological isolation in allopatry.

Speciation theory is still largely descriptive and explanatory despite recent dramatic advances in molecular biology, cytogenetics, ethological metholologies, and preliminary efforts at correlating genetic systems and speciation, as well as establishing a mechanistic taxonomy of speciation. To become more predictive and falsifiable, speciation theory must involve extensive survey of unrelated taxa in different life zones and ecological regimes, involving multidisciplinary approaches. Of paramount importance are critical multifactorial tests incorporating different taxonomic and genetic systems, population and migration structures and ecologies, as well as historical patterns in actively speciating populations. The subterranean ecotope provides a fruitful field for investigating cardinal unresolved problems of speciation.

ACKNOWLEDGMENTS

Some sections appearing in this article as well as Table I were reproduced from my review article "Adaptive Convergence and Divergence of Subterranean Mammals," Annual Review of Ecology and Systematics, 10:269–308 (1979), with permission from Annual Reviews, Inc.

My deep gratitude is extended to M. Avrahami and G. Heth, long-time collaborators in studying Spalax, and to many colleagues who participated in the endless but challenging adventure of deciphering the evolutionary patterns of adaptation and speciation of Spalax. Special thanks go to A. Beiles and G. Heth for their stimulating discussions concerning Spalax and to D. Adler for reading the manuscript.

This study was partly supported by a grant from the United States-Israel Binational Science Foundation, BSF, Jerusalem, Israel.

REFERENCES

1. Ayala F: Genetic differentiation during the speciation process. Evol Biol 8:1, 1975.
2. White MJD: Modes of Speciation. San Francisco: Freeman, 1978, p 455.
3. Littlejohn MJ: Reproductive isolation—a critical review. In Atchley D, Woodruff DS (eds): Essays on Evolution and Speciation in Honor of MJD White. Cambridge: Cambridge University Press, 1980, p 64.
4. Maynard-Smith J: Sympatric speciation. Am Natur 100:637, 1966.
5. Endler J: Geographic Variation, Speciation and Clines. Monographs in Population Biology, No. 10. Princeton: Princeton University Press, 1977, p 246.
6. Nei M: Mathematical models of speciation and genetic distance. In Karlin S, Nevo E (eds): Population Genetics and Ecology. New York: Academic Press, 1976, p 723.
7. Felsenstein J: Skepticism towards santa rosalia, or why are there so few kinds of animals? Evolution 35:124, 1981.
8. Bush GL: Modes of animal speciation. Annu Rev Ecol Syst 6:339, 1975.
9. Bush GL: Stasipatric speciation and rapid evolution in animals. In Atchley D, Woodruff DS (eds): Essays on Evolution and Speciation in Honor of MJD White. Cambridge: Cambridge University Press, 1980, p 17.
10. Dover G, Strachan T, Brown SDM. The evolution of genomes in closely related species. In Scudder GGE, Reveal JL (eds) Evolution Today, Proc 2nd Congr Syst Evol Biol. Pittsburgh: Carnegie-Mellon University, 1981, p 337.
11. Flavell RB, Rimpau J, Smith DM, Odell M, Bedbrook JR: The evolution of plant genome structure. In Leaver C (ed): Genome Organization and Expression in Plants. New York: Plenum Press, 1980, p 35.
12. Wilson AC: Evolutionary importance of gene regulation. Stadler Symp 7:117, 1975.

13. Templeton AR: Modes of speciation and inferences based on genetic distances. Evolution 19:115, 1980.
14. Templeton AR: The theory of speciation via the founder principle. Genetics 94:1011, 1980.
15. Mayr E: Ecological factors in speciation. Evolution 1:263, 1947.
16. Mayr E: Population, Species, and Evolution. Cambridge: Harvard University Press, 1970, p 453.
17. Dobzhansky TH: Species of *Drosophila*. Science 177:664, 1972.
18. Nevo E, Cleve H: Genetic differentiation during speciation. Nature 275:125, 1978.
19. Nevo E: Adaptive convergence and divergence of subterranean mammals. Annu Rev Ecol Syst 10:269, 1979.
20. Anderson S, Jones JK Jr: Recent Mammals of the World: A Synopsis of Families. New York: The Ronald Press, 1967, p 453.
21. Pearson OP: Biology of the subterranean rodents Ctenomys in Peru. Mem Mus Hist Nat Javier Prado 9:1, 1960.
22. Webb SD: A history of savanna vertebrates in the new world. Part I. Annu Rev Ecol Syst 8:355, 1977.
23. Webb SD: A history of savanna vertebrates in the new world. Part II. Annu Rev Ecol Syst 9:393, 1978.
24. Romer AS: Vertebrate Paleontology. Chicago: University of Chicago Press, 3rd Ed., 1966, p. 468.
25. Simpson GG: The nature and origin in superspecific taxa. Cold Spring Harbor Symp Quant Biol 24:255, 1959.
26. Wood AE: Are there rodent suborders? Syst Zool 7:169, 1959.
27. Wood AE: Grades and clades among rodents. Evolution 19:115, 1965.
28. Colwell RK: Predictability, constancy and contingency of periodic phenomena. Ecology 55:1148, 1974.
29. McNab BK: The metabolism of fossorial rodents: A study of convergence. Ecology 47:712, 1966.
30. Arieli R, Ar A, Shkolnik A: Metabolic responses of a fossorial rodent (Spalax ehrenbergi) to simulated burrow conditions. Physiol Zool 50:61, 1977.
31. Withers PC: Models of diffusion mediated gas exchange in animals burrows. Am Natur 112:1101, 1978.
32. Roughgarden J: Niche width: Biogeographic patterns among Anolis lizard population. Am Natur 108:429, 1974.
33. Schoener TW: Theory of feeding strategies. Annu Rev Ecol Syst 2:369, 1971.
34. MacArthur RH, Wilson EO: The Theory of Island Biogeography. Princeton: Princeton University Press, 1967, p 203.
35. Nevo E, Naftali G, Guttman R: Aggression patterns and speciation. Proc. Natl Acad Sci (USA) 72:3250, 1975.
36. Pianka ER: On r- and k-selection. Am Natur 104:592, 1970.
37. Pianka ER: Niche overlap and diffuse competition. Proc Natl Acad Sci (USA) 71:2141, 1974.
38. Wright S: Evolution in Mendelian populations. Genetics 16:97, 1931.
39. Wright S: Isolation by distance. Genetics 28:114, 1943.

40. Karlin S: Population subdivision and selection migration interaction. In Karlin S, Nevo E (eds): Population Genetics and Ecology. New York: Academic Press, 1976, p. 617.

41. Yates TL, Schmidley DJ: Karyotype of the eastern mole (Scalopus aquaticus) with comments on the karyology of the family Talpidae: J Mammal 30:36, 1975.

42. Reig OA, Spotorno AO, Fernandez DR: A preliminary survey of chromosomes in populations of the Chilean burrowing octodont rodent Spalacopus cyanus Molina (Caviomorpha, Octodontidae): Biol J Linn Soc London 4:29, 1972.

43. Heth G, Nevo E: Origin and evolution of ethological isolation in subterranean mole rats: Evolution 35:254, 1981.

44. Nevo E, Bar-El H: Hybridization and speciation in fossorial mole rats. Evolution 30:831, 1976.

45. Patton JL, Dingman RE: Chromosome studies of pocket gophers, genus Thomomys. I. The specific status of Thomomys umbrinus (Richardson) in Arizona. J Mammal 49:1, 1968.

46. Savic' IR, Soldatovic' B: Contribution to the knowledge of the genus Spalax (Microspalax) karyotype from Asia Minor. Arh Biol Nauka, Beograd 31:1P, 1979.

47. Thaeler CS, Jr: Karyotypes of sixteen populations of the Thomomys talpoides complex of pocket gophers (Rodentia: Geomyidae). Chromosoma 25:172, 1968.

48. Thaeler CS Jr: Chromosome polymorphism in Thomomys talpoides agrestis Merriam (Rodentia: Geomyidae) Southwest. Nature 21:105, 1976.

49. Reig OA, Kiblisky P: Chromsome multiformity in the genus Ctenomys (Rodentia, Octodontidae). Chromosoma 28:211, 1969.

50. Patton JL, Yang SY: Genetic variation in Thomomys bottae pocket gophers: Macrogeographic patterns. Evolution 31:697, 1977.

51. Nevo E, Shkolnik A: Adaptive metabolic variation of chromosome forms in mole rats Spalax. Experientia 30:724, 1974.

52. George W: Conservatism in the karyotypes of two African mole rats (Rodentia, Bathyergidae). Z Saugetierk 44:278, 1979.

53. Bush GL, Case SM, Wilson AC, Patton JL: Rapid speciation and chromosomal evolution in mammals. Proc Natl Acad Sci (USA) 74:3942, 1977.

54. Nevo E: Mole rat Spalax ehrenbergi: Mating behavior and its evolutionary significance. Science 163:484, 1969.

55. Nevo E, Heth G: Assortative mating between chromosome forms of the mole rat Spalax ehrenbergi. Experientia 32:1509, 1976.

56. Nevo E, Sarich V: Immunology and evolution in the mole rat Spalax. Isr J Zool 23:210, 1974.

57. Thaeler CS Jr: Four contacts between the ranges of different chromosome forms of the Thomomys talpoides complex (Rodentia: Geomyidae). Syst Zool 23:343, 1974.

58. Patton JL: An analysis of natural hybridization between the pocket gophers Thomomys bottae and Thomomys umbrinus in Arizona. J Mamm 54:561, 1973.

59. Thaeler CS Jr: An analysis of three hybrid populations of pocket gophers (genus Thomomys). Evolution 22:543, 1968.

60. Tchernov E: Succession of Rodent Faunas During the Upper Pleistocene of Israel. Hamburg/Berlin, Verlag Paul Parey, 1968, p. 152.

61. Nevo E, Kim YJ, Shaw C, Thaeler CS Jr: Genetic variation, selection and speciation in Thomomys talpoides pocket gophers. Evolution 28:1, 1974.
62. Russell RF: Evolution and classification of the pocket gophers of the subfamily Geomyinae. University Kansas Pub Mus Nat Hist 16:473, 1968.
63. Miller RS: Patterns and process in competition. Adv Ecol Res 4:1, 1967.
64. Grant PR: Interspecific competition among rodents. Annu Rev Ecol Syst 3:79, 1972.
65. Miller RS: Ecology and distribution of pocket gophers (Geomyidae) in Colorado. Ecology 45:256, 1964.
66. Vaughn TA, Hansen RM: Experiments on interspecific competition between two species of pocket gophers. Am Midl Nat 72:444, 1964.
67. Thaeler CS Jr: Taxonomic status of the pocket gophers, Thomomys idahoensis and Thomomys pygmaeus (Rodentia: Geomyidae). J Mammal 53:417, 1972.
68. Niethammer J: Zur taxonomie europaischer Zwergmaulwurfe (Talpa "mizura"). Bonn Zool Beitr 4:360, 1969.
69. Nevo E, Guttman R, Haber M, Erez E: Habitat selection in evolving mole rats. Oecologia 43:125, 1979.
70. Wahrman J, Goitein R, Nevo E: Mole rat Spalax: Evolutionary significance of chromosome variation. Science 164:82, 1969.
71. Nevo E: Variation and evolution in the subterranean mole rat Spalax. Isr J Zool 22:207, 1973.
72. Nevo E, Heth G, Bodmer M: Olfactory discrimination as an isolating mechanism in speciating mole rats. Experientia 32:1511, 1976.
73. Capranica RR, Moffat J, Nevo E: Vocal repertoire of a subterranean rodent Spalax. Presented at Acoust Soc Am Ann Meeting, Los Angeles, 1973.
74. Nevo E, Shaw C: Genetic variation in a subterranean mammal Spalax ehrenbergi. Biochem Genet 7:235, 1972.
75. Nei M: Genetic distance between populations. Am Natur 106:283, 1972.
76. Nevo E: Observations on Israeli populations of the mole rat Spalax ehrenbergi Nehring 1898. Mammalia 25:127, 1961.
77. Nevo E, Guttman R, Haber M, Erez E: Activity patterns of evolving mole rats. J Mammal (in press) 1982.
78. Enright JT: Ecological aspects of endogenous rhythmicity. Annu Rev Ecol Syst 1:221, 1970.
79. Rosenzweig M: On the evolution of habitat selection. Proc 1st Intl Cong Ecol, 1974, p 401.
80. Lay DM, Nadler CF: Cytogenetics and origin of North African Spalax (rodentia, Spalacidae). Cytogenetics 11:279, 1972.
81. Haim A, Heth G, Nevo E: Adaptive thermoregulatory patterns in speciating mole rats. Proc. 3rd Intl Cong Theriology, 1982, Helsinki (in press).
82. Nevo E, Heth G, Beiles A: Population structure in subterranean mole rats. Evolution (in press) 1982.
83. Nevo E, Heth G, Beiles A: Differential survivorship of evolving chromosomal species of mole rats, Spalax: An unplanned laboratory experiment. Evolution (in press) 1982.

Mechanisms of Speciation, pages 219-240
© 1982 Alan R. Liss, Inc., 150 Fifth Avenue, New York, NY 10011

Advances in Speciation of Cave Animals

Valerio Sbordoni

INTRODUCTION

Cave populations of troglophilic and troglobitic animals have traditionally been of special interest to evolutionary biologists for several reasons. Caves have been regarded as simple natural laboratories [1]. The cave climate is stable and easily defined. The cave communities are simple and can be studied in toto. Cave populations are reduced in size and are easily measurable. Certain adaptive features such as structural rudimentation in eye, pigmentation, and wing, improvement of chemosensory structures, elongation of body and appendages, lowered metabolic rate, and other features typical of K-strategists such as increase in egg size, decrease in egg number and population regulation via longer life span and infrequent reproduction, are typical of most troglobites (obligate cavernicoles) and are highly predictable events which, step by step, evolve convergently in different lineages [1–4].

In addition, like islands, caves are discontinuous habitats, separated by limestone rocks, which limit or prevent gene flow among populations. In this respect, caves stand as a favored subject for the analysis of allopatric or vicariant speciation processes.

In the past the difficulty of breeding troglobitic organisms and the duration of their biological cycle prevented the gathering of substantial data on the genetic variation of cave populations—information necessary to investigate microevolutionary problems. However, in recent years, the development of electrophoretic techniques applied to population studies and more research into population ecology of cavernicoles enabled this obstacle to be overcome, and now a fairly large amount of data on the genetic structure and variability of cave populations is becoming available.

The aim of this paper is to present and discuss some new experimental data relevant to the understanding of the process of speciation and adaptation to the cave environment. Most of the work to which I shall refer was carried out in my laboratory over these last few years; and additional genetic information is still to be revealed as may be seen from the frequent quotations of unpublished studies. For this reason this paper should be regarded as a progress report rather than as a review and, even if some general conclusions may be advanced, it appears still to be premature to organize the material in the form of an exhaustive and well assessed theory of speciation of cavernicoles.

SPECIATION OF TROGLOBITES: THE CLASSICAL MODEL

The theory of speciation of troglobites rests on some firmly assessed observations: 1) The ancestors of troglobites have colonized caves, to which they were probably preadapted, from epigean habitats, namely soil, freshwater, and the sea [2,5–7]. 2) Most troglobites are relict species, epigean descendants of the ancestral stock being extinct or surviving only in areas geographically remote from the cave region [8]. 3) Troglobites are far more abundant in temperate zones than in the tropics; the relatively few tropical troglobites are almost exclusively aquatic. This situation suggests a role of past "anisothermic" conditions of temperate zones alternating cold, wet periods with hot, dry periods, in cave isolation and evolution of troglobites. On the other hand the constant "isothermic" climate of the tropics, prevailing for millions of years, apparently has not promoted cave isolation and faunistic evolution [9–13]. 4) In several different taxonomic groups, assemblages of closely related species occur, showing varying degrees of reduction of structures such as eyes, wings and melanic pigment, indicating a gradual evolution from the troglophile (facultative cavernicoles, able to exist either in caves or in similar epigean habitats) to the troglobite stage [12, 14–17]. 5) Closely related species of troglobites are commonly found in different cave systems in the same geographic region, suggesting descent from a common ancestor which inhabited caves of the region as a troglophile.

Schematically, the classical model of speciation in cavernicoles [3] may be summarized as follows.

Stage I) Cave populations of a troglophilic species are genetically connected with each other via the surface populations; they still belong to the same gene pool and their degree of variability is relatively high depending on the overall effective population size.

Stage II) Following extinction of the surface members of the species, the cave populations become isolated and breed only within the caves. Gene flow among different populations is possible only if there are interconnecting passages or crevices through which dispersal by cavernicoles can take place. But most populations become fully isolated, and their degree of variability is expected to be low due to sampling error associated with the small size of the founder population. Later on, variability would be even lower due to the effects of severe selection, and homozygosity resulting from inbreeding. At this stage the probability of survival of the colony would be rather low.

Stage III) Those populations which succeed in overcoming the critical phase are expected to face an extensive reorganization of the epigenotype, adapting the colony more closely to the environment, and increasing variability at a new level depending on the new population size which is now adjusted to the carrying capacity of the environment. According to the general model by Mayr [18], reproductive isolation should occur at this stage, as a byproduct of the extensive genetic reorganization of the cave population.

DEGREE OF VARIABILITY OF CAVE POPULATIONS: MODELS AND DATA

As seen in the preceding section Barr's model hypothesizes increased levels of genetic variability in the cave population following its genetic revolution and accompanying the cave adaptation process.

In contrast Poulson and White [1] believed that low levels of genetic variability are consistently maintained as a result of strong stabilizing selection in the cave environment. Low levels of polymorphism have been also predicted to evolve as the result of selectively mediated responses to a monotonous subterranean niche [19,20]; however, this last assumption appears to be inappropriate for caves, where spatial heterogeneity occurs in several features including food resources [21,22].

Prediction of the evolution of the amount of genetic variability in cave animals may also be derived from the neutralist models of molecular evolution. According to these models heterozygosity levels depend on population size, evolutionary time, and the occurrence of bottlenecks. If equilibrium conditions and neutral alleles are assumed, a dependence of heterozygosity on N_e (effective population size) and μ (the mutation rate to neutral alleles) is expected according to the relation [23]: $H_e = 4N_e\mu/(4N_e\mu + 1)$. A similar formula [24], $H_e = 1 - 1/\sqrt{1 + 8 N_e\mu_a}$, considers stepwise production of neutral and very slight deleterious mutations (in this formula μ_a is the apparent mutation rate).

One advantage of studying cave populations is the possibility of estimating N_e values to a fairly good degree by combining population censuses and data on sex ratio and age structure. For this purpose, the cave crickets are among the most suitable organisms. These were extensively studied in our laboratory both from the population ecology and the population genetics point of view. Starting from censuses of N_e in nine populations of Dolichopoda cave crickets and of H_e at 15 enzyme loci, we tested the population size—heterozygosity hypothesis by two approaches. The first was to search for clear-cut relationships: A significant correlation was found between heterozygosity and population size; however, this is confused with a similarly significant relationship with latitude. The two independent variables were not correlated to each other [25, unpublished data]. In the second approach we calculated the expected N_e values from our H_e values and obtained realistic figures (ie N_e expected values of the same order of magnitude as empirical values) only assuming an average mutation rate to neutral alleles $\mu_a \simeq 10^{-5}$ [unpublished data]. A similar study on Troglophilus [26] gives comparable results, and one should expect even faster mutation rates ($\mu_a \simeq 10^{-4}$) to cope with actual heterozygosity values. In the context of these analyses, neutralist models correctly predict variation of H_e among populations as a function of population size; however, these models fail to justify the observed high levels of average heterozygosity unless mutation rates are assumed to be unrealistically high.

The possible effects of time and bottleneck on heterozygosity have also been explored theoretically [27,28]. Heterozygosity should increase positively with evolutionary time, but bottlenecks have contrary effects, lowering the degree of heterozygosity. In most epigean habitats subjected to climate changes, population size would not stay constant for a long time, and thus the average heterozygosity would generally be lower than expected from the time dependent neutralistic model. In cave populations an initial bottleneck occurs when a small founder population becomes isolated in a cave. Following this stage, population size should increase to reach the carrying capacity of the environment. From then on, it is unlikely that further bottlenecks would occur because of the notable stability of the environment and the density-dependent control of the population size, a feature typically occurring in K-strategist organisms, such as troglobites [29].

A number of data enable the bottleneck-time hypothesis to be tested at different levels. We studied genetic variability and divergence at 17 enzyme loci in populations of two species of Troglophilus cave crickets: T cavicola and T andreinii [26]. Most cavicola populations in northern Italy are recent colonizers of caves in the Prealps, and one of the populations studied dates back less than 15,000–10,000 years since it actually lives in a cave which was covered by a glacier until the last Ice Age (Würm). On the other hand Apulian populations of T andreinii have been isolated in caves presumably some hundreds of thousands of years. Divergence estimates between populations support this view. Variability estimates strongly differ between the two groups of populations and they are in good agreement with the bottleneck-time hypothesis.

On a much broader scale, the set of data now available on the genetic variability of cave organisms belonging to different taxa (Table I) shows that cave populations of long standing display high levels of polymorphism. Some early papers reporting very low levels of heterozygosity for cave fishes (Astyanax) [30], Ptomaphagus beetles [31], and Ceuthophilus cave crickets [32] refer indeed to cases of young troglobites the evolution of which started in the Pleistocene. Further genetic studies of a selection of paradigmatic troglobites in several groups such as amphipods [33,34], isopods [35], bathysciine [36] and trechine beetles [37] clearly demonstrated the occurrence of high levels of heterozygosity both in aquatic and in terrestrial cave populations of long standing.

From a careful analysis of data reported in Table I it may be concluded that evolutionary time, namely the age of the cave population, is the best predictor of the heterozygosity level in cave animals (Fig. 1). Most significantly the role of gene flow in enhancing variability may be ignored since the opportunities for it decrease as the process of cave adaptation (in isolation) proceeds.

The intriguing problem is that old aged troglobites are much more variable than expected according to their very reduced population size. In most cases old troglobites are more variable than their epigean relatives and some of them, particularly certain Crustaceans (ie, Niphargus, Monolistra) are placed among the most polymorphic organisms ever reported.

It seems to me that, apart from stochastic factors, some form of balancing selection might concurrently be at work in late stages of subterranean evolution, increasing heterozygosity and enhancing coarse grained adaptive strategies as a response to the temporal stability of the environment (see [4] and [38] for detailed discussion of this topic). In addition the small population size characteristic of cave organisms may allow species to expand their niches, with an associated increase in genetic variability [34].

In summary it should be remarked that the present evidence contradicts the Poulson and White hypothesis predicting low variability in troglobites and shows that the narrow niche-variation hypothesis is inapplicable to cave populations. The data are not in contrast with the predictions of Barr; however, a deeper insight is needed to better understand the relative role of the determinants of variability in cave populations.

THE BOTTLENECK EFFECT ON DIVERGENCE OF CAVE POPULATIONS

Let me discuss briefly another aspect of the bottleneck concerning its role in enhancing genetic differentiation of cave isolates. Chakraborty and Nei [28] showed that genetic distance increases rapidly. This is because, under conditions of inbreeding, the genes segregating in the original population are rapidly fixed. Thus the rate of increase of genetic distance is higher for small bottlenecks than for large ones. Chakraborty and Nei [28] have shown that these predictions are in agreement with the infinite allele model and the stepwise mutation model:

TABLE I. Average Estimates of Genic Variability in Cave Populations

	Number of populations	Average sample	Number of loci	P	H̄	Ā
Crustacea						
Crangonyx antennatus [34]	6	150	8	0.25	0.118	—
Gammarus minus [58]	28	—	13	0.23	—	—
Niphargus orcinus[a]	1	15	11	0.27	0.104	1.45
N longicaudatus [33]	6	539	16	0.69	0.298	2.24
N romuleus[a]	1	188	16	0.81	0.301	2.25
N stefanellii[a]	1	110	15	0.67	0.271	2.06
Typhlocirolana moraguesi[a]	2	122	18	0.17	0.030	1.17
Monolistra boldorii [35]	1	51	14	0.71	0.303	2.00
M berica [35]	1	110	14	0.71	0.321	2.00
M caeca [35]	1	53	14	0.79	0.318	2.14
Spelaeomysis bottazzii [59]	3	204	12	0.38	0.080	1.51
Troglocaris anophthalmus (s.l.) [55]	4	47	23	0.10	0.032	1.10
Diplopoda						
Scoterpes copei [60]	2	12	10	0.15	0.032	1.01
Araneae						
Nesticus eremita [45]	13	253	19	0.35	0.111	1.47
N menozzii [45]	1	13	18	0.22	0.081	1.39
N sbordonii [45]	1	10	19	0.32	0.106	1.32
Meta menardi [31]	—	—	15	0.10	0.027	1.31

				P	H̄	Ā
Orthoptera						
Dolichopoda schiavazzii[a]	6	175	15	0.40	0.155	1.59
D baccettii[a]	3	102	15	0.41	0.152	1.60
D laetitiae[a]	5	358	15	0.51	0.163	1.72
D geniculata[a]	9	535	15	0.53	0.191	1.83
Troglophilus cavicola [26]	2	95	16	0.47	0.082	1.59
T andreinii [26]	3	213	16	0.65	0.163	1.88
Ceuthophilus gracilipes [32]	7	330	26	0.11	0.026	1.10
Coleoptera						
Duvalius lepinensis[a]	2	44	20	0.32	0.101	1.32
D jurececkii rasettii[a]	1	37	20	0.35	0.141	1.45
Rhadine subterranea [30]	2	148	3	1.00	0.222	2.67
Bathysciola derosasi (s.l.) [41]	5	250	13	0.37	0.100	1.46
Speonomus delarouzeei (s.l.) [48]	8	260	12	0.49	0.101	1.68
Speonomus lostiai[a]	2	59	16	0.37	0.134	1.69
Ptomaphagus hirtus [31]	6	200	13	0.14	0.047	1.19
Orostygia doderoi [36]	1	20	11	0.73	0.239	1.82
O meggiolaroi [36]	1	20	11	0.64	0.182	1.64
Leptodirus hohenwarti [36]	1	22	14	0.57	0.168	1.86
Neaphaenops tellkampfii [37]	5	99	12	0.36	0.121	1.40
Pisces						
Astyanax mexicanus [30]	3	136	17	0.14	0.031	1.22
Caudata						
Euryceea lucifuga [34]	—	331	12	0.012–0.14	0.022	—

[a]Sbordoni and co-workers, in preparation.
P, frequency of polymorphic loci (those with a frequency of the commonest allele < 0.99); H̄, average heterozygosity (expected under Hardy-Weinberg equilibrium); Ā, mean number of alleles per locus.

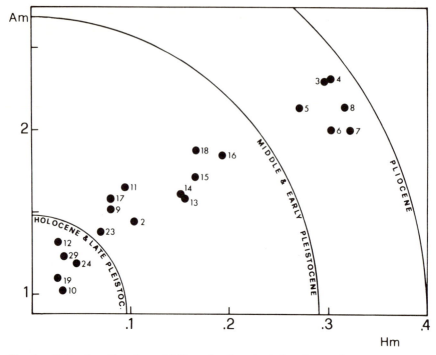

Fig. 1. Estimates of genic variability related to the inferred age of cave populations. Am, mean number of alleles per locus, Hm, average expected heterozygosity. 2. Niphargus orcinus, 3. N longicaudatus, 4. N romuleus, 5. N stefanellii, 6. Monolistra boldorii, 7. M berica, 8. M caeca, 9. Spelaeomysis bottazzii, 10. Scoterpes copei, 11. Nesticus eremita, 12. Meta menardi, 13. Dolichopoda schiavazzii, 14. D baccettii, 15. D laetitiae, 16. D geniculata, 17. Troglophilus cavicola, 18. T andreinii, 19. Ceuthophilus gracilipes, 23. Speonomus delarouzeei, 24. Ptomaphagus hirtus, 29. Astyanax mexicanus.

The accelerated increase of genetic distance ceases as soon as the population size becomes close to the equilibrium value K. With the growth rate of r = 0.1, N becomes close to K in about 100 generations. After this stage the rate of increase of genetic distance is lower than that in a constant population. This reduced rate of increase is maintained until generation 10^6–10^7. It is worth noting that although genetic distance increases very rapidly in the early generations, the initial gain is lost in the later generations, so that in the long run there is no gain.

We had a good opportunity to test this model in the cave beetle Bathysciola derosasi, a cave species endemic to Mount Argentario in Tuscany. Thirty years ago Marquis Patrizi, a keen biospeleologist, introduced 50 individuals of this species into a cave in Latium, where no Bathysciola were living, from the Argentario cave locality [39]. This population is now quite well adapted to the new cave [40], and mark-release-recapture techniques allowed us, last year, to estimate the population size at 15,000 individuals. The calculated rate of population growth is about 0.12. On a set of 13 loci we calculated genetic distance between the original and the introduced population, which is 0.012, a figure that is predicted by the model [41]. The former example refers to quite a large bottleneck. In nature we would expect founder populations of caves to be much more reduced, and the sampling error could play a major role in the differentiation of cavernicoles. In addition, the bottleneck associated with the founder population may induce shifts in the genetic environment, leading to altered selective conditions which, in turn, promote genetic transilience and speciation [42]. This last view, incorporated in a model recently advanced by Templeton, may be considered to better illuminate the speciation processes in caves.

DEGREE OF DIFFERENTIATION BETWEEN POPULATIONS AND SPECIES AND THE POPULATION STRUCTURE OF CAVERNICOLOUS SPECIES

Table II shows the available data on genetic distance of taxa at different stages of evolutionary divergence. The sample included a wide spectrum of organisms such as amphipods (Crangonyx, Niphargus), isopods (Monolistra, Typhlocirolana), mysidacea (Spelaeomysis), shrimps (Troglocaris), millipedes (Scoterpes), spiders (Nesticus), crickets (Troglophilus, Ceuthophilus, Dolichopoda), carabid beetles (Neaphaenops, Duvalius), catopid beetles (Speonomus, Orostygia, Ptomaphagus) and fishes (Astyanax). For some taxa included in the "sibling species" category such as those belonging to the genera Troglocaris and Speonomus, evidence of reproductive isolation is available. The remaining taxa have been put tentatively in this category according to their degree of genetic differentiation which is of the same order of magnitude as detected between reproductively isolated species. As expected, genetic distance between populations is much lower than between sibling species and, in turn, between morphologically differentiated species; however, even geographically close cave populations are fairly different genetically. On average, conspecific populations of troglophilic organisms such as cave crickets and Nesticus spiders are genetically closer than populations of troglobitic animals, which represent the majority of the sample.

Population structure of the species varies to some extent as is shown by the following examples. A notable genetic uniformity occurs between cave populations of the troglophilic crickets Ceuthophilus gracilipes in central Pennsylvania ([32] and Table II) suggesting some actual gene flow among caves through epigean forest populations. Much wider genetic differentiation occurs between populations of other cave crickets such as Troglophilus and Dolichopoda (Table II). Dolichopoda species are all allopatric, but in some areas, as in central Italy, the limits of the range of different species almost touch [43], and if they were

TABLE II. Estimates of Mean Levels of Nei's Genetic Distance at Different Stages of Evolutionary Divergence in Cave Animals

	Number of loci	Number of pairwise comparisons, average D values and ranges									Reference
		between populations			between sibling species			between morphologically differentiated species			
Crangonyx antennatus	8	21	0.091	0.032–0.153							[34]
Niphargus, seven species	18	28	0.221	0.055–0.458	27	0.435	0.247–0.725	36	0.698	0.333–1.192	[33], [a]
Monolistra, three species	22							3	0.752	0.551–0.859	[57] [a]
Typhlocirolana moraguesi group	18	1	0.000		2	0.691					[a]
Spelaeomysis bottazzii	12	6	0.063	0.011–0.116							[59]
Troglocaris anophthalmus group	22	2	0.102	0.072–0.132	4	0.397	0.375–0.420				[55]
Scoterpes copei group	10				1	1.576					[60]
Nesticus, three species	21	78	0.092	0.006–0.294				27	0.916	0.467–1.126	[45]
Troglophilus, two species	17	4	0.060	0.015–0.109				6	0.374	0.307–0.467	[26]
Ceuthophilus gracilipes	26	28	0.001	0.000–0.004							[32]
Dolichopoda, four species	15	47	0.098	0.014–0.240				143	0.281	0.060–0.501	[61]
Neaphaenops tellkampfii	12	10	0.029	0.005–0.071							[37]
Duvalius, two species	20	1	0.028					2	1.15	1.13–1.17	[a]
Bathysciola derosasi	13	6	0.068	0.012–0.169	4	0.304	0.245–0.356				[41], [a]
Speonomus, four species group	12	11	0.063	0.002–0.157	17	0.451	0.333–0.588	2	2.418	2.294–2.542	[48], [a]
Orostygia, two species	12							1	0.947		[36]
Ptomaphagus hirtus	13	15	0.219	0.004–0.478							[31]
Astyanax mexicanus	17										[30]
cave populations		3	0.140	0.027–0.208							
surface populations		15	0.006	0.002–0.014							
cave vs. surface		18	0.142	0.062–0.218							

[a]Sbordoni and co-workers, in preparation.

epigean species, they would be regarded as parapatric. Even if at some enzyme loci a clinal variation of allele frequencies was to occur ([44] and unpublished data), adjacent populations belonging to different species would not be genetically closer than geographically more distant populations. Frequency distribution of interpopulational genetic distance overlaps with the distribution of interspecies values, but the specific groups are, in spite of that, well separated (Fig. 2) as is also indicated by the results of several multivariate analyses of allele frequencies and genetic distance data (unpublished data). Utilizing the Shannon-Weaver information index applied to gene diversity, it has been shown that the mean proportion of the total species gene diversity that resides within populations is 68.6% (see Table III for comparison with other organisms.).

The situation in the cave spiders of the genus Nesticus in Italy is somewhat different from Dolichopoda [45]. One widely distributed species (N eremita) overlaps in certain areas of its range with local, endemic species, coexisting with one of them (eg N eremita/N menozzii) or being competitively excluded (as in the case of N sbordonii, a species known only by one population in the center

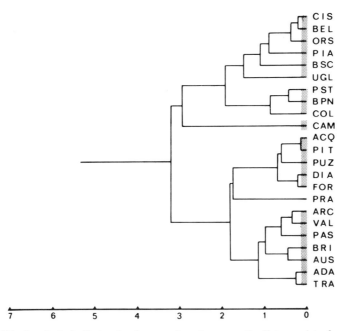

Fig. 2. Biochemical similarity dendrogram based on genetic distance data for 23 populations of Dolichopoda cave crickets from central Italy. Shaded areas cluster conspecific populations. From top to bottom: D schiavazzii, D baccettii, D laetitiae, and D geniculata. Nei's genetic distance values are indicated in the scale.

of the range of N eremita). Interspecies distance values are much higher than in Dolichopoda, but interpopulational values within N eremita are quite the same as in Dolichopoda (Table II and Fig. 3). These are significantly correlated to the geographic distances (r = 0.371; P < 0.01; df = 90). The association between genetic distance and a geographic gradient, suggesting some past or present gene flow, is also strongly supported by the results of the principal component analysis applied to gene frequency data. By analyzing distribution of loci with respect to I values, it has been found that more than 70% of loci have a genetic identity higher than 0.95. This finding is roughly consistent with the values recorded in other organisms between conspecific populations; however, the L-shaped distribution of the loci approaches the figures obtained in comparisons between semispecies. The apportionment of the whole species diversity to single populations of N eremita is quite low; only the 58% of the genetic diversity resides within populations, while 42% is due to between populations variation. It is worthwhile to compare these data with those of other cave and non cave dwelling organisms such as humans and plants (Table III) [46,47].

Different again is the situation of amphipods in the Niphargus longicaudatus group [32], where differentiation between populations is increasingly high, reaching the species level (Fig. 4). Even geographically close populations may show high divergence values. The genetic distance between Beatrice Cenci (CEN) and Verrecchie (VER) populations in the Abruzzi Appennines is 0.12, indicating an almost complete absence of migration in spite of the closeness of the two caves, which are separated by a distance of 300 meters. The absence of a hydrologic

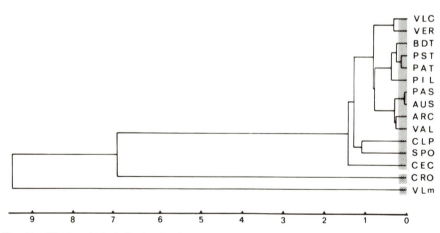

Fig. 3. Biochemical similarity dendrogram based on genetic distance data for 15 populations of Nesticus cave spiders belonging to three species as grouped by the shaded areas. From top to bottom: N eremita, N sbordonii, and N menozzii. Nei's genetic distance values are indicated in the scale.

TABLE III. Mean Amount of Genetic Diversity Within Population (H_{pop}) and Species (H_{sp}), and the Proportion Accounted for Within Population (H_{pop}/H_{sp}) in Four Cave Organisms Compared with a Plant (Phlox) and Homo.

	Number of loci	H_{pop}	H_{sp}	H_{pop}/H_{sp}
Dolichopoda (four species)	12	0.192	0.291	0.686
Nesticus eremita	15	0.093	0.166	0.583
Niphargus longicaudatus[a]	14	0.230	0.416	0.559
Speonomus delarouzeei	10	0.092	0.282	0.433
Phlox drummondii[b]	7	0.464	0.634	0.769
Homo sapiens[c]	17	0.750	0.876	0.849

[a]Sbordoni et al [33]; [b]Levin [47]; [c]Lewontin [46].
Genetic diversity estimates are calculated following Lewontin's [46] application of the Shannon-Weaver information index, at single loci. The values reported result from means over all polymorphic loci.

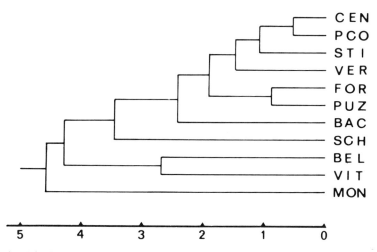

Fig. 4. Biochemical similarity dendrogram based on genetic distance data for 11 populations belonging to the Niphargus longicaudatus group. The scale refers to Nei's genetic distance.

connection between Beatrice Cenci and Verrecchie caves has been now established by water dying experiments.

The situation in the Pyrenean cave beetle Speonomus delarouzeei is similar to that of Niphargus [48]. We studied eight cave populations ranging from 470–1190 m on the southwestern side of the Mount Canigou (eastern Pyrenees). Genetic distance data, based on 12 loci, suggested the existence of three species

(Fig. 5), which are separated by geographic distances of less than 4 km. This supposition has now been confirmed by laboratory experiments which produced evidence of reproductive isolation (ie, gametic mortality) in crosses between populations belonging to different groups (unpublished data). Interestingly enough, the barriers to gene flow are not related to the geology (S delarouzeei cave populations occur in limestone islands separated by granite rocks) or to the topography, but appear to depend on the type of vegetation associated with a particular bioclimate [48].

In summary, according to the present evidence, the population structure of most cavernicoles appears to fit with the Wright's Island model, in which the species is subdivided into isolated demes with little or no gene flow among them. The extent of gene flow depends chiefly on the occurrence of epigean populations and on the existence of interconnecting passages among caves. Potential barriers to migration are represented by vegetation and soil types not suitable for epigean dispersal [48], discontinuities in carbonate rock outcrop, streams, stream and karst drainage divides [49], and may depend on the direction of flow of the water tables [31]. However, a problem still open is represented by the existence of locally abundant populations of troglobite-like trechine and bathysciine beetles and other small arthropods, morphologically close to cave species, occurring in deep soil and micro-crevice habitats [50]. Further research is needed to determine whether these populations are reproductively isolated from cave populations, or whether they may contribute at some extent to the cave gene pools.

SECONDARY SIMPATRY AND OPPORTUNITIES FOR HYBRIDIZATION

It is usually difficult to provide evidence of hybridization between related cavernicolous species, because in most cases they are allopatric. Only occasionally may this phenomenon be observed as the result of natural or artifactual colonization in already-inhabited caves. Similarly, evidence of secondary contact between a cave-adapted population and its epigean ancestor has seldom been reported. Consequently, such instances are obviously of great interest in assessing evolutionary relationships between related taxa.

In the fish Astyanax mexicanus, epigean populations with normal pigmentation and well-developed eyes in the Cueva Chica (Sierra de El Abra, Mexico), intergrade with extreme individuals that are completely blind and unpigmented [51]. This hybrid swarm is likely to be a long-term phenomenon that will probably persist as long as present ecological conditions in the cave (eg, large energy input) and inflow of epigean fish are maintained. Electrophoretic studies have not enabled appreciable introgression to be ascertained [30]. Other mixed fish populations, with more or less reduced hybridization, have also been discovered in the Sierra de El Abra. However, most caves in this area harbour pure pop-

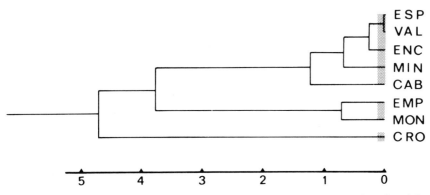

Fig. 5. Biochemical similarity dendrogram based on genetic distance data for eight populations of Speonomus delarouzeei group. Shaded areas cluster populations belonging to three distinct species. Nei's genetic distance values are indicated in the scale.

ulations of blind fish. Of special interest is the occurrence of both blind and surface fish in two oligotrophic caves and the absence of any apparent hybridization [52,53]. This seems to indicate that nutrient availability affects both the relative fitness of the two forms and the probability of their successful interbreeding [52].

Hybridization between Dolichopoda cave cricket species was revealed both by transplant experiments and by occasional coexistence due to passive dispersal [54]. A release experiment was carried out by introducing a sample of D laetitiae into a small natural cave in central Italy already inhabited by D geniculata, whose population ecology had been thoroughly studied in previous years. The two populations investigated were fixed for alternative alleles at the phosphoglucomutase (Pgm) locus so that hybrids could easily be recognized by their heterozygous pattern at this locus. Soon after release, the species ratio was 1:2 (sex ratio 1:1). The introduced population of D laetitiae rapidly decreased until it was extinct after a few months. However, a relatively high frequency of F_1 hybrids was observed. Hybrids were intermediate in some morphological features and did not exhibit any reduction in viability. Research is still in progress to ascertain whether any introgression may occur.

Interspecific hybridization has also been determined for Dolichopoda baccettii and D schiavazzii by using gene-enzyme markers [54]. These two species currently cohabit in a small artificial cave of the Monte Argentario promontory in Tuscany. D baccettii is endemic of this area, whereas the presence of allochthonous D schiavazzii is probably due to passive dispersal. The recent opening of the cave, dating back to the Second World War, indicates recent colonization by both Dolichopoda species, which are thought to have coexisted there for several generations now. The relative frequency of hybrids in the overall population is somewhat high, ie nearly 10%. Nevertheless, hybrid frequency is significantly lower than to be expected in a panmictic population. No positive evidence of introgression has been detected.

These hybridization studies suggest that premating isolation mechanisms between Dolichopoda species are weak. They are also consistent with electrophoretic divergence data (Fig. 2) indicating both slight genetic differentiation between the two pairs of species concerned and relatively recent speciation.

Secondary coexistence of even closely related cave forms does not necessarily lead to hybridization. The troglobite cave shrimp Troglocaris anophthalmus is widely distributed in groundwater systems (including caves) of the Italian, Slovenian, and Croatian Karst. A genetic study at 23 gene loci of the Italian populations enabled us to detect two electrophoretically distinguishable forms. The former, located in the Goritian Karst, belongs to the Vipacco-Isonzo basin, and the latter occurs in caves belonging to the Timavo system, near Trieste [55]. The two forms are morphologically very close and cannot be differentiated on this basis; genetic distance, however, is high, averaging a value of 0.4. The two forms have been proved to coexist near Duino (Trieste) in a cave belonging to the Timavo watershed which occasionally receives flood waters from the Vipacco-Isonzo aquifer. No hybridization seems to occur between the two forms, thereby suggesting that they are, in fact, distinct species separated by reproductive barriers.

AGE OF CLADOGENETIC EVENTS IN CAVE SPECIES

The foregoing discussion and the data presented in Table II indicate that reproductive isolation may occur between cave populations with genetic distances as low as 0.2. This threshold might even be lowered to 0.15 if reproductive isolation should be proved to occur between certain epigean and cave populations of Astyanax mexicanus. On the other hand, lower genetic distances have been recorded between interfertile conspecific populations as shown by laboratory crosses in Bathysciola, Speonomus, Ptomaphagus, and Astyanax. Higher values of genetic distance are, in the great majority of cases, attributable to separate species.[1] Since a raw correlation seems to have been established between genetic

---

[1] A notable exception seems to occur in the case of Catopid beetle Ptomaphagus hirtus, studied by Laing and co-workers [31], where laboratory F_2 adults were obtained even from crosses between populations with a genetic distance of more than 0.4. However, in my opinion, these results require further investigation, because wild captured "males were isolated from females for three months thereby allowing any sperm stored in the female's spermathecae either to be utilized or to be inviable" [31]. According to my experience, this procedure may not be sufficient for the female to lose the sperm stored from previous matings. In Speonomus delarouzeei we found that females could store live sperm for as long as six months from the previous copulation.

distance and evolutionary time [62,63], the data reported in Table II suggest that the time required for speciation in caves is rather long, possibly not less than $5 \times 10^5{-}10^6$ years.

Because troglobites are confined to caves and have no opportunity to disperse over long distances, they are excellent palaeogeographic monitors and their present distribution generally reflects particular palaeogeographic events which have been dated by geologists. One such example is the occurrence in Sardinia of Speonomus cave beetles, a genus otherwise limited to the Pyrenees [56]. Current distributions appear to be due to an earlier connection between the Pyrenees and the Sardinian-Corsican plate which rotated towards the Italian peninsula in the early Miocene to reach its present location [64]. According to palaeomagnetic and deep-sea drilling evidence, the two areas separated 22–25 million years ago. We studied genetic divergence between Sardinian species S lostiai and Pyrenean S delarouzeei, finding an average genetic distance value of 2.418, which, calibrated with albumin immunological distance for 60% of fast-evolving loci, gives an approximately 25 Myr estimate for time divergence [unpublished data].

Monolistra is a similar case [35,57]. The occurrence of Monolistrini (isopods of the typically marine family Sphaeromidae) in continental Europe is a remarkable example of relict distribution. Allozymic variation in proteins encoded by 20 loci was analyzed in three cavernicolous Monolistra and two epigean Sphaeroma species, ie, S serratum, a marine species, and brackish water S hookeri. High levels of heterozygosity and divergence time ($\simeq$ 7 Myr) calculations based on genetic distance support the hypothesis that Monolistra diverged from its Sphaeroma-like marine ancestor during the Messinian Age in connection with the Mediterranean salinity crisis.

CONCLUSIONS

From this summarized report of a fairly extensive set of data the following conclusions may be drawn.

1. As expected, speciation processes in caves are strictly allopatric. Processes may be categorized according to White's scheme [66] as being "strictly allopatric with a narrow population bottleneck," gene flow being interrupted as the result of intermediate (epigean) populations becoming extinct.
2. Experimental data on a limited sample of taxa suggest that speciation leads chiefly to hybrid sterility (and perhaps inviability). However, postmating prezygotic isolation due to gamete mortality has also been detected in the laboratory.
3. Genetic differentiation at structural loci seems to progress steadily after an initial acceleration due to bottleneck. In troglobitic species, a large amount

of genic differentiation may not necessarily be accompanied by divergence at the morphological level at all; consequently, many sibling species may be thought to exist due to convergent evolution in similar environments. On the other hand, strong morphological divergence between epigean and cave populations in adaptive characters, such as structural reduction of eye and pigmentation, may occur over a relatively short period and be associated with only minor genetic change at enzyme loci (eg Astyanax).

4. According to evidence on reproductive isolation and to genetic distance data, it may be argued that the time required for speciation in caves is rather long, possibly not less than $5 \times 10^5 - 10^6$ years.

5. Genetic variability in cave populations appears to be positively related to evolutionary time, a trend which is suggested both by Barr and neutralist models. However, the high degree of heterozygosity reached by old troglobitic populations despite their small population size is higher than heterozygosity in their epigean relatives of considerably larger populations, thereby suggesting a concurrent role of natural selection.

ACKNOWLEDGMENTS

Most of the data discussed in this paper originate from a series of investigations in progress carried out by the author in collaboration with G. Allegrucci, A. Caccone, C. Carchini, D. Cesaroni, M. Cobolli Sbordoni, E. De Matthaeis, M. Lucarelli, M. Mattoccia, and M. Rampini. The manuscript was kindly read and revised for correct English by Jeremy Holloway and Robert Maberry. The investigations were supported by several grants from the Italian National Research Council (CNR) and from the Faculty of Sciences, University of Rome.

REFERENCES

1. Poulson TL, White WB: The cave environment. Science 165:971, 1969.
2. Vandel A: Biospéologie. La Biologie des Animaux cavernicoles. Paris: Gauthier-Villars, 1964.
3. Barr TC Jr: Cave ecology and the evolution of troglobites. Evol Biol 2: 35, 1968.
4. Sbordoni V: Strategie adattative negli animali cavernicoli: uno studio di genetica ed ecologia di popolazione. Acc Naz Lincei, contr Centro Linceo Interdisc Sc Mat e loro appl 51: 61, 1980.
5. Packard AS Jr: On the origin of the subterranean fauna of North America. Am Nat 28: 727, 1894.
6. Racovitza EG: Essai sur les problèmes biospéologiques. Arch zool exp et gèn 36: 371, 1907.
7. Jeannel R: Faune cavernicole de la France, avec une étude des conditions d'éxistence dans le domaine souterrain. Encycl Entom 7, Paris: Lechevalier, 1926.
8. Jeannel R: Les fossiles vivants des cavernes. Paris: Editions Gallimard, 1943.

9. Jeannel R: Situation géographique et peuplement des cavernes. Ann Spéléol 14: 333, 1959.
10. Ruffo S: Su alcuni problemi relativi allo studio degli insetti cavernicoli. Atti Acc Naz Ital Entom Rendiconti 8: 269, 1961.
11. Mitchell RW: A comparison of temperate and tropical cave communities. Southwestern Nat 14: 73, 1969.
12. Sbordoni V, Cobolli Sbordoni M: Aspetti ecologici ed evolutivi del popolamento di grotte temperate e tropicali: osservazioni sul ciclo biologico di alcune specie di Ptomaphagus (Coleoptera, Catopidae). Int J Speleol 5: 337, 1973.
13. Sbordoni V, Argano R, Vomero V, Zullini A: Ricerche sulla fauna cavernicola del Chiapas (Messico) e delle regioni limitrofe: grotte esplorate nel 1973 e nel 1975. Criteri per una classificazione biospeleologica delle grotte. In: Subterranean Fauna of Mexico. Parte III. Acc Naz Lincei, Quaderni 171: 5, 1977.
14. Poulson TL: Cave adaptation in amblyopsid fishes. Am Midl Nat 70: 257, 1963.
15. Jeannel R: Sur l'évolution des coléoptères aveugles et le peuplement des grottes dans les monts Bihor en Transylvanie. C R Acad Sci Paris 176: 1670, 1923.
16. Jeannel R: Monographie des Trechinae. Morphologie comparée et distribution géographique d'un groupe de Coléoptères. L'Abeille 32:221; 33:1; 34: 59; 35: 1, 1926–1930.
17. Lattin G de: Über die Evolution der Höhlentiercharaktere. Sitzber Ges Naturf Freunde 32: 11, 1939.
18. Mayr E: Change of genetic environment and evolution. In Huxley J (ed): Evolution as a Process. London: Allen & Unwin, 1954, p 157.
19. Nevo E, Kim YJ, Shaw CR, Thaeler CS Jr: Genetic variation, selection and speciation in Thomomys talpoides pocket gophers. Evolution 28: 1, 1974.
20. Nevo E: Genetic variation in natural populations: pattern and theory. Theor Pop Biol 13: 121, 1978.
21. Poulson TL, Culver DC: Diversity in terrestrial cave communities. Ecology 50: 153, 1969.
22. Barr TC Jr: Observations on the ecology of caves. Am Nat 101: 475, 1967.
23. Kimura M, Crow JF: The number of alleles that can be maintained in a finite population. Genetics 49: 725, 1964.
24. Maruyama T, Kimura M: Theoretical study of genetic variability, assuming stepwise production of neutral and very slightly deleterious mutations. Proc Nat Acad Sci USA 75: 919, 1978.
25. Sbordoni V, Cesaroni D, Allegrucci G, Caccone A, Cobolli Sbordoni M, De Matthaeis E: Variabilità genetica e dimensione di popolazione in Ortotteri cavernicoli: un test della teoria neutralista. Boll Zool 48 (suppl):100, 1981.
26. Sbordoni V, Allegrucci G, Caccone A, Cesaroni D, Cobolli Sbordoni M, De Matthaeis E: Genetic variability and divergence in cave populations of Troglophilus cavicola and T andreinii (Orthoptera Rhaphidoph). Evolution 35: 226, 1981.
27. Nei M, Maruyama T, Chakraborty R: The bottleneck effect and genetic variability in populations. Evolution 29: 1, 1975.
28. Chakraborty R, Nei M: Bottleneck effects on average heterozygosity and genetic distance with the stepwise mutation model. Evolution 31: 347, 1977.

29. Margalef R: Paralelismo entre la vida de las cavernas y la de las grandes profundidades marinas. Bol Soc Hist Nat Baleares 21: 10, 1976.
30. Avise JC, Selander RK: Evolutionary genetics of cave-dwelling fishes of the genus Astyanax. Evolution 26: 1, 1972.
31. Laing C, Carmody GR, Peck SB: Population genetics and evolutionary biology of the cave beetle Ptomaphagus hirtus. Evolution 30: 484, 1976.
32. Cockley DE, Gooch JL, Weston DP: Genetic diversity in cave dwelling crickets (Ceuthophilus gracilipes). Evolution 31: 313, 1977.
33. Sbordoni V, Cobolli Sbordoni M, De Matthaeis E: Divergenza genetica tra popolazioni e specie ipogee ed epigee di Niphargus (Crustacea, Amphipoda). Lavori Soc Ital biogeogr (ns)6: 329, 1979.
34. Dickson GW, Patton JC, Holsinger JR, Avise JC: Genetic variation in cave-dwelling and deep-sea organisms, with emphasis on Crangonyx antennatus (Crustacea, Amphipoda) in Virginia. Brimleyana 2: 119, 1979.
35. Sbordoni V, Caccone A, De Matthaeis E, Cobolli Sbordoni M: Biochemical divergence between cavernicolous and marine Sphaeromidae and the mediterranean salinity crisis. Experientia 36: 48, 1980.
36. Sbordoni V, Allegrucci G, Caccone A, Cesaroni D, Cobolli Sbordoni M, De Matthaeis E: A preliminary report of the genetic variability in troglobitic Bathysciinae: Leptodirus hohenwarti and two Orostygia species (Coleoptera, Catopidae). Fragm Entomol 15: 327, 1980.
37. Giuseffi S, Kane TC, Duggleby WF: Genetic variability in the Kentucky cave beetle Neaphaenops tellkampfii (Coleoptera, Carabidae). Evolution 32: 679, 1978.
38. Valentine JW: Genetic strategies of adaptation. In Ayala FJ (ed): Molecular Evolution. Sunderland, Massachusetts: Sinauer Ass. Inc., 1978, p 78.
39. Patrizi S: Introduzione ed acclimatazione del coleottero Catopide Bathysciola derosasi Dod. in una grotta laziale. Le Grotte d'Italia (3)1: 303, 1955–1956.
40. Lucarelli M, Sgrò G, Sbordoni V: Ciclo biologico in laboratorio di tre popolazioni cavernicole di Bathysciola derosasi Jeann. (Coleoptera, Bathysciinae). Mém Biospéol 7: 319, 1980.
41. De Matthaeis E, Cobolli Sbordoni M, Mattoccia M, Allegrucci G, Caccone A, Cesaroni D, Rampini M, Sbordoni V: Struttura genetica di due popolazioni cavernicole di Bathysciola derosasi (Coleoptera Catopidae): bottlenecks e divergenza genetica. Atti XII Congr Naz Entom Roma (in press).
42. Templeton AR: The theory of speciation via the founder principle. Genetics 94: 1011, 1980.
43. Baccetti B, Capra F: Notulae orthopterologicae XXVII. Nuove osservazioni sistematiche su alcune Dolichopoda italiane esaminate anche al microscopio elettronico a scansione (Orthoptera, Rhaphidoph.). Mem Soc Entomol Ital 48: 351, 1970.
44. Sbordoni V, De Matthaeis E, Cobolli Sbordoni M: Phosphoglucomutase polymorphism and natural selection in populations of the cave cricket Dolichopoda geniculata. Z Zool Syst Evolut -forsch 14: 292, 1976.
45. Cesaroni D, Allegrucci G, Caccone A, Cobolli Sbordoni M, De Matthaeis E, Di Rao M, Sbordoni V: Genetic variability and divergence between populations and species of Nesticus cave spiders. Genetica 56: 81, 1981.

46. Lewontin RC: The Genetic Basis of evolutionary Change. New York: Columbia University Press, 1974.
47. Levin DA: The organization of genetic variability in Phlox drummondii. Evolution 31: 477, 1977.
48. Delay B, Sbordoni V, Cobolli Sbordoni M, De Matthaeis E: Divergences génétiques entre les populations de Speonomus delarouzeei du Massif du Canigou (Coleoptera, Bathysciinae). Mém Biospéol 7: 235, 1980.
49. Gooch JL, Hetrick SW: The relation of genetic structure to environmental structure: Gammarus minus in a karst area. Evolution 33: 192, 1979.
50. Juberthie C, Delay B, Bouillon M: Extension du milieu souterrain en zone non-calcaire: description d'un nouveau milieu et de son peuplement par les Coléoptères troglobies. Mém Biospéol 7: 19, 1980.
51. Breder CM Jr: Descriptive ecology of la Cueva Chica, with especial reference to the blind fish, Anoptichthys. Zoologica 27: 7, 1942.
52. Mitchell RW, Russell WH, Elliott WR: Mexican eyeless characian fishes, genus Astyanax: environment, distribution and evolution. The Museum Special publications 12: 3, 1977.
53. Sbordoni V, Argano R: Introduction: caves studied during the 1st mission to Mexico (1969). In: Subterranean Fauna of Mexico. Parte I, Acc Naz Lincei, Quaderno 171: 5, 1972.
54. Allegrucci G, Caccone A, Cesaroni D, Cobolli Sbordoni M, De Matthaeis E, Sbordoni V: Natural and experimental interspecific hybridization between populations of Dolichopoda cave crickets. Experientia 38: 96, 1982.
55. Allegrucci G, Caccone A, Cesaroni D, Cobolli Sbordoni M, De Matthaeis E, Sbordoni V: Prime ricerche sulla genetica di popolazioni naturali del gamberetto cavernicolo Troglocaris anophthalmus. Boll Zool 48:(suppl): 12, 1981.
56. Jeannel R: Biospeologica. L. Monographie des Bathysciinae. Arch Zool Expér Gén. 63: 1, 1924.
57. Caccone A, Cobolli Sbordoni M, De Matthaeis E, Sbordoni V: Una datazione su base genetico- molecolare della divergenza tra specie cavernicole e marine di Sferomidi (gen. Monolistra e Sphaeroma, Crustacea, Isopoda). Lavori Soc. Ital. Biogeogr. (ns) 7 (in press).
58. Hetrick SW, Gooch JL: Genetic variation in populations of the freshwater Amphipod Gammarus minus (Say) in the Central Appalachians. Natl Speleol Soc Bull 35: 17, 1973.
59. De Matthaeis E, Colognola R, Cobolli Sbordoni M, Pesce L, Sbordoni V: Divergenza genetica tra popolazioni di Spelaeomysis bottazzii Caroli (Crustacea, Mysidacea) della regione pugliese. Lavori Soc Ital Biogeogr (ns) 7 (in press).
60. Laing C, Carmody RG, Peck SB: How common are sibling species in cave-inhabiting invertebrates? Am Nat 110: 184, 1976.
61. Sbordoni V, Allegrucci G, Cesaroni D, Cobolli Sbordoni M, De Matthaeis E, Rampini M, Sammuri G: Problemi di biogeografia e genetica di popolazioni nel genere Dolichopoda (Orthoptera, Rhaphidoph.). Lavori Soc Ital Biogeogr (ns) 7 (in press).
62. Nei M: Molecular Population Genetics and Evolution. Amsterdam: North-Holland, 1975.

63. Sarich VM: Rates, sample size, and the neutrality hypothesis for electrophoresis in evolutionary studies. Nature 265: 24, 1977.
64. Alvarez W: The application of plate tectonics to the Mediterranean Region. In Tarling DH, Runcorn SK (eds): Implications of Continental Drift to the Earth Sciences. New York: Academic Press, 1973, p 893.
65. Hsu KJ, Montadert L, Bernoulli D, Cita MB, Erickson A, Garrison RE, Kidd RB, Mèlierés F, Müller C, Wright R: History of the Mediterranean salinity crisis. Natura 267: 399, 1977.
66. White MJD: Modes of Speciation. San Francisco: Freeman, 1978.

Mechanisms of Speciation, pages 241-264
© 1982 Alan R. Liss, Inc., 150 Fifth Avenue, New York, NY 10011

Genetic, Ecological, and Ethological Aspects of the Speciation Process

Luciano Bullini

INTRODUCTION

The electrophoretic study of protein variation in natural populations initiated a new era in the understanding of the genetic basis of speciation processes. The significance of the results obtained has been recently revised and discussed by various authors [1–6].

The aim of this paper is to summarize some results we have obtained with this approach on a number of phylogenetically unrelated species. These data appear to be of interest both for the general theme of this meeting on mechanisms of speciation and for the discussion of some genetic, ecological, and ethological aspects of speciation processes in different animal groups, such as Ascarid worms, mosquitoes, moths, stick-insects, and snails.

SIBLING SPECIES AND THEIR INTEREST FOR THE STUDY OF SPECIATION

The term sibling species was introduced by Mayr [7] to indicate "morphologically similar or identical natural populations that are reproductively isolated." The number of sibling species is very high in many animal groups and their recognition is often difficult. This frequently caused serious misunderstandings and confusion not only in taxonomy, but also in genetics, ecology, and evolutionary biology (eg, the case of many "biological races," which in fact are distinct sibling species). The importance of sibling species in evolutionary biology does not come from speciation processes different from those of the other species, but from the fact that they generally represent degrees of evolutionary divergence inferior to those of morphologically differentiated species, often being of recent origin.

Differences in habits, vocalization, host preferences, etc, allowed the discovery, or at least suspicion of, the existence of a certain number of sibling species. Their recognition, however, was in the past chiefly performed by laboratory hybridization tests and comparative study of karyotypes. Now, electrophoresis makes generally possible an easy detection of sibling species and their hybrids, both at the adult and larval stage [2,5,6,8].

Electrophoretic markers can also be used to perform experimental mixed infections with sibling species of endoparasites. Some results of electrophoretic study on sibling species of endoparasite Ascarid worms are presented in the next paragraph.

SPECIATION OF ENDOPARASITE ASCARID WORMS

The evidence of "host races" in endoparasite helminths brought various authors—eg, Baylis [9]—to hypothesize sympatric processes of speciation and the existence of incipient species in this group. We studied, among others, the "univalens" and "bivalens" forms of Parascaris equorum, the man and pig ascarids (Ascaris lumbricoides and A suum), and populations of Anisakis simplex, collected on different hosts. The research on the "univalens" and "bivalens" forms of P equorum has shown the following chief points [10–12]: 1) "Univalens" and "bivalens" are in fact two distinct sibling species, for which we have utilized the names P equorum (Goeze) and P univalens (Hertwig). 2) Natural hybrids are present (one male and 18 females out of the more than 2,000 adult specimens tested), but they do not allow gene flow between the two species. 3) Genetic distance (D) between these two species is exceptionally high: Nei's D = 1.96, a value never before recorded for sibling species. 4) P equorum and P univalens are the product of a very ancient process of speciation which affected a large part of the genome; a simple process of tetraploidization is ruled out both by the chromosome structure and the electrophoretic data. 5) P equorum and P univalens occur sympatrically in Europe and mixed natural infections are frequent in both horses and donkeys. 6) The two species live in the same portions of equine intestines and apparently have the same niche requirements. 7) Some evidence of interspecific competition is available and P univalens appears to be the prevailing species. 8) Some indications exist that P equorum was the most common species in Europe up to the beginning of this century; today its frequency in various European countries is about ten times less than that of P univalens. These data suggest that speciation of P univalens and P equorum occurred in different geographic areas and probably in different host species. Their coexistence is presumably recent and perhaps geographically limited. A precise map of the present distribution of these two species appears to be much needed, together with further ecological data on their trophic niche, and detailed studies on competition, performed by means of experimental mixed infections.

The process of speciation in man and pig Ascaris was much more recent. Average genetic distance between A lumbricoides and A suum was found to be 0.27 [12,13]. The two species occur sympatrically in most regions, but mixed natural infections appear to be rare. However a few cases of human infections with A suum are reported in the parasitological literature. The survival of A suum genotypes in man (no data exist, up to now, on pig infections by A lumbricoides), together with the fact that the parasite's reproduction is performed in the host, and the coexistence of the two hosts in many regions, indicate that in this case a sympatric process of speciation might have occurred.

The electrophoretic study of various Mediterranean populations of Anisakis simplex collected on different host fishes [14] has shown the existence of two distinct species (indicated as A and B); their genetic distance (D = 0.33) is similar to that found between A lumbricoides and A suum. These two Anisakis species have different intermediate hosts, consisting in fishes of the genera Trachurus and Micromesistius for A simplex A and of the genus Scomber for A simplex B. No data are available on the definitive host (whales), and on the first host (gammarids). The life cycle needing a number of hosts to be completed apparently makes it more difficult for sympatric speciation to take place—geographic speciation is the more likely hypothesis.

Single-host parasites like Ascaris, Parascaris, Neoascaris, Toxocara, Toxascaris, and Ascaridia show a very low mean heterozygosity: Observed mean heterozygosity per locus is from 2–8% [12, and unpublished data]. Much higher values of heterozygosity have been observed in Anisakis species, with an average of 16%. The higher genetic variability of multiple-host Ascarid species is in contrast with the higher probability of bottle-neck events; it can be accounted for as being related with the adaptation to a more time-varying environment.

SIBLING SPECIES AND SPECIATION IN MOSQUITOES

Mosquitoes are the first animal group, together with Drosophila, in which sibling species were detected. In particular, the Anopheles maculipennis complex is historically important as its study led to the demonstration that good biological species could exist without the development of significant morphological differences [15]. The presence in the genus Anopheles of good polytene chromosomes allowed the identification of various sibling species complexes [16]. A further improvement in the understanding of mosquito speciation processes came from the use of multilocus electrophoretic techniques [17–19].

Genetic distances found between mosquito sibling species range from 0.1 to 0.6 (Table I, Fig. 1). No apparent correlation exists between values of genetic distance and extent of chromosomal rearrangements. Relatively high values of genetic distance can be found between homosequential species, like An labranchiae and An atroparvus, while lower values of D are found to coexist with fairly high levels of chromosome divergence, as between An gambiae and An arabiensis. When the species of the An gambiae and An maculipennis complexes are compared, a higher number of chromosome rearrangements is found among

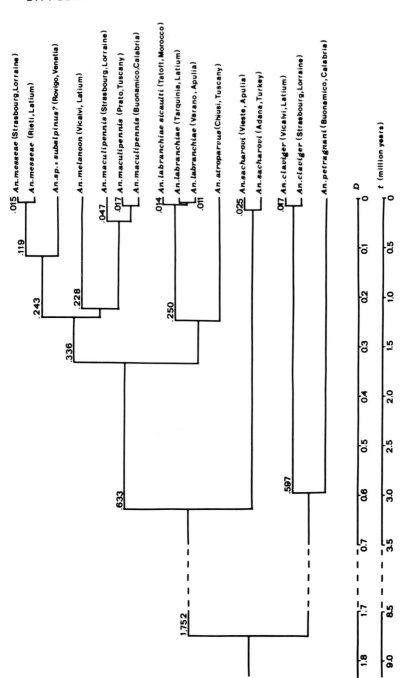

Fig. 1. Dendrogram of the presumed phylogenetic relationships among palearctic species of the Anopheles maculipennis and An claviger complexes. D, Nei's genetic distance; t, time of evolutionary divergence [from 22].

TABLE I. Genetic Differentiation Between Sibling Species of Mosquitos

Species	I	D
Anopheles messeae vs An subalpinus	0.89	0.12
Anopheles gambiae vs An arabiensis	0.88	0.13
Anopheles melas vs An arabiensis	0.87	0.13
Anopheles subalpinus vs An melanoon	0.86	0.15
Anopheles subalpinus vs An maculipennis	0.85	0.16
Aedes caspius A vs Ae caspius B	0.84	0.17
Anopheles melanoon vs An maculipennis	0.80	0.23
Culex pipiens vs C torrentium	0.78	0.25
Anopheles labranchiae vs An atroparvus	0.78	0.25
Anopheles gambiae vs An melas	0.76	0.27
Aedes mariae vs Ae zammitii	0.75	0.29
Aedes caspius vs Ae dorsalis	0.73	0.31
Aedes caspius vs Ae mariae	0.64	0.44
Anopheles melanoon vs An sacharovi	0.59	0.53
Anopheles claviger vs An petragnanii	0.55	0.60

I, Nei's standard genetic identity; D, Nei's standard genetic distance. Data from [19–23, and unpublished].

the former, whereas gene differentiation is higher in the latter [24]. This can be correlated with differences in the speciation processes. In the An maculipennis complex speciation occurred over a longer period of time and in areas with greater opportunities of geographic isolation. This kind of speciation does not necessarily have to be accompanied by chromosomal rearrangements. On the contrary, in the speciation process of An gambiae complex, which occurred in the Afrotropical region in areas with fewer opportunities of geographic isolation, chromosomal rearrangements seem to have played a fundamental role in favoring the preservation of coadapted multilocus complexes selected in marginal niches [25 and Coluzzi, this volume]. Species of both complexes generally present marked ecologic differences and very efficient premating barriers, contrasting with the various degrees of hybrid sterility shown by interspecific experimental crosses.

PREMATING REPRODUCTIVE ISOLATING MECHANISMS AND COMPETITIVE EXCLUSION IN THE AEDES MARIAE COMPLEX

In most cases precopulatory isolating mechanisms (RIMs) between allopatric forms can be studied reliably only by release experiments, enzyme variants offering good natural markers to identify the two forms and their possible hybrids.

This approach was applied for the first time in the study of premating isolating mechanisms in mosquitoes of the Aedes mariae complex [8]. The results of this research will be reported in some detail.

The Ae mariae complex includes at least three forms (mariae, zammitii, and phoeniciae) apparently allopatric, slightly differentiated morphologically, and showing various degrees of hybrid sterility. The larvae of these mosquitoes have an apparently identical adaptation to breeding in rock pools along the Mediterranean coasts. The distribution is Western Mediterranean for mariae and Eastern Mediterranean for zammitii and phoeniciae, without overlapping of the ranges (Fig. 2). A complete sterility barrier was demonstrated between phoeniciae and the other two members of the complex, while only partial sterility limited to F_1 males was found between Ae mariae and Ae zammitii. In all crosses vigorous adult hybrids were obtained. On the other hand, no precopulatory barrier to gene flow was revealed in the laboratory.

A series of release experiments was performed, involving populations of the three species monomorphic for alternative Pgm alleles. Regular sampling in the release areas was taken from 1970 to 1975 (more than 70 generations). The results showed an almost complete absence of hybrids, which never exceeded 2% (Table II), indicating that highly efficient precopulatory RIMs are operating [8,26]. These mechanisms appear related to differences in the premating behavior of the three species, as indicated by the electrophoretic test of nuptial swarms in the release areas. Heterogeneous swarms including more than one species were observed at the beginning of the nuptial flight. Later on only homospecific swarms were present. In Ae mariae, nuptial swarms remain near the rocks along the shore, while in Ae zammitii they generally take place at a height of 2 m from the top of the rocks. The latter behavior appears to be the most primitive, being similar to that of Ae caspius, a possible ancestor of the Ae mariae complex [27].

The present distribution of the species of the Ae mariae complex suggests that the process of speciation originated from a disjunction of a primitive circum-Mediterranean range. Assuming such a process of speciation as truly allopatric, the precopulatory barriers existing between these sibling species, as evidenced by the release experiments, would appear to be a by-product of their genetic divergence. As an alternative hypothesis, premating isolation would be the product of selection against hybridization. This would have operated in areas where the species ranges came into contact.

Each release experiment led, after 1–2 yr, to the survival of a single species, indicating that competitive exclusion takes place between the members of the Ae mariae complex. This agrees well with the lack of sympatric areas between the three species, in spite of their sometimes closely contiguous ranges (for instance, Ae mariae and Ae zammitii are separated in Sicily by only 2–3 km of sandy coasts). Competitive exclusion is presumably present also between An

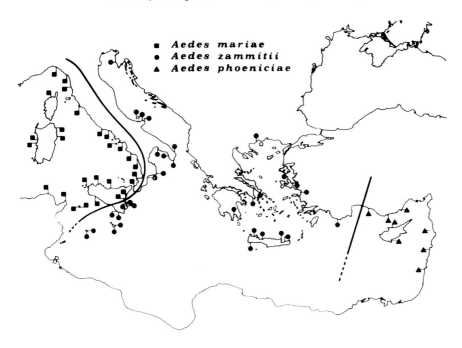

Fig. 2.   Map of distribution of the Aedes mariae complex in the Central and Eastern Mediterranean. Aedes mariae (squares), Ae zammitii (circles), and Ae phoeniciae (triangles).

labranchiae and An atroparvus in central Italy, where their ranges come into contact. The role of this phenomenon in limiting species spread requires further investigation.

PREMATING RIMs AND NICHE DIFFERENTIATION IN THE EUROPEAN POPULATIONS OF THE CULEX PIPIENS COMPLEX

An approach similar to that used in the study of premating RIMs in the Ae mariae complex was applied by us to test the possible role of stenogamy in preventing gene flow between Culex pipiens pipiens and C pipiens molestus [27,28]. The type form C pipiens pipiens (the so-called "rural" form) is the most primitive one, breeding in unpolluted water, showing pronounced ornithophily and winter diapause at the adult stage (eterodinamy), needing a blood meal for egg maturation (anautogeny), and performing nuptial flights in wide sites (eurigamy); this form does not adapt to man and to his environment, while the inverse is true of C pipiens molestus (the so-called "urban" form). The latter is

TABLE II. Absolute Frequencies of PGM Genotypes in Samples of the Aedes mariae Complex Collected at Different Times After the Release Dates*

	PGM genotype						Total tested
	AA	AB	AC	BB	BC	CC	
June 20, 1971	83	1	—	109	1	105	299
July 4, 1971	162	1	—	145	—	116	424
July 16, 1971	117	—	1	161	1	108	388
July 28, 1971	96	—	—	116	1	86	299
August 9, 1971	61	—	1	81	—	50	193
August 21, 1971	71	1	1	94	2	62	231
September 5, 1971	50	1	—	65	1	28	145
September 19, 1971	54	1	—	90	—	60	205
June 24, 1973	161	1	—	99	2	106	369
July 8, 1973	121	—	—	214	1	127	463
July 20, 1973	141	1	1	228	1	128	500
August 2, 1973	86	1	—	174	2	122	385
August 18, 1973	47	—	1	73	1	52	174
September 1, 1973	42	1	—	121	—	35	199
Total tested	1292	9	5	1770	13	1185	4274

*June 6, 1971 and June 10, 1973 (Ae zammitii:Pgm^A/Pgm^A; Ae mariae:Pgm^B/Pgm^B; Ae.phoeniciae:Pgm^C/Pgm^C).

highly adapted to man; its preimaginal stages breed in polluted water drains and ponds, or in septic tanks; adult females take their blood meal from man (anthropophily) or even need no blood meal for egg maturation (autogeny); nuptial flights take place in narrow sites (stenogamy). At the morphological level, two larval characters, siphon shape and the number of mentum teeth, are the most effective in differentiating C pipiens pipiens and C pipiens molestus. Laboratory crosses show that no postmating RIMs between these two forms exist. The hybrids, obtained in the expected numbers, are completely fertile and perfectly vital.

Field experiments, carried out releasing electrophoretically detectable samples of C pipiens molestus in areas occupied by C pipiens pipiens, indicated a very low rate of hybridization between these two forms (about 0.01). Moreover, the electrophoretic assay of males from nuptial flights taking place at different heights from the ground showed the presence of homogeneous swarms, consisting of molestus males near the ground, and of pipiens males near the foliage of trees, at a height of 2–3 m [28]. The stenogamy of C pipiens molestus represents, therefore, a very efficient premating RIM. A certain amount of gene flow between the two forms, preventing their complete reproductive isolation, is however allowed by intermediate populations, having both stenogamous and eurygamous individuals. On the other hand, these "intermediate" populations appear to have a higher fitness in the very heterogeneous environments found in industrialized countries. The existence of "intermediate" populations, together with the well-differentiated "urban" and "rural" ones, was confirmed by the analysis of the genetic structure of several C pipiens populations from Italy, variously adapted to man and his environment (Figs. 3, 4). Nei's average genetic distance between the typical pipiens and molestus populations was found to be 0.06 [28].

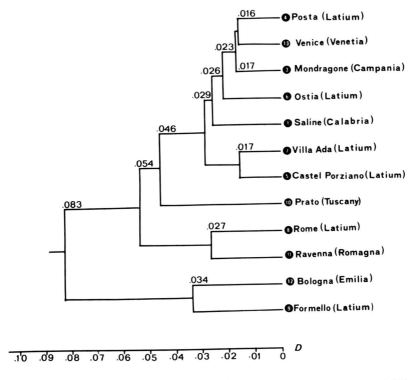

Fig. 3. Dendrogram of the genetic relationships among Italian populations of Culex pipiens. D, Nei's genetic distance. Populations 9, 12, "rural" form; 8, 11, "urban" form; from 1 to 7 and 10, "intermediate" form [from 28].

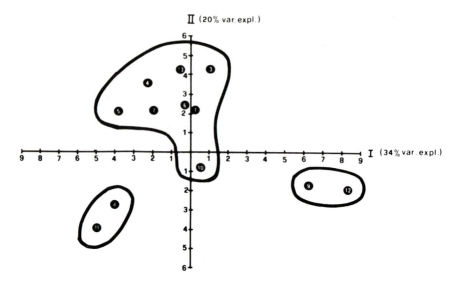

Fig. 4.   Multivariate analysis of Culex pipiens populations (clustering method by Jancey); the numbers representing the populations correspond to those in Figure 3 [from 28].

From the ecological point of view, the "urban" form of C pipiens has achieved a shift from the primitive niche of the "rural" form to a new one, trophically much richer. This evolutionary process apparently took place without relevant changes at the structural genes level, as indicated by the low values of genetic distance. Obviously, classes of genes other than those studies by electrophoresis, like regulatory genes, might be involved in this process to a greater extent [3].

EVOLUTION OF PREMATING RIMs IN THE EUROPEAN CORN BORER, OSTRINIA NUBILALIS

As we have seen, premating behaviors like nuptial flight may play an important role in reproductive isolation of sibling or incipient mosquito species. The nuptial flight pattern may evolve as a by-product of genetic divergence or of ecologic adaptation (eg, in Culex pipiens), but more often it is an ad hoc product of natural selection. Little is known about the genetic basis of the evolution of premating behavior. In the case of C pipiens the shift from eurigamy to stenogamy was found to require changes in two distinct loci, where polymorphism is extremely widespread.

Another genetic polymorphism of a character directly implied in premating isolation was observed in the European corn borer, Ostrinia nubilalis. This is a cosmopolitan moth, originally distributed in Europe and from there introduced into America. O nubilalis has a pheromone communication system. In the range of the species two forms were found, characterized by a different composition of the pheromone secreted by the females [29,30]. Average genetic distance between these two forms, generally allopatric, is 0.02 [31].

Cross experiments showed that the isomer composition of female pheromone is inherited as a monofactorial character; hybrids are fully fertile and obtained in expected numbers; hybrid females have an intermediate pheromone blend [32]. We observed a polymorphic case near Bologna [unpublished data]; heterozygote females were at the expected frequency and attracted chiefly hybrid males. A quite different situation was found in Pennsylvania where the two pheromone forms occur sympatrically, apparently without interbreeding [33].

The understanding of the possible incipient speciation of O nubilalis obviously requires more detailed studies of these two pheromone strains—their geographic distribution, their areas of sympatry, and polymorphic localities such as the one reported near Bologna. Changes in mating pheromones certainly represent possible speciation mechanisms, which can evolve either sympatrically or allopatrically.

PREMATING RIMs IN THE AMATA PHEGEA COMPLEX

We observed premating RIMs completely different from O nubilalis mating pheromones in Ctenuchid moths of the Amata phegea complex. Four species of this complex were found in Italy: A phegea, A marjana, A ragazzii, and A kruegeri, which were confused by systematics up to recently [34]. These species presumably diverged through geographic isolation during Pleistocene glaciations. The electrophoretic study of genetic divergence showed that the species most related are A phegea, A marjana, and A ragazzii (Figs. 5, 6). A phegea and A ragazzii were found together in a number of areas of Central and Southern Italy, while A phegea and A marjana are sympatric in Istria and Dalmatia. According to our research the more effective isolating mechanism in these sympatric areas is represented by a seasonal displacement in the flight period [35]. A phegea, the most widespread species of the complex, emerges from April to August, according to the localities. Where this species occurs sympatrically with A ragazzii, its emergence regularly precedes by about a month that of A ragazzii (Fig. 7). But where A phegea and A marjana ranges overlap, A marjana is the first to fly, and also, in this case the emergence times of the two species are separated by about 1 month. Premating behavior of homospecific and heterospecific pairs, investigated in the laboratory, showed the existence of isolating mechanisms also at the ethological level that lower the mating success of het-

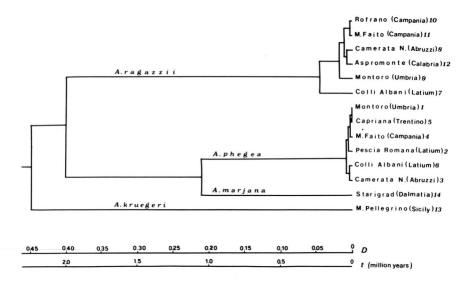

Fig. 5. Dendrogram of the presumed phylogenetic relationships among species of the Amata phegea complex. D, Nei's genetic distance; t, time of evolutionary divergence.

erospecific pairs. Rare natural hybrids (identifiable by electrophoresis) were observed between A phegea and A ragazzii in some sympatric areas, with a frequency ranging from 0 to 5% according to the locality; the highest rates of hybridization were found in biotopes disturbed by man [35].

CHARACTERIZATION OF THE TROPHIC NICHE IN AMATA PHEGEA AND A RAGAZZII

The occurrence of overlap areas between the ranges of A phegea and A ragazzii allowed us to investigate their trophic niche both in sympatric and allopatric conditions [36]. The analysis of the ecological niche of partially sympatric sibling species is expected to show possible correlations between niche differentiation and speciation.

A phegea and A ragazzii larvae are extremely polyphagous; their trophic fundamental niche is very similar. In nature the larvae of both species are active and carry on feeding from autumn until late spring. Field observations, involving quantitative estimates, showed that in sympatric conditions A phegea and A ragazzii larvae subdivide trophic resources following a vertical transect. The former feed mainly on grasses and dried leaves in the forest bed, while the latter

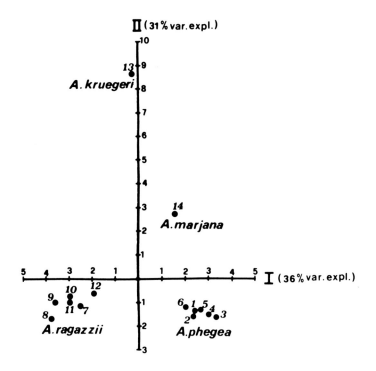

Fig. 6. Multivariate analysis of species of the Amata phegea complex (rq factor analysis). The numbers representing the populations correspond to those in Figure 5.

feed mainly on the bark, buds, and new leaves of trees and shrubs. The percentage of the two species in the two parts of this transect (from 0 to 20 cm and from 20 cm up to 2 m from the ground) is significantly different (Table III). This does not seem to happen in allopatric areas, where the vertical distribution is not dependent on the species, but only on local ecological conditions. Competition appears then to be the cause of this case of niche displacement.

TROPHIC NICHE OF THE MEDITERRANEAN STICK-INSECTS

A second example of trophic niche subdivision is shown in the three Mediterranean stick-insects—Clonopsis gallica, Bacillus rossius, and B atticus [37]. The trophic fundamental niche of the first two species, both highly polyphagous, largely overlaps, while that of the third one, B atticus, is more specialized. The hybrids between B atticus and B rossius show an intermediate degree of poly-

TABLE III. Relative Frequencies of Amata phegea and A ragazzii Larvae Observed in Sympatric Areas of Latium (Italy) Along a Vertical Transect* [from 36]

Species	San Cesareo (1975)			San Cesareo (1976)			Alban Hills (1975)		
	0–20 cm	20 cm–2 m	Total observed larvae	0–20 cm	20 cm–2 m	Total observed larvae	0–20 cm	20 cm–2 m	Total observed larvae
A. phegea	0.70	0.16	73	0.61	0.15	280	0.60	0.07	140
A ragazzii	0.30	0.84	80	0.39	0.85	216	0.40	0.93	334
Total observed larvae	90	63	153	444	52	496	199	275	474
	$\chi^2 = 41.38$ $P < 0.005$			$\chi^2 = 38.01$ $P < 0.005$			$\chi^2 = 155.99$ $P < 0.001$		

*From 0 to 20 cm and from 20 cm up to 2 m from the ground.

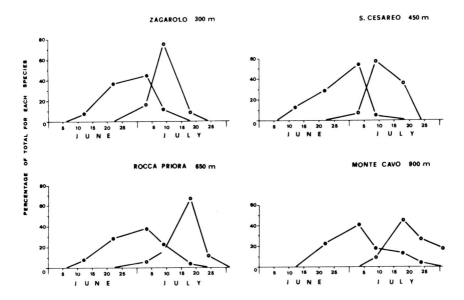

Fig. 7. Phenologies of adult populations of Amata phegea (black circles) and A ragazzii (doughnuts) in four localities south east of Rome, showing temporal displacement in the flight period [from 35].

phagy. In sympatric areas Cl gallica and B rossius subdivide trophic resources, the former feeding mainly on Rosa, Crataegus, and Prunus, the latter on Rubus. Also B rossius and B atticus, when sympatric, show a differentiated trophic niche, B rossius feeding on Rosaceae, B atticus mainly on Pistacia lentiscus. In the latter case niche subdivision represents also an efficient ecological premating barrier.

GENETIC DIFFERENTIATION AND SPECIATION IN MEDITERRANEAN STICK-INSECTS

The electrophoretic study of genetic divergence, together with crossing experiments and field observations, allowed us to show interesting speciation processes, already achieved or occurring at present in stick-insects of the Mediterranean region. The level of vagility in these apterous insects is exceptionally low; they are scattered in many small or medium-sized populations and their reproduction can be both bisexual and parthenogenetic. Two chief modes of speciation seem to occur in Mediterranean stick-insects: parapatric, as in the B rossius complex, and through hybridization, as in B atticus and B whitei.

Six taxa, provisionally treated as subspecies, have been recognized up to now in the B rossius complex [38]: rossius, occurring in Spain, France, northwestern and central-western Italy; redtenbacheri, found in southwestern Italy, in Sicily, on the Italian Adriatic coasts, in Yugoslavia, Albania, Greece, and Turkey; tripolitanus A, present in western Tunisia; tripolitanus B, of northeastern Algeria; lobipes, of eastern Algeria; montalentii, of northcentral Algeria. Their ranges are often contiguous, but no overlap areas have been found to this time. Many populations of B rossius rossius and B rossius redtenbacheri reproduce by constant thelitoky that evolved independently in different geographic areas. The other populations of the complex are bisexual, but unfertilized females can reproduce by parthenogenesis [39]. The members of the B rossius complex are rather well differentiated at the genic level but not at the karyological one (2n = 36 in the females, 35 in the males). A dendrogram showing their presumed phylogenetic relationships is given in Figure 8. North African populations appear

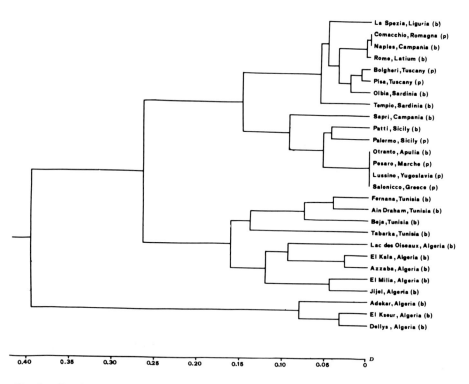

Fig. 8.   Dendrogram of the presumed phylogenetic relationships among bisexual (b) and parthenogenetic (p) populations of the Bacillus rossius complex. D, Nei's genetic distance [from 38].

to have the highest variability, probably being of more ancient origin. European populations, especially the eastern ones, show a much lower variability. Complete homozygosity is found in the parthenogenetic populations; this is due to the diploidization mechanism, which involves endomitotic doubling of the chromosome number in the nuclei of the blastoderm [40]. In the field, gene flow apparently occurs only at the borders between adjacent populations. Partial postmating barriers among members of the B rossius complex have been observed in the laboratory.

The first case of speciation by hybridization in stick-insects was demonstrated in the B atticus complex, on the basis of both allozyme and chromosome studies [38,41, and unpublished data]. One of B atticus bisexual ancestors, B grandii, was recently discovered in southern Sicily [42]. Four parthenogenetic taxa are included in this complex, differing both in external morphology and genetic structure [38,43,44]: B atticus atticus from Attica and Peloponnese, B atticus baccettii from Epirus, B atticus mülleri from Dalmatia and Istria, and B atticus caprai from southern Italy (Fig. 9). All of them are diploid, with the same chromosome number found in B grandii females (2n = 34), and they reproduce by thelitoky. About one-third of the enzyme-loci tested show fixed heterozy-

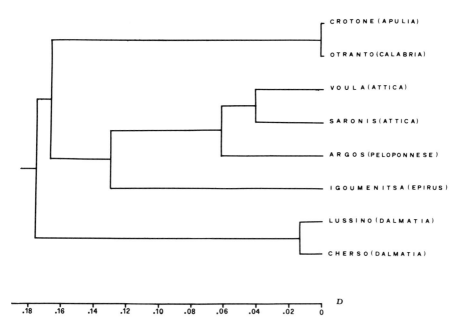

Fig. 9. Dendrogram of the presumed phylogenetic relationships among populations of the Bacillus atticus complex. D, Nei's genetic distance [from 38].

TABLE IV. Genotype Frequencies of 24 Enzyme Loci in the Isidora truncata Complex

Loci	Genotypes	Gyza (Egypt)	Cairo (Egypt)	Alessandria (Egypt)	Assuan (Egypt)	Tai'zz (Yemen)	Poakwe (Ghana)	Akuase (Ghana)	Dakar (Senegal)
Odh	100/100	1.00	1.00	1.00	1.00	1.00	1.00	1.00	1.00
Ldh	100/100	1.00	1.00	1.00	1.00	1.00	1.00	1.00	1.00
Mdh-1	100/100	1.00	1.00	1.00	1.00	1.00	1.00	1.00	1.00
Mdh-2	100/100	1.00	1.00	1.00	1.00	1.00	1.00	1.00	1.00
Sdh	100/100	1.00	1.00	1.00	1.00	1.00	1.00	1.00	1.00
Xdh-1	100/100	—	0.10	—	—	—	—	—	1.00
	100/104	1.00	0.87	—	1.00	1.00	1.00	1.00	—
	104/104	—	0.03	1.00	—	—	—	—	—
Xdh-2	100/100	1.00	1.00	1.00	1.00	1.00	1.00	1.00	1.00
α-Gpdh-1	100/100	1.00	1.00	1.00	1.00	1.00	1.00	1.00	1.00
α-Gpdh-2	100/100	1.00	1.00	1.00	1.00	1.00	1.00	1.00	1.00
Hbdh-1	100/108	1.00	1.00	1.00	1.00	1.00	1.00	1.00	1.00
Hbdh-2	100/100	1.00	1.00	1.00	1.00	1.00	1.00	1.00	1.00
Idh-1	100/100	—	—	—	—	—	—	—	—
	100/106	1.00	1.00	1.00	1.00	1.00	1.00	1.00	1.00
Idh-2	100/100	1.00	1.00	1.00	1.00	1.00	1.00	1.00	1.00
6-Pgdh	94/94	0.20	—	1.00	—	—	—	—	—
	89/102	—	—	—	—	—	—	—	—
	89/94/100	0.30	—	—	—	—	—	—	—
	94/94	—	—	—	—	1.00	—	—	—
	94/100	0.50	—	1.00	1.00	—	1.00	1.00	1.00
	100/100	—	—	—	—	—	—	—	—
Ao	100/100	1.00	1.00	1.00	1.00	1.00	1.00	1.00	1.00
Pgm-2	95/100	—	—	—	—	1.00	—	—	—
	100/100	1.00	1.00	1.00	1.00	—	1.00	1.00	1.00
Got	100/100	1.00	1.00	1.00	1.00	1.00	1.00	1.00	1.00
Adk	100/105	1.00	1.00	1.00	1.00	1.00	1.00	1.00	—
	100/110	—	—	—	—	—	—	—	1.00
Est-2	100/108	1.00	1.00	1.00	1.00	—	1.00	1.00	1.00
	108/108	—	—	—	—	1.00	—	—	—
Est-3	96/100	—	—	—	—	1.00	—	1.00	—
	100/100	1.00	1.00	1.00	1.00	—	1.00	—	1.00
Np	100/100	1.00	1.00	1.00	1.00	1.00	1.00	1.00	1.00
Pgi	100/100	1.00	1.00	1.00	1.00	1.00	—	1.00	1.00
	100/104	—	—	—	—	—	—	—	—
	100/106	—	—	—	—	—	1.00	—	—
Mpi	100/100	1.00	1.00	1.00	1.00	1.00	1.00	1.00	1.00
Sod	100/100	1.00	1.00	1.00	1.00	1.00	1.00	1.00	1.00

bha bya)	Rabat (Morocco)	S Teodoro (Sardinia)	Posada (Sardinia)	S Giovanni (Sardinia)	Tortoli (Sardinia)	Escalaplano (Sardinia)	Padria (Sardinia)
00	1.00	1.00	1.00	1.00	1.00	1.00	1.00
00	1.00	1.00	1.00	1.00	1.00	1.00	1.00
00	1.00	1.00	1.00	1.00	1.00	1.00	1.00
00	1.00	1.00	1.00	1.00	1.00	1.00	1.00
00	1.00	1.00	1.00	1.00	1.00	1.00	1.00
—	—	1.00	—	1.00	1.00	0.33	1.00
00	—	1.00	—	—	0.67	—	
—	—	—	—	—	—	—	—
00	1.00	1.00	1.00	1.00	1.00	1.00	1.00
00	1.00	1.00	1.00	1.00	1.00	1.00	1.00
00	1.00	1.00	1.00	1.00	1.00	1.00	1.00
00	1.00	1.00	1.00	1.00	1.00	1.00	1.00
00	1.00	1.00	1.00	1.00	1.00	1.00	1.00
00	—	—	—	—	—	—	—
—	1.00	1.00	1.00	1.00	1.00	1.00	1.00
00	1.00	1.00	1.00	1.00	1.00	1.00	1.00
—	—	—	—	—	—	—	
00	—	—	—	—	—	—	—
—	—	—	—	—	—	—	—
—	—	—	—	—	—	—	—
—	1.00	1.00	1.00	—	1.00	1.00	1.00
—	—	—	—	1.00	—	—	—
00	1.00	1.00	1.00	1.00	1.00	1.00	1.00
—	—	—	—	—	—	—	—
00	1.00	1.00	1.00	1.00	1.00	1.00	1.00
00	1.00	1.00	1.00	1.00	1.00	1.00	1.00
00	1.00	—	—	—	—	1.00	1.00
—	—	1.00	1.00	1.00	1.00	—	—
00	1.00	1.00	0.20	1.00	1.00	0.67	1.00
—	—	0.80	—	—	0.33	—	—
—	—	1.00	—	1.00	1.00	0.40	1.00
00	1.00	—	1.00	—	—	0.60	—
00	1.00	1.00	1.00	1.00	1.00	1.00	1.00
—	1.00	1.00	1.00	—	1.00	1.00	1.00
00	—	—	—	—	—	—	—
—	—	—	—	1.00	—	—	—
0	1.00	1.00	1.00	1.00	1.00	1.00	1.00
0	1.00	1.00	1.00	1.00	1.00	1.00	1.00

gosity, apparently maintained by the fusion of the products of the first meiotic division. A normal meiosis seems to take place, on the contrary, when B atticus females are fertilized by males of bisexual species, as indicated by the fact that the produced offspring is diploid and shows Mendelian segregation. This occurs, for instance, in experimental crossings with B rossius: in this case the F_1 males appear to be fertile, while females are sterile (atrophic ovaries). Natural hybrids between these two species were occasionally found in southern Italy, where B atticus caprai coexists with B rossius redtenbacheri [41,43].

Another case of speciation by hybridization is that of Bacillus whitei [42]. The bisexual ancestors of this diploid thelitokous species are B rossius and B grandii: their haploid chromosome sets are both recognizable in B whitei karyotype (2n = 35). Also at the electrophoretic level B whitei shows the coexistence of B rossius and B grandii alleles, with more than 50% of loci fixed in heterozygous condition [38]. B whitei is presently widespread in southern Sicily, where it is apparently eliminating its parental species.

HYBRID ORIGIN OF THE ISIDORA TRUNCATA COMPLEX

The snails of the Isidora truncata complex (previously included in the genus Bulinus) are important vectors of schistosomiasis in Africa and in the Mediterranean basin; they were separated by some authors into different species (I truncata, I contorta, I yemenensis, I rohlfsi, and I rivularis), while being considered by others as a single species [45–47]. Electrophoretic studies of samples with different geographic origin (from Egypt, Yemen, Libya, Senegal, Ghana, Morocco, Sardinia, and Corsica, etc) showed a continuous range of genetic variation existing in this complex, with a number of loci fixed in heterozygous condition [48]. Geographic populations of the I truncata complex differ in the number of loci fixed in heterozygosity, and in the alleles implied (Table IV). All populations are tetraploid, many being completely or partially aphallic.

On the basis of electrophoretic data, speciation of the I truncata complex seems to have occurred through hybridization and tetraploidization [48]. This process gave origin to individuals which, taking advantage of their initial heterosis due to their hybrid origin, were able to colonize successfully wide areas, unfavorable to their diploid ancestors (possible ancestors appear to be I natalensis and I tropica, two diploid species showing many of the alleles found in the I truncata complex). The subsequent splitting of the I truncata complex in an indefinitely large number of different biotypes appears to be related to the low vagility of these snails and to their reproduction by self-fertilization.

CONCLUDING REMARKS

The need for more information in order to reach an adequate knowledge of speciation mechanisms has been recently stressed by various authors, such as Mayr [49], Bush [50], and White [4]. It is, for instance, amazing how little is still known about the ecology and ethology of widely studied species like Anopheles mosquitoes, Drosophila flies, etc. White, in his recent book, Modes of Speciation, after listing the data that are necessary to a full understanding of any speciation event, concludes: "It is perhaps unnecessary to state that there is probably no single case of speciation for which all of this information exists." The need for these gaps to be filled is represented by the fact that "again and again we encounter evidence that should warn us against generalizations on mechanisms of speciation based on single groups of organisms, however thoroughly studied."

A number of biologists now begin to share the opinion that general speciation models are to be considered with caution, while many particular cases need a thorough analysis, now made possible by the new techniques available. The exponential increase of research in evolutionary biology during the last 20 years provides hope that this work will be achieved.

REFERENCES

1. Lewontin RC: The Genetic Basis of Evolutionary Change. New York: Columbia University Press, 1974.
2. Avise JC: Systematic value of electrophoretic data. Syst Zool 23:465, 1974.
3. Ayala FJ: Genetic differentiation during the speciation process. Evol Biol 8:1, 1975.
4. White MJD: Modes of Speciation. San Francisco: Freeman 1978.
5. Bullini L, Sbordoni V: Electrophoretic studies of gene-enzyme systems: Microevolutionary processes and phylogenetic inference. Boll Zool 47 (Suppl):95, 1980.
6. Ferguson A: Biochemical Systematics and Evolution. Glasgow: Blackie & Son, 1980.
7. Mayr E: Systematics and the Origin of Species. New York: Columbia University Press, 1942.
8. Coluzzi M, Bullini L: Enzyme variants as markers in the study of precopulatory isolating mechanisms. Nature 231:455, 1971.
9. Baylis HA: Helmints and evolution. In: Evolution: Essays on Aspects of Evolutionary Biology Presented to Professor ES Goodrich. Oxford: Clarendon Press, 1938, p 249.
10. Biocca E, Nascetti G, Iori A, Costantini R, Bullini L: Descrizione di Parascaris univalens (Hertwig, 1890) parassita degli equini e suo differenziamento da Parascaris equorum (Goeze, 1782). Acc Naz Lincei Rend Cl Sc Fis Mat Nat 65:133, 1978.
11. Bullini L, Nascetti G, Ciafrè S, Rumore F, Biocca E: Ricerche cariologiche ed elettroforetiche su Parascaris univalens (Hertwig, 1890) e Parascaris equorum (Goeze, 1782). Acc Naz Lincei Rend Cl Sc Fis Mat Nat 65:151, 1978.

12. Bullini L, Nascetti G, Grappelli C: Nuovi dati sulla divergenza e sulla variabilità genetica delle specie gemelle Ascaris lumbricoides—A. suum e Parascaris univalens—P.equorum. Parassitologia 23 (in press).

13. Nascetti G, Grappelli C, Bullini L: Ricerche sul differenziamento genetico di Ascaris lumbricoides e Ascaris suum. Acc Naz Lincei Rend Cl Sc Fis Mat Nat 67:457, 1979.

14. Nascetti G, Paggi L, Orecchia P, Mattiucci S, Bullini L: Divergenza genetica in popolazioni del genere Anisakis del Mediterraneo. Parassitologia 23 (in press).

15. Kitzmiller JB, Frizzi G, Baker RH: Evolution and speciation within the maculipennis complex of the genus Anopheles. In Wright JW, Pal R (eds): Genetics of Insects Vectors of Disease. Amsterdam: Elsevier, 1976, p 151.

16. Kitzmiller JB: Genetics, cytogenetics and evolution in mosquitoes. Adv Genet 18:315, 1976.

17. Bullini L, Coluzzi M: Electrophoretic studies on gene-enzyme systems in mosquitoes (Diptera, Culicidae). Parassitologia 15:221, 1973.

18. Bullini L, Coluzzi M: Applied and theoretical significance of electrophoretic studies in mosquitoes (Diptera, Culicidae). Parassitologia 20:7, 1978.

19. Bullini L, Coluzzi M: Evolutionary and taxonomic inferences of electrophoretic studies in mosquitoes. In Steiner WWM, Tabachnik WJ, Rai KS, and Narang S (eds): Recent Developments in the Genetics of Insect Disease Vectors. Stipes Publishing Co, Champaign, Illinois (in press).

20. Cianchi R, Urbanelli S, Coluzzi M, Bullini L: Genetic distance between two sibling species of the Aedes mariae complex (Diptera, Culicidae). Parassitologia 20:39, 1978.

21. Bullini L, Bianchi Bullini AP, Cianchi R, Sabatini A, Coluzzi M: Tassonomia biochimica del complesso Anopheles maculipennis. Parassitologia 22:290, 1980.

22. Cianchi R, Sabatini A, Bullini L, Coluzzi M: Differenziazione morfologica e genetica nei complessi Anopheles maculipennis e An.claviger. Parassitologia 23 (in press).

23. Urbanelli S, Sabatini A, Bullini L: Tassonomia morfologica e biochimica di Culex pipiens e C. torrentium. Parassitologia 23 (in press).

24. Bianchi Bullini AP, Bullini L, Cianchi R, Coluzzi M, Todini M: Differenziamento genetico nel complesso Anopheles maculipennis. Atti Ass Genet Ital 25:39, 1980.

25. Coluzzi M, Sabatini A, Petrarca V, Di Deco M: Chromosomal differentiation and adaptation to human environments in the Anopheles gambiae complex. Trans R Soc Tr Med Hyg 73:483, 1979.

26. Coluzzi M, Bianchi Bullini AP, Bullini L: Speciazione nel complesso mariae del genere Aedes (Diptera, Culicidae). Atti Ass Genet Ital 21:218, 1976.

27. Bullini L, Coluzzi M: Ethological mechanisms of reproductive isolation in Culex pipiens and Aedes mariae complexes (Diptera-Culicidae). Monitore Zool Ital (NS)14:99, 1980.

28. Urbanelli S, Cianchi R, Petrarca V, Sabatinelli G, Coluzzi M, Bullini L: Adaptation to the urban environment in the mosquito Culex pipiens (Diptera, Culicidae). In Moroni A, Ravera O, Anelli A (eds): Ecologia, Atti I Congr Naz SITE. Parma: Zara 1981, p 305.

29. Klun JA, Robinson JF: European corn borer moth: Sex attractant and sex attraction inhibitors. Ann Entomol Soc Am 64:1083, 1971.

30. Cardè RT, Roelofs WL, Harrison RC, Vawter AT, Brussard PF, Mutuura A, Munroe

E: European corn borer: Pheromone polymorphism or sibling species? Science 199:555, 1978.

31. Cianchi R, Maini S, Bullini L: Genetic distance between pheromone strains of the European corn borer, Ostrinia nubilalis: Different contribution of variable substrate, regulatory and non regulatory enzymes. Heredity 45:383, 1980.

32. Klun JA, Maini S: Genetic basis of an insect chemical communication system: The European corn borer. Environ Entomol 8:423, 1979.

33. Cardè RT, Kochansky J, Stimmel JF, Wheeler AG Jr, Roelofs WL: Sex pheromone of the European corn borer (Ostrinia nubilalis): Cis and trans responding males in Pennsylvania. Environ Entomol 4:413, 1975.

34. Bullini L, Cianchi R, Stefani C, Sbordoni V: Biochemical taxonomy of the Italian species of the Amata phegea complex (Ctenuchidae, Syntominae). Nota Lepid 4:125, 1981.

35. Sbordoni V, Bullini L, Bianco P, Cianchi R, De Matthaeis E, Forestiero S: Evolutionary studies on Ctenuchid moths of the genus Amata. 2. Temporal isolation and natural hybridization in sympatric populations of Amata phegea and A. ragazzii. J Lepid Soc (in press).

36. Sbordoni V, Bianco P, Bullini L: Trophic niche displacement in coexisting populations of Amata phegea and A.ragazzii (Lepidoptera, Ctenuchidae). In Moroni A, Ravera O, Anelli A (eds): Ecologia, Atti I Congr Naz SITE. Parma: Zara, 1981, p 303.

37. Bullini L: Trophic niche of the Mediterranean stick-insects Clonopsis gallica, Bacillus rossius and B. atticus (Cheleutoptera, Bacillidae). In Moroni A, Ravera O, Anelli A (eds): Ecologia, Atti I Congr Naz SITE. Parma: Zara, 1981, p 263.

38. Nascetti G, Bullini L: Differenziamento genetico e speciazione in fasmidi dei generi Bacillus e Clonopsis (Cheleutoptera, Bacillidae). Atti XII Congr Naz Entomol, Roma (in press).

39. Bullini L: Ricerche sulle caratteristiche biologiche dell'anfigonia e della partenogenesi in una popolazione bisessuata di Bacillus rossius (Rossi). Riv Biol 58:189, 1965.

40. Pijnacker LP: Automictic parthenogenesis in the stick insect Bacillus rossius Rossi (Cheleutoptera, Phasmidae). Genetica 40:393, 1969.

41. Goday C, Bianchi Bullini AP, Nascetti G, Bullini L: Chromosome studies on Bacillus atticus, B rossius, and their hybrids (Cheleutoptera, Bacillidae). Acc Naz Lincei Rend Cl Sc Fis Mat Nat 71 (in press).

42. Nascetti G, Bullini L: Bacillus grandii n. sp. and B. whitei n. sp.: Two stick insects from Sicily (Cheleutoptera, Bacillidae). Boll Ist Ent Univ Bologna 36:245, 1982.

43. Nascetti G, Bullini L: A new phasmid from Italy: Bacillus atticus caprai (subsp. n.) (Cheleutoptera, Bacillidae). Fragmenta Entomol. 16:143, 1982.

44. Bullini L, Nascetti G: Morphological and genetic differentiation in the Bacillus atticus complex (Cheleutoptera, Bacillidae). Boll Ist Ent Univ Bologna 37 (in press).

45. Mandahl-Barth G: Intermediate hosts of Schistosoma: African Biomphalaria and Bulinus. World Health Organization, Monograph Series 37, 1958.

46. Biocca E, Bullini L, Chabaud A, Nascetti G, Orecchia P, Paggi L: Suddivisione su base morfologica e genetica del genere Bulinus in tre generi: Bulinus Müller, Physopsis Krauss e Mandahlbarthia gen. nov. Acc Naz Lincei Rend Cl Sc Fis Mat Nat 66:275, 1979.

47. Biocca E, Bullini L, Chabaud A: Classification of the Subfamily Bulininae and of the Isidora truncata Complex on Morphogenetic Criteria. In Cunning EU (ed): Parasitological Topics. A Presentation Volume to PCC Garnham SRS. Lawrence, Kansas: Allen Press, 1981, p 34.
48. Nascetti G, Bullini L: Genetic differentiation in the Mandahlbarthia truncata complex (Gastropoda: Planorbidae). Parassitologia 22:269, 1980.
49. Mayr E: Populations, Species and Evolution: An Abridgement of Animal Species and Evolution. Cambridge, Massachussets: The Belknap Press of Harvard University, 1970.
50. Bush GL: Modes of animal speciation. Ann Rev Ecol Systemat 6:339, 1975.

Mechanisms of Speciation, pages 265–305
© 1982 Alan R. Liss, Inc., 150 Fifth Avenue, New York, NY 10011

The Evolutionary Genetics of a Unisexual Fish, Poecilia formosa

Bruce J. Turner

INTRODUCTION

There are at present fewer than 50 known species of vertebrate parthenoforms.[1] Given that there are more than 20,000 species of teleosts alone, and probably an equal number of other verterate species, why pay any attention at all to such a minute, indeed trivial, proportion of seemingly aberrant forms? The answer to this reasonable question is severalfold: First, unisexual species may not be as rare as they seem. The relatively recent discovery that the European edible frog, Rana esculenta, a species first described by Linnaeus, is hybridogenetic [1,2] does not inspire confidence in the efficiency with which biologists have detected parthenoforms among vertebrates. The very recent discovery [3] of a unisexual species of the atherinid fish genus Menidia further erodes that confidence, for Menidia has been the object of lavish attention by American ichthyologists, attention that has included both traditional morphological and allozyme studies (reviews in [4,5]). Vertebrate parthenoforms may be more common, perhaps far more common, and their detection may be more difficult, than hitherto realized. The unisexual Menidia was discovered during an intensive allozyme survey of local populations of sympatric sibling species, one in part designed to detect

---

[1]The term "parthenoform" is used throughout to denote organisms with thelytokous breeding systems [6]. The thelytoky is presumably automictic, but may not be so in all cases. Parthenoforms may be either parthenogenetic, gynogenetic or hybridogenetic. As used here, "unisexual" is a synonym for "parthenoform" (even for Rana esculenta, where males are known). The interesting question of whether parthenoforms are really "species" (in either the biological or taxonomic sense [6,7]) will not be dealt with here. The term "bisexual" refers to conventionally gonochoristic animals with sexual reproduction.

small amounts of hybridization between them; it is extremely doubtful if the unisexual could have been detected by morphological analysis *at all*. Secondly, parthenoforms are ideal organisms with which to probe the adaptive significance of heterozygosity, for, by virtue of their hybrid ancestry, they are extremely heterozygous. In fact, the average heterozygosity per individual in most parthenoform populations is nearly an order of magnitude greater than that of their bisexual congeners. These enormous levels of heterozygosity (sometimes approaching 50% of all loci surveyed) are virtually without equivalent among bisexual animals. Third, the existence of successful parthenoforms, animals with no sexual recombination and clonal population structure, forces us to ask again the question: "What use is sex?" [8]. The answer may lie, at least in part, in some yet-to-be-made comparison of unisexuals and their bisexual relatives. Fourth, the reproductive biology of parthenoforms might ultimately tell us much about critical events in meiosis. Thelytoky obviously involves modifications (perhaps suppression) of at least some events that occur during normal meiosis. The hybrid ancestry of the vertebrate parthenoforms clearly implies a causality between thelytoky and the interaction of independantly differentiated genomes in the same organism. If this casuality could be resolved, important insight into meiotic mechanisms at the organelle or molecular levels might be gained. This paper reviews the biology of the first known unisexual vertebrate, Poecilia formosa,[2] with special reference to its clonal variation, the elucidation of its hybrid ancestry and attempts to "synthesize" the parthenoform in the laboratory.

Although Poecilia formosa (Girard) was described as a species in 1859, its reproductive biology was not known to be at all unusual until the initial report of its unisexuality by Hubbs and Hubbs [10].[3] The Hubbses (whose report was greeted with some incredulity—some biologists were reluctant to accept the existence of fish that reproduced like insects [11])—indicated that P formosa was an all-female species that could be propagated in the laboratory by matings with males of two different bisexual Poecilia species. The progeny of these matings were identical to their mothers in appearance, ie, reproduction was matroclinous. The Hubbs' observations, and their suggestion that P formosa was gynogenetic, were confirmed and extended by Meyer [12], Kallman [13], and

---

[2]P formosa is referred to as "Mollienesia" (or earlier, "Mollienisia") formosa in the older literature. "Mollienesia" is nowadays regarded as a junior synonym, or at most a subgenus (see [9]) of Poecilia.

[3]Three Hubbses have been involved in work with unisexual fishes: Carl L. Hubbs (1894–1979; see [21] for festschrift), his wife Laura C. Hubbs, and their son Clark. The latter is referenced in the bibliography as "C. Hubbs," the late senior Hubbs as "C. L. Hubbs."

their own extensive subsequent experiments [14–17]. Poecilia is now known to be one of three teleost genera that contain one or more unisexual species. The other two are Carassius (gynogenetic goldfish or "ginbuna" have been studied by Japanese and Soviet workers, see review in [18]) and Poeciliopsis (several parthenoforms from northwestern Mexico; reviews in [18,19]). Of the three, the unisexual Poeciliopsis have been the subject of by far the most research. Consequently, though they were discovered later, more is known of certain aspects of their biology than is the case with P formosa. The biology of the unisexual Poeciliopsis and Poecilia formosa are similar in some major respects, and appropriate parallels are discussed below. They differ, however, in an important way: Hybridogenesis (sensu [20]) is the predominant mode of reproduction in unisexual Poeciliopsis (gynogenesis occurs only in triploids), but, so far as is known, all unisexual Poecilia are gynogenetic.

The following are the salient features of the biology of P formosa:

Nomenclature and General Systematics

Poecilia formosa (Fig. 1) is a member of a speciose and widely distributed genus of the family Poeciliidae (order Atherinomorpha), a family of roughly 150 livebearing killifish-like fishes that include guppies (Poecilia reticulata), mosquitofish (Gambusia), platies, and swordtails (Xiphophorus) as well as aquarists' "mollies." The most recent monographic treatment of the family is that Rosen and Bailey [22]; reproductive biology is reviewed by Thibault and Schultz [23]. The recognized common name [24] of P formosa is the "Amazon molly" (the allusion in "Amazon" is, of course, Classical, not geographic; the name is apt, but unfortunately I can find little evidence of its use among contemporary biologists (except by writers of reviews on unisexual fishes).[4]

Geographic Range

The known range of P formosa (Fig. 2) extends from the mouth of the Rio Tuxpan, Estdo: Veracruz, Mexico, north to the Rio Grande basin in southern Texas in both coastal lagoons and freshwater streams [25,26]. Of the three range

---

[4]The existence of the common name may have seriously misled S. Ohno, for there is no evidence whatsoever in support of his statement ([29], p 6) that "One can purchase the Amazon molly (Poecilia formosa) from almost any pet shop that has a sizeable aquarium section, for this small (fish) has a rather pleasing coloration." In fact, the fish is one of the least prepossessing of the poeciliids (Girard's choice of the specific name "formosa" remains a minor ichthyological mystery) and enjoys little or no popularity among aquarists. It is at present not commercially available at all (the single fish farmer who bred the fish for research purposes ceased production in 1978) and it is next to impossible for experimental biologists to obtain material short of collecting it in the field.

Fig. 1. A) Poecilia formosa, specimen from 8 miles south of San Rafael, Tamaulipas, Mexico (UMMZ 184396, 56 mm standard length (SL)). B) F_1 laboratory hybrids, P latipinna male × P m mexicana female (see [75]); female above, (37 mm SL) sexually mature male below (32 mm SL). Note the close similarity of the female hybrid to P formosa.

extensions attributed to human activity by Darnell and Abramoff [25], two, the San Marcos River at San Marcos, Texas and the Rio Mante at Cd. Mante, Tamaulipas, Mexico, are still extant (B. J. Turner, field observations), and the San Marcos population is apparently thriving. The third presumptive introduction, in the lower Nueces River near Corpus Christi, Texas, actually may have

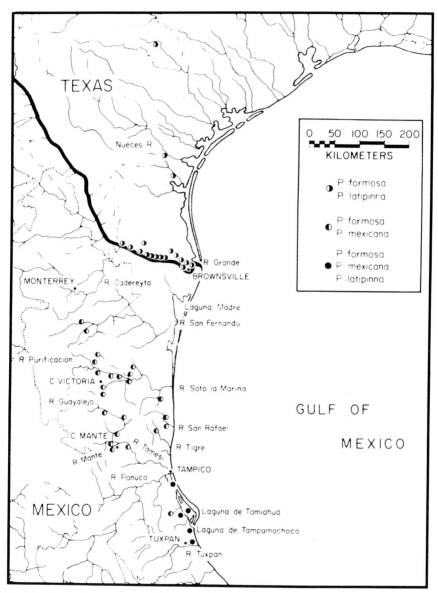

Fig. 2. Approximate geographic range of Poecilia formosa (modified from [25]). Host species as indicated.

been the fluctuating northern limit of the natural range of the species; to my knowledge, no P formosa have been taken in the Nueces since 1977. Darnell and Abramoff [25] suggested that the distribution of P formosa "can be explained on the basis of present-day biological and geographic factors." Biochemical genetic data will be presented below that are consistent with this suggestion and also argue that the distribution was achieved relatively recently from one or a few centers of origin.

Unisexuality

Throughout its range (roughly 90,000 km^2) very nearly all the P formosa that have been collected are females. Exceptions (mostly nonfunctional males) are discussed in [25,27,28].

Reproduction

So far as is known, the reproductive system of P formosa is exclusively gynogenesis or "pseudogamy." Sperm are required to "activate" embryogenesis [12], but there is apparently no syngamy. The mechanisms by which either activation of the egg, or exclusion of sperm pronuclei occur in P formosa are totally unknown, and neither process has been directly observed. In the laboratory, sperm from a wide variety of poeciliid species can serve to activate embryogenesis [14,15]. In all cases, with the exception of rare syngamies that result in triploids (see below), progeny are female genetic replicates of their mothers [13].

Reproductive Ecology

In nature, P formosa is sympatric with one of two bisexual Poecilia species, P latipinna (the sailfin molly) mostly but not exclusively in brackish or marine habitats at or near the coast, and P mexicana (the shortfin molly) in freshwater habitats more inland. Both of these bisexual species are strongly implicated in the hybrid ancestry of P formosa, and their biology is discussed in more detail below. Males of these species provide sperm for P formosa as well as their female conspecifics. This relationship has been termed "sexual parasitism" [28], but it has not been convincingly shown that the reproductive rate or potential of the "host" bisexual species, or indeed any other component of its biology, is affected at all by the presence of sympatric parthenoforms. Sperm per se may not be a limiting resource in these populations [30]. The ecological interrelationships of the gynogenetic "sexual parasite" and the bisexual "host" are under current scrutiny [30–32]; it appears that mechanisms have evolved which may promote asynchrony of breeding and minimize competition for space and other resources. The ecological relationships of the Poeciliopsis parthenoforms with each other and their bisexual host species have been the subjects of a penetrating series of investigations [33–36], but the hypotheses which have emerged have not yet been broadly applied to studies of Poecilia.

A third bisexual species, Poecilia latipunctata, may serve, along with sympatric P mexicana, as a host species for P formosa in some upstream tributaries of the Rio Tamesi system, but direct data, other than the sympatry itself, are lacking.

Meiosis and Thelytoky

Resolution of the question of meiosis and thelytoky in P formosa is complicated by technical difficulties. Ovaries with large, yolk-laden eggs and oocytes are rather poor material for detailed cytological studies. Nevertheless, a variety of indirect but powerful evidence, including isogenicity of mother and progeny, argues that the germinal vesicle of P formosa is diploid (eg [37]). In the absence of direct data, it has been assumed by most workers in the field that diploid or 2C eggs in P formosa are produced by premeiotic enduplication, a mechanism well known in other vertebrate parthenofroms [38–43] and now recently found to occur in the Australian grasshopper, Warramaba virgo [44]. The process is as follows: During the last oogonial mitosis, prior to the first meiotic division, an endomitotic duplication occurs without an ensuing cytokinesis, so that primary oocytes are 8C. Presumably during zygotene or pachytene pseudobivalents form from the synapsis of identical, replicated chromosomes. This renders chiasma, recombination, and segregation non-functional in a genetic sense. The two meiotic divisions that follow are normal, resulting in genetically identical diploid gametes that are exact copies of their mother.

However, recent data from E. Rasch's laboratory [45,46], which I am priviledged to mention here, indicate that neither premeiotic doubling of chromosome number nor synapsis occurs in P formosa. During meiotic prophase, oocytes of both P formosa and its bisexual congeners never contain more than a 4C amount of DNA, not the 8C that would be predicted by premeiotic endoduplication. Oogenesis in P formosa may well be ameiotic. Previous data from triploids related to P formosa, which suggested that a premeiotic endoduplication occurs in these biotypes [47] based on DNA cytophotometry of oocytes in serial sections, may have been subject to unavoidable systematic errors (P. Monaco and E. Rasch, personal communication). It has now been shown that oogenesis in triploid Poecilia, as in P formosa, is ameiotic [46]. Thus, there is no compensatory doubling of chromosomes in diploid or triploid unisexual Poecilia, as is reported to occur in certain unisexual forms of Poeciliopsis [48].

Triploidy

Triploidy resulting from the actual syngamy of diploid egg nuclei and host sperm is a significant feature of the biology of P formosa. In the laboratory, roughly 1% of all broods from diploid unisexual females contain at least one triploid ([37]; J. Balsano, personal communication). Triploids express paternal as well as maternal traits, and have been detected when either commercial "black molly" or Poecilia (Limia) vittata males have been used as laboratory "pseudofathers" [37,49]. Nearly all laboratory-produced triploids are sterile. The discovery of triploidy in laboratory-maintained populations was unexpected; the phenomenon is a reasonable alternative explanation for the observations which engendered certain early hypotheses that now seem unlikely [50,51].

Triploids were first reported in natural unisexual Poecilia populations more than a decade ago [52] and have been a major focus of research in Poecilia ever since. Triploids are most frequently encountered in populations where P mexicana limantouri is the bisexual host species. They have been most extensively studied in the Rio Soto la Marina system, most notably in the Rio Purificacion and its affluents. In these streams triploids form a signficant though fluctuating proportion of the total poeciliid fauna; at some localities the seasonal peak abundance of triploids may exceed 90% of all female Poecilia present [53]. In general, triploids appear more common at downstream localities, bisexual females at upstream ones [31]. Natural triploids (in contrast to laboratory-produced ones) are fertile and, so far as is known, all are gynogenetic and produce triploid female progeny [53–55]. Triploids are rather variable in morphology ([56]; J. S. Balsano, personal communication], and in the Rio Soto la Marina, when populations of Poecilia are sampled, an almost complete morphological continuum is observed [31]. A combination of electrophoretic and cytophotometric data [57,58] is needed to reliably distinguish among triploids, P formosa and P mexicana females. A variety of biochemical genetic data [59,60] is consistent with the hypothesis that the natural triploids are produced in a manner analogous to those observed in the laboratory: syngamy of P formosa eggs and P mexicana sperm. Additional data will be presented below that seem to remove all reasonable doubt of this ancestry, at least for the triploids in the Rio Soto la Marina. Accordingly, the term, "P formosa-limantouri" will be used as a name for the triploid gynogens that occur sympatrically with P mexicana limantouri in the Soto la Marina.[5] Small numbers of triploids have recently been encountered among populations of unisexuals sympatric with P latipinna at a locality in Texas [61]. Much less is known about these than the triploids in the Rio Soto la Marina. They presumably contain a paternal genome from host P latipinna males and, by analogy, could be called "P formosa-latipinna."

Heterozygosity

As determined by allozyme techniques, Poecilia formosa is heterozygous at 31–33% of its genetic loci [62]. However, as is the case with other parthenoforms

---

[5]This name is an obvious partial extension of the successful analytical terminology developed by R. J. Schultz for the unisexual Poeciliopsis. The name has the advantage of clearly denoting the ancestry of the triploids, and seems more convenient than the large labels previously used (eg, "naturally occurring triploid fish related to P formosa"). "Limantouri" is chosen because it is the most specific term available; the "mexicana" genome in P formosa itself is probably not that of P m limantouri (see below).

of hybrid origin, the adaptive significance of this level of heterozygosity is not understood. To some biologists, the high levels of heterozygosity encountered in vertebrate parthenoforms almost automatically imply heterosis. To others, the heterozygosity is irrelevant except as it preadapts the parthenoforms to a "hybrid niche" between that of their parental species or enables them to coexist ecologically with a single parental species as host (see [19] for a discussion of these alternatives as they pertain to the Poeciliopsis parthenoforms). A study of thermal tolerances in Poeciliopsis monacha-lucida by Bulger and Schultz [63] produced some evidence of heterosis, but the generality of these results, especially as they pertain to fishes from less extreme environments, is open to question.

Evaluation of the significance of heterozygosity in P formosa is made especially difficult by the nature of its presumptive ancestral (and present day host) bisexual species. Both P latipinna and P mexicana are broadly distributed euryhaline and eurythermal species with ranges that seem more limited by geographic or hydrographic features than directly by biotic or physiological factors. It is by no means clear that P formosa is more physiologically plastic or any more fit than are these species. It has been suggested that heterozygosity, perhaps by virtue of increasing developmental homeostasis (canalization), and thereby enhancing reproduction in novel environments, has promoted the colonization of new habitats by P formosa [62]. However, while it is quite probable that P formosa gained its present range by expansion from one or a few places of origin ([25]; also see below), it is not certain that its colonizing ability has had anything to do with its heterozygosity. Increased colonizing ability is an obvious probable consequence of the greater intrinsic rate of increase ("r") associated with all-female reproduction. Colonization in gynogenetic forms is necessarily limited by the presence of bisexual species to provide sperm.

Population Genetics

Perhaps no aspect of the evolutionary biology of the vertebrate parthenoforms has attracted wider attention than their population genetics, for the unisexuals have a clonal population structure that sets them well apart from bisexual animals. As pointed out first by Kallman [13,64], clonal population structure is a direct consequence of a parthenogenetic reproductive system. In the complete absence of recombination, changes in the genetic structure of a unisexual population reflect only changes in the relative frequencies (reproductive success) of clonal genotypes. New clonal genotypes are generated by mutations within existing clones, or by the advent of new unisexual lineages by repetitive hybridization of progenitor bisexual species. Presumably, all clonal genotypes do not have equivalent fitness in a particular environment, so that the environment can be viewed as imposing "clonal selection."

Clonal genotypes in unisexual fishes have been distinguished by histocompatibility (scale or organ transplants) and allozyme techniques. Kallman [64] detected four clones in a single population of P formosa from a stream at Olmito,

Texas using histocompatibility techniques; he eventually detected 12 clones in this population [K. Kallman, personal communication]. Two clones were predominant and, in a nearby drainage ditch, these clones accounted for 80% of the population. Darnell et al [65] detected at least two histocompatibility clones in the Rio Guayelejo. Using allozyme techniques, this laboratory [62] detected two clones by adh phenotype in P formosa collected at a locality adjacent or indentical with Kallman's. (In our original report [62] the heterozygous adh phenotype was attributed to a mutation that had occured after the evolutionary origin of the unisexual. This was probably incorrect, for we have since detected the secondary adh allele in some populations of P mexicana, including one population that is otherwise well qualified as a potential ancestor of P formosa.) Most recently, our laboratory has detected three additional "allozyme clones" within several Mexican populations of P mexicana [B. Turner and P. Monaco, unpublished data]. Interestingly, each of these clones was found in at least two river systems, and the clonal diversity of populations geographically close to the area of probable origin of the unisexual was not greater than elsewhere (an observation most consistent with a small number of origins, or even a unique origin, of the unisexual. See below). These data must be interpreted cautiously, however, for allozyme techniques used alone may seriously underestimate the clonal diversity of P formosa populations; it is quite clear that the histocompatibility analyses are a more sensitive tool for detecting clonal diversity in unisexual Poeciliopsis [66,67]. In general, the allozyme clonal variation within P formosa populations seems to be considerably less than that in most unisexual Poeciliopsis populations (references in [68]) but, in the absence of histocompatibility data, the biological significance of this observation is difficult to interpret. A detailed study of clonal diversity in P formosa, using both histocompatibility and allozyme techniques, is badly needed.

If histocompatibility measures of clonal diversity in unisexuals are uniformly more sensitive than allozyme determinations, the following questions will need to be answered: Is the difference between the measures actually related to a greater underlying allelic diversity of histocompatibility loci, or does it mean that many alleles of enzyme loci are simply not detected by routine electrophoretic surveys? If the difference is genetically real, is this because the histocompatibility loci more readily mutate (or mutate more frequently in ways not affecting clonal fitness) than do those loci encoding the allozymes? Or, does the difference simply reflect the fact that the progenitor bisexual populations are (or were) richer in polymorphisms for histocompatibility genes (such that multiple origins are more frequently genetically distinctive) than in enzyme polymorphisms?

Leslie and Vrijenhoek [68] have presented a picture of clonal evolution in unisexual Poeciliopsis that probably pertains, at least in part, to P formosa as well:

1. Unisexuality per se does not influence natural mutation rates. Mutation rates

in unisexual lineages and in related bisexual species are probably not significantly different.

2. Most mutations lead to the loss of biological function; they are therefore deleterious or potentially lethal.

3. Clonally reproducing parthenoforms lack a means of purging themselves of deleterious mutations. Therefore, in the absence of recombination, the proportion of deleterious genes in a unisexual lineage can only increase, an example of "Muller's ratchet." Over the long term therefore, the fitness of a unisexual population must decline. This is true even if the mutations are sheltered in heterozygous condition, unless they are completely recessive. How then, do unisexual populations escape being "ratcheted" out of existence? In the hybridogen Poeciliopsis monacha-lucida, Leslie and Vrijenhoek suggest that rare matings with P monacha males yield bisexual progeny (as they do in the laboratory) and thus "unlock" a previously "hemiclonal" monacha genome for recombination and therefore for the elimination of its load of deleterious lethals. They also suggested that while some clones might ultimately become extinct, the pool of unisexual clones could be refurbished by the addition of new clones resulting from fresh interspecific hybridizations.

Neither of the escapes from Muller's ratchet suggested by Leslie and Vrijenhoek can work in Poecilia formosa. So far as is known, there is no mechanism by which recombination can occur in natural populations. Neither is there any evidence that new P formosa clones are being produced at any appreciable rate. How does P formosa avoid the fate of gradual but ultimately complete clonal extinction due to the accumulation of deleterious mutations? The following considerations might be pertinent:

1. The heterozygosity of a new parthenoform clone, even if it does confer some initial heterosis (and this is far from certain), may represent an embarassment of riches. The new parthenoform, though diploid, is in an evolutionary position similar to that of a newly evolved allopolyploid. The genes it inherits from each of its bisexual parents have been separately selected into two different genomes. In combination, these homoeologous genes may not function at the same level of efficiency as each one does in its "own" genome. This may be especially true of genes encoding molecules that regulate major developmental or metabolic pathways.

2. Given that homoeologous genes may function suboptimally together, a mutation that eliminates the function of one of them may increase the fitness of the clone in which it occured. In other words, provided that dosage compensation can occur, clones carrying mutations to functional hemizygosity of previously heterozygous loci might be favored.

3. The mutation that confers adaptive hemizygosity to a unisexual lineage is of the same type that would ordinarily be considered deleterious or lethal in a

bisexual population. Thus, mutations of the sort that must accumulate in a unisexual lineage can actually increase heterosis (hybrid fitness) while decreasing functional heterozygosity. They are involved in "fine tuning" the interaction of the separate components of a hybrid genome.

Therefore, clones of a parthenoform such as P formosa may well accumulate "deleterious" mutations but, considered in the context of hybridity, these mutations could confer increased fitness; they may lead to long term survival of clones, not to extinction.

At present, it is doubtful if it is possible to directly compare a sufficient number of genes to test this hypothesis. However, as a greater proportion of genomes becomes accessible to examination, it should be possible to test the hypothesis because it contains an inherent prediction of asymmetery of hemizygosity. To a certain extent at least, the individual genes in each of the specific genomes of the hybrid parthenoform are coadapted to one another. An increase in fitness due to adaptive hemizygosity would most probably result in a restoration of this coadaptation by progressively eliminating the alleles characteristic of one species. For example if, in a given clonal lineage, the first mutation that led to increased fitness did so by eliminating the "A" allele at a heterozygous "AB" locus, a mutation that left the "B" allele functional at another locus would result in additional fitness. A mutation that left a functional "A" allele at the second locus would not restore the original coadaptation, and the clone carrying it would likely be less fit. If mutation to adaptive hemizygosity is a significant factor in clonal evolution, some clones in a population of P formosa might be enriched in alleles characteristic of P latipinna, others in those of P mexicana. If, on the other hand, accumulation of deleterious genes is at random, no such asymmetry should be evident.

The above hypothesis appears to be a more general statement of one made in a behavioral ecology context by Vrijenhoek and Schultz [69] and Leslie and Vrijenhoek [70]. They pointed out that male host bisexuals in Poeciliopsis prefer to mate with female conspecifics rather than sympatric parthenoforms. This mating preference might impose a selection pressure on unisexual clones in favor of mutations which enable them to resemble host females (ie by silencing the "heterologous" allele at heterozygous loci that influence critical morphological features). Thus, the unisexual comes to mimic the female host by a process of adaptive hemizygosity and clonal selection. This more specialized hypothesis is quite testable at present with P formosa, as some populations live only with P latipinna, others live only with P mexicana. Though the presence of triploids in the populations that live with P mexicana might complicate the issue somewhat, direct morphological and/or ethological comparisons of both host bisexual species and the P formosa that live with them might provide an extremely interesting study.

Clonal Diversity of Triploids

Clonal diversity in populations of the unisexual triploid P formosa-mexicana is currently being surveyed by J. Balsano's laboratory using histocompatibility techniques. Recently, in collaboration with that group, our laboratory completed a preliminary survey of allozyme variation among 48 triploids collected over a two-year period in the Rio Soto la Marina [Turner, Monaco, Balsano, and Rasch, unpublished data]. Results are summarized in Table I. This survey, which has not yet been coordinated with histocompatability data, must be regarded as somewhat preliminary. With that caveat however, the following points seem of interest:

1. The overall picture is one of one or two major clones and many minor ones. Twenty-seven out of 48 triploids, for example, are members of clone I. Even if this clone is proven to be heterogeneous by other criteria, the distribution of clonal frequencies is clearly skewed in favor of a few clones.
2. Apparently there are also differences among the clones in their geographic extent. Clone I is encountered in 5 or 6 localities, while clone II, though fairly common (n = 13) is clustered at only two. All three members of clone VIII are found only at a single locality.
3. The clonal diversity at any single locality seems only moderate. The average locality has only two clones, two localities have only one, and no one has more than four (five of the six localities surveyed are on a 70 km continuous sketch of the Rio Purificacion).

At present, the factors influencing clonal success and regulating clonal diversity in triploids are not known. However, the existence of poeciliid populations that contain both diploid and triploid gynogens presents an intriguing question: If, as postulated by in Poeciliopsis [69,70], clones that most closely resemble females of the bisexual host species are awarded increased fitness, then triploids with the two P mexicana genomes (albeit rather different ones; see below) should be more fit than diploids. In fact, the frequency of triploids in the Rio Sota la Marina fluctuates markedly and, in some cases, diploids are far more prevalent. This fluctuation seems to indicate that mechanisms other than male mating choice are important regulators of the relative abundance of diploid and triploid unisexuals. It is tempting to speculate that diploids, more morphologically distinct from host bisexuals, are able to use resources or habitat somewhat differently, and therefore accrue relative fitness at times of the year when resources become limiting.

THE EVOLUTIONARY ORIGINS OF POECILIA FORMOSA

In announcing the discovery of unisexuality in Poecilia formosa, the Hubbses [10] suggested that the species was of hybrid origin. In so doing, they seem to

TABLE I. Genotypes and Geographic Distribution of Members of Eight Allozyme Clones of the Triploid P formosa-limantouri in the Rio Soto la Marina

Clone no.	Genotypes at informative loci[a]				Localities[b]						Total individuals per clone
	ADA	Est5	Got1	XDH	1	2	3	4	5	6	
1	86/100	83/83	89/89	100/100	5	6	2	1	13	—	27
2	86/100	83/83	89/89	91/100	6	—	7	—	—	—	13
3	86/100	83/100	89/89	91/100	—	—	1	—	—	—	1
4	86/100	83/100	89/100	100/100	—	—	—	—	—	1	1
5	77/86/100	83/83	89/89	91/100	—	—	1	—	—	—	1
6	86/100	83/100	89/89	100/100	—	—	—	—	1	—	1
7	86/100	100/100	89/89	91/100	1	—	—	—	—	—	1
8	86/100	83/83	89/100	100/100	—	—	—	—	—	3	3
No. of clones per locality					3	1	4	1	2	2	$\bar{X} = 2$
Total n of triploids per locality					12	6	11	1	14	4	48

[a]Alleles are presented in terms of relative electrophoretic mobilities.

[b]Locality codes (see map in [31]): 1, Rio Purificacion, 7.3 mi E. Hidalgo; 2, Rio Purificacion at earthern dam; 3, Rio Purificacion at Barretal; 4, Rio Purificacion at Nuevo Padilla; 5, Rio Purificacion, 16 mi W. Nuevo Padillo; 6, Arroyo Raton.

have been particularly impressed with the morphologic intermediacy of the parthenoform when compared to two bisexual species they know well, the sailfin molly (P latipinna) and the shortfin molly (then known only as a single taxon, P sphenops). The idea of a causal link between interspecific hybridization and parthenogenesis or apomixis did not originate with the Hubbses, but can be traced, at least among botanists, to 1918 and possibly earlier (see [71] for review).

The Hubbses attempted to confirm the hypothetical ancestry of P formosa by synthesizing it by direct hybridization of the presumptive parental species. Their crosses, however, and those of Meyer [12], who obtained his material from them, failed to produce gynogens. Instead, they yielded vigorous, fertile, bisexual progeny of both sexes (eg [72]). A potential explanation for the failure of the P latipinna × P sphenops crosses to produce gynogens became apparent when the biological complexity of "P sphenops" began to be appreciated in the late 1950s [73]. The "P shenops" of earlier workers is a complex of sibling and near-sibling species, many of which are rather well differentiated genetically (see below). Of these species, the one most likely, on zoogeographic grounds, to have been an ancestor of P formosa was P mexicana, a widespread Atlantic slope species. The shortfin molly material used by the Hubbses stemmed largely from collections made on the Pacific coast of Mexico near Acapulco, and thus was probably P sphenops or P butleri (senu [74]) or inadvertant laboratory hybrids of one of these species with the other or with P mexicana [75]. Thus, the Hubbses' experiments used at least one incorrect species or species hybrid. Their failure to produce a unisexual Poecilia hybrid has been taken as critical evidence against the theory of the hybrid origin of thelytoky by some biologists. In fact, their crosses probably do not address the issue at all.

Other attempts to directly synthesize a unisexual Poecilia were made by the Darnell–Balsano group in the late 1960s [52: Table II] and more recently our laboratory [75]. In both cases, though reciprocal combinations of the presumably correct species were used, the crosses produced progeny of both sexes, and all fertile progeny were uniformly bisexual.

Assuming that the parental species have been correctly identified, the failure of laboratory hybridizations to produce at least some unisexual progeny could be viewed as a rather serious challenge to the hybrid theory of the origin of thelytoky, especially as that theory has been developed by its early proponents [76–78]. Such a challenge, however, would represent a rather narrow view of the problem for, among the three animal groups in which diploid parthenoforms have been synthesized by laboratory hybridization, in only one, the hybridogenetic Rana, [79] does the thelytoky appear to be a completely direct result of hybridization per se. In the others, the results are clearly variable, and variable in biologically suggestive ways. This variability can be resolved into three separate components: an individual component, a sex-related component, and a geographic component. Examples of each follow:

TABLE II.* Heterozygous "Allozyme" Loci in Poecilia Formosa**,†

Locus[a]	Protein	Comments
Aco-1	aconitase; liver	Type II. No variation detected in P formosa. Polymorphic, apparently with the same alleles. in P latipinna and P mexicana.
Ada	adenosine deaminase	Type II. See Aco-1.
Adh	alcohol dehydrogenase	Type II. Polymorphic in P formosa. Adh-b former "orphan" allele detected in scattered samples of P m mexicana[b] only. *Not* a Type IV locus as previously suggested [62].
Alb	serum albumin	Type II.
Es-5	major liver carboxylesterase	Type II. See Adh; "orphan" allele Es-5b detected in widely scattered samples of both subspecies of P mexicana.
Got-2	glutamate-oxaloacetate transaminase (aspartate aminotransferase); liver specific cytoslic enzyme	Type II. Got-2b orphan allele detected thus far only in P m mexicana samples from the Lagunas des Tamiahua and Tampamachoco[c] ($q = 0.7$–0.8).
Ldh-2	lactate dehydrogenase "muscle" enzyme	Type II. See Aco-1. In P mexicana, polymorphism largely restricted to samples of P m limantouri. P latipinna samples frequently multiallelic.
Lpp	leucyl-prolyl peptidase	Type II. Monomorphic.
Mdh-2	malate dehydrogenase (presumptive mitochondrial enzyme)	Type II. Low level of variability in some P m mexicana populations.
Pgd	6-phosphogluconate dehydrogenase	Type II; see Mdh-2.
Phi-1	phosphohexose isomerase	Type II; see Aco-1
Xdh	xanthine dehydrogenase	Type II; see Adh. Polymorphic in P formosa Xdh-b allele detected in scattered samples of both subspecies of P mexicana. Polymorphic also in P latipinna. Polymorphisms not reported in [62].

*Revises Table 3 in [62].

**Based on several ongoing unpublished surveys by B. L. Brett, P. Monaco, and B. J. Turner.

†The range of "Nei identities" (I_N) For P latipinna vs P mexicana comparisons is 0.62–0.70 ($\bar{x} = 0.68$).

[a]Parvalbumin loci are not included in this summary.

[b]Including samples from the Rio Tamesi and the Lagunas de Tamiahua and Tampamachoco.

[c]Not yet detected in Rio Tamesi samples.

Individual variability. In attempting to synthesize the parthenogenetic grasshopper Warramaba virgo in the laboratory by hybridizing its ancestral species "P169" and "P196" (designations for undescribed bisexual forms) White et al [80] noted that the offspring produced were largely sterile; only a few apparently were fertile parthenoforms. Crosses of Poeciliopsis lucida males with P monacha females usually have lethal results; only a small proportion of broods survive to become "synthetic" P monacha-lucida hybridogens [81]. Most crosses of P monacha females with P occidentalis males yield hybridogenetic female P monacha-occidentalis progeny, but some produce sterile male progeny instead [R. Vrijenhoek, personal communication]. The naturally occurring all-female hybridogen P monacha-latidens has never been synthesized by direct hybridization of its ancestral bisexual species, for these crosses yield mostly sterile males ([19]; the cross has recently produced a few hybridogenetic *males,* R. Vrijenhoek, personal communication). However, crosses of the P monacha-lucida females with P latidens sometimes do result in female P monacha-latidens parthenoforms.

Sex-related variability. In reciprocal crosses of Warramaba grasshoppers, White et al [80] found that the cross "P196" (female) × "P169" (male) was essentially lethal; the reciprocal produced at least some parthenoforms. In Poeciliopsis, crosses of P lucida females with P monacha males do not result in progeny, while the reciprocal produces some unisexuals [81].

Geographic variability. White and Contreras [82] found that if the parents were taken from an area where the two species were sympatric, crosses of the grasshoppers "P169" (female) × "P196" (male) fail to produce offspring at all, while sterile progeny and parthenoforms are produced if the parents are taken from "allopatric areas." In Poeciliopsis, the cross P monacha-lucida × P latidens produces hybridogenetic female P monacha-latidens (as well as some sterile males) if the parents are taken from the Rio Sinaloa or Rio Mocorito. However, if the parents are taken from the Rio Fuerte, the cross produces 100% sterile males [19].

The brief review above leads to the realization that the original concept of the relationship of hybridization to the advent of unisexuality is incomplete. If thelytoky resulted from the wholesale interaction (meiotic or otherwise) of two disparate genomes, then the variability that exists would not be expected: Particular specific hybrid combinations, regardless of how or from where they were put together, would be expected to result in a reasonable proportion of thelytokous progeny. The sex-related component of the variability suggests that the origin of unisexuality is related to genetic (or chromosomal) sex determination systems or, equally likely, that specific kinds of nucleocytoplasmic interaction are involved, perhaps including cytoplasmically inherited components. The individual and geographic components of the variability imply that the differences between thelytokous and nonthelytokous (sterile or bisexual) hybrid genotypes may be small in magnitude, perhaps even sometimes allelic differences at single loci.

These considerations lead to the following restatement of the relationship between hybridization and the origin of unisexuality: Thelytokous reproductive systems are a set of parameiotic phenotypes associated with some, but not necessarily all, of the genotypes that result from certain interspecific hybridizations. They are not ordinarily present in either bisexual species alone, but are the result of the interaction of particular combinations of genes in a novel hybrid genetic environment.

If this restatement is correct, the inability to synthesize P formosa by laboratory hybridization is a consequence of the failure thus far to produce very specific hybrid genotypes. Stated another way, hybridization experiments have missed critical "progynogenetic" genotypes within P latipinna and/or P mexicana. These genotypes may be in low frequency[6] or may be heterogeniously distributed geographically, and have thus far escaped detection. This hypothesis has, in turn, an interesting implication: If the actual progenitor populations, the ones that hybridized to produce P formosa in nature, could be identified, the progynogenetic genotypes might still be present, and might produce gynogenetic progeny upon hybridization in the laboratory. The success of future attempts to synthesize P formosa may well be a function of the precision with which its ancestral populations (as well as species) have been recognized.

Ancestral Species Complexes

Morphologically, P formosa falls rather neatly between two important species group in the genus Poecilia, the shortfin mollies of the "P sphenops complex"[7]

---

[6]The possibility exists that these critical genotypes are exceedingly rare indeed, and that P formosa represents an essentially accidental, much exaggerated, amplification of a unique hybrid genotype that can never be precisely duplicated by subsequent hybridizations in nature or the laboratory.

[7]"Complex" is used here as it is used in contemporary ichthyology—as an informal taxonomic category that somewhat overlaps both "superspecies" and "subgenus" and may include taxa of uncertain or controversial specific status. This usage is different from that of Balsano et al [31] who writes of a "breeding complex" of P mexicana, P formosa and triploids in the Rio Soto la Marina, or of [83] who uses the term "P formosa complex" for P latipinna, P formosa, and P mexicana. By convention, taxonomic complexes are named for their oldest included taxon.

(dorsal fin with 7–10 rays) and the sailfin mollies of the P latipinna complex (12–19 dorsal fin rays).[8]

The P sphenops complex represents a major component of the freshwater fish fauna of Mexico and middle America. Formerly, the complex was considered by many to be a single, variable, polytypic species [22], but morphological and genetic studies [9,73,74,83] demonstrated that more than a single species was almost certainly involved.[9] Recent allozyme data from this laboratory [84,85] have provided fairly convincing evidence of the genetic distinctiveness and specific status of many of the Mexican and insular Caribbean components of the complex. The major species of the complex in Mexico are P mexicana (Atlantic slope), P butleri (Pacific slope), and P sphenops (both slopes, but at least some Pacific populations may be specifically distinct). These are essentially sibling species. P mexicana and Atlantic coast populations of P sphenops are extremely difficult to distinguish morphologically, except by the nature of the inner jaw teeth (tricuspid in P sphenops, unicuspid in P mexicana), but even when sympatric are well differentiated allozymically and are rather unequivocally distinctive [Brett, in preparation]. Though the inner jaw teeth of P sphenops are tricuspid,

---

[8]The two complexes together are not coextensive with the genus Poecilia, but constitute a seemingly natural grouping which Miller [9] refers to as the subgenus Mollienesia. Poecilia contains other such apparent natural groups, most notable of which are Lebistes (guppies and their close relatives) and the insular Limia, as well as some species which seem intermediate to the major complexes (eg, P vivipara).

[9]Lack of understanding of the systematic of the P sphenops complex has had important biological consequences, for shortfin mollies are in wide use as both experimental animals and as articles of commerce in the ornamental fish industry. For example, the cavernicolous "P sphenops" studied by Parzefall and his associates (eg [86]) from a cave in Tabasco are in fact P mexicana [B. L. Brett, personal communication]; surface streams in the area of the cave may contain both P sphenops and P mexicana and, if these were not recognized, cave/surface comparisons could be utterly confounded. The "P sphenops" of the American pet trade and of experimental biologists (eg [87]) has unicuspid inner jaw teeth, and is probably P mexicana or a P mexicana hybrid. Since many of the species in the complex are rather distinctive in ways that are biologically potentially important, experimental biologists are urged to use material of known taxonomic identity and geographic origin. The name "P sphenops" is also sometimes used in the pet trade and pet literature for totally melanistic "black mollies." The black molly was in fact originally bred from partially melanistic P latipinna, though melanistic forms of other sailfins may have been involved, as well as hybridization with shortfin species.

laboratory P sphenops × P latipinna hybrids have unicuspid inner teeth [75], as do P latipinna and P formosa itself, and P sphenops is otherwise virtually indistinguishable morphologically from P mexicana. It is therefore impossible to disqualify P sphenops as a potential ancestor of P mexicana on morphological criteria alone, and the allozymic distinctiveness of P shenops is critical evidence eliminating it from consideration as a progenitor of P formosa.

P mexicana, P butleri, and P sphenops are essentially freshwater species, though the former two do enter brackish to marine estuarine and mangrove habitats near the coasts, sometimes quite extensively. However, another sibling species, P orri, the "mangrove molly," is essentially marine, and another member of the complex, P vandepolli has been taken from completely marine "rifwater" habitats on Curacao [B. J. Turner, field observations, 1974]. Other members of the complex include morphologically well marked species such as P chica and P latipunctata (despite its morphological distinctiveness, hybrids of P latipunctata and P mexicana have been taken in the Rio Guayelejo [R. R. Miller, personal communication]), large lacustrine derivatives such as P catemaconis and P gracilis (each morphologically distinct, but allozymically difficult if not impossible to distinguish from P sphenops and P mexicana respectively), and differentiated local forms that may not be distinct at the species level. The relationships and systematics of the complex south of Mexico are not clear, and will not be dealt with any further here.

Poecilia mexicana, the most likely component of the P sphenops complex to have been a progenitor of P formosa, is itself a widespread, morphologically variable species. Menzel and Darnell [88] have dealt with its general distribution and systematics: They recognized two subspecies, P m mexicana (so called "southern mexicana") and P m limantouri ("northern mexicana"). These subspecies are distinguished largely by color pattern, adult size, and some minor meristic and morphometric traits. Menzel and Darnell suggested that the subspecies involved in the ancestry of P formosa was most probably P m limantouri. However, the coastal regions of contemporary sympatry between P latipinna and P mexicana are supposedly a region of intergradation between the two subspecies.

Much less is known of the sailfin mollies of the P latipinna complex. Largely through the unpublished work of R. R. Miller, three species are currently recognized: P latipinna, P petenenis, and P velifera. The three are allopatric and differ somewhat in meristic traits, adult size, and color pattern. P latipinna ranges coastally from North Carolina to just north of the outlet of the Rio Tuxpan in the Laguna de Tampamochoco, Veracruz, Mexico; the southern terminus of its range is coincident with that of P formosa. It is typically found in brackish and marine coastal habitats, but has successfully colonized freshwater in Texas (notably in the Rio Grande) and elsewhere. There is significant local variation in the number, size, and aspects of the color pattern of reproductive males (and also in the frequency of black-speckled individuals), but no attempts to order this variation into formal subspecies appear to have been made. Simanek [83,89] has described the allozymic variability of some P latipinna populations, and the relationship of that variability to population structure (see also [90]).

P petenensis is a somewhat larger species than P latipinna. As presently understood, it occurs in the freshwaters of the vast Rio Grijava-Rio Usumacinta system of eastern Mexico and nearby Guatemala (including Lago Peten). The precise limits of its range are not well established. In eastern Mexico the species ranges northwestward from the Rio Grijalva itself coastwise (including brackish water habitats) to the vicinity of Cardenas, Tabasco. In these most westward extensions of its range the species appears to some workers to more strongly resemble P latipinna than it does elsewhere, and before the recognition of their relationship to Rio Grijalva fish, these populations were sometimes thought to be conspecific with P latipinna [25,91]. P petenensis also ranges eastward from the Rio Grijalva to the base of the Yucatan peninsula. It may intergrade there with P velifera, for sailfin samples taken in the Laguna de Teriminos are intermediate between the two species in some respects [R. R. Miller, personal communication]; the two may also intergrade along the eastern coast of the Yucatan peninsula or coastal Belize. P velifera, a rather large form with a very well developed sailfin, is known only from the Yucatan peninsula in marine and brackish waters; it has apparently been introduced into interior "cenotes."

Though all components of the P latipinna complex appear to be well differentiated on morphological criteria, future work may show the three to be components of one or two polytypic species. Preliminary data, based on very limited sampling, suggests that the sailfins are not nearly so well separated from one another allozymically as are the components of the P sphenops complex.

Various components of the P sphenops and P latipinna complexes are sympatric (but not necessarily synoptic) in different parts of their ranges. Thus P latipinna and P mexicana at least sometimes occur together in coastal lagoons near Tampico; P petenensis occurs with P mexicana and P sphenops in the Rio Grijalva, with P gracilis in Lago Peten, and possibly with P orri (as well as P mexicana) in Belize; and P velifera sometimes occurs with P mexicana (and, in Quintana Roo, possibly P orri as well). Unisexual Poecilia have thus far been associated only with P latipinna and P mexicana, but this may change as other faunas become better known. Hubbs [91] has described and illustrated ([91]: plate 8, Fig. 4) a natural hybrid between P velifera and a component of the P sphenops complex (his "M sphenops altissima"—probably P mexicana).

Cytogenetics

Research on the cytogenetics of P formosa and its congeners has been hampered to some extent by difficulties common to most fish chromosome work: Small chromosome size and the lack of readily detectable morphological differentiation within the complement. Poeciliid chromosomes are frequently difficult or impossible to distinguish from one another. These difficulties have apparently discouraged active research on the cytogenetics of Poecilia, for even in the current relative "boom" in fish cytogenetics, the most recent treatment of Poecilia is that of Prehn and Rasch [52].

As in most other poeciliids, the diploid complement of P formosa (2 n = 46) consists of a more or less continuously graded series of acrocentric or telocentric chromosomes. The only chromosome pair that could be reliably distinguished from the others is the largest, and that pair is heteromorphic: one member is the same relative size as the largest component of the complement of P mexicana: Its presumptive homoeologue, slightly larger, is the same relative size as the largest chromosome in the genome of P latipinna. However, the largest chromosome in the genome of P latipinna, at least in Prehn and Rasch's preparations, is telocentric, but the P formosa chromosome is clearly acrocentric or subtelocentric (Fig. 3). Prehn and Rasch termed the large P formosa chromosome "1'" and suggested that "1'" and "1" (the large chromosome of P latipinna) were related by a "centric shift," most likely a small pericentric inversion. This interpretation is reasonable, but an important question remains: Did the "1" to "1'" shift occur after the advent of the parthenoform or was it part of its ancestry? Was it established during early clonal evolution, or was it present, perhaps as a polymorphism, in the sailfin progenitor of P formosa?

Ongoing research in this laboratory, using improved techniques [92], has confirmed the existence of the "1'" chromosome in P formosa collected from near Brownsville, Texas, and from the Rio Soto la Marina. In addition we find the following [B. J. Turner, unpublished]: 1) The short arm of the "1'" chromosome is not heterochromatic. 2) The "1'" chromosome in the P latipinna we have studied (material mostly from the Nueces River in Texas) is distinctly subtelocentric (not telocentric), but the short arm is shorter than that in P formosa. At present we do not know if the discrepancy between our results and those of Prehn and Rasch [1969] reflect true geographic variation (their material was from the Rio Grande) or simply differences in technique (in work with fish chromosomes, even small increments in resolution can yield major changes in interpretation eg, see [93]).

It is reasonable to suggest that the "1'" chromosome represents a minor variant of the "1" chromosome of P latipinna, a variant that might be common in some local populations. This will be checked when Mexican populations of P latipinna become available for analysis.

Alternatively, the "1'" chromosome may be a component of an as yet undetected heterogametic sex chromosomal system in P latipinna. Geographic variation in cytological heterogamety is known in another poeciliid, the widely distributed Gambusia affinis [94], and has by no means been conclusively ruled out in P latipinna. Interestingly, though their figures are difficult to interpret due to poor reproduction in the journal, Rishi and Gaur [95] claim to have detected heterogamety in aquarium stocks of the "jet black molly," which they identify

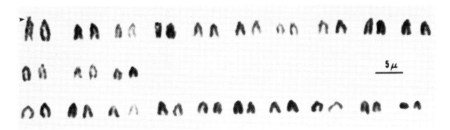

Fig. 3. Karyotype of Poecilia formosa prepared from gill epithelium (E Rasch, unpublished). The largest subtelocentric (arrowhead) is the 1' chromosome of [52].

as P sphenops. Most black mollies which we have analyzed allozymically have had largely latipinna-like genomes.

Electrophoretic Studies

The earliest electrophoretic study of P formosa and its relatives was from C. P. Haskins' laboratory [96]. Their paper is a pioneering application of electrophoretic techniques to problems in the evolutionary biology of fishes, but the resolution of the paper electropherograms that they used is so poor compared to that of gel-based techniques that it must now be regarded to be of strictly historical interest.

J. S. Balsano studied the serum proteins of P formosa and its congeners by electrophoresis in polyacrylamide gels [59,60,97,98]. P formosa clearly has the two-banded phenotype of a heterozygote for the structural gene locus encoding serum albumin (the "albumin 3 + 4" phenotype); the faster of the two albumins is characteristic of P mexicana, the slower of P latipinna. The staining intensities of the bands in the P formosa pattern are approximately equal, presumably indicating equality of gene dosage. In triploids from the Rio Soto la Marina the faster band stains more intensely, (reflecting a "2 mexicana: 1 latipinna" genome dosage). Balsano's work provided the first substantial nonmorphological evidence of the hybrid ancestry of P formosa and of the probable ancestry and genomic composition of the triploids. The albumin phenotypes also provided a means by which P formosa, triploids, and P mexicana could be distinguished in field collections.

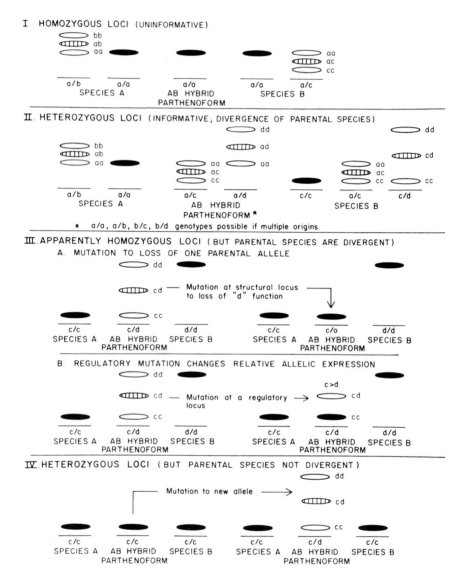

Fig. 4. Diagram of four types of allozyme electrophoretic patterns that might be encountered in comparing a parthenoform of hybrid origin with its presumptive acestral bisexual species. The patterns are drawn as if the allozyme locus encoded a dimeric protein with freely combining subunits. Genotypes are given under the origin in each case, and the subunit composition of each heterozygote pattern is presented to its right. Staining intensities of bands: black > stippled > white.

Electrophoretic comparison of gene products (eg, allozymes) from a parthenoform of hybrid origin and its presumptive progenitor bisexual species have predictable results. The following kinds of gene loci are possible outcomes of such a comparison (see Fig. 4):

1. Loci with identical homozygous phenotypes in the parthenoform and both progenitor species (type I loci). The progenitor populations were not detectably divergent at this locus, and no mutations have occurred in the parthenoform. Note: this does not necessarily imply monomorphism in all populations of both bisexual species.
2. Loci with heterozygous phenotypes in parthenoforms due to divegence of progenitor species or populations (type II loci). The serum albumin phenotype in P formosa is encoded by a type II locus. If one or both of the parental species are polymorphic at this locus and there have been multiple origins of the parthenoform, then this locus can have more than a single heterozygous phenotype. If the same allele is part of the polymorphism in both parental species, then parthenoforms with homozygous phenotypes could also be generated by hybridization.
3. Loci with homozygous phenotypes in the parthenoform but the parental genotypes are divergent (type III loci). The loci could most probably arise from a mutation to functional silence (hemizygosity) at one allele of an original type II locus. If clones with the new phenotype are favored, then selection may have eliminated the originals in favor of the hemizygotes. Included in this category also are loci with basically heterozygous genotypes but whose expression has been modified in favor of one parental allele. Modification of expression can be by regulatory mutation or may be a direct consequence of interspecific hybridization. Asymmetric expression of parental loci is well known in experimentally produced interspecific fish hybrids (reviewed in [99]).
4. Loci that are nondivergent in parental species but have a heterozygous phenotype in the parthenoform (type IV loci). These are due to mutation in a clonal lineage after the parthenoform had evolved. The result of such mutations could be polymorphism in the parthenoform or, if the new genotype is successful (or "fixed" by chance), a population of parthenoforms with an invariant phenotype that is hetrozygous for an allele not present in either parental species.

Thus far, an almost perfect knowledge of the genetic structure and variation of both progenitor species and the hybrid parthenoform has been assumed, and the four categories appear clear-cut. However, if knowledge is less complete, the categorization can be far more difficult. For example, type II and type IV loci can be particularly difficult to distinguish if samples of the parental species are limited. Is an allele present in all unisexuals but absent in parental samples (previously termed an "orphan allele" by Turner et al [62]) the result of mutation and selection in clonal lineages, or does it indicate that the parentals have been inadequately sampled, and that a bisexual population exists (or existed) somewhere with the orphan allele at reasonable frequency? Resolution of this question is important: 1) If the supposed orphan alleles result from mutation and subsequent clonal selection, they then provide clues as to the specific kinds of evolutionary changes that are important in clonal evolution. 2) If they are the result of hybridization of differentiated populations, then they are important clues for locating the precise ancestral bisexual progenitor populations of the parthenoform.

The "l'" chromosome discussed above seems to be such a troublesome but potentially important orphan "allele" in P formosa. Allozyme surveys have detected others.

Our laboratory began allozyme surveys of P formosa and its congeners in 1977 in collaboration with J. S. Balsano and E. Rasch; we were aided substantially by R. R. Miller and his extensive live collections of Poecilia at the University of Michigan. While our initial report [62] was being prepared, a doctoral dissertation on the same subject [83] appeared from another laboratory.[10]

Our first surveys were of P formosa and triploids from field collections in the Rio Soto la Marina and the Rio Grande, field samples of P latipinna and P m limantouri and laboratory stocks of P m mexicana. We detected 11 heterozygous loci in P formosa, only one of which, Adh, was variable in phenotype [62:Table 3]. Alleles at two other loci, Es-5 and Got-2, appeared to be "orphans." They were present (as heterozygotes) in P formosa, but could not be detected in our samples of P mexicana or P latipinna. More recent surveys, some still in progress, have revealed a very substantial amount of geographic variation in P mexicana, variation that our first surveys barely hinted at. Table II here revises and updates Table 3 in our earlier paper [62]. Note that our previous orphan alleles have acquired parents: We can now account for all of the heterozygous allozyme

---

[10]It is instructive to compare the initial data in [62] with that of [83]. Our survey included the products of a relatively large number of loci, but relatively few populations. Consequently, we detected potentially important phenomena like the supposed orphan alleles, but substantially underestimated the geographic variation in the parental species, especially in P mexicana. The other study surveyed fewer loci but many more populations. It obtained a far better appreciation of the geographic variation present in the system than we did, but missed the Got-2b orphan allele and the Adh variability. Thus, the entire system appeared more straightforward than it did to us. The nature of the geographic variation at allozyme loci in P mexicana requires extensive surveys of both large numbers of loci and many populations.

TABLE III. Frequency of Parvalbumin M₂ Null Phenotypes in Coastal Samples of P m mexicana

Locality	Nulls	N	q^a
Laguna de Tamiahua at Tamiahua	1	4	0.5
Laguna de Tampamachoco at Rio Tuxpan outlet	2	15	0.37
Rio Tamesi at Tampico	2	9	0.47
Rio Vinasco (R. Tuxpan drainage)	1	10	0.32
all	38	6	0.40 ± 0.17

q = frequency of presumptive null allele, assuming null phenotypes are homozygous (q^2 = nulls/total); confidence intervals for q in individual populations are not given due to small sample sizes. If the populations are in Hardy-Weinberg equilibrium at the locus, then the average frequency of heterozygote carriers of the null allele is 0.35–0.49 (95% confidence interval).

genotypes in P formosa by variation within or among P m mexicana populations. Most of the geographic variation we have detected in P mexicana requires no special comment, but the following points deserve emphasis:

1. The best fit to the genotypes of P formosa occur in P mexicana from the Rio Tamesi and the large coastal Lagunas de Tamiahua and Tampamochoco. The Got-2b orphan allele of P formosa has been detected only in the latter two localities. These populations, from brackish to marine habitats, are P m mexicana (or at best, integrades between the two subspecies) in phenotype. They are not P m limantouri. The data therefore contradict earlier indications [25,83] that P m limantouri was the ancestral subspecies of P formosa.
2. Population samples of P m mexicana and P m limantouri are very distinctive genetically. In fact, there is more differentiation in between P m limantouri from Monterrey or the Rio Soto la Marina and P m mexicana from Tampico than there is between P m mexicana from Tampico and the same species sampled from Belize or Costa Rica. This observation leads immediately to the speculation that P m mexicana and P m limantouri may in fact be distinct at the species, not the subspecies, level. Collections that might settle this question remain to be made (note: if this speculation is correct, the triploids in the Rio Soto la Marina are in fact *tri*specific).
3. Some of the variation is intensely local in nature. For example, in a survey of P m limantouri from the Soto la Marina, a particular allele of the Got-3 locus was encountered at reasonable frequency (q = 0.4) in a small tributary, the Arroyo la Presa, but was absent in much larger samples from localities on either side of the confluence of the small stream and the larger river. This

extreme local variation at one or a few loci would be lost if comparisons were based solely on some averaging "genetic identity" statistic. Variation of this sort may be critical in attempts to understand the evolutionary origin of the unisexual. However, local variation of this intensity within P mexicana appears to be a potentially significant phenomenon that merits study in its own right.

Parvalbumin Phenotypes

In our original study, parvalbumins were surveyed on routine starch gel systems of only moderate resolving power. Unlike most of the other heterozygous loci we detected in P formosa, the parvalbumins displayed what appeared to be gene dosage-dependent phenotypes in the triploids. We were interested in developing an electrophoretic "screen" for triploids that did not involve bleeding specimens in the field (for serum albumin typing and for cytophotometry), a relatively troublesome procedure. We had already developed high resolution survey techniques for these low molecular weight Ca^{++}-binding proteins in other fishes, and we therefore applied these to parvalbumins in Poecilia species.

There are three distinguishable parvalbumins in muscle homogenates prepared from most cyprinodontoid fishes and many other teleosts, apparently encoded by three paralogous gene loci. In aggregate, the parvalbumins almost always provide species-specific electrophoretic phenotypes. In addition, two kinds of polymorphisms are known: The most frequent type appears to be due to the segregation in a population of two codominant alleles, much as in a typical allozyme polymorphism. Parvalbumin polymorphisms of this type in fishes have frequently been reported, generally without reference to the identity of the proteins concerned (eg, [100]). Our laboratory has studied them in walleye [101] and in Jordanella floridae and Poecilia reticulata [unpublished]. A second kind of polymorphism, termed "presence/absence," was first detected in the goodeid fish Skiffia multipunctata [102], and has now been found in other goodeids as well. There are only two phenotypes, typically one with three parvalbumins, the other with only two. We attribute these polymorphisms to the segregation of a "null" allele at one of the parvalbumin loci. The presumptive null allele is recessive, and the two-parvalbumin phenotypes are considered to comprise the homozygous recessive class; heterozygotes for the null allele are usually not distinguishable, even by densitometry.[11]

P mexicana and P latinna have distinctive parvalbumin phenotypes on polyacrylamide gels (Fig. 5); that of P latinna essentially invariant. The P formosa phenotype, as one might predict, was identical to that obtained by mixing P latipinna and P mexicana samples, with one very conspicuous exception. Par-

---

[11]It is recognized that the null phenotypes may have other genetic bases, including regulatory variants at other loci.

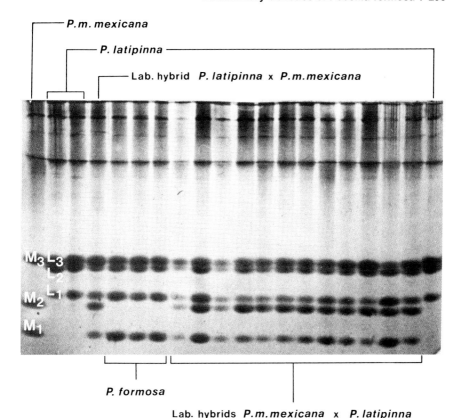

Fig. 5. Parvalbumin phenotypes in Poecilia latipinna, P m mexicana, their reciprocal laboratory hybrids [75], and P formosa. The parvalbumins of each bisexual species are given species-specific designations in order of their mobility ("1" is the fastest component in each). As expected, the laboratory hybrids have completely additive parvalbumin phenotypes, but note that the component "M_2" is completely absent in P formosa. This discrepancy between the phenotypes of laboratory hybrids and P formosa was initially attributed to a mutation to functional silence of the "M_2" structural locus in P formosa. In this and subsequent figures, parvalbumins are analyzed in 16% total monomer (4% crosslinked) gels at pH 8.9 in the Ornstein-Davis buffer system. The bands identified as parvalbumins have mobilities identical to that of authentic carp parvalbumin when extirpated from these gels and reanalyzed in gels containing sodium dodecyl sulfate (B. J. Turner, unpublished).

valbumin "M_2," present in P mexicana and in mixtures, was completely absent in P formosa. The complete absence of M_2 was a characteristic of all the P formosa we surveyed, from the Rio Grande to the Rio Tuxpan. Parvalbumin M_2 is present without exception in all P latipinna $\times$ P m mexicana F_1 laboratory

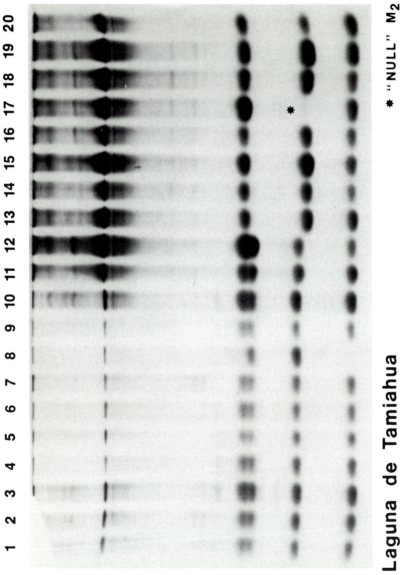

Laguna de Tamiahua

* "NULL" M₂

Fig. 6. Survey of parvalbumin phenotypes from mollies taken in the Laguna de Tamiahua. Specimens 1–7) P formosa. 8) Unidentified (juvenile P latipinna?). 9–12) P formosa. 13–20) P m mexicana. Note that specimen 17 has a "null" phenotype for parvalbumin M₂; the phenotype is interpreted as that of a homozygote for a null allele at the M₂ structural locus. (The very slight amount of staining visible at the M₂ position in this sample is due to minor crosscontamination from the samples loaded into adjacent wells.)

hybrids of both reciprocal combinations. In other words, the parvalbumin phenotypes of F_1 laboratory hybrids are what would be predicted from a combination of codominant parental alleles; that of P formosa lacks a P m mexicana component, M_2. For some time, we believed that the discrepancy between the P formosa and F_1 hybrid phenotypes was due to a mutation in the early clonal history of P formosa, a mutation of loss of M_2 gene function. The gene locus encoding parvalbumin M_2 would then be the only type III locus known in P formosa. However, subsequent surveys of paravalbumin phenotypes in additional P m mexicana populations provided an alternative, potentially more significant explanation of the P formosa phenotype: Nulls for parvalbumin M_2 were detected in a few coastal populations (Figs. 6, 7). The P formosa parvalbumin phenotype could have resulted from the mating of one of these presumptive homozygous nulls (or, more likely, a heterozygous carrier) with a P latipinna. Table III gives the frequencies of these nulls and the gene estimated frequencies of the presumptive null alleles in the coastal populations.

The parvalbumins have thus provided another set of distinctive phenotypes by which we can recognize the bisexual populations that most likely were the progenitors of P formosa. At present, the most "qualified" potential progenitor populations (Got-2b allele present, appropriate alleles at other loci present at high frequency, null for parvalbumin M_2 present) are those in the coastal Laguna de Tamiahua and its outlet, the Laguna de Tampamochoco. However, the area around the coast and the Rio Tamesi near Tampico have not yet been well surveyed, and other populations may be equally qualified. No other P m mexicana populations, and certainly none of P m limantouri, that we have surveyed outside this coastal region are nearly as qualified.

It is therefore reasonable to suggest that P formosa originated by hybridization of P m mexicana and P latipinna in the coastal region around Tampico or the Laguna de Tamiahua itself. More speculatively, I hypothesize that, from its area of origin in brackish and marine water, P formosa spread northwards coastwise, utilizing both P m mexicana and P latipinna as hosts. From an essentially coastal base the parthenoform invaded rivers and moved into freshwater habitats, utilizing resident populations of P m limantouri as hosts. The relative paucity of allozyme variation in P formosa suggests that there have been relatively few independent origins of the unisexual (theoretically, all known allozyme clones could be genotypes found among the progeny of a single mating of an appropriate P m mexicana from the Laguna de Tamiahua with P latipinna). In aggregate, the genetic data do not suggest that P formosa is an especially old parthenoform. They certainly do suggest that it is a rather good colonist, spreading to occupy a significant range from a relatively small area of origin.

P m limantouri and the Soto la Marina Triploids

The P m limantouri in the Rio Soto la Marina have a variant parvalbumin phenotype that is different from that of P m mexicana (Fig. 8); the difference

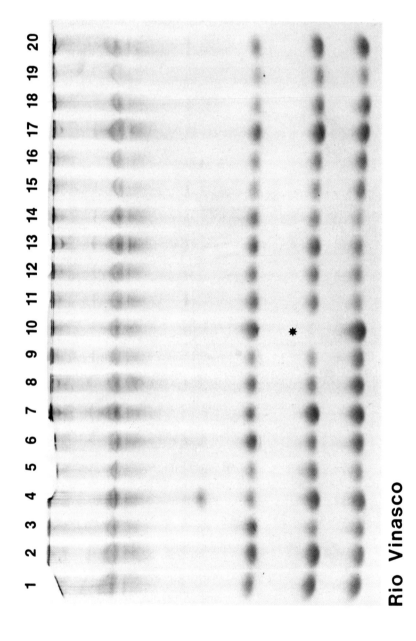

Rio Vinasco

Fig. 7. Survey of 20 P m mexicana taken from the Rio Vinasco (Rio Panuco system). Note the null for parvalbumin M_2 in specimen 10, and the heterozygote (apparently for the locus encoding M_1) in specimen 4.

involves, at least, a fixed allelic substitution the locus encoding parvalbumin M_1. Triploids in the Rio Soto la Marina have parvalbumin phenotypes with all P formosa components and those of the host, thus providing direct evidence that they result from syngamy of P formosa eggs with host sperm. The different "M_1" parvalbumin phenotype is completely characteristic of P m limantouri from both the Rio Soto la Marina and from the Rio Grande drainage near Monterrey, and may well be characteristic of the entire "sub" species. Survey of parvalbumin phenotypes in P mexicana populations from the presumptive zone of intergradation of the subspecies [88] should be especially interesting.

PROSPECTUS FOR FUTURE RESEARCH

1. Obviously, the populations of P m mexicana genetically the best qualified to be the ancestors of P formosa should be used in future attempts at synthesizing unisexuals in the laboratory by interspecific hybridization. The P latipinna that live with these populations, rare and/or difficult to collect though they may be, should also be used. Such hybridization experiments might be facilitated by the following: Restriction nuclease patterns of mitochondrial DNA might be used to establish which of the parental species is the female component in the cross, as has been done with unisexual Cnemidophorus [103]. In addition, artificial insemination [104], instead of natural breeding should be used in future experiments for both the production of the F_1 hybrids and their genetic evaluation.

2. Clonal diversity in P formosa across its range needs to be measured by both allozyme and histocompatibility techniques. In addition, mitocohondrial DNA polymorphisms, if they can be detected in these small fish, might provide a new tool for clonal analysis of potentially superlative sensitivity (eg, [105]).

3. The evolution of gene expression in P formosa clones (and in triploids) has been largely unexplored. The direct comparison of synthetic laboratory hybrids (made with the correct genotypes) with P formosa clones reared under identical conditions in the laboratory might provide a rather powerful experimental approach to the area. The technique of two dimensional ("isodalt") electrophoresis, particularly in the manifestation developed by Celis and Bravo [106], in which over 1100 polypeptides can be separated, would allow the detection, and ultimately the quantitation, of sufficient numbers of different gene products to make such comparisons worthwhile.

4. Parthenoforms such as P formosa offer unique systems for studying the evolutionary characteristics of "selfish" DNA (literature referenced in [107]). When a new parthenform lineage is started by hybridization, do the selfish sequences characteristic of one of the parental species "take over" the genome? If so, is the same sequence always the successful one, or does the ratio of parental selfish sequences vary among different clonal lineages? In the absence

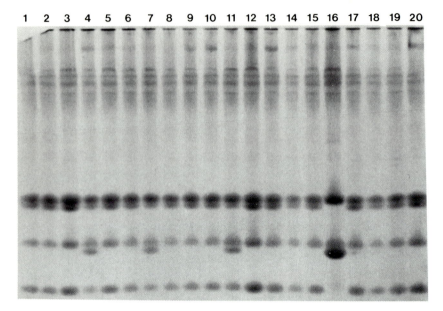

Arroyo Raton

Fig. 8. Parvalbumin survey of two localities in the Rio Purificacion (Rio Soto la Marina system). Specimens are as follows: P formosa: 1, 2, 3, 5, 6, 8, 9, 10, 12, 13, 14, 15, 17, 18, 19, 20, 21, 22, 24, 25, 27, 28, 29. P m limantouri: 16, 23, 30–37 (note that the P m limantouri in these samples lack the parvalbumin "M_1" present in P m mexicana and P formosa, but instead have a band identical in position to the "L_1" component of P latipinna.) Triploids: 5, 7, 11, 26. The phenotypes of hosts, P formosa, and triploids

of meiosis in P formasa, do selfish sequences evolve at all? Do new sequences evolve by mutation and take over the genomes of some clones? If they do, are the successful sequences different from those that evolve in bisexual species (where the need for functional meiosis might impose different selection pressures on newly evolved sequences)? These questions are perhaps more speculative than others that have been raised here, but they are no less interesting.

ACKNOWLEDGMENTS

The work of my colleagues J. S. Balsano and E. Rasch has been a constant source of inspiration. In addition, I thank the following individuals for field assistance and/or helpful discussions over the years: R. Andrews, R. Angus, J. W. Atz, R. M. Bailey, A. Bulger, J. Cranford, B. Eisenbrey, D. Grosse, M. Howell, C. L. Hubbs, L. C. Hubbs, K. Kallman, J. Leslie, J. Maynard Smith, F. H. Miller, R. R. Miller, W. Moore, R. J. Schultz, M. Smith, T. Uzzell, R. Vrijenhoek, and B. Wallace. P. Monaco and B. L. Brett allowed me to mention

21 22 23 24 25 26 27 28 29 30 31 32 33 34 35 36 37

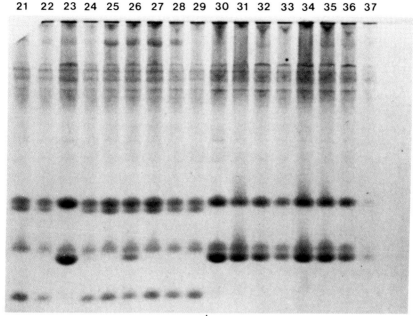

Arroyo Raton ─────────────┘ Barretal (bridge)

can be reliably distinguished, making practical a survey of the allozyme variation in each biotype without resort to cytophotometry or bleeding specimens in the field. The phenotypes of the triploids are compatible with their presumptive ancestry by syngamy of a diploid P formosa egg and host sperm.

our unpublished genetic data on P mexicana and also generously permitted me to cite material in their doctoral dissertations now in preparation. I thank my colleagues at the Rome meeting on Mechanisms of Speciation for some very stimulating discussions, particularly F. Ayala, J. Beardmore, and H. Carson (geographic variation), G. Dover (selfish sequences), R. Ehrendorfer (meiotic mechanisms), E. Mayr (clonal evolution), L. Stebbins (practically everything), and M. J. D. White (Warramaba parthenogenesis). I acknowledge the generous financial support of the National Geographic Society and the National Science Foundation, and the expert technical assistance of Ms Carolyn Harris. Field work in Mexico was made possible by permits issued by the Direccion General de Regulacion Pesquera, Instituto Nacional de Pesca, for which I am extremely greatful.

REFERENCES

1. Berger L: Systematics and hybridization in European green frogs of the Rana esculenta complex. J Herpetol 7:1, 1973.

2. Uzzell T. Berger L: Electrophoretic phenotypes of Rana ridibunda, Rana lessonae, and their hybridogenetic associate, Rana esculenta. Proc Acad Natur Sci Philadelphia 127:13, 1975.

3. Echelle AA, Mosier DT: All-female fish: A cryptic species of Menidia (Atherinidae). Science 212:1411, 1981.

4. Mickevich MF, Johnson MS: Congruence between morphological and allozyme data in evolutionary inference and character evolution. Syst Zool 25:260, 1976.

5. Chernoff B, Conner JV, Bryan CF: Systematics of the Menidia beryllina complex (Pisces: Atherinidae) from the gulf of Mexico and its tributaries. Copeia 1981:319, 1981.

6. White MJD: Animal Cytology and Evolution. 3rd ed, Cambridge: Cambridge University Press, 1973.

7. Maslin TP: Taxonomic problems in parthenogenetic vertebrates. Syst Zool 17:219, 1968.

8. Maynard Smith J: The Evolution of Sex. Cambridge: Cambridge University Press, 1978.

9. Miller RR: Five new species of Mexican poeciliid fishes of the genera Poecilia, Gambusia, and Poeciliopsis. Occ Pap Mus Zool Univ Michigan 672:1, 1975.

10. Hubbs CL, Hubbs LC: Apparent parthenogenesis in nature, in a form of fish of hybrid origin. Science 76:628, 1932.

11. Howell AB: The involved genetics of fish. Science 77:389, 1933.

12. Meyer H: Investigations concerning the reproductive behavior of Mollienisia formosa. J Genet 36:329, 1938.

13. Kallman KD: Gynogenesis in the teleost Mollienesia formosa (Girard), with discussion of the detection of parthenogenesis in vertebrates by tissue transplantation. J Genet 58:7, 1962.

14. Hubbs CL, Hubbs LC: Breeding experiments with the invariably female, strictly matroclinous fish, Mollienisia formosa. Genetics 31:218, 1946.

15. Hubbs CL, Hubbs LC: Experimental breeding of the Amazon Molly. Aquarium J 17:4, 1946.

16. Hubbs CL: Species and hybrids of Mollienisia. Aquarium 1:263, 1933.

17. Hubbs CL: Double-crossing molly. Home Aquarium Bull 4:5, 1934.

18. Schultz RJ: Role of polyploidy in the evolution of fishes. In Lewis WH (ed): Polyploidy: Biological Relevance. New York: Plenum p 313, 1980.

19. Schultz RJ: Evolution and ecology of unisexual fishes. In Hecht MK, Steere WC, Wallace B (eds): Evolutionary Biology. New York: Plenum, vol 10, p 277, 1977.

20. Schultz RJ: Hybridization, unisexuality and polyploidy in the teleost Poeciliopsis (Poeciliidae) and other vertebrates. Am Nat 108:605, 1969.

21. Horn MH (ed): [Festschrift] In honor of Carl L. Hubbs. Bull So Cal Acad Sci 75:57, 1976.

22. Rosen DE, Bailey RM: The poeciliid fishes (Cyprinodontiformes), their structure, zoogeography and systematics. Bull Am Mus Nat Hist 126:1, 1963.

23. Thibault RE, Schultz RJ: Reproductive adaptations among viviparous fishes (Cyprinodontiformes: Poeciliidae). Evolution 32:320, 1978.

24. Bailey RM, Fitch JE, Herald ES, Lachner EA, Lindsey CC, Robins CR, Scott WB: A list of common and scientific names of fishes from the United States and Canada. 3rd ed. Am Fish Soc Special Publ 6:1, 1970.

25. Darnell RM, Abramoff P: Distribution of the gynogenetic fish, Poecilia formosa, with remarks on the evolution of the species. Copeia 1968:354, 1968.
26. Miller RR: Geographical distribution of Central American freshwater fishes. Copeia 1966:773, 1966.
27. Hubbs C, Drewry GE, Warburton B: Occurrence and morphology of a phenotypic male of a gynogenetic fish. Science 129:1227, 1959.
28. Hubbs C: Interactions between a bisexual fish species and its gynogenetic sexual parasite. Bull Tex Mem Mus No 8:1, 1964.
29. Ohno S: The development of sexual reproduction. In: Austin CR, Short RV (eds): The Evolution of Reproduction. Cambridge: Cambridge University Press, p 1, 1976.
30. Monaco PJ, Rasch EM, Balsano JS: Sperm availability in naturally occuring bisexual-unisexual breeding complexes involving Poecilia mexicana and the gynogenetic teleost Poecilia formosa. Env Biol Fish 6:159, 1981.
31. Balsano JS, Kucharski K, Randle EJ, Rasch EM, Monaco PJ: Reduction of competition between bisexual and unisexual females of Poecilia in northeastern Mexico. Env Biol Fish 6:39, 1981.
32. Kucharski K: Habitat preferences and associations of a unisexual-bisexual complex of Poecilia (Pisces: Poeciliidae) from northeastern Mexico. MS thesis, Florida Atlantic Univ., Boca Raton, Florida.
33. McKay FE: Behavioral aspects of population dynamics in unisexual-bisexual Poeciliopsis (Pisces: Poeciliidae). Ecol 52:778, 1971.
34. Moore W, McKay FE: Coexistence in the unisexual-bisexual species complexes of Poeciliopsis (Pisces: Poeciliidae). Ecol 52:791, 1971.
35. Thibault RE: Ecological and evolutionary relationships among diploid and triploid unisexual fishes associated with the bisexual species, Poeciliopsis lucida (Cyprinodontontiformes: Poeciliidae). Evolution 32:613, 1978.
36. Vrijenhoek RC: Factors affecting clonal diversity and coexistence. Am Zool 19:787, 1979.
37. Schultz RJ, Kallman KD: Triploid hybrids between the all-female teleost Poecilia formosa and Poecilia sphenops. Nature 219:280, 1968.
38. Macgregor HC, Uzzell TM: Gynogenesis in salamanders related to Ambystoma jeffersonianum. Science 143:1043, 1964.
39. Uzzell TM: Meiotic mechanisms of naturally occuring unisexual vertebrates. Am Nat 104:433, 1970.
40. Cimino MC: Meiosis in triploid all-female fish (Poeciliopsis: Peociliidae). Science 175:1484, 1971.
41. Cuellar O: Reproduction and the mechanism of meiotic restitution in the parthenogenetic lizard Cnemidophorus uniparens. J Morphol 133:139, 1971.
42. Cuellar O: Cytology of meiosis in the triploid gynogenetic salamander *Ambystoma tremblayi*. Chromosoma 58:355, 1976.
43. Uzzell TM, Darevsky IS: Biochemical evidence for the hybrid origin of the parthenogenetic species of the Lacerta saxicola complex (Sauria: Lacertidae), with a discussion of some ecological and evolutionary implications. Copeia 1975:204, 1975.
44. White MJD: Meiotic mechanisms in a parthenogenetic grasshopper species and its hybrids with related bisexual species. Genetica 52/53:379, 1980.

45. Monaco PJ, Rasch EM, Balsano JS: Nucleoprotein cytochemistry during oogenesis in a unisexual fish, Poecilia formosa. Hist J 13:747, 1981.
46. Rasch EM, Monaco PJ, Balsano JS: Cytophotometric and autoradiographic evidence for functional apomixis in the gynogenetic fish, Poecilia formosa, and its related triploid unisexuals. Histochem 73, 1982.
47. Rasch EM, Balsano JS: A mechanism for meiotic restitution during oogenesis in a triploid, unisexual fish. J Cell Biol 63:280a, 1974.
48. Cimino MC: Egg production, polyploidization, and evolution in a diploid all-female fish of the genus Poeciliopsis. Evolution 26:294, 1972.
49. Rasch EM, Darnell RM, Kallman KD, Abramoff P: Cytophotometric evidence for triploidy in hybrids of the gynogenetic fish, Poecilia formosa. J Exp Zool 160:155, 1965.
50. Haskins CP, Haskins EF, Hewitt RE: Pseudogamy as an evolutionary factor in the poeciliid fish Mollienisia formosa. Evolution 14:473, 1960.
51. Kallman KD: Homozygosity in a gynogenetic fish. Genetics 50:260, 1964.
52. Prehn LM, Rasch EM: Cytogenetic studies of Poecilia (Pisces): I. Chromosome numbers of naturally occuring poeciliid species and their hybrids from eastern Mexico. Can J Genet Cytol 11:888, 1969.
53. Rasch EM, Balsano JS: Biochemical and cytogenetic studies of Poecilia from eastern Mexico. II. Frequency, perpetuation, and probable origin of triploid genomes in females associated with Poecilia formosa. Rev Biol Trop 21:351, 1974.
54. Rasch EM, Balsano JS: Cytogenetic studies of Poecilia (Pisces). III. Persistence of triploid genomes in the unisexual progeny of triploid females associated with Poecilia formosa. Copeia 1973:810, 1973.
55. Strommen CA, Rasch EM, Balsano JS: Cytogenetic studies of Poecilia. V. Cytophotometric evidence for the production of fertile offspring by triploids related to Poecilia formosa. J Fish Biol 7:1, 1975.
56. Menzel BW, Darnell RM: Morphology of naturally occuring triploid fish related to Poecilia formosa. Copeia 1973:350, 1973.
57. Rasch EM, Prehn LM, Rasch RW: Cytogenetic studies of Poecilia (Pisces). II. Triploidy and DNA levels in naturally occuring populations associated with the gynogenetic teleost Poecilia formosa (Girard). Chromosoma 31:18, 1970.
58. Strommen CA, Rasch EM, Balsano JS: Cytogenetic studies of Poecilia (Pisces). VI. Use of epithelial cell biopsies to identify triploid females associated with Poecilia formosa. Copeia 1975:568, 1975.
59. Balsano JS, Darnell RM, Abramoff P: Electrophoretic evidence of triploidy associated with populations of the gynogenetic teleost Poecilia formosa. Copeia 1972:292, 1972.
60. Balsano JS, Rasch EM: Biochemical and cytogenetic studies of Poecilia from eastern Mexico. I. Comparative microelectrophoresis of plasma proteins of seven species. Rev Biol Trop 21:299, 1974.
61. Rasch EM, Monaco PJ, Balsano JS: Identification of a new form of triploid hybrid fish by DNA-Feulgen cytophotometry. J Histochem Cytochem 26:218, 1978.
62. Turner BJ, Brett BH, Rasch EM, Balsano JS: Evolutionary genetics of a gynogenetic fish, Poecilia formosa, the Amazon molly. Evolution 34:246, 1980.
63. Bulger AJ, Schultz RJ: Heterosis and interclonal variation in thermal tolerance in unisexual fishes. Evolution 33:848, 1979.

64. Kallman KD: Population genetics of a gynogenetic teleost, Mollienesia formosa. Evolution 16:497, 1963.

65. Darnell RM, Lamb E, Abramoff P: Matroclinous inheritance and clonal structure of a Mexican population of the gynogenetic fish, Poecilia formosa. Evolution 21:168, 1967.

66. Angus RA, Schultz RJ: Clonal diversity in the unisexual fish Poeciliopsis monacha-lucida: a tissue graft analysis. Evolution 33:27, 1979.

67. Moore WS, Eisenbrey AB: The population structure of an asexual vertebrate, Poeciliopsis 2 monacha-lucida (Pisces: Poeciliidae). Evolution 33:563, 1979.

68. Leslie JF, Vrijenhoek RC: Consideration of Muller's ratchet mechanisms through studies of genetic linkage and genomic compatibilities in clonally reproducing Poeciliopsis. Evolution 34:1105, 1980.

69. Vrijenhoek RC, Schultlz RJ: Evolution of a trihybrid unisexual fish (Poeciliopsis, Poecililidae). Evolution 28:306, 1974.

70. Leslie JF, Vrijenhoek RC: Genetic dissection of clonally inherited genomes of Poeciliopsis. I. Linkage analysis and preliminary assessment of deleterious gene loads. Genetics 90:801, 1978.

71. Stebbins GL Jr: Variation and Evolution in Plants. New York: Columbia University Press, 1950.

72. Hubbs CL: Natural hybridization in fishes. Syst Zool 4:1, 1955.

73. Hubbs CL: Isolation mechanisms in the speciation of fishes. In Blair WF (ed): Vertebrate Speciation. Austin: University of Texas Press, 1961, p 5.

74. Schultz RJ, Miller RR: Species of the Poecilia sphenops complex in Mexico. Copeia 1971:282, 1971.

75. Turner BJ, Brett BH, Miller RR: Interspecific hybridization and the evolutionary origin of a gynogenetic fish, Poecilia formosa. Evolution 34:917, 1980.

76. Lowe CH, Wright JW: Evolution of parthenogenetic species of Cnemidophorus (whiptail lizards) in western North America. J Ariz Acad Sci 4:81, 1966.

77. Schultz RJ: Gynogenesis and triploidy in the viviparous fish Poeciliopsis. Science 157:1564, 1967.

78. Uzzell TM, Goldblatt SM: Serum proteins of the salamanders of the Ambystoma jeffersonianum complex, and the origin of the triploid species of this group. Evolution 21:345, 1967.

79. Berger L: Inheritance of sex and phenotype in F_1 and F_2 crosses within Rana esculenta complex. Genetica Polonica 12:517, 1971.

80. White MJD, Contreras N, Cheney J, Webb GC: Cytogenetics of the parthenogenetic grasshopper Warramaba (formerly Moraba) virgo and its bisexual relatives. II. Hybridization studies. Chromosoma 61:127, 1977.

81. Schultz RJ: Unisexual fish: laboratory synthesis of a "species." Science 179:180, 1973.

82. White MJD, Contreras N: Cytogenetics of the parthenogenetic grasshopper Warramaba (formerly Moraba) virgo and its bisexual relatives. III. Meiosis of male "synthetic virgo" individuals. Chromosoma 67:55, 1978.

83. Simanek DE: Population genetics and evolution in the Poecilia formosa complex (Pisces: Poeciliidae). PhD Dissertation, Yale University, New Haven, CT, 1978.

84. Brett BH, Turner BJ, Miller RR: Allozymic divergences among the shortfin mollies of the Poecilia sphenops species complex. Isozyme Bull 13:104, 1980.

85. Brett BH, Turner BJ, Miller RR: Genetic divergence in the Poecilia sphenops complex in Middle America (Pisces: Poeciliidae). Copeia 1982 (in press).
86. Parzefall J: Zur Genetik und biologischen Bedeutung des Aggressionsverhaltens von *Poecilia sphenops* (Pisces, Poeciliidae). Z Tierpsychol 50:399, 1979.
87. Kriebel RM: The caudal neurosecretory system of Poecilia sphenopss (Poeciliidae). J Morphol 165:157, 1980.
88. Menzel BW, Darnell RM: Systematics of Poecilia mexicana in northern Mexico. Copeia 1973:225, 1973.
89. Simanek DE: Genetic variability and population structure of Poecilia latipinna. Nature 276:612, 1978.
90. Snelson FF, Wetherington JD: Sex ratio in the sailfin molly, Poecilia latipinna. Evolution 34:308, 1980.
91. Hubbs CL: Fishes of the Yucatan peninsula. Carnegie Inst Wash Publ 457:157, 1936.
92. Klingerman AD, Bloom SE: Rapid chromosome preparations from solid tissues of fishes. J Fish Res Bd Canada 34:266, 1977.
93. Chen RR: A comparative study of twenty killifish species of the genus Fundulus (Teleostei: Cyprinodontidae). Chromosoma 3:436, 1971.
94. Black DA, Howell WM: The North American mosquitofish, Gambusia affinis: a unique case in sex chromosome evolution. Copeia 1979:509, 1979.
95. Rishi KK, Gaur P: Cytological female heterogamety in jet-black molly, Mollienesia sphenops. Curr Sci 45:669, 1976.
96. Hewitt RE, Word LW, Haskins EF, Haskins CP: Electrophoretic analysis of muscle proteins in several groups of poeciliid fishes, expecially the genus Mollienesia. Copeia 1963:269, 1963.
97. Balsano JS: Systematic relations of fishes of the genus Poecilia in eastern Mexico based upon plasma protein electrophoresis. PhD Dissertation, Marquette University, Milwaukee, Wis. 1968.
98. Abramoff P, Darnell RM, Balsano JS: Electrophoretic demonstration of the hybrid origin of the gynogenetic teleost Poecilia formosa. Am Nat 102:555, 1968.
99. Whitt GS: Developmental genetics of fishes: Isozymic analyses of differential gene expression. Am Zool 21:549, 1981.
100. Morgan RP II, Ulanowicz NI: The frequency of muscle protein polymorphism in Menidia menidia (Artherinidae) along the Atlantic coast. Copeia 1976:356, 1976.
101. Murphy BR, Nielsen LA, Turner BJ: Electrophoretic assessment of walleye (Stizostedion vitreum) stock structure in Claytor Lake, Virginia. Isozyme Bull 13:100, 1980.
102. Turner BJ, Miller RR, Rasch EM: Significant differential gene duplication without ancestral tetraploidy in a genus of Mexican fish. Experientia 36:927, 1980.
103. Brown WM, Wright JW: Mitochondrial DNA analyses and the origin and relative age of parthenogenetic lizards (genus Cnemidophorus). Science 203:1247, 1979.
104. Clark E: A method for artificial insemination of viviparous fishes. Science 112:722, 1950.
105. Avise JC, Lansman RA, Shade RO: The use of restriction endonucleases to measure mitochondrial DNA sequence relatedness in natural populations. I. Population structure and evolution in the genus Peromyscus. Genetics 92:279, 1979.
106. Celis, JE, Bravo R: Cataloguing human and mouse proteins. Trends Biochem Sci 6:197, 1981.
107. Ohta T: Population genetics of selfish DNA. Nature 292:648, 1981.

NOTE ADDED IN PROOF

Field collections and laboratory work done immediately after this review was submitted have further clarified the ancestry of P formosa. Sammples of P mexicana from the upper Rio Tamesi (including the Rio Guayelejo) and the Rios Tigre and San Rafael (small rivers between the much larger Tamesi-Panuco and Soto la Marina systems) are *fixed* for the "null M_2" parvalbumin phenotype. These populations are thus prime candidates for the ancestry of P formosa. Hybrids between some of these populations and P latipinna have been made in the laboratory and are now being evaluated for gynogenesis. The relationship of these populations to those of the coastal lagoons around Tampico and the Laguna de Tamiahua (where the null M_2 phenotype was originally discovered) is not yet clear. In this review, the latter have been considered to be straight-forward polymorphic populations, but they could also be mixtures of previously unresolved sibling species, or part of a "suture zone" between well differentiated subspecies. These alternatives are now being examined. For the present, the following "forms" of P mexicana can be recognized: 1) "Extreme northern"— Rio Soto la Marina and Rio Grande. This form, referred to as P m limantouri in the present review (probably erroneously), is genetically very distinct from all southern populations, and may well merit recognition as a distinct species. 2) "Northern"—upper Rio Tamesi,, Rio Tigre, and Rio San Rafael. This form has the fixed null M_2 parvalbumin phenotype, and is best qualified for the ancestry of P formosa. 3) "Transitional"—Rio Tamesi near Tampico and coastal lagoons south to the mouth of the Rio Tuxpan. This form contains the null M_2 parvalbumin phenotype, apparently as a polymorphism. Until the discovery of the fixed null M_2 phenotype in the "northern form," these populations were regarded as the best qualified for the ancestry of P formosa. 4) "Southern"—Rio Panuco and southward into Belize (or further). This form is traditional P m mexicana. "Transitional" may simply be a mixture of "northern" and "southern." The name "limantouri" might be used for the "northern" form, but the original type locality attached to that name was Tampico, a "transitional" locality. Application of the name should therefore await resolution of the relationship of the "northern," "southern," and "transitional" populations.

Mechanisms of Speciation, pages 307–344
© 1982 Alan R. Liss, Inc., 150 Fifth Avenue, New York, NY 10011

Speciation Events Evidenced in Turbellaria

Mario Benazzi

INTRODUCTION

Some groups of Turbellaria have offered many examples of microevolutionary changes revealed by karyological variation, by reproductive strategies, and by the rise of isolating mechanisms representing effective processes of speciation.

Most information concerns Tricladida Paludicola depending, at least to a certain extent, on the fact that among the various turbellarian groups the freshwater planarians are the most suitable for longlasting laboratory cultures. In fact, cytogenetical studies were first realized using these organisms and it was also possible to follow the offspring of interracial crosses. If we take into consideration the 13 points in which, according to White [1], a fully documented history of a single case of speciation might be included, the value of such an experimental approach appears valid. But White himself points out that there is probably no single case of speciation for which all of the required information exists. This assumption is certainly valid also for Turbellaria; however, in several cases evolutionary events at two levels, 1) anagenesis or phyletic evolution and 2) cladogenesis or phyletic branching (of which speciation is the most decisive process), appear supported by plausible evidence.

Many investigations on these topics have been accomplished by our team in Pisa over the course of about 40 years, and the results have been summarized in the preceding papers (cf particularly: Benazzi [2], Benazzi Lentati [3,4], Benazzi and Benazzi Lentati [5]). I now wish to provide a review of that knowledge, placing special emphasis on the speciation mechanisms and especially taking into account some recent data. Unfortunately the contribution brought about by application of biochemical techniques is very scarce. Peter [6], with the disk electrophoretic method, attempted the characterization of nine species of freshwater planarians. The comparison among different populations of a given species yielded higher degrees of similarity than those between different species.

However, the varying interspecific correspondences could not be used for taxonomical grouping. The author concluded that while intraspecific distinctions of paludicolous triclads may be made with much success using the applied method, more extensive investigations with a variety of methods seem to be necessary for a definitive conclusion concerning the applicability of electrophoresis in elucidating relationships above the species level. More recently, only a few turbellarian representatives have been studied from this viewpoint.

The data I have selected as signaling speciation events will be presented separately for each species group according to their taxonomic position. Some general traits of the evolutionary mechanisms will be analyzed in a separate chapter.

TRICLADIDA PALUDICOLA

The freshwater planarians (Probursalia) are now divided [7] into three well-defined families: the Dugesiidae, the Planariidae, and the Dendrocoelidae. In several genera of these families a substantial number of karyological studies have been made during the past three decades, usually combined with data on the geographical distribution of the taxa. Experimental investigations have also been accomplished, particularly on species of the genus Dugesia.

Family Dugesiidae

Genus dugesia. This is the more common freshwater genus with a worldwide distribution. It comprises, however, species differentiated by various characteristics that justifies its division into three subgenera [7], viz Dugesia, Girardia, and Schmidtea. Before the recognition of the three subgenera I pointed out [8] that the genus Dugesia could have been divided into two clearly differentiated sections. One section comprises the species that possess an arrow-shaped anterior end with well developed auricles, namely Dugesia gonocephala together with the allied species of the Old World (now attributed to the subgenus Dugesia), and the species of the New World (now attributed to the subgenus Girardia). All these species have oocytes of small diameter with multipolar spindles (only after fertilization does the normal bipolar spindle appear). Furthermore, these species frequently multiply by fission. The other section comprises the European species of the "lugubris-polychroa group," which show a rounded triangular head and are now attributed to the subgenus Schmidtea which possess oocytes of a great diameter with a bipolar spindle and which reproduce, but for one exception, exclusively sexually.

There are karyological data to support this dichotomy within the genus Dugesia since in the subgenus Schmidtea the basic number is 4 and acrocentric chromosomes are present, while in the species of the subgenera Dugesia and Girardia

TABLE I. Tricladida paludicola: Family Dugesiidae, Genus Dugesia

| Subgenus | Species | Chromosome number | | Distribution |
		2x	x	
Dugesia	D gonocephala s.s. and			
	some allied species[a]	16	8	Europe
	D sicula	18	9	Europe
	D biblica[a]	18	9	Israel
	D japonica:[a]			
	D j japonica	16	8	Far east
	D j ryukyuensis	14	7	Far east
Girardia	D dorotocephala[a]	16	8	USA
	D jenkinsae	8	4	USA
	D arizonensis	8	4	USA
	D tigrina[b]	16	8	New world
	D sanchezi	16	8	Chile
	D anceps	16	8	Argentina
	D cubana	18	9	Cuba
Schmidtea	D polychroa[b]	8	4	Europe
	D lugubris	8	4	Europe
	D mediterranea	8	4	Europe
	D nova (nomen nudum)	6	3	Europe

[a]Polysomic and polyploid forms, and in some species strains with B chromosomes are also present.
[b]Polyploid forms are present.

the basic number is 8, but for a few exceptions, lacking acrocentric chromosomes, and their karyotypes are rather similar (Table I).

Subgenus dugesia. It comprises a great number of taxa, distributed in Eurasia and in Africa, strictly allied to the Dugesia gonocephala (Dugès) from Central Europe but reproductively isolated. All these forms represent species or sometimes semispecies belonging to the "Dugesia gonocephala group" which may be considered a superspecies, ie a monophyletic group of closely related, largely or entirely allopatric species. Within some species polyploid biotypes occur.

The eight members of the haploid complement differ in length but have not displayed marked morphological variations, all chromosomes being meta- or submetacentric (Fig. 1a). It seems likely that only minor chromosome mutations, such as small pericentric inversions, have accompanied speciation. However, the reproductive isolation appears well established at least in most cases, since hybrids may be obtained only among races of single species. In some cases I observed copulation between specimens of different species but with no resulting

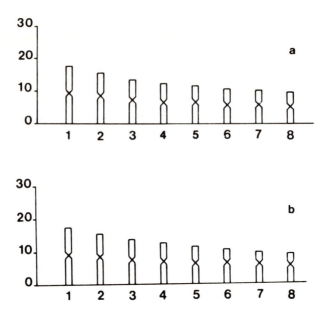

Fig. 1.   Idiograms of a) Dugesia gonocephala s.l., n = 8 and b) D dorotocephala, n = 8. Originals.

offspring. As experimental insemination is not feasible in planarians it is not possible to determine the exact nature of the isolating mechanisms.

Only in a few cases has speciation within the "gonocephala complex" involved numerical chromosome variation. I found 2n = 18, n = 9 in D sicula Lepori, a number recently confirmed for a population from Mallorca, Spain [9]. The same basic number 9 has been found by Bromley [10] in D biblica Benazzi and Banchetti, which presents diploid sexually reproducing populations and triploid (plus 1 to 5 extra chromosomes) generally fissiparous populations. Recently Bromley (unpublished thesis) has attributed the sexual populations to the subspecies monticola, and the fissiparous ones to the subspecies biblica. On the other hand, the basic number 7 has been found by Kawakatsu et al [11] in D japonica Ichigawa and Kawakatsu. The typical subspecies of this Asiatic polymorphic planarian possesses 2n = 16, n = 8, while the subspecies ryukyuensis has diploid and triploid complements with 14 and 21 chromosomes, respectively.

It is finally to be pointed out that fissioning, often correlated with various heteroploid conditions, is of frequent occurrence in many other species of the gonocephala complex (see Table VII).

Subgenus girardia.   Until now the following species have been studied karyologically: D tigrina from North America (included specimens found as immigrants in Europe), D dorotocephala and the two strictly related species, D jenkinsae and D arizonesis, all from the US, D anceps from Argentina, D sanchezi from Chile, and D cubana from Cuba. The basic number 8 has been ascertained in D tigrina [12–14], in D dorotocephala [8], in D anceps [15], in D sanchezi [16]. D tigrina is frequently represented by fissiparous populations with diploid, triploid, or mixoploid karyotypes; a sexual population from Canada resulted in being diploid. In D dorotocephala almost all the populations, both sexual and fissiparous, possess a diploid complement (see Table VII).

In all the species now mentioned the haploid complement looks very similar with the eight members differing in length (the longest being about twice the size of the shortest one). However, the length decreases so gradually that it is very difficult to classify the chromosomes into groups (Fig. 1b). Also the centromeric index varies little, all chromosomes being meta- or submetacentric; however, according to Dutrillaux and Lenicque [14] in a diploid population of D tigrina the chromosome labelled No.5 is acrocentric (centromeric index, c_i, = 12). Even so, the karyological similarity between these species suggests that their specific differentiation has occurred principally by allelic mutations without any pronounced structural repatterning of the chromosomes.

An exception to this is offered by D cubana which posseses a diploid complement of 18 chromosomes [17]. It is to be pointed out that the form examined is not the typical one described by Codreanu and Balcesco, but an unpigmented and micro-oculated strain living in Cuban subterranean waters. Its Karyotype is peculiar also for the presence of a subtelocentric chromosome pair (c_i = 16,44). Gourbault was not able to study the epigean form; therefore, we can not speculate about the possible influence of the hypogean habitat on the karyotype.

However, the most impressive variation in the chromosome complement within the subgenus Girardia is represented by the forms with $2n = 8$, $n = 4$. This complement was found by Benazzi [18] in two populations, one sexual (from San Felipe, Texas) the other asexual (from Sabino, Arizona), both apparently belonging to D dorotocephala. The sexual population was later recognized as a new species (Dugesia jenkinsae Benazzi and Gourbault), while the sexual form represents, with all probability, a fissiparous strain of D arizonensis Kenk, a species which possesses the same complement $2n = 8$, $n = 4$ [19]. In all these forms the four chromosomes of the haploid set look identical. The longest and the two middle-sized are metacentric while the shortest one is submetacentric (Table I). On the basis of these data neither the hypothesis of centric fusion, nor that of chromosome fission can be used to explain the relationship between this karyotype and that of the typical dorotocephala; in fact, all chromosomes are two armed in both karyotypes. It may be suggested that $n = 4$ represents the basic number and the typical dorotocephala is a tetraploid. A comparative analysis has shown that the karyometrical data are not completely consistent with this hypothesis, which in addition meets with the other difficulty that almost all the "arrow-headed" Dugesia from both the Old- and New-World present the haploid number 8, which appears to be the ancestral one.

The relationship between the two complements with $n = 8$ and $n = 4$ respectively, as well as the taxonomic significance of the latter, remain therefore unclear.

Subgenus schmidtea. The taxa included in this subgenus, which are widely distributed in Europe, have been a taxonomic puzzle for many years. O. Schmidt distinguished within Planaria torva (Müller) two new species, Planaria lugubris and P polychroa, which were subsequently accepted by some authors, but not by others who considered the differences between the presumed two new species purely quantitative. The question has been settled by Benazzi, Puccinelli, and Del Papa [20] and by Reynoldson and Bellamy [21] who demonstrated on the basis of karyological and morphological data, the validity of the two species. Moreover, new taxa have been later recognized within this species complex.

The first steps in the solution of the problem were accomplished by myself [22] showing a pronounced karyological differentiation (seven biotypes) within the forms attributable to lugubris or polychroa. This, together with reproductive isolation in some cases, offered a new approach to the taxonomic problem. The seven biotypes, indicated by the first seven letters of the alphabet, do not have the same taxonomic rank.

In fact, the first four biotypes (A,B,C,D) form a polyploid series starting with biotype A which is diploid ($2n = 8$, $n = 4$) and amphigonic, while the three polyploid biotypes (B,C,D) are pseudogamic. The haploid set of biotype A is made up of four chromosomes, one large metacentric, and three acrocentrics of different lengths. The basic set of biotypes B,C,D is the same as biotype A, thus they must be considered as autopolyploids which are capable of interbreeding, although in some cases the fertilization of diploid biotype eggs by sperm of polyploid biotypes is difficult to be realized. These four biotypes, therefore, represent a single species which, on the basis of the structure of the copulatory system, must be considered as corresponding to D polychroa [20,21] (Table I).

Biotypes E,F,G are, on the contrary, chromosomally differentiated and reproductively isolated. Biotype E is diploid ($2n = 8$, $n = 4$), but its haploid set differs from that of biotype A, being formed by three large acrocentrics of different lengths and by a very small submetacentric chromosome—besides which the oocyte and spermatocyte bivalents do not correspond to those of biotype A, in either size and chiasma frequency. On the basis of the morphology of the copulatory system, Benazzi et al [20] and Reynoldson and Bellamy [21] have concluded that this biotype corresponds to D lugubris (O. Schmidt).

Biotype F has $2n = 6$, $n = 3$, and its haploid set comprises a long metacentric, an acrocentric, and a very small metacentric. This biotype certainly originated from biotype E through a Robertsonian translocation (centric fusion) and on the basis of a karyometrical analysis, Benazzi and Puccinelli [23] have shown that the centric fusion occurred between the first and the third acrocentric chromosomes of biotype E (Table II) (Fig.2). The formation of the new karyotype seems to have been completed by a pericentric inversion of the tiny submetacentric in the E biotype, so giving rise to the tiny metacentric of the F biotype. The close genetic relationship between the two biotypes suggests the possibility of considering them as a single species. However, I was not able to obtain hybrids between them, with the exception of one very dubious case. Moreover, Reynoldson and Bellamy [21] have described the presence of a small protrusion or nipple in the distal end of the penial papilla for the specimens of biotype E, a feature that is not present in the biotype F (Benazzi et al [23]). Therefore, the two biotypes are probably separate species, perhaps species in statu nascendi.

Manfredi Romanini of Pavia University suggested that it was worth delving into the analysis of this model of karyotype evolution through an in situ cytochemical evaluation of the DNA content of the whole genome in each of the two biotypes, and of their single chromosomes and chromosome arms. The research accomplished by her staff (cf Benazzi et al [24]) confirms the hypothesis of centric fusion, since there are no significant differences between the DNA content of the nucleus in the two biotypes. The DNA content of isolated chromosomes and of both arms of the large metacentric allow us to assign to each chromosome, or chromosome arm, a percent value of the total genome. These results are in very good agreement with the karyometric data by Benazzi and Puccinelli. In regenerative blastemas of both biotypes the Feulgen-DNA content distribution of interphasic nuclei shows the presence of several nuclear classes, whose DNA contents follow a series of successive doubling of the lowest (2c) value found. Some findings could be accounted for by a different compactation of chromosome material between the two biotypes, without a true difference of DNA content. It is also to be pointed out that a difference in chromosome C

TABLE II.

Chromosome no.		Biotypes A–D (D polychroa)	Biotype G (D mediterranea)	Biotype E (D lugubris s.s.)		Biotype F (D nova n nudum)
1	rl	38.73	39.53	rl	33.70 →	33.35 + 26.73
	ci	44.35	42.33			↗
2	rl	25.60	32.54	rl	31.45	31.68
	ci		49.60			↗
3	rl	20.68	18.06	rl	27.08 ↗	
	ci		26.84			
4	rl	15.00	9.93	rl	7.76	8.23
	ci		43.06	ci	30.77	42.21

rl, relative length; ci, centromeric index.

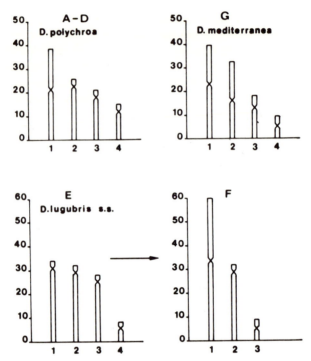

Fig. 2. Idiograms of Dugesia polychroa (biotypes A–D), D lugubris s.s. (biotype E), D mediterranea (biotype G), all with n = 4, and of biotype F with n = 3 (Dugesia nova nomen nudum). Demonstration of the origin of the biotype F from biotype E, through a Robertsonian translocation (modified from Benazzi and Puccinelli [23]).

banding in the two biotypes has been observed. These cytochemical data may support the view that biotypes E and F are two separate species: I propose now for biotype F the name Dugesia nova nomen nudum.

The seventh biotype (G) was previously found in Corsica, Sardinia, and Sicily. Its haploid complement is characterized by two longer chromosomes, both metacentric but the first more heterobrachial than the second. The third chromosome is at the boundary between submeta- and subtelocentric and the fourth is metacentric but more or less heterobrachial. This karyotype seems to be the most similar to that of biotype A, from which it perhaps originated through chromosome changes such as pericentric inversions and translocations. The cytogenetic data and the morphology of the copulatory system have shown that biotype G represents a bona species to which Benazzi et al [25] have given the name Dugesia mediterranea due to its peculiar geographical distribution.

A new contribution to this field has been brought by Ball [26] who collected from the Greek island of Corfu a sexual planarian with a diploid complement comprising one submetacentric and three metacentric chromosome pairs. This karyotype is most similar to biotype G, however, there are subtle karyological and morphological differences between this form and D mediterranea. This is especially the case for relative chromosome lengths in that the elements of the Corfu population decrease far less steeply than the ones of mediterranea. The distinctness of the Corfu population is also indicated by the morphology of the male copulatory apparatus. According to Ball, there can be little doubt that this population belongs to D polychroa s.s.; its karyotype would be derived from biotype A quite independently of the derivation of biotype G. Ball points out that further questions as to the origins of the peculiarities of the Corfu population must be left until such time as additional populations of D polychroa from Greek and Albanian mainlands have been investigated.

Family Planariidae

Although the cytogenetical and cytotaxonomical studies accomplished on species belonging to this family are less exhaustive than those referring to the genus Dugesia, some data appear interesting and suggest new investigations (Table III).

Genus planaria. I wish to present a short account on the phyletic status of this genus, which originally comprised all known Turbellaria: Its extent was gradually narrowed and Kenk selected the European Planaria torva (Müller) as the type-species. Besides torva, only two North American species dactyligera Kenk and occulta Kenk have been attributed to the genus. This amphiatlantic distribution was deemed by Ball to be incompatible with his hypothesis of the historical biogeography of freshwater planarians, and karyological studies by Ball and Gourbault [27] have suggested the possibility that the genus is not strictly monophyletic. The diploid complement of torva, first examined by Benazzi and Puccinelli [28] and by Melander [29] consists of nine chromosome pairs: the 1st pair is sharply differentiated by its length, thereafter the elements decrease gradually in size. With respect to centromere position, five pairs (the 1st, the 5th, and the three smallest ones) are metacentric, the other pairs are submetacentric (Fig. 3a). The karyometrical data by Ball and Gourbault [27] on this species are in agreement with the previous ones. On the contrary, P dactyligera possesses 19 chromosome pairs (Fig. 3b). Moreover in spermatocytes I of this species most of the 19 bivalents are ring shaped because of the complete chiasma terminalization, while in P torva, Benazzi and Puccinelli found both in male and female lines nine bivalents showing numerous chiasmata. Ball and Gourbault conclude that the karyotype of P dactyligera is unlike that of P torva, bears little resemblance to that of other European Planariidae, and is almost identical with that of several North American Planariidae, mostly of which belong to the genus Phagocata. According to the authors, the American species of

TABLE III. Tricladida paludicola: Family Planariidae

Genus	Species	Chromosome number		Distribution
		2x	x	
Planaria	P torva	18	9	Europe
Paraplanaria	P dactyligera	38	19	North America
Phagocata	P vitta	21 to 70	7	Europe
		34		Europe
	P dalmatica	32		Europe
	P paravitta	34		Europe
	P albissima	36		Europe
	P morgani	14	7	North America
	P velata	~60		North America
		(60 to 96)		North America
	P fawcetti	38		USA
	P gracilis	38		USA
	P papillifera	24		Japan
	P suginoi	24		Japan
	P kawakatsui	24		Japan
	P teshirogi	24		Japan
	P vivida	36 + 1B		Japan
Atrioplanaria	A racovitzai	52		Europe
	Atrioplanaria sp	46		Europe
	A delamarei	95 to 104		Europe
	A morisii	96 to 124		Europe
Polycelis	P nigra[a]	16	8	Europe
	P tenuis[a]	12	6	Europe
		14	7	Europe
	P felina[a]	18		Europe

[a]Also with polyploid forms.

Planaria are merely specialized species of the group characterized by a particular karyotype (2n = 38), all of which, no doubt, have a common origin in, and are endemic to, North America. Therefore, they have proposed the new generic name Paraplanaria to contain the two American species already attributed to the genus Planaria.

Genus phagocata. This genus, which also represents a puzzling taxonomic problem, was first proposed by Leidy for the American species Planaria gracilis Haldeman, 1840, while the generic name Fonticola was proposed by Komárek for the European species Planaria vitta Dugès, 1830. Hyman [30] admitted that all Neartic and Paleartic species, referred to by these two genera, are congeneric and that the name Phagocata, having priority over Fonticola, is the valid name.

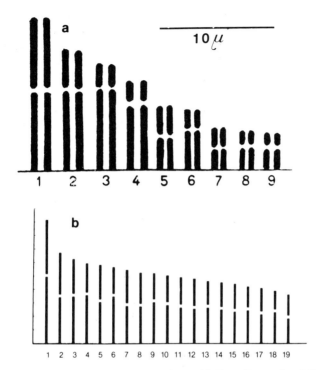

Fig. 3.   a) Karyogram of Planaria torva, 2n = 18 (from Benazzi and Puccinelli [28]); b) idiogram of P dactyligera, n = 19 (from Ball and Gourbault [27]).

However, the affinity between the Nearctic and the Paleartic species is as yet undecided. Cytological studies have been carried out by Dahm [12] on some Northern European populations of P vitta; he found eight chromosome numbers, viz 21,28,35,42,49,56,63,70, all of which are multiples of the number 7, and concluded that the specimens studied with all probability represented 3–10 ploids. Most of the populations were asexual; a complement of 28 chromosomes has been found for the population with sexual reproduction (see Table VI).

Benazzi and Gourbault [31] in three sexual populations of the same species originating from subterranean waters of Southern France have found 2n = 34, n = 17, the first chromosome pair being formed by markedly longer metacentric elements. This complement may be considered as corresponding to a diploid condition; therefore, these three populations cannot be included in the polyploid series outlined by Dahm for the Northern European populations of the same species (Fig. 4a,b).

Dahm [32] studied the karyotype of three other European species, viz Phagocata albissima (Vejdovsky) with 2n = 36, Ph paravitta (Reisinger) with 2n = 34, Ph dalmatica (Stanković and Komárek) with 2n = 32, and admitted that there is evidence that polyploidy has occurred in the evolution of these taxa.

With regards to the American species of the genus, the karyological data are very scarce. Whitehead [33] found in Phagocata morgani (Stevens and Boring) two biotypes, one diploid and sexual with 14 chromosomes and seven bivalents indicated in meiosis, the other fissiparous, possibly polyploid with 35 chromosomes. The same author recorded for Ph velata (Stringer) an approximate mitotic number of 60 chromosomes, which may indicate polyploidy and noted irregular meiosis. She also suggested that this was a recently evolved species and that pseudogamy may be involved in the reproductive cycle.

Phagocata velata has been reexamined by Ball and Goubault [34] who have found a varied chromosome number of 60–96 elements corresponding to different degrees of polyploidy. These authors have also studied the complement of Phagocata fawcetti sp. nov. from California, consisting of 38 elements in which the

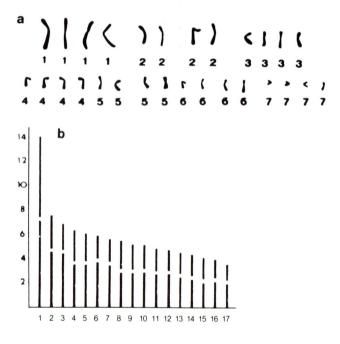

Fig. 4. a) Karyogram of Phagocata vitta from North Europe, tetraploid set (from Dahm [32]); b) idiogram of Ph vitta from South France, n = 17 (from Benazzi and Gourbault [31]); c) idiogram of Ph fawcetti from California, n = 19 (from Ball and Gourbault [34]); d) metaphase plate of Atrioplanaria delamarei (from Gourbault and Benazzi [39]).

karyotype exhibits one evident peculiarity, the presence of a pair of very large chromosomes which are more than 1.5 times the size of the second pair (Fig. 4c). The authors point out that whereas Ph fawcetti might be expected to be morphologically allied with Ph velata, the two species differ greatly in their karyotypes. In contrast, the karyotype of fawcetti is quite similar to that of gracilis, although the two species are morphologically distinct. Ball and Gourbault point out that in the genus Phagocata specific differentiation seems to be linked with chromosomal variations, a statement which appears corroborated also by the data referring to the European species.

In this context it is noteworthy that recent karyological data have also been collected on five Japanese species of Phagocata, namely: teshirogii Ichikawa and Kawakatsu, papillifera Ijima and Kaburaki, suginoi, kawakatsui Okugawa, vivida Ijima and Kaburaki. In the first species, Teshirogi and Sasaki [35] have found 2n = 24, n = 12 where the karotype consists of five pairs of metacentric, six pairs of submetacentric, and one pair of subtelocentric chromosomes. Chromosome No.1 is particularly large, No.2 is medium sized, and all the others are

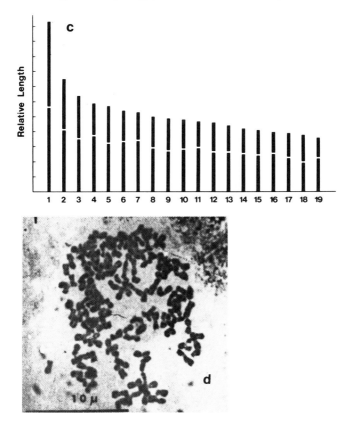

distinctly small. The karyotypes of papillifera and suginoi, according to Sugino et al [36] and of kawakatsui, according to Teshirogi et al [37], also possess 2n = 24, n = 12, and are very similar to one another as the karyotypes of kawakatsui and teshirogii are distinguishable only because chromosomes No.2 and No.12 are respectively submetacentric and metacentric in the first species while in the latter one are respectively subtelocentric and submetacentric. On the contrary, Ph vivida possesses a very different karyotype: According to Teshirogi et al [38] its complement consists of 36, or 36 + 1B chromosomes in the somatic cells and 18, or 18 + 1B, in germ cells, and chromosome No. 11 shows a satellite at the end of the long arm. The only common trait of the karyotype of vivida and of the other four species is that chromosome No.1 is extremely large and metacentric.

Summing up, it appears that the karyotypes of the species included in the genus Phagocata differ remarkably, in agreement with the assumption of Ball and Gourbault [34].

Another interesting problem referring to speciation within Phagocata s.l. concerns the individuality of the taxa attributed to Atrioplanaria. This genus was created by de Beauchamp for Planaria racovitzai de Beauchamp, 1928 and includes some European unpigmented forms inhabiting subteranean waters previously attributed to the genus Fonticola. The validity of the genus Atrioplanaria, denied by some authors, has been substantiated by karyological data. Gourbault and Benazzi [39] in A delamarei and A morisii have found a very peculiar complement, consisting of a high number of mostly metacentric tiny chromosomes (this number is liable to fluctuate from 95 up to 124 in the same animal (Fig. 4d)). Already Dahm [32] in A racovitzai (see Table VII) and in Atrioplanaria sp. from Sardinia (Benazzi 1938) observed a fairly high number (apparently 52 and 46 respectively) of tiny chromosomes lacking conspicuous diagnostic features. These data confirm full taxonomical individuality for this genus and suggest several hypotheses on the origin of its karyological characteristics, such as polyploidy, or chromosome fragmentation.

Genus polycelis. The species belonging to this genus, which is largely distributed in Eurasia, offer interesting evolutionary perspectives. First of all, it is to be pointed out that the European species may be distinguished into two sections, well differentiated from both the morphologic and ecologic viewpoints. The first section comprises two allied species P nigra (Müller) and P tenuis Ijima, commonly found in stagnant or slowly running waters and reproducing only sexually. The other section is represented by P felina (Dalyell), a cold running-water dweller which frequently multiplies by fission. The karyological data emphasize the evolutionary diversification of the two sections. Another peculiar species, Polycelis benazzii de Beauchamp, was found in an Italian cave but, unfortunately, no karyological studies were accomplished.

The "nigra-tenuis group" is the most valuable as a speciation model. The two species, although clearly related, are morphologically as well as karyologically distinguishable. P nigra has 2n = 16, n = 8 (Fig. 5a). This diploid complement has been ascertained by many authors for different European populations. There are, however, polyploid biotypes also (Lepori [40,41]) (see Table VII). P tenuis appears to have 2n = 12, n = 6 in specimens from Sweden (Melander [29,42]) and from the Finnish Gulf (Lepori [43]), 2n = 14 in specimens from Italy and Germany. In specimens from Central Europe, on the other hand, Lepori found 19 to 22 chromosomes in mitoses and 19 to 22 bivalents in the oocytes; he assumed that tenuis possesses the basic number of six chromosomes and that the number seven recorded in some populations may probably represent a tetrasomic condition. In support of this interpretation, he noticed that in specimens from Monfalcone (Italy) with 2n = 14, the karyotype shows two pairs of chromosomes similar both in size and centromere position (Fig. 5b). Populations with higher chromosome numbers were interpreted as polyploid or polysomic strains. In

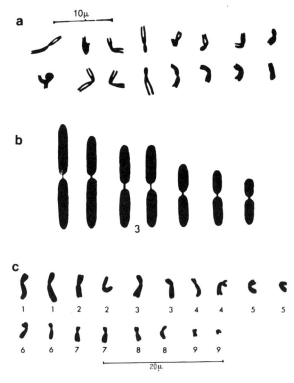

Fig. 5. a) Karyogram of Polycelis nigra, 2n = 16 (from Le Moigne [90]); b) karyogram of P tenuis, n = 7 (from Lepori [43]); c) Karyogram of P felina, n = 9 (from Dahm [32]).

1954 Hansen-Melander et al [44] created a new species in the "nigra-tenuis group" which they named hepta because this planarian, considered by them morphologically distinguishable from both nigra and tenuis, has n = 7. According to these authors, the haploid number is eight in *nigra*, six in *tenuis*, and seven in *hepta*. However, Benazzi [45] pointed out that, according to the data now available, P hepta cannot be considered a valid species.

A biochemical approach to the phyletic relation between nigra and tenuis has been accomplished by Biersma and Wijsman [46] using electrophoretic technique (iso-electric focusing). No difference was found in the isozyme pattern of malate dehydrogenase and tetrazolium oxidase, indicating a close relationship between the two species. I wish to recall that there are indications that nigra and tenuis may interbreed in the laboratory, although this question is not finally settled (Benazzi [45]). Biersma and Wijsman's investigations on banding patterns agree with the recent findings that structural proteins are especially conservative. In fact, the sodium dodecyl sulphate extracted proteins of Polycelis nigra-tenuis, Planaria torva and Phagocata vitta were very similar, while their soluble proteins were not. The authors point out that this technique may be of great help in taxonomic studies of higher taxa.

Different cytogenetical conditions are proper to Polycelis felina. Dahm [12] found that mitotic chromosome numbers varied from 18 to 27 and deduced that the basic number is nine and that 18 and 27 represent the diploid and triploid set respectively, while cells with 20 to 26 chromosomes were considered as aneuploids (Fig. 5c). Dutrillaux and Lenicque [15] found in a French population a diploid complement of 18 chromosomes. I have studied (unpublished data) three populations, two of which from Italy are sexual and diploid, the third from England is fissiparous and triploid.

The differentiation between felina and the "nigra-tenuis group" is, therefore, very marked and must be considered phylogenetically ancient. In this context, it is interesting to make a comparative analysis with the Asian species of the genus Polycelis, which appear from the morphologic, reproductive and ecologic points of view, more stricly related to felina than to the "nigra-tenuis group." Unfortunately, data referring to speciation processes are very meager. However, recently Teshirogi and Ishida [47] have studied for this purpose the Japanese species P auricolata Ijima and Kaburaki, which shows various chromosome polymorphisms or karypotypic variations. Electrophoretic analysis of the constitutive proteins extracted from specimens with different karyotypes (2n = 6, 2n = 12, 3x = 6) shows some dissimilarities. The species exhibits population polymorphism also in many morphologic features, and the authors admit that P auricolata is still progressing in speciation or chromosomal evolution.

Besides auricolata, three other Polycelis species have so far been found in Japan: sapporo (Ijima and Kaburaki), akkeshi Ichikawa and Kawakatsu, and schmidti (Zabusov) in which their diploid numbers are 42,40,32, respectively.

No data are available at present in order to interpret these chromosomal differences.

In North America the genus Polycelis is represented by two species only: coronata (Girard) and sierrensis Kenk, inhabiting the western half of the continent. Both species are restricted to fairly cold streams and reproduce more frequently by fission. They appear, therefore, related to the European P felina.

As far as I know, no cytological studies have been accomplished on the American species. It is, however, worthy to note research by Nixon and Taylor [48] on genetic similarities among 22 populations of P coronata determined from electrophoretically obtained protein data. Similarity values were lower than expected, being significantly below values reported for many vertebrate and invertebrate species. The authors state that this difference is partially explained by various ecological and behavioral isolating mechanisms. Lack of a morphological variation between populations which does not correspond with the biochemical divergence may imply different rates of evolution at these two levels. Polycelis has its center of origin in Europe or Asia and its eastward migration into Alaska may have occurred across the Bering Land Bridge. Evolutionary divergence times, based on the protein data, are used as possible indices of the migratory pattern of worms in the Pacific Northwest and correlate with the theory of land bridges between North America and Asia.

Family Dendrocoelidae

This family contains many taxa diffused in Eurasia and in North America. The classification at the generic level has not reached a definitive assessment; however, at present most species are attributed to the genera Dendrocoelum, Dendrocoelopsis, and Bdellocephala. Several species are adapted to underground waters. Although experimental investigations are lacking, some evolutionary problems have emerged (Table IV).

Genus dendrocoelum. D lacteum (Müller), the most common representative of the family, was the first to be studied cytogenetically. It possesses a diploid complement of 14 chromosomes, while three other European species, ie, D album (Steinmann), D infernale (Steinmann), D tubuliferum de Beauchamp have 28, 32, and 32 chromosomes, respectively. Aeppli [49,50] admitted that D infernale (a blind hypogeous form) is an autotetraploid of lacteum for which he assumed a diploid complement of 16 chromosomes. But this hypothesis does not hold because the diploid number 14 for lacteum, first recognized by Gelei [51], has been confirmed by many authors. Dahm [52], on the basis of the cytological patterns of the four species now mentioned, suggested the existence of two ancestral basic numbers, viz. 7 and 8. The first is present in the diploid D lacteum and the probable tetraploid D album, the second is shown by D infernale and D tubuliferum, which are considered as probably being originally tetraploid. According to Dahm, several cytological features suggest that the forms

TABLE IV. Tricladida paludicola: Family Dendrocoelidae

| Genus | Species | Chromosome number | | Distribution |
		2x	x	
Dendrocoelum	D lacteum	14	7	Europe
	D nausicaae	14	7	Europe-Asia minor
	D album	28		Europe
	D infernale	32		Europe
	D tubuliferum	32		Europe
	D tuzetae	32		Europe
	D coiffaiti	32		Europe
Neodendrocoelum	Neodendrocoelum species from Lake Ohrid	32		Europe
Procotyla	P fluviatilis	14	7	North America
		12	6	
Dedrocoelopsis	D spinosipenis	10	5	Europe
	D piriformis	10	5	Canada–Alaska
	D lactea	16	8	Japan
	D ezensis	20		Japan
	D bessoni	20		Europe
	D americana	20		North America
	D beauchampi	30		Europe
	D chattoni	≈30		Europe

interpreted as originally tetraploid species have undergone a "diploidization," a process which is likely to be of considerable importance during evolution as it sets the stage for further cycles of evolution. However, it is to be pointed out that from the taxonomic viewpoint the relationships among these four species do not seem to be congruent with those suggested by the karyotypes. In fact, lacteum and infernale are attributed to Dendrocoelum s.s., while album belongs to the well differentiated subgenus Polycladodes, and tubuliferum to the subgenus Eudendrocoelum.

Complements of 32 chromosomes have also been found in D tuzetae Gourbault (Benazzi and Gourbault [53]) and in D coiffaiti de Beauchamp (Gourbault and Benazzi [54]). These two species are closely related and their karyotypes can readily be compared (Fig. 6).

An interesting problem of speciation has been advanced for the species attributed to Neodendrocoelum, a genus established by Komárek but now considered by some authors (cf Kenk [55]) a subgenus of Dendrocoelum. Most of these species are endemic of lake Ohrid in Macedonia, and Stanković [56,57] has put forth the hypothesis that all species of Neodendrocoelum have a common origin

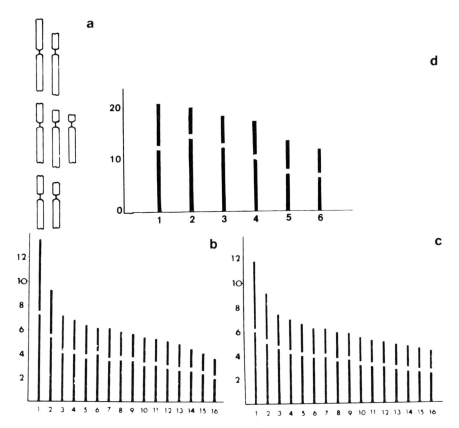

Fig. 6. a) Idiograms of Dendrocoelum lacteum, n = 7 (from Paunović [58]); b) of D tuzetae, n = 16 (from Benazzi and Gourbault [53]); c) of D coiffaiti, n = 16 (from Gourbault and Benazzi [54]); d) of Protocyla fluviatilis, n = 6 (from Gourbault [61]).

in that lake. The ancestral species would be Dendrocoelum nausicaae Schmidt, a planarian first found in the Jonian islands of Corfu and Cephalonia, and later proved to be widely distributed in Southeastern Europe and in Asia Minor. Stanković and Paunović [58] have shown that this species is diploid with 2n = 14, n = 7 and seven bivalents with 2–3 chiasmata are present in oocytes and spermatocytes. According to Paunović [59], the species of Neodendrocoelum of Lake Ohrid are derived from this diploid species. She has studied nine of the Neodendrocoelum species (five from sublitoral and four from litoral regions of the lake) and found that all have a complement of 32 chromosomes and show a marked resemblance in their karyotypes. However, variation in chiasma frequency and in the number of quadrivalents in metaphase I indicate that their

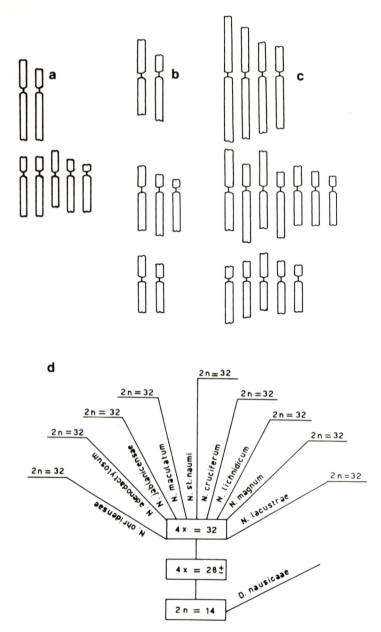

Fig. 7. a,b,c,) Idiograms, respectively, of Dendrocoelum lacteum n = 7, D nausicaae n = 7, and Neodendrocoelum ohridense, n = 16; d) scheme of evolution of the genus Neodendrocoelum (from Paunović [59]).

meiotic systems are different. Paunović suggests that the speciation process of this genus was manifested in the diploidization of autotetraploid species. She found that chromosome numbers are duplicated and in their morphology are almost identical with the chromosomes of D nausicaae. The hypothesis is advanced that during evolution this complement becomes an autotetraploid that was probably an aneuploid at the beginning, which implies that it had a number of 28 chromosomes (Fig. 7).

A last question regarding the species related to Dendrocoelum concerns the North American Procotyla fluviatilis which is the only common dendrocoelid in that region. For many years this white planarian was confused with and regarded as the European Dendrocoelum lacteum; its taxnomic individuality was established by Hyman. Pennipaker [60] in specimens from Illinois and New Jersey found $2n = 14, n = 7$, ie, the same complement of D lacteum, and admitted that there is also a similarity in the form of the chromosomes in the two species. However, Gourbault [61] in specimens of P fluviatilis from the St. Lawrence River (Ontario) has observed that the diploid complement is of 12 chromosomes (Fig. 6d) and six bivalents occur in the oocytes. New research would be opportune in order to explain the origin of this numerical variation and also to set the bases for an analysis of the relationship between the American and the European species.

Genus dendrocoelopsis. This genus, which is distributed in the Holarctic region, also deserves attention because the chromosome numbers found in the different species vary greatly and the causes of this fact are not clear.

Up until now, out of fifteen species described, eight have been studied karyologically and the results are as follows:

1. D spinosipenis Kenk (the type-species), European: $2n = 10$ (Dahm [62]).

2. D piriformis Kenk, North American: $2n = 10$ (Holmquist [63]).

3. D ezensis Ichikawa and Okugawa, Japanese: $2n = 20$ (Dahm [62]).

4. D lactea Ichikawa and Okugawa, Japanese: $2n = 16$ (Dahm [62], Teshirogi et al [64]).

5. D beauchampi Gourbault, European: $2n = 30$ (Benazzi and Gourbault [65]).

6. D chattoni (de Beauchamp), European: $2n = 30, 28$, or infrequently 32 (Gourbault and Benazzi [66]).

7. D bessoni Gourbault et al, European: $2n = 20$ (Gourbault et al [67]).

8. D americana (Hyman), North American: $2n = 20$ (Gourbault [68]).

In all species the chromosomes are metacentric or submetacentric (Fig. 8), only in populations of D lactea is the chromosome pair No. 5 according to Teshirogi et al [64] subtelocentric. Therefore, a Robertsonian mechanism may not be accounted to the variation in chromosome complements. A polyploid evolution could be suggested by the fact that D ezensis, bessoni, americana have

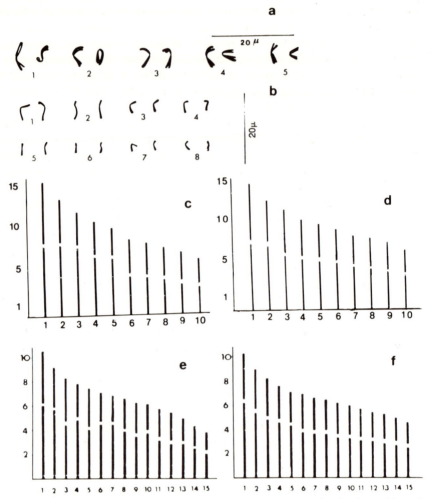

Fig. 8. a,b) Karyograms, respectively, of Dendrocoelopsis spinosipenis, 2n = 10, and D ezensis, 2n = 16 (from Dahm [62]); c) idiograms of D bessoni, n = 10 (from Gourbault et al. [67]); d) of D americana, n = 10 (from Gourbault [68]); e) of D beauchampi, n = 15; and f) of D chattoni, n = 15 (from Benazzi and Gourbault [65]).

a chromosome number which is twice that of spinosipenis and piriformis, while beauchampi and chattoni have a number which is three times as much. However, the karyotypes of the different taxa do not seem to support this hypothesis. Dahm [62], on the basis of the data referring to the three species studied by himself, admitted that "it is unlikely that they are closely related, thus strengthening the same view suggested from purely morphological examination." Therefore, Gourbault and Benazzi [66] have put forward the question of the taxonomic affinity among the species within the genus Dendrocoelopsis. They have also taken into consideration another aspect of the question, ie, the possibility of a relationship between chromosome numbers and adaptation to subterranean waters. In fact, the first four species with 2n = 10,10,20, and 16, respectively, are clearly epigean, while beauchampi and chattoni with 2n = 30 are true cave animals. Dendrocoelopsis americana with 2n = 20 must be considered a troglophile rather than an obligate troglobite, while bessoni also with 2n = 20 is a blind cave dweller. Although the data are not all concordant, it is perhaps possible to assume that speciation concomitant with adaptation to underground habitats frequently implied an increase in chromosome number. This assumption appears valid also for species of Dendrocoelum, if we remember that the epigean lacteum is diploid (14 chromosomes), while the subterranean species infernale, tubuliferum, tuzetae, coiffaiti all have 32 chromosomes. It should also be noticed that Plagnolia vandeli de Beauchamp and Gourbault, which is the most hypogean representative of the family Planariidae, possesses a complement of 44 chromosomes (Benazzi and Gourbault [69]). The hypogeous forms are narrowly distributed inside geographic areas making it reasonable to suppose that the high chromosome number, with eventual increased heterozygosity, may confer an adaptive superiority in colonizing new unfavorable niches.

SPECIATION PROCESSES IN SOME MICROTURBELLARIANS

At the beginning of this paper I stated that most data referring to speciation processes in Turbellaria concern freshwater planarians. This is certainly true; however, in some species belonging to other groups such as Dalyelloida or Typhloplanoida, populations differentiated by morphological and/or ecological characters, by variations in chromosome complements due to polypoidy or other genome mutations, and by different reproductive strategies, are present. Frequently this differentiation appears to be restricted within the intraspecific level, but the discrimination between intra- and supra-specific levels may be difficult and in some cases a more thorough analysis would likely show the occurrence of effective speciation events.

The genus Mesostoma (Typhloplanidae) offers clear examples in this respect. M ehrenbergi (Focke) has been reported from Europe, Asia, North and South America. The European form shows 2n = 10,n = 5 in all populations (Bresslau and many subsequent authors) and the same chromosome set occurs in the South American form (Marcus [70]). By contrast, the North American form has 2n = 8,n = 4 (Husted and Ruebush [71]). Besides this karyological variation the North American form differs from the European one in size and many characteristics of the copulatory system. Because of these differences, Husted and Ruebush divided the species into two subspecies: M e ehrenbergi (the European form) and M e wardi (the North American one). The reliability of this taxonomic conclusion, ie, the attribution of both forms to the same species, is perhaps disputable and new investigations would be helpful.

Many other Mesostoma species have been studied but the research carried out by Heitkamp and Schrade-Mock [72] on M lingua deserves particular attention. In eight populations from North, Middle, and South Europe and North Africa (all with $2n = 8, n = 4$) morphological, ecological, reproductive, and developmental differences could be demonstrated. Defined on morphological-ecological data and on reproductive isolation, M lingua s.l. appears to be a species-group with three sibling species represented, respectively, by the populations from: 1) France and Italy, 2) Göttingen (West Germany), 3) Schleswig-Holstein (West Germany), and Finland. The two latter populations, although lacking reproductive isolation, show great differences in ecology, and the authors suggest that they may be interpreted as species in statu nascendi.

Another turbellarian in which the occurrence of speciation processes has been admitted is Gyratrix hermaphroditus Ehrenberg (Kalyptorhinchia), an animal widely distributed throughout the world in fresh, brackish, and salt waters. Reuter [73] examined specimens from many localities in the environs of the Zoological Station Twärminne (Finland) and found three different chromosome sets: $2n = 4, n = 2$; $2n = 8, n = 4$; $2n = 6$. These three complements were regarded by Reuter as corresponding to a diploid, a tetraploid, and a triploid form, respectively. More recently, Heitkamp [74] has examined 16 populations from 15 small ponds in lower Saxonia. Two types of life cycles (polyvoltine cycles with facultative dormance periods, and univoltine cycles with a period of nine months or more survived in the egg capsules by a diapause) are recognized. Morphologically it is possible to distinguish small, medium-sized, and large specimens, with significant differences in the form and largeness of the egg capsules and the male cuticolar organs. Reproductive isolation exists among populations; therefore G hermaphroditus is considered to be a group of closely related species forming at least five sibling species. The biota of the sibling species are either isolated ponds or neighbouring ones, but in one case two species are living sympatric in the same pond. Considering the possibility that further sibling species from marine, brackish, and limnetic biota will be discovered, Heitkamp has given up on taxonomic terms and the species group Gyratrix hermaphroditus is provisionally carried on under this name. The reproductive isolation is primarily caused by a sexual substance inhibiting copulation, which is different in all populations. The origin of speciation processes is referred to ecogeographical differentiation, autopolyploidy, and chromosome mutations. The karyologycal data, however, are not fully convincing. Heitkamp assumes that one population is tetraploid with $2n = 8, n = 4$, but the description of the karyotype does not allow any conclusions. Precise data on the germ lines are also lacking. All other populations possess $2n = 4, n = 2$. In the first population the production of sperm is very low, while it is abundant in the others. In the general discussion Heitkamp observes that polyploidy, parthenogenesis, or pseudogamy may be significant in the speciation processes.

SOME REFERENCES TO CYTOGENETICAL VARIATION CORRELATED WITH PARTICULAR REPRODUCTIVE STRATEGIES

This final chapter does not concern speciation in its achievement, but is intended to offer some indications as to the possible ways followed by turbellarian species in their evolution (for general references cf [5]).

About 250 species have been karyologically studied, although in many cases only one of the two germ lines, generally the male, has been examined. Moreover, the haploid number has been frequently deduced from the diploid number rather than from observations of pairing at meiosis. In spite of these limitations, various cytogenetic processes have shown their value as a microevolutionary strategy in a particularly clear manner. The following events may be considered in connection with unusual gametogenetic mechanisms and reproductive modalities: Male and female asynapsis in diploid forms; polyplody with synaptic or asynaptic ovogenesis; parthenogenesis in diploid and polyploid forms; pseudogamy in polyploid forms; polysomy; fissioning in diploid, polyploid, or aneuploid strains.

Before dealing with these deviant patterns in the forms in which they are definitely assessed, it should be pointed out that variation in chromosome number, and especially asynapsis, may appear sporadically in diploid and synaptic species belonging to various turbellarian families. This fact suggests that the genetic factors responsible for these deviant patterns are widespread, but only in determined species or races do their phenotypic manifestations become normal. In Tables V and VI are listed the species showing asynapsis and polyploidy which, in some cases, are harmoniously correlated with regular gametogenetic processes.

In many other species differentiation has occurred at the level of races (or biotypes) as shown in Table VII. The Typhloplanidae Tetracelis marmorosa represents a case of polyploid evolution. Luther [77] in a Finnish population found $2n = 8, n = 4$ with complete absence of spermatozoa in the male organs as well as the absence of subitaneous eggs. Papi [78] confirmed this number for the Finnish population, but reported it in an Italian population $2n = 4, n = 2$. He believes that the two populations represent two distinct races: The Finnish tetraploid parthenogenetic with winter eggs only and the Italian diploid amphimictic with the two types of eggs.

In the Polycistidae Gyratrix hermaphroditus, as already mentioned, three karyological biotypes have been found with $2x = 4, 4x = 8, 3x = 6$, respectively (the third race probably reproduces partenogenetically, since sperm were never found in the seminal vesicles and in the copulatory bursa).

TABLE V. Asynapsis in Turbellarian Species

Family and species	Ploidy level	Maturation divisions			Reproduction modalities
		synapsis	partial asynapsis	total asynapsis	
Typhloplanidae					
Castrada		regular	in the ♂		
cristatispina	2x	meiosis[a]	line		amphimixis
Mesostoma					
ehrenbergi	2x	"	"		amphimixis
M wardi	2x	"	"		amphimixis
M benazzii	2x	"	"		amphimixis
M lingua	2x	"	"		amphimixis
Mesostoma sp.I	2x	?	"		?
Mesostoma sp.II	2x	?	"		?
Mesostoma sp.III	2x	?	"		?
Rynchomesostoma					
rostratum	2x	regular meiosis	"		amphimixis
Opistomum pallidum				ameiosis[b]	
	2x			in the ♂ line	amphimixis
Pseudostomidae					
Pseudostomum					
coecum	2x	"	"		amphimixis
Bothrioplanidae					
Bothrioplana				ameiosis[c]	
semperi	2x			in the ♀ line	parthenogenesis
Planariidae:					
Atrioplanaria				ameiosis[d]	
racovitzai	?			in the ♀ line	parthenogenesis
A delamarei	?		"		?
A morisii	?		"		sexual ? fission

[a]Essentially the male line is known.
[b]Peculiar ameiosis, with segregation of the homologous elements (Papi [75]).
[c]Peculiar equational division (Reisinger [76]).
[d]Ameiotic mechanism unknown.

However, the most impressive examples of intraspecific differentiation have been found in some freshwater planarians. Dugesia gonocephala s.l., D lugubris s.l., and the Polycelis of the "nigra-tenuis group" are the taxa having been the most thoroughly studied. D benazzii (of the gonocephala complex) and D polychroa (of the lugubris complex) present, besides the diploid amphimictic biotype, different polyploid biotypes which are correlated with pseudogamous de-

TABLE VI. Turbellarian Polyploid Species

Family and species	Ploidy level	Maturation divisions		Reproduction modalities
		regular meiosis	irregular meiosis	
Macrostomidae				
Macrostomum				
hustedi	4x	with biv. only	with biv. and multiv.	amphimixis
Typhloplanidae				
Mesocastrada				
fürmanni	6x	?	?	?
Planariidae				
Phagocata vitta	3x to 10x	with 4x[a]		amphimixis
			with other ploidy levels	sexual ? fission
P velata	?[b]			?
Atrioplanaria				
(some species)	?[c]			mainly asexual
Crenobia				
alpina	4x	with 4x[a]		amphimixis
	6x	with 6x[a]		"
	hyperploid	?	?	fission

[a]The data now disposable do not allow outlining the cycles with confidence.
[b]60 to 96 chromosomes.
[c]High chromosome numbers (see Table III).

velopment (gynogenesis). One biotype is triploid in the somatic line and hexaploid in the female line, owing to a chromosome set doubling which occurs in the oogonia; the oogenesis is synaptic. Other biotypes are triploid or tetraploid in both the somatic and female lines where oocytes are asynaptic and maturation takes place with a single equational division. In these polyploid biotypes the male line is, except in a few cases, diploid as a result of chromosome elimination. A peculiar biotype, triploid in both the somatic and female lines and amphimictic, has been found in D benazzii. In the Polycelis of the "nigra-tenuis group" asynaptic biotypes are not present, rather only synaptic biotypes with various ploidy levels occur; their male line is also polyploid and development is pseudogamic.

In D Benazzii and in D polychroa an experimental approach through crosses between diploid amphimictic biotypes and polyploid pseudogamic biotypes have been realized, while in the Polycelis species such crosses have not proven to be conclusive because frequently these species practice self insemination. It is to

334 / Benazzi

TABLE VII. Polyploidy, Asynapsis, and Polysomy Within Single Species of Turbellaria

Family and species	2x soma	polysomic. soma	polyploid soma	female germ lines synaptic	asynaptic	Reproduction modalities
Typhloplanidae						
Tetracelis						
marmorosa	4					amphimixis
			4x = 8			parthenogenesis
Polycistidae						
Giratrix						
hermaphroditus	4					amphimixis
			3x = 6			parthenogenesis
			4x = 8			?
Dugesiidae						
Dugesia						
gonocephala s.l.	16					amphimixis and fission[a]
		16 to 30[b]				amphimixis and fission
D benazzii	16					amphimixis and fission[a]
		16 to 30[b]				amphimixis and fission
			3x = 24	24[c]		amphimixis
			3x = 24[b]	24[c]		amphimixis and fission
			3x = 24	48		pseudogamy
			4x = 32		32	pseudogamy
D biblica	18					—
			3x = 27	27?		amphimixis ? and fission

be pointed out that these breeding experiments are possible because the egg of an amphimictic diploid individual may be fertilized by the spermatozoon of an individual belonging to a pseudogamous biotype.

The experiments allowed us to outline the possible modes by which the characteristics proper to the polyploid biotypes are realized. It has been established that these characteristics are genetically controlled by multifactorial com-

TABLE VII. (continued)

Family and species	Chromosome number					Reproduction modalities
			polyploid			
				female germ lines		
	2x soma	polysomic. soma	soma	synaptic	synaptic	
D dorotocephala	16					← amphimixis and fission
			3x = 24			fission
D tigrina	16					amphimixis and fission
			3x = 24			fission
Dugesia lugubris s.l.						
D polychroa	8					amphimixis
			3x = 12	24		← pseudogamy
			3x = 12		12	pseudogamy
			4x = 16		16	pseudogamy
D mediterranea	8					amphimixis
D lugubris s.s.	8					amphimixis
D nova n.nud.	4					amphimixis
Planariidae Polycelis nigra	16					amphimixis
			3x = 24	48		pseudogamy
P tenuis	12					
	14					amphimixis
			≈20ᵈ	≈40		pseudogamy
P felina	18					amphimixis
		20 to 26				fission
			3x = 27			fission

ᵃRarely.
ᵇIn some cases with B chromosomes.
ᶜBivalents and univalents.
ᵈHyperploid.

plexes which are inherited independently of one another, whereby they may occur in various combinations in the offspring. For instance, the following anomalous situations may be found: diploid and polyploid asynaptic oocytes in which the equational division is irregular or sometimes completely hindered; or synaptic oocytes with a doubled chromosome set which develop by amphimixis rather than pseudogamically. In fact, most of the diploid oocytes of the hybrids are amphimictic and even in the polyploids of the first generation pseudogamy is very rare. On the contrary, in the successive generations the multifactorial complexes may join in such a manner so as to give balanced chromosome cycles, thus the experimental reconstitution of all natural biotypes of the two species was obtained (Benazzi Lentati [3,4], Benazzi Lentati and Puccinelli [80]).

The results now summarized suggest that the genetic factors controlling the various manifestations of the chromosome cycles are present even in the diploid amphimictic biotypes. This assumption is corroborated also by the sporadic appearance of asynapsis and polyploidy in diploid forms of other turbellarian groups, as already stated. The frequency of these factors in the diploid biotypes is indeed scant, therefore, only in very few cases may their phenotypic effect become manifest.

For instance, investigations on the descendents from consecutive incrosses of two diploid populations revealed the occasional production of tetraploid oocytes.

More demonstrative results have been obtained by Benazzi Lentati [81] on a strain derived from a single diploid individual of D benazzi reared in the laboratory for about 30 years. A few specimens with both diploid and tetraploid synaptic oocytes as well as specimens with diploid asynaptic oocytes appeared in the last years. A preliminary study of their offspring indicates that the tetraploid synaptic oocytes after reductional division and amphimixis produce triploid offspring, while the chromosome set doubling may occur in diploid asynaptic oocytes giving tetraploid asynaptic oocytes. Therefore, already at these stages the characteristics of the three polyploid biotypes of the species have appeared.

Specimens with pseudogamic oocytes have not yet been obtained, but this fact is in agreement with the results offered by crosses among different biotypes, showing that pseudogamy generally appears in the successive generations and is correlated with particular ploidy levels.

On the basis of these results, Benazzi Lentati suggests that polyploidy represents a preliminary condition for the rise of pseudogamy, an assumption which does not agree with the hypothesis advanced about the evolution of parthenogenesis, according to which polyploidy is successive to parthenogenesis.

The deviant oocytes with either synapsis, or synapsis and chromosome set doubling, found in individuals of the diploid synaptic biotype represent the first step of the evolution towards asynapsis and polyploidy. Successively, strains with constant degrees of ploidy may originate. The laboratory crosses have shown, moreover, that the spreading of the polyploid individuals may be favored by their capacity to bachcross with individuals of the diploid original biotype.

Chromosome variation pertaining to polysomy have also been evidenced within species of D gonocephala s. l. In some races, originally diploid and amphimictic, an increase of the chromosome number, up to a high value, may take place, with adjustment of the gametogenesis to the new karyological conditions. A typical example has been offered by D etrusca Benazzi in which the chromosomes in addition are transmitted through the egg only, being expelled during spermatogenesis. The development is amphimictic and the chromosome number of the zygote corresponds to the one of the mother (Benazzi Lentati [79]).

It is to be emphasized that the polysomic specimens gradually lose their capacity of sexual reproduction, giving rise first to strains in which both sexual and asexual individuals occur, and finally to strains almost exclusively fissiparous. These facts, observed in laboratory cultures, have enabled the interpretation of the origin of the fissiparous races showing variable chromosome numbers which frequently occur in nature. In races with a high chromosome number, B chromosomes may also appear, and this fact has been observed both in nature and in the laboratory (Deri [82,83], Benazzi Lentati and Deri [84,85]).

With regards to the origin of asexual reproduction, it is to be pointed out that it is favored but not directly determined by heteroploidy, since diploid fissiparous races are known in many planarian species (Benazzi [86]).

GENERAL REMARKS

The aim of my contribution was to give a collection of data reflecting recent achievements on speciation in Turbellaria. I hope to have attained this goal even if, in many cases, in a rather descriptive fashion. In fact, cladogenesis is a much more complex phenomenon than was recognized in the past, and general rules of its happening are not established. Also the central question of the respective importance of karyological and genetical variation still remain debatable, even after the great advancement realized through the methods of molecular biology. The models evidenced in planarians show that also in these organisms speciation may take place through different mechanisms, and in different ways. For instance, within the "Dugesia gonocephala group" and the Dugesia of the subgenus Girardia it appears likely that in most cases only minor chromosome mutations, such as small pericentric inversions, have accompanied speciation. On the contrary, in other genera such as Phagocata, specific differentiation seems to be linked with marked numerical chromosome variation. I therefore agree with White's statement ([1]: 324) that "at the present time, it would seem that sweeping generalizations about methods of speciation should be avoided as far as possible."

I would like to conclude my talk with an excursus in the field of the phylogenetic classification, referring to the speculations outlined by Ball [87] about the evolutionary history in time and space of some planarian taxa. The author takes into consideration morphological, karyologycal, and biogeographical characters of the following species: 1) Cura foremanii (Girard) from North America and Cura pinguis (Weiss) from Australia. The first is diploid with $2n = 12, n = 6$, the second probably hexaploid having 36 chromosomes which are easy to assemble in six groups of six elements each. Cura is considered an old and

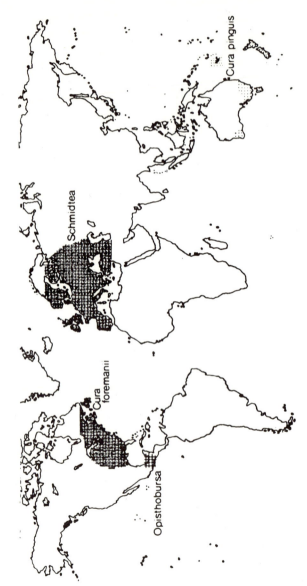

Fig. 9. Distribution of taxa carrying particular character states relating to subtelocentric chromosomes (squares) and usually large oocytes (dots) (from Ball [87]).

primitive genus of the family Dugesiidae. 2) Opisthobursa mexicana Benazzi, a retrobursal planarian living in freshwater but certainly of marine origin. 3) The "Dugesia lugubris complex," namely the species of the subgenus Schmidtea previously analyzed in this paper.

Ball points out that Gourbault and Benazzi [88] have found in the two species of Cura a basic chromosome number of 6, as in Opisthobursa mexicana (according to Benazzi and Giannini [89] and some marine planarians. More importantly, however, Gourbault and Benazzi found that in Cura foremanii subtelocentric chromosomes are present as in species of Schmidtea and that both Cura species possess unusually large oocytes that are otherwise found only in Schmidtea. The geographical distribution of the taxa carrying these two characters is shown in Figure 9.

In an early paper Ball [7] had regarded Schmidtea as being a derived group sharing a recent ancestor with the "Dugesia gonocephala group." But he admits that the distribution of the two species of Cura (in North America foremanii, in Australasia pinguis) and of the species of Schmidtea in Europe, is rather puzzling. The difficulties are increased by the awareness that all the species considered possess preovarial vitellaria, a primitive feature, and all but Opisthobursa show an unusual feature of the copulatory system, namely an expanded chamber of the bursal canal receiving the shell glands. Moreover C pinguis shares marked derived features of the anterior sensory organs with all the other endemic Australian species and also with Eviella hynesae Ball (a primitive retrobursal planarian from Australian freshwater), but neither with C foremanii, nor with the Opisthobursa and Schmidtea species.

According to Ball, an explanation for all these apparently conflicting data would involve the vicariance paradigm on a global scale. If it is assumed that the progenitors of the Paludicola, with many of the features of the extant species of Cura and Opisthobursa, were widespread (or cosmopolitan) in Pangea, then the present systematic and biogeographical relationship of these forms is best understood by involving a series of vicariant events related to plate tectonics. According to this hypothesis the Australasian isolate diverged rapidly to give the wealth of species found there today. The North American isolate contracted so as to be represented by a single species, Cura foremanii, that although primitive does show some derived features. The Holarctic isolate persisted, with much change, as the Schmidtea group. Ball points out that it is as parsimonious to derive the latter from a Cura-like ancestor as from the "D gonocephala group," and remembers that also de Beauchamp was of the opinion that the affinities of C foremanii laid with the "D lugubris group," although his views have been largely ignored.

These conjectures are certainly highly speculative, as recognized by Ball himself and are perhaps not directly correlated with speciation problems. However, I do not consider this type of approach devoid of interest, because, as emphasized by Ball, it has the further advantage of directing attention to a problem area where further comparative anatomical, karyological, and perhaps electrophoretic studies might prove to be very profitable. I would also like to point out that the vicariance models imply the assumption of allopatric speciation, which certainly represents the most diffused mechanism of cladogenesis.

ACKNOWLEDGMENTS

I am grateful to my wife, Dr. Giuseppina Benazzi Lentati for her redaction of the section on cytogenetical variation and reproductive strategies.

REFERENCES

1. White MID: Modes of speciation. San Francisco: W.H. Freeman, 1978, p 11.
2. Benazzi M: Evoluzione cromosomica e differenziamento razziale e specifico nei Tricladi. Acc Naz Lincei Quaderno 47: 273, 1960.
3. Benazzi LG: Amphimixis and pseudogamy in freshwater triclads: Experimental reconstitution of polyploid pseudogamic biotypes. Chromosoma 20: 1, 1966.
4. Benazzi LG: Gametogenesis and egg fertilization in planarians. Inter Rev Cytol 27: 101, 1970.
5. Benazzi M, Benazzi LG: Platyhelminthes. In John B (ed): Animal Cytogenetics I. Berlin-Stuttgart: Gebrüder Borntraeger, 1976.
6. Peter R: Disk-Electrophoretishe Untersuchungen zur Frage der Artcharakterisierung paludicoler Tricladen (Platyhelminthes: Turbellaria). Z zool Systematik u Evolutionforshung 9: 263, 1971.
7. Ball IR: A contribution to the phylogeny and biogeography of the freshwater triclads (Platyhelminthes: Turbellaria). In Riser NW, Morse MP (eds): Biology of the Turbellaria. New York: McGraw-Hill, 1974, p 339.
8. Benazzi M: Cariologia della planaria americana Dugesia dorotocephala. Rend Acc Naz Lincei Ser VIII, 40:99, 1966.
9. Gourbault N, Benazzi M: Une nouvelle espèce ibérique du "groupe Dugesia gonocephala" (Turbellariés, Triclades). Bull Mus nat Hist nat Paris, 4 ser 1: 329, 1979.
10. Bromley HJ: Morpho-Karyological types of Dugesia (Turbellaria, Tricladida) in Israel and their distribution patterns. Zoologica Scripta 3: 239, 1974.
11. Kawakatsu M, Oki I, Tamura S, Sugino M: Studies on the morphology, karyology and taxonomy of the Japanese freshwater planarian Dugesia japonica Ichikawa and Kawakatsu, with a description of a new subspecies, Dugesia japonica ryukyuensis. Bull Fuji Women's Colleg, ser II 14: 81, 1976.
12. Dahm AG: Taxonomy and ecology of five species groups in the family Planariidae (Turbellaria Tricladida Paludicola). Nya Litografen Malmö, 1958.
13. Benazzi M, Giannini-Forli E, Puccinelli I: Cariologia della planaria americana Dugesia tigrina. Boll Zool 38: 439, 1971.
14. Dutrillaux B, Lenicque P: Analyse du caryotype de cinque espèces de planaires par la méthode du choc hypotonique. Acta Zool 52: 241, 1971.

15. Durán Troise G, de Lustig ES: Cariologia della planaria Dugesia anceps. Caryologia 23: 455, 1970.

16. Benazzi M: Karyological and genetic data on the planarian Dugesia sanchezi from Chile. Rend Ac Naz Lincei Ser VIII, 64: 299, 1978.

17. Gourbault N: Données biologiques et cytotaxonomiques sur un triclade de l'ile de Cuba. Arch Zool Expér et Génér 120: 131, 1979.

18. Benazzi M: A new karyotype found in the American fresh-water planarian Dugesia dorotocephala. System Zool 23: 490, 1975.

19. Gourbault N: Karyology of Dugesia arizonensis Kenk (Turbellaria, Tricladida). Caryologia 30: 63, 1977.

20. Benazzi M, Puccinelli I, Del Papa R: The planarians of the Dugesia lugubris-polychroa group: taxonomic inferences based on cytogenetic and morphologic data. Rend Acc Naz Lincei Ser VIII, 48: 369, 1970.

21. Reynoldson TB, Bellamy LS: The status of Dugesia lugubris and D. polychroa (Turbellaria, Tricladida) in Britain. J Zool London 162: 157, 1970.

22. Benazzi M: Cariologia di Dugesia lugubris (O. Schmidt). Caryologia 10: 276, 1957.

23. Benazzi M, Puccinelli I: A Robertsonian translocation in the freshwater triclad Dugesia lugubris: karyometric analysis and evolutionary inferences. Chromosoma 40: 193, 1973.

24. Benazzi M, Formenti P, Manfredi Romanini MS, Pellicciari C, Redi CA: Feulgen-DNA content and C-banding of Robertsonian transformed karyotypes in Dugesia lugubris. Caryologia 34:129, 1981.

25. Benazzi M, Baguñà J, Ballester R, Puccinelli I, Del Papa R: Further contribution to the taxonomy of the "Dugesia lugubris–polychroa group" with description of Dugesia mediterranea n. sp. (Tricladida Paludicola). Boll Zool 42: 81, 1975.

26. Ball IR: The karyotypes of two Dugesia species from Corfu, Greece (Platyhelminthes, Turbellaria). Bijdragen tot de Dierkunde 48: 187, 1979.

27. Ball IR, Gourbault N: The phyletic status of the genus Planaria (Platyhelminthes, Turbellaria, Tricladida). Bijdragen tot de Dierkunde 48: 29, 1978.

28. Benazzi M, Puccinelli I: Cariologia di Planaria torva (Müller). Caryologia 16: 653, 1963.

29. Melander Y: Cytogenetic aspects of embryogenesis in Paludicola Tricladida. Hereditas 49: 119, 1963.

30. Hyman LH: The two species confused under the name Phagocata gracilis, the validity of the generic name Phagocata Leidy 1847, and its priority over Fonticola Komárek 1926. Trans Am microsc Soc 56: 298, 1937.

31. Benazzi M, Gourbault N: Etude caryologique de quelques populations hypogées de la planaire Phagocata (Fonticola) vitta (Dugès, 1830). Caryologia 27: 467, 1974.

32. Dahm AG: The taxonomic relationships of the European species of Phagocata (? = Fonticola) based on karyological evidence. Arkiv Zool 16: 481, 1964.

33. Whitehead MM: The triclads of Cattarangus County, New York. PhD Thesis, St Bonaventure Univ, NY, 1965.

34. Ball IR, Gourbault N: The morphology, karyology and taxonomy of a new freshwater planarian of the genus Phagocata from California. Life Sci Contr R Ont Mus Number 105, 1975.

35. Teshirogi W, Sasaki S: Karyotype of a freshwater planarian, Phagocata teshirogii, and chromosomal variations in neoblasts of regenerating pieces. Jap J Genet 52: 387, 1977.
36. Sugino H, Murayama H, Horikoshi I: Karyological studies of two Phagocata species in Japan. Zool Mag Tokyo 87: 535, 1978.
37. Teshirogi W, Ishida S, Nimura F: Karyotype of a freshwater planarian, Phagocata kawakatsui and chromosomal variations in neoblasts of regenerating pieces. Annot Zool Japanenses 52: 191, 1979.
38. Teshirogi W, Hasebe K, Ishida S: Karyotype of a Japanese freshwater planarian, Phagocata vivida. Jap J Genet 55: 1, 1980.
39. Gourbault N, Benazzi M: Etude caryologique du genre Atrioplanaria (Triclade Paludicole). Arch Zool Exp Génér 118: 53, 1977.
40. Lepori NG: Ricerche sulla ovogenesi e sulla fecondazione della planaria Polycelis nigra Ehr. con particolare riguardo all'ufficio del nucleo spermatico. Caryologia 1: 280, 1949.
41. Lepori NG: Nuova mutazione genomica in Polycelis nigra Ehr. Caryologia 6: 90, 1954.
42. Melander Y: Accessory chromosomes in animals, especially in Polycelis tenuis. Hereditas 36: 19, 1950.
43. Lepori NG: Prime ricerche cariologiche su alcune popolazioni europee di Polycelis tenuis Jjima. Caryologia 6: 103, 1954.
44. Hansen-Melander E, Melander Y, Reynoldson TB: A new species of freshwater triclad belonging to the genus Polycelis. Nature 173: 354, 1954.
45. Benazzi M: Il problema sistematico delle Polycelis del gruppo nigra-tenuis alla luce di ricerche citologiche e genetiche. Monit Zool Ital 70–71: 288, 1963.
46. Biersma R, Wijsman HJW: Studies on the speciation of the European freshwater planarians Polycelis nigra and Polycelis tenuis based on the analysis of enzyme variation by means of iso-electric focusing. In Shockaert E, Ball IR (eds): Developments in Hydrobiology 6, Turbellaria. The Hague: Junk, 1981, p 79.
47. Teshirogi W, Ishida S: Studies on the speciation of Japanese freshwater planarian Polycelis auricolata based on the analysis of its karyotypes and constitutive proteins. In Schockaert E, Ball IR (eds): Developments in Hydrobiology 6, Turbellaria. The Hague: Junk, 1981, p 69.
48. Nixon SE, Taylor RJ: Large genetic distances associated with little morphological variation in Polycelis coronata and Dugesia tigrina (Planaria) Syst Zool 26: 107, 1977.
49. Aeppli E: Die Chromosomen verhaltnisse bei Dendrocoelum infernale (Steinmann). Ein Beitrag zur Polyploidie im Tierreich. Rev suisse Zool 58: 511, 1951.
50. Aeppli E: Natürliche Polyploidie bei den Planarien Dendrocoelum lacteum (Müller) und Dendracoelum infernale (Steinmann). Z ind Abstammungs—Vererbungslehre 84: 182, 1952.
51. Gelei J: Uber die Ovogenese von Dendrocoelum lateum. Arch Zellf 11: 51, 1913.
52. Dahm AG: Cytotaxonomical analyses of four Dendrocoelum species. Lund Univer Arsskrift N F Avd 2, 57: 1, 1961.
53. Benazzi M, Gourbault N: Recherches caryologiques sur quelques Dendrocoelidae hypogés. C R Ac Sc Paris 278: 1051, 1974.

54. Gourbault N: Etude caryologique des triclades paludicoles hypogés: Dendrocoelum coiffaiti de Beauchamp. Ann Spéléol 30: 427, 1975.

55. Kenk R: The planarians (Turbellaria: Tricladida Paludicola) of lake Ohrid in Macedonia. Smithsonian Contrib Zool, N.280, 1978.

56. Stanković S: The Balkan lake Ohrid and Its Living World. Den Haag: Dr W. Junk, 1960, p 357.

57. Stanković S: Turbellariés triclades endémiques nouveaux du lac d'Ohrid. Archiv Hydrobiologie 65: 413, 1969.

58. Paunović D: Karyological analysis of Dendrocoelum nausicaae (Tricladida Paludicola). Arch Sc Biol Beograd 27: 3, 1977.

59. Paunović D: A cytogenetic analysis of the genus Neodendrocoelum from lake Ohrid. Chromosoma 63: 161, 1977.

60. Pennypaker MI: The meiotic chromosomes of the triclad turbellarian Procotyla fluviatilis with attention to the "lampbrush" phase. J Morphol 84: 365, 1949.

61. Gourbault N: Une donnée nouvelle sur la garniture chromosomique de la planaire américaine Procotyla fluviatilis Leidy. C R Ac Sc Paris 279: 1171, 1974.

62. Dahm AG: The karyotypes of some freshwater triclads from Europe and Japan. Arkiv Zool 16: 41, 1963.

63. Holmquist C: Dendrocoelopsis piriformis (Turbellaria, Tricladida) and its parasites from Northern Alaska. Arch Hydrobiol 52: 453, 1967.

64. Teshirogi W, Ishida S, Hasebe K: Karyological studies of a Japanese freshwater Planarian, Dendrocoelopsis lactea. Zool Magazine 89: 41, 1980.

65. Benazzi M, Gourbault N: Recherches caryologiques sur quelques Dendrocoelidae hypogés. C R Ac Sc Paris 278: 1051, 1974.

66. Gourbault N, Benazzi M: Etude caryologique du triclade hypogé Dendrocoelopsis chattoni (de Beauchamp). Ann Spéléol 29: 621, 1974.

67. Gourbault N, Benazzi M, Hellouet MN: Triclades obscuricoles des Pyrénées. V. Etude morphologique et cytotaxonomique de Dendrocoelopsis bessoni n. sp. Bull Mus Nat Hist Nat N.406, Zool 283: 1095, 1976.

68. Gourbault N: Karyology of the troglophile planarian Dendrocoelopsis americana (Hyman). Ann Spéléol 30: 125, 1975.

69. Benazzi M, Gourbault N: Données préliminaires sur la caryologie de la planaire hypogée Plagnolia vandeli de Beauchamp et Gourbault. C R Ac Sc Paris 277: 1337, 1973.

70. Marcus E: Sobre turbellaria brasileiros. Bol Fac Ci Letr Univ São Paulo Zool 12: 99, 1947.

71. Husted L, Ruebush TK: A comparative cytological and morphological study of Mesostoma herenbergii herenbergii and Mesostoma herenbergii wardii. J Morphol 67: 387, 1940.

72. Heitkamp U, Schrade-Mock W: Speziationprozesse bei Mesostoma lingua. In Karling TG, Meinander M (eds): The Alex Luther Centennial Symposium on Turbellaria. Acta Zool Fenn 154, 1977 p 47.

73. Reuter M: Untersuchungen über Rassenbildung bei Gyratrix hermaphoditus (Turbellaria Neorhabdocoela). Acta Zool Fenn 100: 1, 1961.

74. Heitkamp U: Speziationprozesse bei Gyratrix hermaphroditus Ehrenberg, 1831 (Turbellaria, Kalyptorhynchia). Zoomorphologie 90: 227, 1978.

75. Papi F: Ricerche cariologiche su Rabdoceli. II. La meiosi nella linea germinale maschile di Opisthomum pallidum O. Schmidt. Caryologia 5: 123, 1952.

76. Reinsinger E: Die cytologische Grundlage der parthenogenetischen Dioogonie. Chromosoma 1: 531, 1940.

77. Luther A: Untersuchungen an rhabdocoelen Turbellarien. IX. Zur Kenntniss einiger Typhloplanidaen. Acta Zool Fenn 60: I, 1950.

78. Papi F: Aspetti del differenziamento razziale e specifico nei turbellari rhabdocoeli. Boll Zool 21: 357, 1954.

79. Benazzi LG: Sul determinismo e sulla ereditarietà della aneuploidia in Dugesia etrusca Benazzi, planaria a riproduzione afigonica. Caryologia 10:352, 1957.

80. Benazzi LG, Puccinelli I: Ulteriori ricerche sugli ibridi fra biotipo diploide e biotipo tetraploide di Dugesia benazzii. Produzione di individui triplo-esaploidi. Caryologia 12: 110, 1959.

81. Benazzi LG: On the appearance of female asynapsis and polyploidy in a population of the diploid synaptic biotype of the planarian Dugesia benazzii. Atti Soc Toscana Sc Nat Ser B 88:83, 1981.

82. Deri P: B-cromosomi in popolazioni polisomiche di Dugesia Benazzii (Tricladida Paludicola) della Corsica. Atti Soc Toscana Sc Nat Ser B 82: 25, 1976.

83. Deri P: Incremento del numero cromosomico e comparsa di B-cromosomi durante l'allevamento in laboratorio di una popolazione diploide di Dugesia benazzii (triclade paludicolo). Rend Acc Naz Lincei Ser VIII, 68: 327, 1980.

84. Benazzi LG, Deri P: Insorgenza di B-cromosomi in individui poliploidi di Dugesia benazzii (triclade paludicolo) allevati in laboratorio. Rend Ac Naz Lincei Ser VIII, 62: 847, 1977.

85. Benazzi LG, Deri P: On the origin of heterogeneous chromosome sets in some fissiparous planarians. Rend Ac Naz Lincei Ser VIII, 68: 318, 1980.

86. Benazzi M: Fissioning in planarians from a genetic standpoint. In Riser NW, Morse MP (eds): Biology of the Turbellaria. L. H. Hyman Memorial Volume. New York: McGraw-Hill, 1974, p 476.

87. Ball IR: On the phylogenetic classification of aquatic planarians. In Karling TG, Meinander M (eds): The Alex Luther Centennial Symposium on Turbellaria. Acta Zool Fenn 154, 1977, p 21.

88. Gourbault N, Benazzi M: Karyological data on some species of the genus Cura (Tricladida, Paludicola). Canad J Genet Cytol 17: 345, 1975.

89. Benazzi M, Giannini E: A remarkable cave planarian: Opisthobursa mexicana Benazzi, 1972. Acc Naz Lincei, Quaderno 171: 47, 1973.

90. Le Moigne A: Etude de formules chromosomiques de quelques Polycelis (Turbellariés Triclades) de la région parisienne. Bull Soc Zool France 87: 259, 1962.

Mechanisms of Speciation, pages 345-376
© 1982 Alan R. Liss, Inc., 150 Fifth Avenue, New York, NY 10011

Genetic Differentiation and Speciation in the Brine Shrimp Artemia

F. A. Abreu-Grobois and J. A. Beardmore

INTRODUCTION

Artemia an anostracan branchiopod, was first described in 1755. It is widely distributed throughout the world, partly by natural colonisation and partly by deliberate or accidental spread by man. Characteristically Artemia is found in areas of solar salt production where the high levels of salinity provide an insurmountable obstacle to predators. It is also interesting to note that while Artemia is associated with warm or tropical areas of the world this is probably only because most salt production is in such areas. In fact the type specimen was taken from the South Coast of England at Lymington where a small salt industry based on cheap coal was in operation until 1865 [1].

The brine shrimp has a striking ability to tolerate extremely high salinity levels, the physiological basis for which depends upon a highly efficient hypoosmotic regulation in both adults and nauplii [2,3]. While reproduction is probably not possible in a saturated solution, the adults appear to survive reasonably well if food is available. The animal is able to exist and reproduce at normal sea water salinity (28–32‰) but, because in most circumstances predators will also be present, Artemia is generally found in nature only in waters of high salinity (>70‰). These include both sodium chloride waters, much like concentrated seawater (also called thalossohaline), and other high salinity lakes and ponds which are quite different from seawater in relative ionic content (often with chloride ions replaced by sulphate and/or carbonate ions, and termed athalassohaline). An additional adaptation is seen in haemoglobin synthesis which is increased with increasing salinities to counteract the decrease in oxygen tension [4,5]. The animal can also adapt to temperature changes extremely rapidly [6] and dry cysts are able to tolerate a range of temperatures between $-273°$ and $100°C$ [7].

Under unfavourable conditions, the brine shrimp is able to form a resistant stage when the embryo reaches the gastrula by the development of a thick and highly resistant shell which leads to the formation of a cyst (so-called oviparous reproduction) although the embryos under other conditions will develop within the mother's ovisac to larvae which are then released (ovoviviparous reproduction), [8]. Artemia is thus an example of a life form restricted in its distribution to a limited range of ecological conditions but within this range being extremely successful, achieving colossal population sizes and tolerating large environmental variations as well. It is thus of special interest to those biologists concerned with adaptation and adaptability and the origin of differentiated populations.

Artemia is, however, of interest for other reasons. It is widely used for physiological and toxicological studies. It has considerable value and even more potential as food both in larval and adult stages for fish and shellfish, and it is interesting to note that it is said to be particularly efficient in the conversion of algae into animal protein. Its general biology has been described in two recent papers [7,8].

It is peculiarly appropriate that the brine shrimp should be one of the organisms to be discussed in this symposium in Rome for one of the first geneticists to take a serious interest in this organism was Professor Barigozzi who was working on it almost half a century ago [9].

Barigozzi [10] produced a comprehensive review of the genetically interesting aspects of Artemia biology. In this he pointed out the salient features of parthenogenesis and polyploidy together with the fact that a short life cycle, tractability under laboratory conditions, and favourable qualities for cytological study predispose it to productive use as a sort of marine Drosophila.

The Species Problem in Artemia

As Barigozzi [10] notes, although the existence of bisexual and parthenogenic forms was recognised many years ago by Artom [11,12], there has been a tendency to regard Artemia as a single species. A number of studies [13–16] have established the existence of at least three bisexual forms which are reproductively isolated from each other. Table I shows the results of crosses carried out by Clark and Bowen [14]. The European and North African populations are clearly cross fertile and these we would refer to as Artemia salina, the Argentinian population is of Artemia persimilis, and the SFB population is A franciscana. The Urmia populaltion (which unfortunately we have so far not been able to assay) is described as A urmiana by Bowen et al [17]. A population in Mono Lake is also reported [14] not to be crossable with other North American populations because of the inability of the two types to tolerate the same ionic composition of the water. (Mono Lake has a mixture of Cl^-, SO_4^- and Co_3^- ions). This population is therefore thought to be effectively debarred from exchanging genes with A franciscana and has been called by Bowen A monica.

TABLE I. Results of Single-Pair Crosses in Artemia*

	♀				
	SB	TUNIS	HIDALGO	SFB[a]	URMIA
♂					
SB (Sardinia)	15/24	9/13	0/4	n.t.	n.t.
TUNIS	8/10	21/33	0/8	0/24	0/18
HIDALGO (Argentina)	0/14	0/25	19/28	0/29	n.t.
SFB[a] (USA)	n.t.	0/21	0/19	b	0/26
URMIA (Iran)	n.t.	0/25	n.t.	0/31	6/15

Figures are fertile/total matings.
*From Clark and Bowen [14].
[a]San Francisco Bay or an equivalent population.
[b]Many successful crosses in other experiments.
nt, Not tested

The anatomical variation displayed by Artemia is quite small and it seems proper to regard these species as sibling species. Table II summarises what is known of the five bisexual species with respect to anatomical features, chromosome number, chromocentres, and distribution.

The parthenogenetic forms of Artemia, while known for a considerable time, present a confused picture which this study, we believe, helps to resolve. A summary of views of various workers including his own is given by Barigozzi [10] who raises the question as to which of the bisexual forms gave rise to a parthenogenetic form and by implication whether the derivation of parthenogenetic forms was monophyletic or not.

Despite this evidence the many published studies of biochemical, mutational, toxicological, and aquaculture work involving Artemia are notable for the general lack of concern with the origin and specific status of the form of Artemia employed, and indeed the name Artemia salina is very freely employed as a label for any stock. In addition, the existence of a range of ploidies with several different types of egg maturation, within the parthogenetic types, further complicates the pictures and leads to problems both of nomenclature and of inferences concerning origins.

The aim of the work reported here was, by the use of electrophoretic assay of a large number of samples of Artemia from known localities, to calculate genetic distances by which it would be possible to determine relationships within and between bisexual species and parthogenetic species, and by combining these data with cytological and available ecological information to infer something of the pattern of evolutionary descent in this interesting genus.

TABLE II. Bisexual Species of Artemia

Species	Distribution	$2n^a$	Chromocentres[b]	Anatomical characters[c]	
1 A salina (or tunisiana) [17]	Europe and N. Africa	42	—	Furca two lobed many setae	Clasper knob sub conical
2 A franciscana (or gracilis)	N. America and Caribbean	42	+ + +, + +	Furca two lobed many setae	Clasper knob subspherical
3 A persimilis	Argentina	44	+	Furca rudimentary few setae	Clasper knob subspherical
4 A monica	Mono Lake, California	42	+ + +	?	?
5 A urmiana	Lake Urmia, Iran	?	?	?	?

[a]1–3, Barigozzi [10]; 1–4, this study.
[b]1–3 Barigozzi (personal communication); 1–4, this study.
[c]1–3, Barigozzi [10].
column 4: —, absent; +, 1–2 per nucleus; + + (Caribbean + Mexico), 5–8; + + + (N. America), 17–22.

MATERIALS AND METHODS

In total, samples from some 48 different geographic areas have been assayed (Table III). These were obtained in the form of cysts largely from the International Artemia Reference Centre in Ghent.

Populations are kept in modified raceway systems and fed on a suspension of micronised rice bran [18]. Some strains which are difficult to keep under these conditions are kept in smaller cultures and fed on live Dunaliella tertiolecta. Cultures were kept at 26–28°C and 32–35‰ salinity. Single pair cultures set up to determine the inheritance of specific allozyme loci [19] are kept at the same temperature but at 65‰ salinity and fed exclusively on live D teriolecta.

Electrophoresis

Full details of the electrophoretic techniques employed are given in Abreu-Grobois and Beardmore [20]. Sample size was normally 40–48 individuals.

In total, 16 systems representing 23 loci were scored (Table IV).

Nei's [21,22] coefficient of gene differentiation (G_{ST}) was used to study the degree of inter-populational substructuring at single loci and overall. It is measured by: $G_{ST} = D_{ST}/H_T$. Where D_{ST} is the average gene diversity between subpopulations (including the comparisons of the individual subpopulations with themselves,) and H_T is the average gene diversity in the total population [see 22].

G_{ST} varies from 0 to 1 and is similar to Wright's well-known F_{ST} statistic [22]. Sampling standard errors for the overall G_{ST} values were approximated according to the formula of Chakraborty [23].

For within-locus studies of differentiation, the test of agreement in distribution of alleles in two populations was carried out using the χ^2 method described by Nei and Roychoudhury [24]. χ^2 is computed as:

$$\chi^2 = n_x n_y \sum_i \frac{(x_i - y_i)^2}{x_i n_x + y_i n_y}$$

where ns are the number of genes sampled from populations x and y, and x_i and y_i are the allele frequencies at the ith allele in populations x and y, respectively. The number of degrees of freedom is equal to the number of alleles minus one.

Estimates of Nei's D statistic and the corresponding standard errors were calculated according to Nei [25] and Nei and Roychoudhury [24]. Allele frequencies in polyploid Artemia where the gene dosage was clearly asymmetrical were counted accordingly [26]. Cluster diagrams were constructed using a computer program by D. Wishart [27] based on the unweighted pair-group method with arithmetic means (UPGMA) procedure by Sokal and Michener [28].

Genetic variation in the populations was calculated as mean expected heterozygosity per locus ($\overline{H}_e$. The formulae used were according to Nei [22] for $\overline{H}_e$ and Nei and Roychoudhury [24] for sampling variances.

TABLE III. Localities from Which Artemia was Assayed and Results of Cytological Observations

Locality	Chromosome numbers	Chromocentres present	Mixed populations	Abbreviations
Bisexuals:				
Artemia franciscana				AUS-R
Bahía Salinas, Puerto Rico	42	+	—	BS
Cabo Frio, Brazil	42	+	—	CF
Chaplin Lake, Sask., Canada	42	+	—	CHA
Great Salt Lake, Utah, USA 1966	42	+	—	G66
Great Salt Lake, Utah, USA 1977	42	+	—	G77
Green Pond, Arizona, USA	42	+	—	GP
Jesse Lake, Nebraska, USA	n.s.	n.s.	—	JE
Little Manitou, Sask., Canada	42	+	—	LM
Macau, Brazil	42	+	—	MAC
Manaure, Guajira, Colombia	42	+	—	MAN
Mono Lake, California, USA	42	+	—	MON
Pekelmeer, Bonaire	42	+	—	PEK
Punta Araya, Venezuela	42	+	—	PA
Red Pond, Arizona, USA	42	+	—	RP
Rockhampton, Australia	42	+	—	AUS-R
San Francisco Bay, California, USA 1971	42	+	—	S71
San Francisco Bay, California, USA 1976	42	+	—	S76
San Pablo Bay, California, USA 1976	42	+	—	SPB
Shark Bay, Australia	n.s.	n.s.	—	AUS-B
Yavaros, Mexico	42	+	—	YAV
Zuni Salt Lake, New Mexico, USA	42	+	—	ZUN

			+*	
Artemia persimilis				
Buenos Aires, Argentina	BA	--	+	44
Artemia salina				
Barbarena, Spain	BAR	--	--	42
Cagliari, Sardinia, Italy	CAG	--	--	42
Chott, Ariana, Tunisia	CHO	--	--	42
Larnaca Lake, Cyprus	LAR	--	--	42
San Felix, Spain	SF	d,pg	--	42
San Pablo, Spain	SPA	d,pg	--	42
Santa Pola, Spain	SPO	d,te,pg	--	42
Parthenogenitics:				
Diploids				
Alcochete, Portugal (3)	AL21-III	te,pg	--	42
Burgas Pomerije, Bulgaria	BUR2	p,pg	--	42
Cadiz, Spain (2)	CAI-II	--	--	42
Calpe, Spain (4)	CALI-IV	--	--	42
Lake Techirgiol, Rumania (7)	LTI-VII	--	--	42
Lavalduc, France	LAV2	te,pg	--	42
Margherita di Savoia, Italy	MS2	te,pg	--	42
Salin de Giraud, France (13)	SGI-XIII	--	--	42

TABLE III. continued from previous page

Locality	Chromosome numbers	Chromocentres present	Mixed populations	Abbreviations
San Felix, Spain	n.s.	n.s.	b	SF
San Pablo, Spain	n.s.	n.s.	b	SPA
Santa Pola, Spain	42	—	te,pg,b	SPO2
Sète, France (2)	42	—	te,pg	SEI-II
Tientsin, China (4)	42	—	---	TNI-IV
Triploids				
Eilat, Israel	63	—	---	EIL
Izmir, Turkey	63	—	te,p,pg	IZM3
Tuticorin, India	63	—	---	TUT
Tetraploids				
Alcochete, Portugal	84	—	d,pg	AL4
Comacchio, Italy	n.s.	n.s.	---	COM
Delta del Ebro, Spain	84	—	---	DE
Izmir, Turkey	84	—	d,t,pg	IZM4
Lavalduc, France	84	—	d,pg	LAV4
Margherita di Savoia, Italy	84	—	d,pg	MS4
Saelices, Spain	84	—	---	SAE
Santa Pola, Spain	84	—	d,pg,b	SPO4
Sete, France	84	—	d,pg	SE4
Tierzo, Spain (2)	84	—	---	TZI-II
Pentaploids				
Burgas Pomerije, Bulgaria (2)	105	—	d,pg	BURI-II
Izmir, Turkey	105	—	---	IZM5

In some localities a mixture either of bisexual and parthenogenetic or of various ploidies of parthenogenetic shrimps were found as indicated (d = 2n, t = 3n, te = 4n, p = 5n, pg = parthenogenetic, b = bisexual (A salina)). Figures after pg populations indicate the number of clones present, ns = not scored. * = see Table II.

TABLE IV. Loci Assayed in Bisexual Artemia Populations

System	Number of loci scored	Polymorphic
Amy	1	—
Cat	1	f,p
Est	2	f
Est-D	1	f,p
Got	2	f,p,s
Idh	2	f,p,s
Lap	3	f,s
Ldh	1	—
Mdh	2	f,p,s
Me	1	—
Mpi	1	f,p,s
Pep	2	f,s
6Pgdh	1	f,p,s
Pgi[a]	1–2	f,p,s
Pgm	1	f,s
Sod	1	f

[a]Two loci found only in A persimilis.
f, franciscana; p, persimilis; s, salina.

The same calculations may also be used to determine the extent of genetic variation in asexual organisms, although of course $\overline{H}_e$ does not then relate directly to the frequence of heterozygotes. In order to avoid confusion, Nei [22] called $\overline{H}_e$ "gene diversity" as a general term of universal applicability. It is in this sense that it has been used in the present work.

Cytological observations were made from cell spreads derived from freshly hatched nauplii. The procedure for the preparations is a slightly modified version of Barigozzi's techniques (Barigozzi, personal communication) for chromosome study.

RESULTS

Bisexual Species

Variation within populations. Of the 23 loci studied in all bisexual species, 19 (83%) were found to be variable in the A franciscana populations and 11 (48%) in the A salina populations. The single A persimilis sample showed a unique duplication of the Pgi locus, raising the number of loci studied in this species to 24, of which 12 (50%) were polymorphic (see table IV). Variability is widespread at most polymorphic loci throughout the populations, but some

TABLE V. The Anionic Composition of A franciscana habitats[a] and the Population $\overline{H}_e$ Values.

Population	Relative anionic composition (%)[b]			$\overline{H}_e$
	CO_3^{--}	SO_4^{--}	CL^-	
Bahía Salinas[c]	—	12	88	0.017
Cabo Frio[c]	—	12	88	0.073
Chaplin Lake[d]	1	69	31	0.119
GSL66	—	12	88	0.094
GSL77	—	12	88	0.106
Green Pond	51	1	48	0.133
Jesse Lake[e]	66	26	8	0.059
Little Manitou	1	68	31	0.131
Macau[c]	—	12	88	0.091
Manaure[c]	—	12	88	0.129
Mono Lake	38	22	40	0.176
Pekelmeer[c]	—	12	88	0.119
Punta Araya[c]	—	12	88	0.106
Red Pond	27	1	72	0.129
Rockhampton[c]	—	12	88	0.073
SFB 71	—	14	86	0.069
SFB 76	—	14	86	0.072
SPB	—	14	86	0.072
Shark Bay[c]	—	12	88	0.036
Yavaros[c]	—	12	88	0.026
Zuni	—	5	95	0.072
Sea water	—	12	88	—

[a]From review in [29].
[b]Calculated so that sum of the concentration of major anions (CO_3^{--}, SO_4^{--}, CL^-) = 100%. These make up about 70% of the total dissolved solids.
[c]The CL values in these populations, associated with salt works are *minimum* estimates. Total salinity in the ponds where Artemia is found varies widely between ponds, increasing from typical sea water levels at the intake, up to saturation (>240 ppt) in the crystalisation pans. Above 184 ppt, the SO_4^{--} begins to precipitate out (as $CaSO_4$) thus raising the relative CL^- concentration to 100%.
[d]Inferred from Little Manitou lake values.
[e]Excluded from calculations due to small sample size (N = 5).

of the rare alleles are characteristic of individual populations or of particular groups of populations (see later section).

Mean expected heterozygosities ($\overline{H}_e$, also referred to as gene diversity, [22]) are listed in Table V. The mean H_e for the A franciscana populations is 0.091 ($\pm$ 0.034), while corresponding values for A salina and A persimilis are 0.089

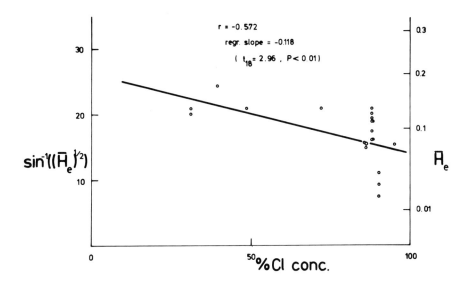

Fig. 1.   The regression of transformed $\overline{H}_e$ on percentage Cl⁻ for populations of A franciscana.

(± 0.039) and 0.130 (± 0.038), respectively. No significant departures from Hardy-Weinberg expectations were seen in any of the populations.

An association between habitat-type and heterozygosity was found in A franciscana populations (for which most of the available ecological studies have been made). Mean H_e for the Cl⁻ populations is 0.077 (± 0.031), while that for the non-Cl populations (ie, Mono Lake, Green and Red ponds, Little Manitou and Chaplin Lakes) is 0.138 (± 0.043).

Table V indicates that a higher level of heterozygosity is seen in non-Cl;-populations regardless of whether the major anion is SO_4^{--} or CO_3^{--}.

A regression line fitting the general expression:

$$\overline{H}_e = b \cdot x_{Cl} + a$$

was obtained using arcsin transformed values of $\sqrt{\overline{H}_e}$. In this equation x_{Cl} represents the Cl⁻ concentration in the medium, regardless of other major constituents. Figure 1 illustrates the relationship between chloride ion concentration and heterozygosity. The regression slope is significantly negative (b = –0.118, P < 0.01). A similar analysis was performed after removing from the data populations which were sampled more than once (GSL and SFB) and those which are thought to have been started from San Francisco Bay material (Rockhampton, Shark Bay, Macau, and Cabo Frio). Results from this latter analysis give a very similar slope and only slightly reduced significance (P <0.02).

Values of $\overline{H}_e$ do not seem to be associated with "area effects," as Mono Lake (mixed ion composition) has the highest $\overline{H}_e$ while being closest to the San Francisco Bay zone (Cl⁻ waters), and Zũni (Cl⁻) has a very low value of $\overline{H}_e$ being very close geographically to Green and Red ponds (Cl⁻/CO₃⁻). Lack of ecological data and sampling from different ecotypes for the A salina populations prevent a similar analysis for that group.

Within locus differentiation. Although the common polymorphic loci remain variable throughout the range of species and populations, some allozymes and some loci were found to be characteristic of particular populations or of one of the species. To demonstrate this, χ^2 values were calculated for all possible comparisons of allele frequencies between populations for each variable locus. The proportion of these comparisons found to be significantly different at the 0.001 level is given for each population for each locus in Figure 2. There appear to be two extreme types of locus: (1) loci which have low levels of variability throughout the range of populations, but in one or more individual population or group of populations have uniquely characteristic alleles which distinguish it or them from the rest of the species; and (2) loci which have very high levels of variability and, in some cases, very many alleles (eg, Pgi) which show considerable discriminatory differences throughout the range of populations. Examples of type (1) loci in the A franciscana group are Cat, Est-4, and Lap-3 (the last two of which contain alleles which are characteristic of Green and Red Ponds, and of Green and Red Ponds and Bahía Salinas, Puerto Rico, respectively). Good examples of type (2) loci are Pgi, Idh-1, and 6Pgdh in this species.

On the other hand, examples of type (1) loci in A salina are Lap-3 and -2, and Mdh-1; and of type (2) Idh-2 and 6Pgdh.

Reflections of this trend can be seen in other genetic parameters of these populations (see Table VI). Type (1) loci tend to have low within-populational gene diversity (H_S) and high coefficient of gene differentiation (G_{ST}), while type (2) loci have high H_S _and_ high G_{ST} values.

There is a general trend for G_{ST} values to vary in a way directly proportional to the $\overline{H}_e$ values, a tendency in agreement with the results of Skibinski and Ward [31] who found that loci with high $\overline{H}_e$ values diverge at a faster rate than loci with low $\overline{H}_e$ values. An exception to this are type (1) loci. The sample sizes used for the present study, however, do not allow the extensive statistical analysis used by Skibinski and Ward. Another general tendency from which type (1) loci are excepted is for enzymes with a higher subunit molecular weight to show higher H_S and higher G_{ST} values, a result that would agree with the work of numerous authors (see [32] for review).

TABLE VI. Genetic Parameters in A franciscana and A salina for Individual Polymorphic Loci

Locus	A. franciscana			A. salina		
	H_S	H_T	G_{ST}	H_S	H_T	G_{ST}
Cat	0.096	0.112	0.141	0.000	0.000	0.000
Est-4	0.049	0.084	0.414	0.000	0.000	0.000
Est-D	0.135	0.286	0.527	0.000	0.000	0.000
Got-2	0.061	0.075	0.186	0.062	0.066	0.062
Got-1	0.129	0.240	0.462	0.000	0.000	0.000
Idh-2	0.207	0.410	0.495	0.181	0.226	0.200
Idh-1	0.166	0.477	0.652	0.009	0.009	0.042
Lap-3	0.042	0.355	0.880	0.023	0.232	0.899
Ldh	0.002	0.002	0.021	0.000	0.000	0.000
Mdh-1	0.025	0.026	0.051	0.059	0.272	0.782
Mpi	0.120	0.259	0.536	0.554	0.581	0.047
Pep-4	0.015	0.016	0.064	0.325	0.420	0.226
Pep-1	0.010	0.010	0.067	0.000	0.000	0.000
Pgi	0.328	0.570	0.425	0.191	0.240	0.203
6Pgdh	0.366	0.529	0.308	0.638	0.736	0.134
Pgm	0.274	0.500	0.453	0.194	0.215	0.098
Sod	0.0002	0.0002	0.006	0.000	0.000	0.000
Mean G_{ST} ($\pm$ SE)	0.237 ($\pm$ 0.045)			0.117 ($\pm$ 0.118)		

Mean of G_{ST} based on both polymorphic and monomorphic loci.

In comparisons of the A franciscana populations with the A persimilis sample, six loci (Amy, Est-1, Got-2, Lap-1, Ldh, and Pep-4) give interpopulational genetic distance values not significantly different from zero. At three other loci (Est-4, Idh-1, and 6Pgdh) there is significant overlap between the two species and these are shown diagrammatically in Figure 3. No overlap was seen in the remaining loci between any pair of populations from these two species. Overlap in gene frequencies between A persimilis and A salina only exists at the Amy and the Me loci, which in both species are homozygous for the same alleles.

Population subdivision. The coefficient of gene differentiation (G_{ST}) can be utilised to analyse the extent of interpopulational subdivision. This measure is equivalent to Wright's F_{ST} statistic if there are only two alleles at a locus, and in the case of more than two alleles, G_{ST} is equal to a weighted mean of F_{ST} for all alleles [22].

Although the known populations of A franciscana and A salina clearly do not form continuous populations in the respective species, it is important to study the degree of subdivision as mixing of populations may occur from birds mi-

(a) A. franciscana

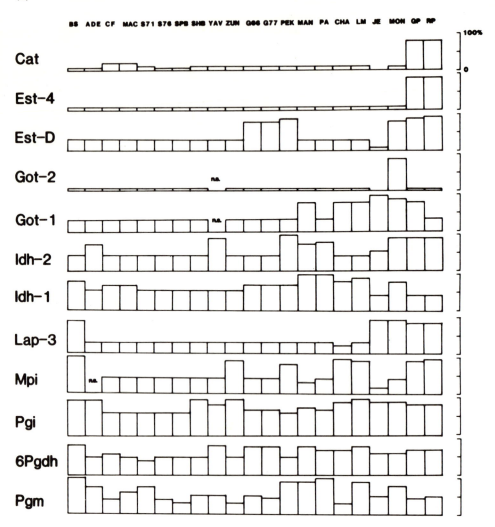

Fig. 2. a) Proportion of comparisons, for single loci, of individual populations with all other populations in A franciscana. b) Proportion of comparisons, for single loci, of individual populations with all other populations in A salina. c) Proportion of comparisons, for single loci, of A persimilis with all populations of A franciscana. The histograms indicate the proportion of comparisons where P < 0.001 for χ^2. (Loci not shown have zero entries; n.s., not sampled).

(b) A. salina

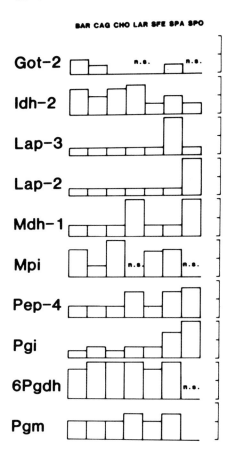

(c) A. franciscana vs. A. persimilis

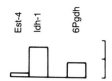

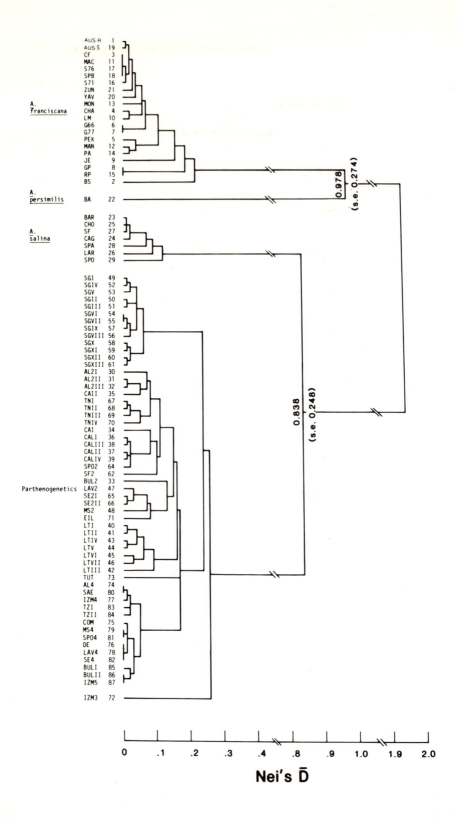

Nei's D̄

grating between saline lakes transporting cysts either adhering to their bodies or in their gut in cases where they feed on the shrimps [33].

The overall mean G_{ST} value for the A franciscana populations is 0.237 with a standard error of 0.0437 [23]. In other words 24% of the genic variation in the species is due to interpopulational gene differences. This value is significantly different from zero (P $<$0.05). On the other hand, the G_{ST} value for the A salina group is 0.117 which, with a standard error of 0.118, is not significantly different from zero. Nevertheless, as shown in Figure 2 considerable differentiation has occurred between some populations at particular loci.

The G_{ST} statistic has been used previously [34] to analyse the significant subdivision existing in populations of the cutthroat trout Salmo clarki in inland streams of the US, although the value for this species was much higher (0.703).

Genetic distances. Genetic distances (D of Nei, [25] are given in Table VII for all possible comparisons of populations in the three bisexual species, together with standard errors [24]. The mean conspecific D value for the A franciscana populations is 0.106 ($\pm$ 0.059), while that for the A salina group is 0.072 ($\pm$ 0.057). These values are quite high for an invertebrate group (Ayala et al [35] found a mean D value of 0.032 for geographical populations of the Drosophila willistoni group, and similar values have been obtained for other invertebrates). It must be borne in mind, however, that high levels of differentiation are seen between the Artemia populations.

UPGMA clustering [36] was used to construct the dendrogram shown in Figure 3. This analysis clusters together all the North American Cl^- A franciscana, the SO_4^- populations (Chaplin and Little Manitou), one carbonate population (Mono Lake), and all of the populations in South America (Cabo Frio and Macau) and Australia (Shark Bay and Rockhampton) where man has introduced cysts [14].

Because of the large standard errors involved, none of the D values between any of these populations is significant, suggesting that considerable gene flow occurs, or has occurred, between these populations. This is not surprising for the SO_4^- populations, for, although SO_4^- ions are slightly toxic to Artemia [37], shrimps hatched from their cysts grow and breed perfectly well in chloride water. The result for the Mono Lake sample is very surprising. Bowen [30] found that Mono Lake water is lethal to Artemia from Great Salt Lake, and Mono Lake shrimps cannot survive in normal chloride waters. Based on this ecological distinction she tentatively suggested that the Mono Lake shrimps deserved full species status as A monica. Our genetic distance data suggest a much smaller degree of genetic differentiation between these two than is usually associated with a pair of sibling species. However, important differences between this population and the other members of its cluster could reside at a very small

Fig. 3.   Dendrogram based on genetic distances between all populations of Artemia (all genotypes for parthenogenetic forms).

TABLE VII. Genetic Differentiation in Artemia

									A. franciscana														*A. persimilis*					*A. salina*	
	ADE	BS	CF	CHA	PEK	Q66	Q77	GP	JE	LH	MAC	MAN	MON	PA	RP	S71	S76	SPB	SHB	YAV	ZUN	BA	BAR	CAQ	CHO	LAR	SF	SPA	SPO
	1	2	3	4	5	6	7	8	9	10	11	12	13	14	15	16	17	18	19	20	21	22	23	24	25	26	27	28	29
1	--	18	02	06	09	07	07	17	19	06	01	10	06	11	17	01	02	02	01	05	03	.982	312	332	322	182	332	262	19
2	09	--	22	16	27	20	20	25	19	14	20	23	20	24	25	22	22	23	23	33	22	.892	352	372	372	192	372	302	19
3	01	10	--	04	09	05	05	16	15	05	00	08	05	10	16	01	00	00	00	03	03	.992	342	362	362	182	362	302	19
4	03	08	02	--	13	07	07	21	10	02	04	09	06	11	21	04	04	05	07	09	07	.882	292	302	312	152	312	272	15
5	05	11	06	--	--	13	12	15	25	13	08	04	13	04	16	07	09	09	11	09	.121	012	322	332	332	152	342	262	17
6	05	09	03	04	06	--	00	17	15	06	05	05	07	13	17	06	06	06	07	08	08	.922	302	302	302	142	302	232	12
7	04	09	03	04	06	00	--	18	15	06	05	13	06	12	17	05	06	06	07	08	08	.922	292	302	302	132	302	232	12
8	08	10	07	08	07	08	08	--	27	09	16	20	17	20	00	15	16	15	21	15	15	.062	312	322	332	172	332	262	17
9	09	09	08	06	10	08	08	10	--	09	15	18	10	20	26	16	15	17	19	19	17	.881	941	962	001	781	962	301	79
10	03	07	02	01	06	04	04	09	05	--	04	09	05	10	24	05	05	06	05	12	06	.882	122	132	141	962	132	142	01
11	00	09	00	02	05	03	03	07	08	02	--	08	05	09	17	00	00	00	02	03	03	.972	332	352	352	172	352	292	18
12	04	10	04	05	02	06	05	08	08	04	04	--	12	02	20	08	08	09	12	11	13	.982	312	332	332	152	332	262	17
13	02	09	02	02	06	03	03	06	05	01	02	05	--	05	17	05	05	05	06	08	05	.932	062	072	091	902	072	191	90
14	05	10	05	05	08	06	06	08	09	05	05	01	05	--	21	09	09	10	13	13	14	.992	322	342	342	162	342	272	17
15	08	10	08	08	07	08	07	00	10	09	08	09	06	09	--	16	16	16	21	16	16	.191	052	312	332	172	332	262	17
16	01	10	00	01	05	02	04	07	08	03	00	04	02	05	07	--	00	01	03	03	03	.031	012	342	362	172	362	292	19
17	01	10	00	01	02	04	04	07	08	02	00	02	02	05	08	00	--	00	02	03	03	.992	342	362	362	182	362	302	19
18	00	10	00	02	05	03	03	08	09	03	00	02	02	06	08	00	00	--	02	--	02	.031	012	342	362	182	362	302	19
19	00	10	02	03	06	06	05	08	09	03	02	06	03	06	09	02	02	02	--	08	08	.932	062	382	382	202	382	312	21
20	03	13	01	03	05	05	05	08	10	06	02	05	03	07	08	01	01	01	05	--	04	.071	122	292	292	212	292	212	21
21	02	10	03	03	07	05	05	09	09	03	02	06	02	07	09	03	03	02	03	04	--	.022	342	362	292	192	362	292	21
22	28	26	27	25	28	26	26	29	25	27	28	28	28	28	29	28	28	28	28	32	28	--1.022	342	362	362	192	362	292	21
23	68	68	68	66	68	66	66	68	54	68	68	68	56	68	68	68	68	68	69	68	68	.921	941	941	941	781	941	881	78
24	69	69	69	66	67	65	65	68	54	68	68	68	56	68	68	68	68	68	69	68	68	.54	02	02	07	01	07	07	09
25	68	69	68	67	68	65	65	68	55	68	68	68	57	68	68	67	67	68	69	68	68	.54	01	--	01	08	02	06	09
26	68	68	68	66	66	66	66	68	54	68	68	68	56	68	68	68	68	68	69	68	68	.54	01	01	--	06	08	16	18
27	69	69	69	66	68	65	65	68	54	68	68	68	57	68	68	67	67	68	69	68	68	.54	00	01	00	06	--	07	10
28	68	68	68	66	68	65	65	68	68	61	68	66	66	68	68	68	68	68	69	68	68	.54	05	05	04	09	05	--	15
29	68	68	68	66	64	65	65	68	54	58	68	68	66	68	68	68	68	68	69	68	68	.54	06	06	07	10	07	09	--

number of loci not observed in normal electrophoresis. Speciation involving small genetic distances has been observed in some groups of animals, for example, in the Hawaiian Drosophila [38,39].

On the other hand, Lenz [40] reported the existence of a population in Fallon, Nevada, which can survive either in carbonate or in chloride water and is interfertile with shrimp from Mono Lake. Existence of populations of this kind could provide a route for genetic interchange between chloride water populations and Mono, thereby preventing marked genetic differentiation.

Caribbean populations of A franciscana (Pekelmeer, Manaure and Punta Araya) have become increasingly differentiated from the rest of the American populations. The mean D between Caribbean and San Francisco clusters is 0.106 (± 0.053).

The remaining alkali lake populations (Jesse, Red, and Green Ponds) have become strikingly differentiated from the other A franciscana populations (mean D = 0.185). Although the Jesse Lake Artemia, similar to the Mono Lake shrimps, cannot survive in chloride waters, the Red and Green pond shrimps thrive in this type of water. It may be that a combination of geographical isolation *and* ecological factors (based on chemical constitution of the medium) are necessary to prevent significant gene flow between these populations.

The differentiation of the Bahía Salinas populations, particularly from the other Caribbean groups is more difficult to explain. It may be necessary to ascertain whether migratory routes between Caribbean islands and the South American north coasts do not cover the island of Puerto Rico.

In the A salina populations, the largest distances found were between Santa Pola and the rest, and between Larnaca and the rest. These differences are due mainly to extensive differentiation at the Lap-2, Mdh-1, and Pgi loci in the first population; and at the Mdh-1, and 6Pgdh loci in the second. These results suggest local population differentiation, in spite of the low G_{ST} values seen.

The mean D between A persimilis and A franciscana is 0.978 ($\pm$ 0.274), while mean D between the A salina and A franciscana cluster is 1.950 (± 0.540) (see Figure 3).

Cytological observations. The chromosome numbers of the bisexual Artemia were counted and the presence or absence of darkly staining areas (chromocentres) visible in resting nuclei, staining with guinacrine and representing highly repetitive regions of DNA (Barigozzi personal communication) was noted. Because of the small size of these chromosomes no chromosomal morphological studies have been made. All A franciscana populations had a 2n value of 42 and all contained numerous chromocentres (see Table II). The 2n values for the A persimilis and the A salina species were 44 and 42, respectively. The latter species showed no chromocentres and A persimilis typically has 1–2 rather small chromocentres.

The genetic differentiation observed between North American populations, Bahiá Salinas and other Caribbean populations is parallelled to some extent by differentiation in number of chromocentres. Most Caribbean populations examined have about eight small chromocentres per nucleus. Typical North American populations have about 19 large chromocentres and Bahiá Salinas resembles the North American type.

Parthenogenetic Artemia

Genetic variation. In the context of parthenogenetic populations two components of genetic variation need to be looked at separately. The first of these is gene diversity, calculated in the same way as expected heterozygosity, but in this case not measuring the number of heterozygotes in a sample [22]. It is a good measure of genetic diversity within a clonal genotype. The second component is clonal, or genotypic diversity (the number of clones found in a sample). Asexual forms may have high values of gene diversity index, but a low number of related genotypes. The reverse is also found. On the other hand, no cases were found with high gene diversity and large numbers of related genotypes.

Electrophoretically detectable variability in the asexual Artemia populations was, in general, found at the same loci as those which are most variable in the bisexual forms. Variation between genotypes can occur as a result of mutation producing new allelic forms, or as a result of recombination. The first mechanism will occur in any of the ploidies of asexual Artemia. However, the second mechanism may only occur in the diploid asexual forms as the polyploids reproduce ameiotically (or apomictically; see [10] for a review.)

Variation between genotypes was due to the presence in polyploid forms of unique alleles at some of the loci (eg, Cat, Est-D, Got-2, Mdh-1, Mpi, Pgi, and Pgm), while in most of the remaining was the result of permutations of genotype composition particularly amongst diploid forms. In keeping with the trend seen amongst the bisexual forms, loci coding for enzymes of high subunit molecular weight (eg, Mdh, Mpi, Pgi, 6Pgdh, and Pgm) showed high levels of intergenotypic variation.

A relationship exists between mean heterozygosity and ploidy level. This is illustrated in Figure 4 which shows a striking and significant correlation between the two factors. The regression slope, calculated using arcsin transformed values of root gene diversity from values given in Table VIII is positive with b = 5.42, t_{56} = 9.0311, P ≪0.001. Such a relationship has been observed in the lizard Cnemodophorus tesselatus [41] and in parthenogenetic insects and is in keeping with the theoretical expectation that polyploids are far better buffered against genetic loads than diploids [42].

Intergenotypic differentiation. Tetraploid and pentaploid Artemia demonstrates a tendency to monoclonality, with very similar or identical genotypes

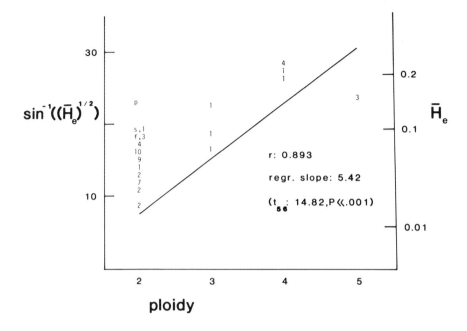

Fig. 4. Regression of transformed $\bar{H}_e$ on ploidy level. s, f, and p represent mean values for the three bisexual species; figures represent the number of entries at that coordinate.

found in distant localities (eg, Margherita di Savoia and Santa Pola; Delta del Ebro, Lavalduc and Sète). The average D between all tetraploids is 0.029 ($\pm$ 0.024). Pentaploids follow the same trend, although only two populations were sampled (one of them, Burgas Pomerije contained two pentaploid genotypes differing in allelic composition at only one locus: 6Pgdh). The mean D for these populations is 0.01 ($\pm$ 0.01). Genetic distance between tetraploids and pentaploids is also very low (0.045 $\pm$ 0.027) suggesting a very close evolutionary relationship (see Discussion).

Triploid populations show relatively high genetic distance between themselves and the remaining asexual Artemia, but also between each other (D Eilat-Tuticorin = 0.170 $\pm$ 0.089, D Eilat-Izmir = 0.136 $\pm$ 0.075, and D Tuticorin-Izmir = 0.234 $\pm$ 0.105).

The polyploid populations are thus relatively similar in some respects, eg, genotype at the Cat locus while at other loci, eg, Mpi there is considerable variation from population to population (Table IX). Data for diploids are not given in this table but these appear to be considerably more heterogeneous than the polyploids with a tendency for there to be two groups. One group consists of populations each containing a relatively large number of clones and the other

TABLE VIII. Values of Gene Diversity ($\overline{H_e}$) and Standard Errors [24] for Parthenogenetic Populations

	Genotype designation	$\overline{H_e}$ (± s.e.)
Diploids	AL2I	0.048(± 0.033)
	AL2II	0.071(± 0.040)
	AL2III	0.048(± 0.033)
	BUL2	0.109(± 0.044)
	CAI	0.071(± 0.040)
	CAII	0.095(± 0.045)
	CALI	0.043(± 0.030)
	CALII	0.065(± 0.036)
	CALIII	0.043(± 0.030)
	CALIV	0.065(± 0.036)
	LTI	0.095(± 0.045)
	LTII	0.071(± 0.040)
	LTIII	0.071(± 0.040)
	LTIV	0.071(± 0.040)
	LTV	0.095(± 0.045)
	LTVI	0.071(± 0.040)
	LTVII	0.071(± 0.040)
	LAV2	0.065(± 0.036)
	MS2	0.071(± 0.040)
	SGI	0.022(± 0.022)
	SGII	0.065(± 0.036)
	SGIII	0.043(± 0.030)
	SGIV	0.043(± 0.030)
	SGV	0.065(± 0.036)
	SGVI	0.065(± 0.036)
	SGVII	0.065(± 0.036)
	SGVIII	0.087(± 0.041)
	SGIX	0.043(± 0.030)
	SGX	0.043(± 0.030)
	SGXI	0.065(± 0.036)
	SGXII	0.065(± 0.036)
	SGXIII	0.087(± 0.041)
	SF	0.087(± 0.041)
	SPA	0.087(± 0.041)
	SPO2	0.022(± 0.022)
	SE2I	0.071(± 0.040)
	SE2II	0.071(± 0.040)
	TNI	0.033(± 0.024)
	TNII	0.054(± 0.031)
	TNIII	0.033(± 0.024)
	TNIV	0.045(± 0.032)
Triploids	ISR	0.099(± 0.040)
	TUR3	0.080(± 0.037)
	TUT	0.142(± 0.046)
Tetraploids	AL4	0.217(± 0.051)
	COM	0.217(± 0.051)
	DE	0.217(± 0.051)
	TUR	0.217(± 0.051)
	LAV4	0.217(± 0.051)
	MS4	0.217(± 0.051)
	SAE	0.217(± 0.051)
	SPO4	0.217(± 0.051)
	SE4	0.217(± 0.051)
	TZI	0.207(± 0.048)
	TZII	0.201(± 0.051)
Pentaploids	BUL5I	0.188(± 0.049)
	BUL5II	0.188(± 0.049)
	TUR5	0.188(± 0.049)

TABLE IX. Genotypes Observed in Some Polyploid Parthenogenetic Artemia

Population	Ploidy	Cat	Est-D		Got-2		Idh-1		Idh-2		Mdh-1		Mdh-2		Mpi				Pgi-2				6Pgdh		Pgm		Sod	
		2/9	6/6	6/10	4/4	4/1	9/4	9/9	9/5	9/9	9/4	3/3	5/5	3/8	10/10	13/10	8/6	10/5	2/2	6/2	3/3	6/1	5/3	5/5	5/2	8/5/2	5/4	5/5
Eilat	3n	X	X		X		X	X	X		X			X		a							X		X			X
Tuticorin	3n	X	X		X		X	X	X		X			X							X			X	X			X
Delta del													X		X				X									
Ebro	4n	X		X		X	X		X			X				X				X			X					
Lavalduc	4n	X		X		X	X		X		X	X					X			X			X			X	X	
Burgas P	5n	X		X		X	X		X	X		X						X				X	X	X		X	X	
Izmir	5n	X		X		X	X		X		X	X						X				X	X			X	X	

[a]No Mpi activity was detected in the Eilat material.

group consists of populations which are mono-, bi-, or triclonal. Abreu-Grobois and Beardmore [20] suggested that the type of egg maturation in these two groups is likely to differ significantly. In the localities containing many genotypes, these tend to be closely related and distinctive of the locality (g, Salin de Giraud) while in the other groups with few genotypes at least some of these are more closely related to genotypes from different localities (eg, Calpe and Santa Pola; Lavalduc and Sète). However, the overall tendency is towards local differentiation and diversification.

Examples of multiclonality in two diploid populations are given in Table X. Such multiclonality has been described in other organisms, notably Solenobia [43]. In Artemia multiclonality appears to be almost always confined to diploid parthenogenetic populations, those of higher ploidy levels being essentially monoclonal though the pentaploid population of Burgas Pomerije and the tetraploid from Saelices are exceptions to this.

The frequencies of the clones vary considerably and many genotypes appear to be unique to a particular locality, a situation found to a lesser extent in diploid Solenobia. Solenobia is, however, different in that the tetraploid form of S. triquetrella is also multiclonal, but here no clone is represented in more than one locality out of the ten studied [45].

It seems likely that multiclonal populations utilise the system of variation in a way something akin to that found in a balanced polymorphism in a bisexual outbreeding form or perhaps a more apt parallel is the variation found in species such as Avena barbata with a very high level of inbreeding [44]. In Avena the frequencies of the different (largely homozygous) genotypes relate strikingly well to environmental variables, particularly moisture.

Cytological studies. Parthenogenetic populations were found in some cases to co-exist with bisexual forms (eg, in Santa Pola and San Pablo), but in other cases samples were found to contain mixtures of different ploidies (eg, the Izmir population contained 3n, 4n, and 5n Artemia, while others such as Santa Pola and Sète contained 2n and 4n parthenogenetic shrimps). Wherever diploid populations with large numbers of closely related genotypes coexisted (eg, Lake Techirgiol, Salin de Giraud and Calpe) no polyploids were found. Generally the chromosomal counts complement extremely well the analysis of clustering derived from the electrophoretic study.

In no cases were chromocentres observed in preparations from parthenogenetic Artemia. Results of the cytological studies are summarised in Table III.

DISCUSSION

We may now try to deal with the mechanisms involved in the production of species. It has been argued that these mechanisms can only be studied in cases where a speciation process is actually taking place. We do not share this view

TABLE X. Genotypes Observed in Some Diploid Parthenogenetic Artemia

Population	Clone	Cat		Idh-2			Mdh-2		Pgi-2			6Pgdh				Pgm		
		2/2	2/9	2/2	5/5	9/9	3/3	3/8	2/2	6/2	6/6	5/3	5/5	7/5	7/7	5/2	6/3	7/7
Salin	I	X				X	X		X				X				X	
de	II		X			X	X		X					X			X	
	III		X			X	X		X						X		X	
Giraud	IV	X				X		X	X					X			X	
	V	X				X		X	X				X				X	
	VI	X				X	X			X				X			X	
	VII	X				X	X			X				X			X	
	VIII		X	X			X				X			X			X	
	IX	X				X	X			X				X			X	
	X	X		X				X	X					X				X
	XI	X		X				X	X					X				X
	XII		X	X				X		X				X				X
	XIII		X	X				X			X			X				X
Calpe	I	X			X		X			X				X		X		
	II	X			X		X			X		X				X		
	III	X			X		X				X	X				X		
	IV	X			X		X			X				X		X		

but it provides a useful pointer as to where to begin. A monica is ecologically distinct from A franciscana yet the genetic distance between it and franciscana is less than the distance between some populations of the latter. The adaptation to the unusual ionic balance of Lake Mono water must clearly reside in a rather small number of genes and yet monica has become sufficiently ecologically differentiated as to be unable to share genes directly with the vast majority, if not all, populations of franciscana. It is also conceivable that the population of Lake Urmia is distinguished from A salina only by a small number of genes and this we hope to test in the near future.

Nevertheless, A franciscana populations have become locally differentiated sufficiently to show physiological differences [45], and this differentiation may be reflected in the electrophoretic data presented here. It is difficult at present, however, to ascertain whether the population subdivision of this species is the effect of genetic drift or whether it is the result of local selection and adaptation. But it does seem that in order to produce significant genetic differentiation in the franciscana group, ecological barriers are not sufficient without additional geographical isolation.

The genetic distance data strongly suggests that persimilis is derived from the franciscana line, particularly as chromocentres are found only in these two species. Data on DNA content in these species are not available.

The differences between A persimilis and A franciscana involve an additional pair of chromosomes [10], and relatively limited natural geographic distribution in persimilis. The additional chromosomes, despite the lack of comparative data on DNA content, seem likely to be true additions to the genome from the otherwise unlikely occurrence of two Pgi loci in persimilis. The presence of chromocentres is thought to indicate relative local abundance of highly repetitive DNA. The evolution of this character in both American lines presumably postdates the separation from the European species. Although the zoogeography of South American Artemia is but poorly known it seems possible that persimilis evolved in geographic isolation from franciscana following a colonisation event which might well have involved only a small founder population. Such conditions are propitious for the generation of new species [46].

The most striking feature of differentiation in the genus Artemia is the large number (in this study no less than 52) of different parthenogenetic forms. This pattern, while unusual in animals, could not reasonably be regarded as an evolutionary dead end particularly as some of these forms are capable of generating new genotypes through oogenesis [20]. Similar arguments have been adduced by Suomalainen [47].

In trying to account for the complex pattern of microspeciation involved in the production of these forms we are led to conclude that the pattern is intimately related to the rather unusual set of ecological circumstances in which this organism lives. In particular the number of other species co-existing in the same water will in most cases be quite small. Thus opportunities for the development of subtle interaction between species and the creation of niches which thereby results are considerably reduced.

The origin of parthenogenesis in parthenogenetic weevils Solenobia and Otiorrhychus has been plausibly attributed to the premium placed on the parthenogenetic state in conditions where populations are very thinly scattered in a habitat newly opened for exploitation as after retreat of a glacier [47].

An analogous pattern can be sketched for Artemia. It seems very probable that the centre of origin is in the region of the Mediterranean and Middle East. The measures of genetic distance which we have obtained can be converted to estimates of periods of time from a common ancestor. Such conversions are, of course, subject to errors and are not necessarily always legitimate. Nevertheless, such methods have been used with some success to compare estimates of D with the time since separation of lineages which can be independently verified from available geological records [48,49]. Two formulae for the conversion of measures of genetic relatedness have been described. The first one due to Nei [22] estimates the divergence time (t) between two lineages from the value of genetic distance (D), $t = D / (2\alpha)$, where α is the rate of electrophoretically detectable mutations for which he suggested the value of 10^{-7} per year. The second measure, originally derived for albumin immunological distances and times of divergence correlations but also convertible to use with Nei's D value [50], holds that one unit of D is approximately equivalent to 18.9×10^6 years.

Utilising these formulae the estimates of time since divergence we have obtained when comparing the A salina with the parthenogenetic populations are 4.2×10^6 years from the first, and 15.83×10^6 years from the second method. It may be dangerous to take either of these estimates at face value, particularly as insufficient geological data on Anostracans is available. Furthermore, generation length (which is considerably shorter in Artemia than in any of the taxa for which the second formula has been found to correlate well with geologic data) will have an important effect on estimates of rates of divergence between lineages [51,52]. In our case it will lead to overestimation of divergence time when compared to a species with much longer generation length. Unfortunately, the specific correction factors for disparate generation lengths are not available.

However, the calculation of time since divergence between the A salina and parthenogenetic lineages, though admittedly controversial and subject to inaccuracies, may allow us to speculate on the possible scenario leading to the establishment of parthenogenesis in this genus. Geological evidence suggests that about 6×10^6 years ago the processes of mountain building in the pre-Mediterranean region led to closure of the straits from the Atlantic Ocean and eventually this led to the severe and extensive desiccation of the Mediterranean basin [53]. This would inevitably have created opportunities for colonisation in a dispersed manner by highly salt-tolerant species such as Artemia. Under such conditions the establishment of a parthenogenetic mode of reproduction would be selectively advantageous [54]. The advantages of parthenogenesis in conditions of low population density and relatively open habitat have been stressed by White [54,55], and Artemia might be added to the examples of terrestrial forms he gives. As mentioned above, an analogous process has been suggested to explain the creation of the parthenogenetic mode of reproduction in weevils.

One conclusion not available from previous work which we may draw from our results [10] is that the parthenogenetic lineage in this genus as a whole is monophyletic rather than polyphyletic. A general trend of monophyletic derivation of parthenogenesis has also been observed from other studies [47].

Further conclusions related to the speciation processes in parthenogenetic Artemia can also be drawn. Genetic distance data for the triploid populations suggest the possibility that they have originated independently at different times and places in the past. Such origins would have occurred either through an aberrant automictic event taking place in a diploid parthenogenetic line [10] (autopolyploidy), or through a successful fusion between a parthenogenetic nucleus and a reduced nucleus from a related male Artemia (allopolyploidy). However, the present data do not allow us to discriminate between these two possibilities. On the other hand, the very close genetic similarity seen between the tetraploid and the pentaploid populations does strongly suggest a derivation of the pentaploids from a tetraploid line. Again one may envisage an allopolyploidisation event involving a male gamete and a parthenogenetic nucleus (tetraploid in this case). Although in most cases of the involvement of male gametes in the production of heteroploid lineages [47] one is dealing with males derived from closely related bisexual species, in the case of Artemia one should also consider the involvement of males which occasionally are produced by parthenogenetic clones [10] and which have been reported even to be fertile with bisexual females [56]. Both the estimates of D and those of t have large standard errors, and we would not wish to attach too great a significance to the relationships between them. However, it does seem that our supposition as to the way in which the parthenogenetic mode may have become established is not entirely unrealistic.

Judging from the data available at present, the pattern of descent in Artemia appears to have involved the following steps in order of sequence:

1. Separation of New World and Old World bisexual lines.
2. Separation of the persimilis and franciscana lines.
3. Origin of parthenogenetic diploids from salina line.
4. Origin of tetraploid from diploid parthenogenetic line.
5. Origin of pentaploid from tetraploid line.

The triploids would appear to have arisen independently from diploids on several occasions but it is not clear whether they preceded the origin of the tetraploid line.

The processes of speciation in Artemia as a whole have evidently involved a number of different processes or modes differing considerably in their effects upon the genomes involved. Despite the unusual aspects of the biology of Artemia (which include the feature of a small number of bisexual species in the genus) there seems little reason to suppose that the brine shrimp is atypical in the variety of speciational modes employed. Hence, we agree with the view of White [55] and Templeton [57] that the mechanisms of speciation are manifold.

SUMMARY

The species problem in Artemia is briefly reviewed. The results of an electrophoretic survey of 48 samples derived from cysts collected over a very wide geographic area are described. Samples included A salina, A persimilis, and A franciscana, and many parthenogenetic forms distributed over four ploidy levels.

A franciscana is characterised by considerable geographic differentiation ($\overline{D}$ = 0.106) compared with A salina ($\overline{D}$ = 0.072). Population subdivision tested by Nei's G_{ST} statistic gave similar results. Values of $\overline{H}_e$ were found to be strongly correlated with habitat type in A franciscana. The mean D value for A franciscana–A persimilis is 0.978 while that for A franciscana–A salina is 1.950. Chromosome counts from the three bisexual species were 2n = 42 (franciscana), 44 (persimilis), 42 (salina) confirming earlier work. A franciscana and A persimilis are the only species to show chromocentres in chromosome preparations.

In parthenogenetic populations genetic variation as measured by number of genotypes (clones) appears to be greater in diploids than in polyploids while genetic diversity as measured by $\overline{H}_e$ increased considerably and significantly with ploidy level. Only euploid numbers of 42, 63, 84, and 105 were found in chromosomal studies representing a polyploid series where n = 21. Genetic distance results suggest 1) that parthenogenetic Artemia have a monophyletic origin, 2) that the ancestral form was derived from a common ancestor with A salina, 3) that tetraploids originated from diploids, 4) that pentaploids are derived from tetraploids, and 5) triploids have arisen in several independent events from diploids. The origin of parthenogenesis is tentatively related to geological events in the Mediterranean basin of perhaps 6 million years ago.

The genetic processes involved in speciation in Artemia are seen to be multimodal despite the small number of bisexual species in the genus. It seems probable that the processes which lead to speciation in other, more richly speciose lines, will also be many and various, although specific groups may tend to utilise a particular mode preferentially.

ACKNOWLEDGMENTS

We are deeply indebted to Sarane T. Bowen, N. Collins, F. A. Domenech, G. MacDonald, R. Nobili, A. Ramirez, P. Sorgeloos, G. Wallis, and R. D. Ward for help of various kinds, particularly in the supply of cysts. Professor

Barigozzi gave us valuable advice and the details of his unpublished technique for chromosome preparation for which we are most grateful. F. A. Abreu-Grobois acknowledges the support of CONACYT, Mexico City in the provision of a scholarship.

The work reported in this paper was carried out within the framework of the International Study on Artemia.

REFERENCES

1. Lloyd AT: The Salterns of the Lymington area. Hants Field Club Arch Soc Proc 24: 1, 1967.
2. Croghan PC: The osmotic and ionic regulation of Artemia salina (L). J Exp Biol 35: 219, 1958.
3. Conte FP: Neck organ of Artemia salina nauplii, a larval salt gland. J Comp Physiol 80: 239, 1972.
4. Bowen ST, Lebhenz HG, Poon M-C, Chow VHS, Grigliatti TS: The haemoglobins of Artemia salina I. Determination of phenotype by genotype and environment. Comp Biochem Physiol 31: 733, 1969.
5. Gilchrist BM: The oxygen consumption of Artemia salina (L) in different salinities. Hydrobiologia 5: 54, 1954.
6. Grainger JNR: First stages in the adaptation of poikilotherms to temperature change. In Prosser CL (ed): Physiological Adaptation. Washington D.C.: American Physiological Society, 1958, p 79.
7. Persoone G, Sorgeloos P: General aspects of the ecology and biogeography of Artemia. In Persoone G, Sorgeloos P, Roels C, Jaspers E (eds): The Brine Shrimp Artemia. Wetteren, Belgium: Universa Press, 3:3, 1980.
8. Sorgeloos P: Life history of the brine shrimp Artemia. In Persoone G, Sorgeloos P, Roels C, Jaspers E (eds): The Brine Shrimp Artemia. Wetteren, Belgium: Universa Press, 1, XIX, 1980.
9. Barigozzi C: Il legame genetico fra i biotipi partenogenetici di Artemia salina. Arch Ital Zool 22: 33, 1935.
10. Barigozzi C: Artemia. A survey of its significance in genetic problems. Evolutionary Biology 7: 221, 1974.
11. Artom C: Richerche sperimentali sul modo di riprodursi dell'Artemia salina Lin. di Cagliari. Biol Zentralbl 26: 26, 1906.
12. Artom C: Analisi comparativa della sostanza cromatica nelle mitosi di maturazione e nelle prime mitosi di segmentazione dell'uovo dell'Artemia sessuata di Cagliari (univalens) e dell'uovo dell'Artemia partenogenetica di Capodistria (bivalens) Arch f Zell 7: 277, 1911.
13. Gilchrist B: Growth and form of the brine shrimp Artemia salina (L). Proc Zool Soc 134: 221, 1960.
14. Clark LS, Bowen ST: The genetics of Artemia salina. VII Reproductive isolation. J Hered 67:385, 1976.
15. Halfer Cervini AM, Piccinelli M, Prosdocimi T, Baratelli Zambruni L: Sibling species in Artemia (Crust. Branchiopoda). Evolution 22: 373, 1968.

16. Piccinelli M, Prosdocimi T: Descrizione tassonomica delle due specie Artemia salina L e Artemia persimilis n sp. Rend Ist Lomb Sci Lett B102: 170, 1968.
17. Bowen ST, Davis ML, Fenster SR, Lindwall GA: Sibling species of Artemia. In Persoone G, Sorgeloos P, Roels C, Jaspers E (eds): The Brine Shrimp Artemia. Wetteren, Belgium: Universa Press, 1:155, 1980.
18. Bossuyt E, Sorgeloos P: Technological aspects of the batch culturing of Artemia at high densities. In Persoone G, Sorgeloos P, Roels C, Jaspers E (eds): The Brine Shrimp Artemia. Wetteren, Belgium: Universa Press, 3:133, 1980.
19. Abreu-Grobois FA, Beardmore JA: Genetic differentiation in bisexual Artemia (in preparation).
20. Abreu-Grobois FA, Beardmore JA: International Study on Artemia II. Genetic characterisation of Artemia populations: An electrophoretic approach. In Persoone G, Sorgeloos P, Roels C, Jaspers E (eds): The Brine Shrimp Artemia. 1: 133–153. Wetteren, Belgium: Universa Press, 1: 133, 1980.
21. Nei M: Analysis of gene diversity in subdivided populations. Proc Natl Acad Sci USA 70: 3321, 1973.
22. Nei M: Molecular Population Genetics and Evolution. Amsterdam: North-Holland, 1975.
23. Chakraborty R: A note on Nei's measure of gene diversity in a substructured population. Humangenetik 21: 85, 1974.
24. Nei M, Roychoudhury AK: Sampling variances of heterozygosity and genetic distance. Genetics 76: 379, 1974.
25. Nei M: Genetic distance between populations. Am Natur 106: 283–292, 1972.
26. Abreu-Grobois FA, Beardmore JA: Evolution in parthenogenetic Artemia. (in preparation).
27. Wishart D: Clustan computer program. Edinburgh: Program Library Unit, University of Edinburgh, 1978.
28. Sokal RR, Michener CO: A statistical method for evaluating systematic relationships. Kansas Univ. Sci Bull 38: 1409, 1958.
29. Cole GA, Brown RJ: The chemistry of Artemia habitats. Ecology 48: 858, 1967.
30. Bowen ST: The genetics of Artemia salina IV. Hybridization of wild populations with mutant stocks. Biol Bull 126: 333, 1964.
31. Skibinski DOF, Ward RD: Relationship between allozyme heterozygosity and rates of divergence. Genet Res Camb Genet Res 38:71, 1981.
32. Koehn RK, Eanes WF: Molecular structure and protein variation within and among populations. Evol Biol 11: 39, 1978.
33. MacDonald GH: The use of Artemia cysts as food by the flamingo (Phoenicopterus ruber roseus) and the shelduck (Tadorna tadorna). In Persoone G, Sorgeloos P, Roels O, Jaspers E (eds): The Brine Shrimp Artemia. Wetteren, Belgium: Universa Press, 3:97, 1980.
34. Loudenslager EJ, Gall GAE: Geographic patterns of protein variation and subspeciation in cutthroat trout Salmo clarki. Syst Zool 29: 27, 1980.
35. Ayala FJ, Tracey ML, Barr LG, MacDonald JF, Perez-Salas S: Genetic variation in natural populations of five Drosophila species and the hypothesis of the selective neutrality of protein polymorphisms. Genetics 77: 343, 1974.
36. Sneath P, Sokal RR: Numerical Taxonomy. San Francisco: W. H. Freeman, 1973.

37. Croghan PC: The survival of Artemia salina (L) in various media. J Exp Biol 35: 213, 1958.

38. Carson HL, Johnson WE: Genetic variation in Hawaiian Drosophila 3. Allozymic and chromosomal similarity in 2 Drosophila species. Proc Natl Acad Sci USA 72: 4521, 1975.

39. Templeton AR: Modes of speciation and inferences based on genetic distances. Evolution 34: 719, 1980.

40. Lenz, PH: Ecology of an alkali-adapted variety of Artemia from Mono Lake, California, USA. In Persoone G, Sorgeloos P, Roels O, Jaspers E (eds): The Brine Shrimp Artemia. Wetteren, Belgium, Universa Press, 3:79, 1980.

41. Parker ED Jr, Selander RK: The organisation of genetic diversity in the parthenogenetic lizard Cnemidophorus tesselatus. Genetics 84: 791, 1976.

42. Lokki J, Saura A: Polyploidy in insect evolution. In Lewis WH (ed): Polyploidy—Biological Relevance. New York: Plenum, 1979, p 277.

43. Lokki J, Suomalainen E, Saura A, Lankinen P: Genetic polymorphism and evolution in parthenogenetic animals II. Diploid and polyploid Solenobia triquetrella (Lepidoptera: Psychidae). Genetics 79: 513, 1975.

44. Hamrick JL, Allard RW: Microgeographical variation in allozyme frequencies in Avena barbata. Proc Natl Acad Sci USA 69: 2100, 1972.

45. Sorgeloos P, Baeza-Mesa M, Benijts F, Persoone G: Research on the culturing of the brine shrimp Artemia salina L. at the State University of Ghent (Belgium). In Persoone G, Jaspers E (eds): 10th European Symp Mar Biol 1, Wetteren, Belgium: Universa Press 473, 1975.

46. Mayr E: Populations, Species, and Evolution. Cambridge, USA: Belknap Press, 1970.

47. Suomalainen E, Saura A, Lokki J: Evolution of parthenogenetic insects. Evol Biol 9:209, 1976.

48. Sarich VM, Cronin VE: Molecular systematics of the primates. In Goodman, Tashian (eds): Molecular Anthropology, New York: Plenum, 1977, p 141.

49. Vawter AT, Rosenblatt R, Gorman GC: Genetic divergence among fishes of the eastern Pacific and the Caribbean—support for the molecular clock. Evolution 34: 705, 1980.

50. Yang SY, Soule M, Gorman GC: Anolis lizards of the eastern Caribbean: A case study in evolution 1. Genetic relationships, phylogeny, and colonization sequence of the roquet group. Syst Zool 23: 387, 1974.

51. Lovejoy CO, Burstein AH, Heiple KG: Primate homology and immunological distance. Science 176: 803, 1972.

52. Korey KA: Species number, generation length, and the molecular clock. Evolution 35: 139, 1981.

53. Hsü KJ, Montadert L, Bernoulli D, Cita MB, Garrison RE, Kidd RB, Melieres F, Müller C, Wright R: History of the Mediterranean salinity crisis. Nature 267: 399, 1972.

54. White MJD: Heterozygosity and genetic polymorphism in parthenogenetic animals. In Hecht MK, Steere WC (eds): Essays in Evolution and Genetics in Honor or Theodosius Dobzhansky. New York: Appleton-Century-Crofts, 1970, p 237.

55. White MJD: Modes of speciation. San Francisco: WH Freeman, 1978.

56. Bowen ST, Durkin JP, Sterling G, Clark W: Artemia haemoglobins—genetic variation in parthenogenetic and zygogenetic populations. Biol Bull 155: 273, 1978.

57. Templeton AR: Mechanisms of speciation—A population genetic approach. Ann Rev Ecol Syst 12: 23, 1981.

acters. Differences, for instance, have been detected in the degree of polymorphism, in its type of genetic control, in physiological adaptations, in certain population dynamics parameters, in sex determination, in adaptive and reproductive strategies. Differences of a similar kind and others such as, for instance, those concerning concealed genetic variability, have also been found between different populations of the same species [7–10]. On the other hand, morphological differentiations are often barely conspicuous; in some cases they are almost absent even between good species.

The other genus considered, Tigriopus, is also widely distributed but represented by a much smaller number of species compared with Tisbe. Moreover, Tigriopus is a narrow ecological specialist, its distribution being exclusively confined to the rock- or tide-pool environment [9].

Cases of Incipient Reproductive Isolation: Tisbe reticulata and Tisbe dobzhanskii

The species T reticulata is characterized by a striking color polymorphism, which is genetically controlled. Observations in the field and laboratory experiments have indicated the adaptive and balanced nature of this polymorphism. Temperature and salinity seem to play a major role as selective agents [11,12].

The only two geographic populations available come from Roscoff, Northern France, and from the Lagoon of Venice, Italy. They differ as to degree of polymorphism, the Atlantic population consisting of seven morphs and that from Venice of only three. Moreover, at Roscoff polymorphism is under the control of a few genes, some of which are independent, whereas at Venice it is more simply controlled by a series of alleles of the same locus.

The two geographic populations are interfertile, but there is evidence of an incipient reproductive isolation: the F_1 hybrids produced by interpopulation crosses exhibit lower fecundity and viability, as well as strong deviations of sex ratio in favor of males [13].

Similar deviations of sex ratio were observed in another species, Tisbe dobzhanskii, in crosses between a population from Anzio, Italy, and one from Tunis, North Africa. Two thirds of these crosses produced only male offspring. In Tisbe, sex ratio alterations are believed to reflect also various levels of genetic incompatibility [13].

The above observations have led us to search for other possible cases of reproductive isolation and to investigate the mechanisms responsible for the arising of barriers.

Tisbe reticulata had to be abandoned, due to its limited geographic distribution and rare occurrence. Other species proved, instead, much better material for this type of research.

The results of an intensive program of cross-breeding experiments, utilizing populations from the Atlantic and Mediterranean coasts of Europe and from the

East coast of North America, have pointed out the existence in Tisbidae of a wide range of isolation levels. In addition to species such as T reticulata, where the isolation is only at an incipient stage, there are others characterized by a complete lack of reproductive barriers even between geographically remote populations. The opposite extreme is represented by species whose populations, even when geographically close to each other, are separated by a high degree of genetic incompatibility.

The case of species lacking reproductive barriers will be discussed first. It is exemplified by the species Tisbe holothuriae.

The case of Tisbe holothuriae: No barriers to gene-flow. This species has an almost cosmopolitan and ubiquitous distribution. In the northern hemisphere, it is perhaps the most widespread species, with a practically continuous distribution along the coasts of the continents where it has been found. The species can occupy a variety of ecological niches: Populations of remarkable size and density can be collected in places which differ considerably with regard to temperature and salinity. The different populations are genetically adapted to the local ecological features. A number of physiological races have been described, characterized by a differential ability to tolerate osmotic shocks [8]. Experiments have shown that these differences have a genetic basis. Genetic differences in rate of development have also been detected in populations from different latitudes [14]. The generation time of T holothuriae is the shortest (ca 12 days) of all the species considered, and females produce a large number of offspring per egg sac.

The species does not show any visible polymorphism, but its genetic variability estimated by electrophoretic techniques is quite high (Table I). In comparing populations from typical marine habitats (eg, Banyuls) with populations collected in brackish water basins (eg, Sigean), differences in gene frequencies have been detected (Table II). Moreover, moving from comparatively stable marine habitats towards progressively unstable, and therefore ecologically demanding, brackish water environments, genetic variability as measured by electrophoretic techniques undergoes a slight drop. This trend becomes more obvious when comparisons concern the concealed genetic variability (Table III).

The above results suggest that the adaptive strategy of Tisbe holothuriae must be based upon a number of mechanisms, among which individual flexibility and genotypic plasticity are likely to play the major role. This strategy is perhaps the optimal compromise for the species to conquer a broad spectrum of ecological niches, spatial as well as temporal, physical as well as biotic. The high genetic load of ecological origin is probably the cost imposed for a strategy of this kind.

All crosses carried out between various Mediterranean populations are perfectly fertile, with signs of heterosis in the F_1. The same applies to crosses between North European or Mediterranean populations and populations from the east coast of North America. In these cases too the viability of interpopulation

TABLE I. Summary of Electrophoretic Variation in Tisbe holothuriae, Tisbe clodiensis, and Tigriopus sp.

Parameter	Species and locality			
	T hol. (Banyuls)	T hol. (Sigean)	T clod. (Banyuls)	Tigriopus sp. (Livorno)
Number of enzymes	10	10	10	13
Number of loci	19	19	15	19
Number of individuals	672	710	590	1893
Mean	132.22	136.16	121.60	339.58
genes sampled per locus	± 10.09	± 7.57	± 6.82	± 26.77
Mean	2.158	2.000	2.000	1.158
alleles per locus	± 0.268	± 0.268	± 0.276	± 0.115
Polymorphic loci (%) (crit of 5%)	57.90	49.99	40.33	10.52
H (gene diversity, according to Nei)	0.241 ± 0.058	0.240 ± 0.061	0.181 ± 0.063	0.054 ± 0.036

TABLE II. Local Distribution of Some Allelic Frequencies in Tisbe holothuriae

Locus	Alleles	Banyuls (marine)	Sigean (brackish)
Aph-1	b	0.897	0.671
Adh-1	b	0.730	0.431
Me-1	a	0.300	0.728

The lagoon of Sigean is situated at about 40 miles north of Banyuls.

TABLE III. Estimates of Genetic Load in Different Populations of Tisbe holothuriae

Locality	b
Split	2.265
Banyuls	1.897
Alberoni	1.779
Malamocco	1.244
Sigean	1.067
Grado	1.065

b, Lethal equivalents.
The localities in the list range from typical marine conditions to progressively more brackish lagoon environments.

F_1 offspring is slightly higher than in intrapopulation controls [15]. Similar results were obtained in transatlantic crosses of other species, namely T bulbisetosa, T lagunaris, and T battagliai.

Speciation in fieri: The superspecies Tisbe clodiensis. The copepod Tisbe clodiensis is a typically coastal marine species. Unlike T holothuriae its distribution, although wide, is rather patchy or discontinuous. This is perhaps due to its lower tolerance of extreme ecological conditions, such as low salinities or salinity and temperature fluctuations. Whereas T holothuriae is the dominant species in brackish water lagoons, the relative abundance of T clodiensis is greater in typical marine habitats. However, the species can occupy niches in lagoon environments, where abundant food can be found, but only in waters physicochemically closer to those of marine habitats.

Its life-cycle parameters and population statistics do not differ significantly from those described for T holothuriae. Genetic variability, estimated in a population from Banyuls-sur-Mer by means of electrophoretic techniques, is slightly reduced compared with the values obtained in T holothuriae.

The species exhibit a colour polymorphism under a simple mendelian control. As in T reticulata, polymorphism is adaptive and balanced. As selective agents, biotic factors are at least equally active as physicochemical factors [16,17]. Adaptive values are strongly affected by interspecific competition, as well as by density-dependent factors [18]. Unlike T holothuriae, the various populations are believed to undergo, from time to time, severe bottlenecks.

Laboratory crosses between several Mediterranean populations of T clodiensis led to results unexpectedly different from those obtained in T holothuriae. In fact, while some crosses were fertile, in other no hybrids were produced. The effect appeared to depend on the geographic origin of the female or male, and it was found that reproductive success within this species ranges from full interfertility to almost complete intersterility [19]. Cross-breeding experiments were then extended to a larger number of strains, obtained by sampling Atlantic populations from Europe and from the east coast of North America (Fig. 1). A summary of the crosses involving four Mediterranean, two Eastern Atlantic, and one Western Atlantic population is given in Table IV. In interpopulation crosses characteristic differences were found showing basically three kinds of results: 1) Adult offspring obtained in both directions; 2) adult offspring only in one direction; 3) no offspring or dying larvae in both directions. In case 2, development of the hybrids was greatly retarded, the number of progeny per female was lower than in the controls, and the sex-ratio strongly shifted towards males. Non-reciprocal incompatibility, or even incompatibility in both directions, was more pronounced in all crosses involving the Venice and Arcachon populations. The Venice strain showed the highest degree of isolation from others when the male was used, whereas in the Arcachon strain the reproductive barrier was

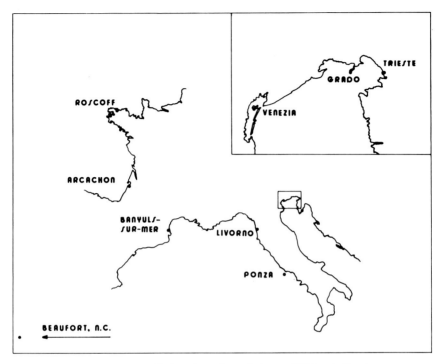

Fig. 1.   Geographic location of populations of Tisbe clodiensis.

TABLE IV. Crossability Between Strains of Different Geographic Origin of Tisbe clodiensis

♀	♂						
	V	B	P	L	R	A	BE
Venezia	+	(+)	(+)	(+)	(+)	(+)	C
Banyuls	O	+	+	+		(+)	O
Ponza	N	+	+	+	+	O	+
Livorno	O	+	+	+			
Roscoff	O		+		+	(+)	+
Arcachon	N	N	N		N	+	N
Beaufort, NC	O	N	+		+	(+)	+

+, Perfect compatibility; (+), non-reciprocal incompatibility; O, no hybrids; N, hybrids dead at nauplius stage; C, hybrids dead at copepodite stage. Empty spaces indicate lacking crosses.

greater utilizing the female. The results of backcross experiments suggest that incompatibility is controlled directly by at least one nuclear gene or, more likely, by a multiple system of incompatibility factors responsible for the various degrees of failure. No ethological barriers were detected between the various strains. Even in those crosses that yielded no viable offspring, mating took place regularly and there was no sign of discrimination between partners of different origin.

It should also be mentioned that all the geographic populations utilized in our experiments are more or less indistinguishable morphologically. Also in the case of completely intersterile populations there are no appreciable differences in conventional taxonomic characters.

On the basis of the above results, it was concluded that the case of Tisbe clodiensis is in many ways similar to that of Drosophila paulistorum. However, the mechanisms responsible for the breaking up of the superspecies in partly reproductively isolated strains, and for their maintenance, must be somewhat different. It seems legitimate to assume that also T clodiensis represents a cluster of semispecies, or species in statu nascendi. This is supported by the results of a biochemical comparison between the various populations carried out by Bisol [20]. The index of biochemical affinity lies between those generally regarded as relating to geographic races and those characterizing sibling species, the figures showing a clear cut differentiation.

Finally, our results indicate that geographic distance per se should not play an important role in the process of isolation and genetic differentiation of T clodiensis. Further evidence has recently been provided by Fava [21], who collected two populations, geographically very close to each other but exhibiting a high degree of genetic incompatibility. They came, respectively, from the brackish lagoon of Grado and from a marine station off Trieste (Fig. 1). The results of crosses were as follows:

♀ Trieste × ♂ Grado ---> from no mating observed to undeveloped egg sacs

♀ Grado × ♂ Trieste ---> from non-extrusion of egg sac to hybrids at copepodite stage

This situation will be further discussed in the last section of this paper.

Reproductive Isolation in the Genus Trigriopus.

The copepod Trigriopus occupies the most demanding and severe habitat which can be found in the marine environment: that of the rock- or tide-pools. The physiological adaptability of this organism is extraordinary. The species can live in conditions ranging from nearly freshwater up to salinity oversaturation; this is made possible through a sort of dormancy from which the animals recover when conditions become more favorable [22].

In Tigriopus from Northern Europe the reproductive stage is reached in about 20 days. The ability to reproduce is maintained for at least three months, and mean longevity (ca 210 days) exceeds that of all the species of harpacticoids reared in the laboratory [23].

The genetic variability of Tigriopus, as indicated by the enzyme data reported in Table I, is the lowest in the species and populations considered.

Observations and experiments carried out in samples from a series of rock pools (Livorno) of various sizes, situated at different sea levels or at the same level but at various distances from each other, indicate the negligible role of drift as a factor responsible for the low genetic variability detected in Tigriopus [9]. It was, therefore, argued that the prevailing adaptive strategy of this species consists in the fixation of alleles conferring the highest possible individual flexibility in environments which are extremely challenging in their short term fluctuations of physicochemical parameters. This agrees with the view, shared by several authors, according to which in challenging habitats little genetic variation may be permitted, natural selection acting in favor of a few genotypes underlying flexible phenotypes.

Cross-breeding experiments were carried out utilizing seven geographic populations from the Atlantic and the Mediterranean Coasts of Europe: Tayvallich (T), Scotland; Isle of Man (M), West Britain; Aber vach (A), Northern France; Livorno (L), Tyrrhenian coast of Italy; Rijeka (R), Upper Adriatic; Gargano (G), Lower Adriatic; Egnazia (E), Lower Adriatic (Fig. 2).

A detailed analysis of the crosses, together with the relevant technical information, will be given elsewhere [24]. A schematic representation of the main results obtained is provided in Figure 3 labeled with the letter abbreviations given above.

In Tigriopus sp, as well as in Tisbe clondiensis, reproductive success between the different populations ranges from full interfertility to complete intersterility. The effect of crosses appears to depend on the geographic origin of the female or male.

Three groups of populations can be distinguished: 1) North European (T, M, and A), with perfect reciprocal compatibility; 2) Adriatic (R, G, and E): perfect compatibility, as in the previous case; 3) Tyrrhenian (L).

The latter population shows complete reproductive isolation from groups 1 and 2; however, the results of crosses are not identical. Let us consider first Mediterranean populations: In crosses ♀ L × ♂ R, G, or E, sterile adults were obtained, whereas the offspring of the reciprocal crosses died at early larval stages (nauplius or, occasionally, copepodite). The degree of compatibility between the Tyrrhenian and the North European populations is considerably reduced: no offspring were yielded in reciprocal crosses L × T. In crosses L × M and L × A, no offspring were produced in one direction, and offspring which died at the nauplius stage, in the other.

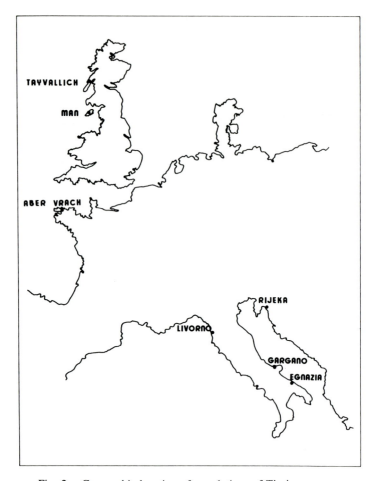

Fig. 2. Geographic location of populations of Tigriopus sp.

The last crosses to consider are those between the Adriatic and the North European populations. Each Northern strain was crossed with each Adriatic strain. All crosses produced, in both directions, offspring which survived to the nauplius or the copepodite stage. These results indicate a reduced degree of incompatibility between the Adriatic and the North European populations compared with that found between the North European and the Tyrrhenian populations. This is supported by the preliminary data of a comparison based on the distributions of enzyme variability. The biochemical results agree with those obtained by cross-breeding experiments and suggest that geographic distance is largely, although not exclusively, responsible for the observed differentiations.

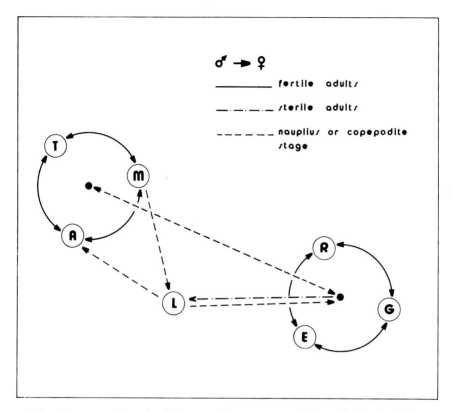

Fig. 3.   Summary schematic of the crosses between geographic populations of Tigriopus sp.

The value of genetic affinity within each population group, Adriatic and North European, is very close to 1 whereas it decreases considerably in the comparison between groups. However, of the three Adriatic strains, Rijeka is the one which exhibits the highest affinity with the Northern group, suggesting that factors other than distance per se must be involved in the differentiation pattern observed.

DISCUSSION AND CONCLUSIONS

Materials suitable for genetic analysis of early stages of speciation are rare. In the terrestrial environment such materials are provided, for instance, by some species of the genus Drosophila, in particular D willistoni and D paulistorum, investigated for a long time by Dobzhansky and his associates. In the marine environment, similar opportunities are offered by certain species of the Harpacticoid Copepods Tisbe and Tigriopus, which are suitable for analysis of reproductive isolation within a species. A series of investigations carried out in recent years on Tisbe has shown that relatively little interspecific morphological differentiation has occurred, which makes the taxonomy of this genus particularly complex [25,26].

The detection of reproductive barriers by means of crossbreeding experiments indicates that certain populations previously considered as belonging to the same species actually belong to different species. In fact, increasing numbers of sympatric or allopatric sibling species groups and closely related species within the genus are now being recognized. On the other hand, it has generally been found that different strains of Tisbe are able to mate and produce fertile offspring only when they belong to the same species, regardless of their geographic origin.

As described in the previous section, in Tisbe, in addition to species characterized by a complete lack of reproductive barriers, there are others whose populations are separated by various levels of genetic incompatibility. This condition is exemplified by Tisbe clodiensis, which has been considered as a cluster of semispecies, or species in statu nascendi [19].

A tentative hypothesis to account for the origin of the situation observed in Tisbe clodiensis could be suggested by the following scenario. Let us imagine a temperate Atlantic species, endowed with a rich genetic variability, which has sometime in the past colonized the Mediterranean Sea. There the species had to cope with a particularly heterogeneous environment, well diversified in time and space as to a wide range of conditions: climatic, hydrographical, geological, and biological. The Mediterranean habitat provides a multiplicity of barriers, physical as well as biological (articulated coast lines, river mouths and estuaries, biotic diversity also due to recurrent invasions by northern and tropical forms), and therefore offers good opportunities for speciation.

The results of our crosses suggest that Beaufort, Roscoff, and Ponza are among the localities where the "ancestral" population is still represented. At least two of the poorly differentiated strains collected there are to a certain extent comparable with the transitional races described by Dobzhansky et al. [27] in Drosophila paulistorum. They may in fact serve as channels through which the gene exchange among all the incipient species may take place.

In the Mediterranean the Banyuls strain, though perfectly compatible with Livorno and Ponza, is separated from Beaufort. The degree of incompatibility increases remarkably in all crosses involving the strain from Venice, that is reproductively isolated from all other Mediterranean populations.

A similar situation can be observed for the two Atlantic populations of Europe, Arcachon and Roscoff, displaying a non-reciprocal incompatibility. Therefore, Arcachon and Venice appear as the more isolated and differentiated populations. What they have in common is a similar brackish lagoon habitat and the alluvial basin character of the localities from which they are derived. In both areas, the presence of rivers and the consequential diversified hydrological properties and currents regimes might have played a significant role in isolating barriers.

In the previous section, we have seen that distance per se does not seem to play an important role in Tisbe clodiensis as an isolating factor. In fact, the population from Grado and that from Trieste, although geographically very close to each other, are reproductively isolated. In this case some signs of premating incompatibility were also detected. Once again the situation recalls that observed in Drosophila, where such an isolating mechanism proved efficient only where different races come together.

Concerning the unpredictable results of certain crosses, it should be remembered that, in contrast to Tisbe holothuriae, Tisbe clodiensis is an organism which very rarely occurs in great numbers inside brackish water lagoons, where it inhabits only the outer parts. It is in fact less tolerant of salinity and temperature variations. Moreover, the species is very sensitive to the selective action of the biotic components of the environment; it has a very discontinuous distribution and is subject to wide fluctuations in population size. This, in absence of gene flow, might induce divergencies independently of distance and, in part, also of ecological conditions. The environment, however, should play an important role considering that the more differentiated populations happen to be those from Venice and Arcachon, both dwelling at the edges of brackish water lagoons with rather similar ecological features.

A further indication that the species T clodiensis is in a rather labile state, with high potentialities for diverging even within a restricted area, emerges from the following result: the equilibrium frequencies of the allele p (a gene controlling color polymorphism) are significantly different in two populations collected from localities geographically very close to each other, namely Venice and Grado [Fava, unpublished]. Both places belong to an area which originally was a continuous system of coastal lagoons. An analysis conducted at a smaller scale in this critical area, where very incompatible populations, such as those from Trieste and Grado, almost come into contact, could be quite promising.

It might also be interesting to search for other "lagoonoid" strains in order to test the hypothesis that the higher levels of differentiation and reproductive incompatibility are features of common occurrence in populations of Tisbe clodiensis inhabiting these areas.

The cases of reproductive isolation detected in Tigriopus, where between geographic populations there are various levels of unilateral or bilateral incompatibility, is not dissimilar from that described in Tisbe clodiensis.

According to Božić [28], the populations from the Northern and Western coasts of Europe and those from the Mediterranean Sea would belong to two different species: Tigriopus brevicornis and T fulvus, respectively. They were considered thus far as belonging to a single species, but the results of cross-breeding experiments, showing a complete reproductive barrier between the two sets of populations, and some morphological comparisons, have induced Božić to classify them as distinct specific units. The situation, however, is complicated by the presence of reproductive barriers between the various Mediterranean strains, which are also quite heterogeneous morphologically.

Our results confirm in part those by Božić, but they do not support his taxonomical conclusions. The finding of a comparatively lower incompatibility between the Northern strains and the Adriatic ones, which are geographically the remotest from each other, and the surprising biochemical affinity between the Rijeka strain and the Northern group of populations suggest that in Tigriopus, as in Tisbe clodiensis, we are probably dealing with a cluster of semispecies rather than with more taxonomic entities having achieved the status of full species. Moreover, the preliminary results of recent investigations seem to indicate, also in Tigriopus, the existence of bridging populations.

As to the greater diversification of the species inside the Mediterranean, this could be ascribed to the same factors responsible for the higher heterogeneity of Tisbe clodiensis in the same area. Another factor could be the possible occurrence of repeated migrations in the Mediterranean by Tigriopus of Northern and Southern origin. Numerous cases are known of species which have repeatedly invaded the Mediterranean Sea during the climatic fluctuations of the Quaternary period. The Upper Adriatic is one of the places where a relic fauna of northern character seems to be still present.

In conclusion, the three species which were the object of the present investigations exhibit different adaptive strategies, which helps to understand their dissimilar distribution patterns. No sign whatsoever of speciation in fieri has been detected in Tisbe holothuriae, characterized by a continuous gene-flow. Tisbe clodiensis and Tigriopus sp. show, instead, phenomena of incipient speciation where ecology, at least for T clodiensis, seems still to play an active role. In the case of Tigriopus, we are probably dealing with a mechanism based on allopatric speciation, where geographic distance acts as the main diversifying factor. Environmental factors, due to the extreme similarity of the Tigriopus habitats, perhaps have played an active role in the very early stages of the species history, when this peculiar habitat specialization occurred originally.

A mechanism responsible for interruption of gene exchange, and whose apparent consequences have permitted the detection of speciation events in fieri, is the observed "intraspecific relative incompatibility", a step which various other species of Tisbe and Tigriopus have probably accomplished. The occurrence of this phenomenon and the remarkable diversity of life cycle parameters and adaptive strategies even within the same genus, provide useful information on the speciation mechanisms operating in the intertidal zones of the sea.

I hope that the cases just described indicate, at least, that also for marine organisms information and tools are becoming more readily accessible contributing to a better approach to the analysis of evolution in action.

REFERENCES

1. The Systematics Association, Publication No. 5, "Speciation in the sea." Harding JP, Tebble N (eds), 1st Ed, 1963.
2. Mayr E: Geographic speciation in tropical echinoids. Evolution 8: 1–18, 1954.
3. Wieser W: Problems of species formation in the benthic microfauna of the deep sea. In Buzzati-Traverso AA (ed): "Perspectives in Marine Biology." Berkeley and Los Angeles, University of California Press, pp 513–518, 1958.
4. Kohn AJ: Problems of speciation in marine invertebrates. In Buzzati-Traverso AA (ed): "Perspectives in Marine Biology." Berkeley and Los Angeles, University of California Press, pp 571–588, 1958.
5. Marshall NB: Diversity, distribution and speciation of deep-sea fishes. In Harding JP, Tebble N (eds): "Speciation in the sea." London, The Systematics Association, Publication No. 5, pp 181–195, 1963.
6. Bocquet Ch: Les problèmes de l'espèce chez quelques crustacés: le genre Tisbe (Copépodes, Harpacticoïdes) et le complexe Jaera albifrons (Isopodes, Asellotes). In Bocquet Ch, Génermont J, Lamotte M (eds): "Les problèmes de l'espèce dans le règne animal". Paris, Socièté Zoologique de France, Memoire No. 38, pp 307–340, 1976.
7. Battaglia B: Advances and problems of ecological genetics in marine animals. In: "Genetics Today." Proc. XI Internat. Congr. of Genetics, London, Pergamon Press, pp 451–463, 1965.
8. Battaglia B: Genetic aspects of benthic ecology in brackish waters. In: "Estuaries". American Association for the Advancement of Science:574–577, 1967.
9. Battaglia B, Bisol PM, Fava G: Genetic variability in relation to the environment in some marine invertebrates. In Battaglia B, Beardmore JA (eds): "Marine organisms. Genetics, Ecology, and Evolution." New York, Plenum Press, pp 53–70, 1978.
10. Parise A, Lazzaretto I: Misure di popolazione sul copepode Tisbe furcata (Baird). Memorie Accad. Patavina Sci. 79:3–11, 1966.
11. Battaglia B: Balanced polymorphism in Tisbe reticulata, a marine Copepod. Evolution 12: 358–364, 1958.
12. Battaglia B, Lazzaretto I: Effect of temperature on the selective value of genotypes of the Copepod Tisbe reticulata. Nature 215: 999–1001, 1967.
13. Battaglia B: Ecological differentiation and incipient intraspecific isolation in marine Copepods. Ann. Biol. 33: 259–268, 1957.
14. Battaglia B: Il polimorfismo adattativo e i fattori della selezione nel copepode Tisbe reticulata Bocquet. Archo Oceanogr. Limnol. 11: 19–69, 1959.
15. Battaglia B, Volkmann-Rocco B: Geographic and reproductive isolation in the marine harpacticoid Copepod Tisbe. Mar. Biol. 19: 156–160, 1973.
16. Battaglia B, Finco G: Fattori biotici e selezione naturale in copepodi del genere Tisbe. Atti Ist. Ven. SS LL AA 127: 363–370, 1969.
17. Fava G: Effetti selettivi della temperatura in Tisbe clodiensis (Copepoda, Harpacticoida) Atti Accad. Naz. Lincei, Rend. Cl. Sci. FF MM NN 53: 22–27, 1972.
18. Fava G: Studies on the selective agents operating in experimental populations of Tisbe clodiensis (Copepoda, Harpacticoida). Genetica 45: 289–305, 1975.

19. Volkmann B, Battaglia B, Varotto V: A study of reproductive isolation within the super-species Tisbe clodiensis (Copepoda, Harpacticoida). In Battaglia B, Beardmore JA (eds): "Marine Organisms. Genetics, Ecology, and Evolution." New York, Plenum Press, pp 617–636, 1978.

20. Bisol PM: Polimorfismi enzimatici ed affinità tassonomiche in Tisbe (Copepoda, Harpacticoida). Atti Accad. Naz. Lincei, Rend. Cl. Sci. FF MM NN 60: 864–870, 1976.

21. Fava G: Personal communication.

22. Issel R: Vita latente per concentrazione dell'acqua (anabiosi osmotica) e biologia delle pozze di scogliera. Pubbl. Staz. Zool. Napoli 22: 191–254, 1914.

23. Bisol PM, Renier M, Tombolan E, Varotto V: Influenza della temperatura e della salinità su alcuni parametri del ciclo biologico del copepode arpacticoide Tigriopus brevicornis. Atti Accad. Naz. Lincei, Rend. Cl. Sci. FF MM NN 66: 214–222, 1979.

24. Varotto V: Patterns of reproductive isolation in Tigriopus (Copepoda, Harpacticoida) (In preparation).

25. Volkmann-Rocco B: Some critical remarks on the taxonomy of Tisbe (Copepoda, Harpacticoida). Crustaceana 21: 127–132, 1971.

26. Volkmann B: A revision of the genus Tisbe (Copepoda, Harpacticoida). Part I. Archo Oceanogr. Limnol. 19 suppl. : 121–283, 1979.

27. Dobzhansky Th, Pavlovsky O, Ehrman L: Transitional populations of Drosophila paulistorum. Evolution 23: 482–492, 1969.

28. Božić B: Le genre Tigriopus Norman (Copépodes, Harpacticoides) et ses formes européennes; recherches morphologiques et expérimentales. Arch. Zool. exp. gén. 98: 167–269, 1960.

Mechanisms of Speciation, pages 393–410

Evolutionary Biology and Speciation of the Stick Insect Bacillus rossius (Insecta Phasmatodea)

Valerio Scali

INTRODUCTION

The Phasmatodea are an order of land insects of medium or large size, living in tropical and subtropical regions. They have an orthopteroid body structure and are characterized by having the first abdominal segment fused with the metathorax to give the so-called "median segment." The overall body appearance is rather squat or compressed in the most primitive morphs, but generally the slender and frail types predominate. Therefore, the Phasmatodea are commonly named stick insects. The Phasmatodea appear to be very ancient and surely date back to the mesozoic age. Actually, some specimens have been found embedded in the baltic sea amber and the appearance of some of them is very similar to that of today's species.

The taxonomy of the order is not settled. The two suborders, Areolatae and Anareolatae, are classified according to the presence or the absence of the "area apicalis," which is a tiny triangular zone of middle and hind tibiae. This characteristic appears to be rather feeble and sometimes produces unnatural groupings. At any rate, even if one accepts such a distinction, the new classifications of the order differ very much from each other, including high-level taxa [1–3].

The cytotaxonomy of the order is far behind that of other insect orders due to the less favourable conditions for chromosomal studies found in the stick insects. Also, a great variability of the chromosome structure and number has been observed, even among species belonging to the same genus. Such variability appears to affect both autosomes and heterochromosomes, sometimes involving complex or even unexplicable rearrangements in closely related species [4–8].

The reproductive biology of stick insects is rather peculiar, because several instances of parthenogenetic reproduction have been observed among them. Actually, it is possible that the unfertilized females of amphigonic species reproduce by facultative thelytokous parthenogenesis. This type of reproduction is so common that it is very likely all the species share this ability to a certain degree. Only a few, however, reproduce by obligatory ameiotic parthenogenesis [8].

The distribution of parthenogenetic morphs compared to the amphigonic relatives sometimes shows such a pattern that a geographic character of their diffusion is recognizable, as it is in the case of the genera Leptynia and Clonopsis [9–11].

In the Phasmatodea the male sex is heterogametic, its model being XX female and XO male. However, instances of a reversion to XY male are not infrequent. This chromosome condition explains perfectly the thelytoky of both kinds of parthenogenesis.

It has been suggested that obligatory parthenogenetic species are polyploid. It may well be so in some instances, but this remains only a suggestion, since positive and precise karyological evidence is still lacking. This fact, however, seems to be true: Obligatory parthenogenetic morphs have, on the average, more chromosomes than bisexual relatives [8,12–16]. Polyploid parthenogenesis is at present well documented for several insects, including the orthopteran Saga pedo (see [17]), but the same kind of evidence has not yet been obtained in the Phasmatodea.

REPRODUCTIVE BIOLOGY

Bacillus rossius is a circummediterranean species ranging from southern Europe to northern Africa and the Middle East. It feeds on a number of spontaneous plants, but more commonly on wild Rosaceae. It is active at night when it climbs on bushes to feed, mate, and lay eggs. The patchy distribution of food plants and the very low mobility favor the isolation of small demes of this apterous insect, even in the absence of real geographic barriers.

B rossius is parthenogenetic in the northern part of its range and amphigonic in the southern one. Early investigations on this geographic parthenogenesis showed that, in France, populations only consist of thelytokous females. For this reason the species was erroneously thought to reproduce by obligatory parthenogenesis. The subsequent findings, however, revealed that in Italy, along the Tuscan coast, there was passage from unisexual to bisexual populations (Fig. 1). It was shown that in the transition area some populations had an altered sex ratio with females greatly outnumbering males, that these populations exist either in a stable or transitory condition, and also that the parthenogenetic demes may be in close vicinity to the bisexual populations and distributed in a very irregular pattern. South of Tuscany bisexual populations are the rule [18–23].

B rossius parthenogenesis is facultative in two respects: First, parthenogenetic females can be fertilized by males of amphigonic populations even if they are of a very distant zone. This has been observed by crossing north African males

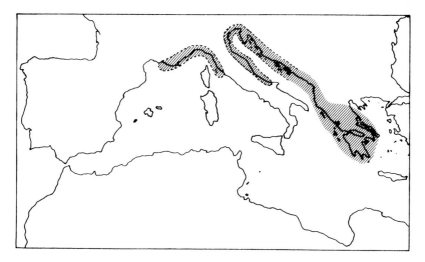

Fig. 1. Map showing main areas where B rossius is parthenogenetic. Along the southern border of the Tirrenian area bisexual populations irregularly intermingle with partheno-genetic demes.

to French females [24,25]; second, virgin females of bisexually reproducing populations lay eggs which develop and give female progeny. Their hatching rate is rather variable, ranging from less than 1% in Sardinian samples [Scali and Masetti, unpublished], to the 68% of continental populations. These figures can be selectively increased up to the levels of unisexually reproducing populations. It is interesting that tuscan bisexual females demonstrate this already at the first generation with a parthenogenetic hatching as high as 80%, which is very close to the fertilized egg rate of about 88% [23].

LIFE CYCLE

Observations on the length of embryonic development have revealed the existence of two kinds of eggs: *rapid,* developing within 50–70 days and *slow,* only hatching after 100–150 days. A second type of slow egg takes up to 200–250 days (Fig. 2). These sharp differences of embryonic development duration are found both in fertilized and unfertilized eggs. The latter, however, take some 12 days longer (on the average), owing to the fact that the diploidization process requires some time, as will be discussed further [23].

Direct and diapaused developments are photoperiodically controlled. Spring and summer long photoperiods experienced by laying females induce non-dia-pause eggs, while successive autumnal short photoperiods cause diapause egg

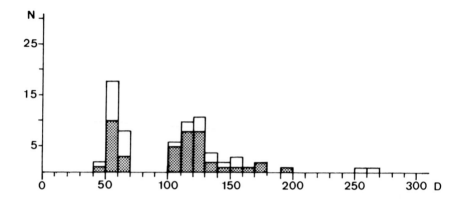

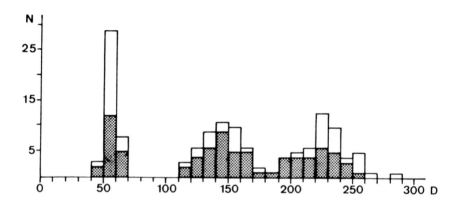

Fig. 2. Histograms of the hatching course in two families of B rossius. Ordinates give the number of individuals (dotted areas, females; white areas, males), while abscisses refer to the days from egg laying to hatching. One or two kinds of diapause eggs are apparent (modified from [23]).

laying (Fig. 3). Shifts from rapid to slow development can be easily obtained experimentally. The reverse is only partially inducible, because the treated females die very soon.

Embryonic diapause is a quite effective adaptive device: It ties up the hatching period with favourable environmental conditions. This allows the B rossius to get over the critical winter time, either as a sufficiently grown larva (third instar), or as a very early embryo, which readily withstands low winter temperatures because it is in a resting stage, well protected within the egg [23].

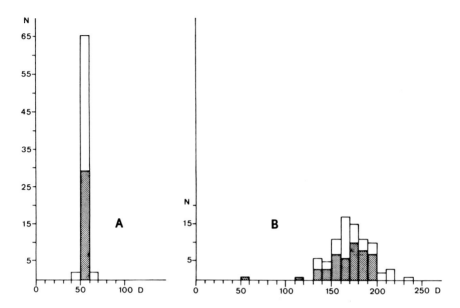

Fig. 3. Durations of embryonic development in eggs laid by a female experiencing varying photoperiods. A) From June 21 to July 24; B) from August 23 to September 22 (modified from [23]).

CYTOLOGY OF PARTHENOGENESIS

The diploid chromosome number of B rossius is 35 for the male and 36 for the female, with a sex-chromosome mechanism XX female–XO male.

The analysis of male mitoses has shown that the chromosome set is made up of a pair of large metacentrics, a pair of medium sized acrocentrics, an unique medium sized subacrocentric, X chromosome,[1] and 15 pairs of small-to-very-small elements. Among these the acrocentric type predominates, but at least a pair of subacrocentrics can be constantly recognized (Fig. 4). The female autosome set is obviously the same, but a pair of medium sized subacrocentric X chromosomes is also present (Fig. 5).

Cytological investigations on the development of parthenogenetic eggs have clearly shown that all eggs have normal meiosis, which is completed within a day from laying, in the ventral side of the egg, opposite to the micropylar

---

[1]That such a chromosome is the X is also supported by the observations of the two kinds of spermatocytes II, with 18 and 17 chromosomes, respectively. The latter type does not possess the medium sized subacrocentric element which is always found in the other.

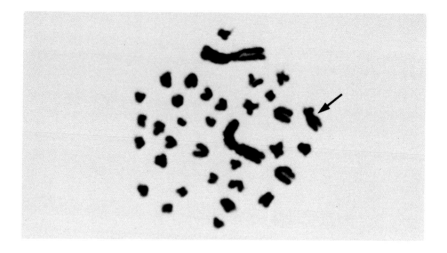

Fig. 4. The 35 chromosome set of male B rossius. Arrow points the unique X chromosome (× 2000).

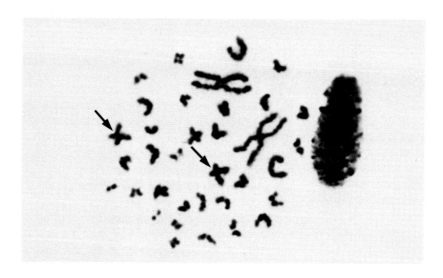

Fig. 5. The 36 chromosome set of female B rossius. Arrows point the two X chromosomes (× 2000).

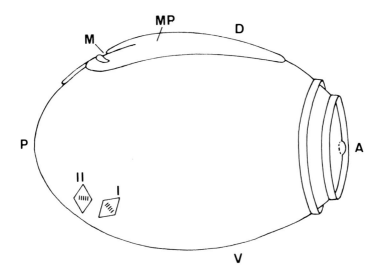

Fig. 6. Schematic drawing of the B rossius egg. A, Anterior pole; P, posterior pole; D, dorsal side; V, ventral side; MP, micropylar plate; M, micropyle. I and II, approximate sites of the first and second meiotic divisions (modified from [44]).

apparatus (Fig. 6). Embryonic development starts with haploid cells which divide rapidly, migrate to the micropylar region and often become aneuploid [26]. In these cells chromosomes appear rather thick and undergo frequent stickiness and lagging, causing unequal repartitions to the cell poles. From the seventh day onwards binucleated cells are produced by anaphase restitution and subsequently diploid or near diploid cells are obtained (Fig. 7). Vast degenerations take place in the blastoderm at this point, and from about a thousand cells only a few diploid cells survive in the micropylar region. From these cells the whole embryo develops, either immediately or after the already described diapause. Diapausing embryos are therefore made of very few cells. Pijnacker [27], when reporting on the same issue, describes a very similar sequence of events, but he finds an invariable number of 18 chromosomes and a diploidization by means of blocked mitoses (C mitoses), in a way similar to that found in Clitumnus extradentatus by Bergerard [13]. Whichever the doubling mechanism may be (anaphasic restitution or blocked mitoses), the facultative parthenogenesis of B rossius is, cytologically, of a very primitive kind. This is in turn confirmed by the observation of the very low success rate in some egg batches when doubling the haploid set, as already discussed.

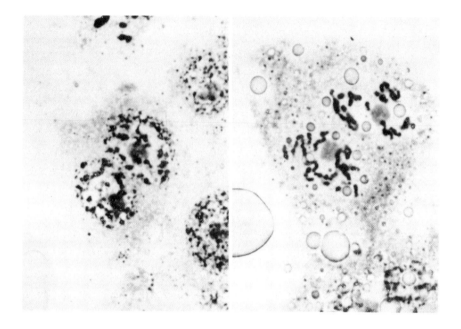

Fig. 7. Embryonic binucleated cells in prophase (left) and pro-metaphase (right). Nuclear evolution of binucleate cells is perfectly synchronous (× 1000) [26].

Aneuploid cells are also found in amphigonic embryos; their chromosome number may range from 28 to 39, ypodiploid cells being more common than the yperdiploid ones. Such a chromosomal variability lowers during larval development. Spermatogonia may show from 32 to 38 chromosomes, but spermatocytes are in fact without variation. The only recurrent anomaly noticed in first metaphases is the loss of the X chromosome in three cases out of 40 [28,29].

MORPHOLOGICAL DIFFERENTIATION

In spite of a noticeable variability of several body characters (such as size, coloration, exoskeleton granulations, shape and size of the subgenital plate and femoral laminae), stable and significant differences between populations do not exist. Some statistic differences between Adriatic, Tirrenian, and Sardinian populations, irrespective of their reproductive mode, have begun to be detected, but a detailed study has not yet been concluded.

Morphological investigations on the egg chorion sculptures of the Phasmatodea are at present carried out by means of the scanning electron microscope. Chorionic sculpturing shows many species-specific characters, so that different taxa can be easily identified. Similarities and differences of the egg chorion appear to reflect phylogenetic relationships with a good precision, making it possible that we can build now an ootaxonomy [30–34]. Furthermore, the SEM approach appears to be useful even at the subspecific level: In C gallica clearly differentiated patterns have been found among Spanish and Italian populations [Scali and Mazzini, unpublished].

Owing to these findings in C gallica, and taking into account the genetic differentiation attained by Tirrenian and Adriatic populations of B rossius (see below), a screening of several differentiated populations of the species is now in progress.

CHROMOSOMAL DIFFERENTIATIONS

The chromosome number appears to be uniform. This has been observed in French, Italian, and North African specimens [24,25,27,28,35,36]. A thorough search for structural rearrangement has not been carried out. However, up to this point the only structural changes reported are those found in an insular population. A survey on the Giglio Island (Tuscan Archipelago) revealed that seven males out of 20 examined showed structural changes in all spermatogonia and spermatocytes [29]. Five males had the same pericentric inversion affecting the unique medium sized pair of acrocentric chromosomes, either in heterozygous (four males) or in the homozygous condition (one male) (Fig. 8).

However, chromosome number variation has occurred during speciation within the genus Bacillus. For instance, the B atticus, a parthenogenetic species found in Greece, Yugoslavia, and in a narrow belt of the eastern coasts in Southern Italy, has been found to have 34 chromosomes (Fig. 9). It is apparent that B atticus karyotype is very similar to that of B rossius, the most obvious difference being the presence, in B atticus, of two unequal pairs of large metacentrics. It is very likely that the shorter metacentric chromosomes of B atticus were the outcome of a Robertsonian fusion between two acrocentrics of the B rossius. This fusion obviously reduced the number from 36 to 34 chromosomes, producing a new metacentric pair. It can also be observed that one of the fusing chromosomes must have been the largest acrocentric of the B rossius complement. Such a pair is no longer found in the chromosome set of B atticus, and it is of the same size as the long arm of the new metacentric. In this context it has been assumed that the 36 chromosome condition is the oldest one, but a better chromosomal knowledge of the northern African species of Bacillus is needed before we can give a definite interpretation of the B atticus karyotype. At any rate, the fact that there is the chromosome number variation in the genus Bacillus remains.

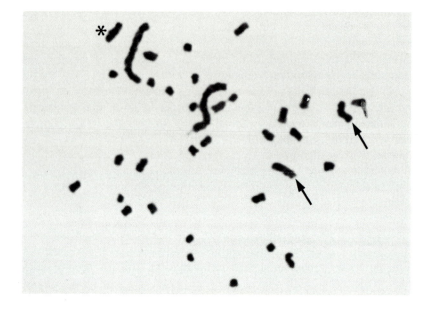

Fig. 8. Spermatogonial mitosis showing the heterozygous pericentric inversion of the medium sized pair of autosomes (arrows). Asterisk marks the unique X chromosome (× 2000) [29].

REPRODUCTIVE ISOLATION

As has been already briefly reported, no reproductive isolation exists between parthenogenetic French females and North African males [24]. Their offspring was kept for 10 years and showed a stable amphigonic reproduction [25]. Further experiments, however, revealed a more complex reproductive condition. Bullini [21] obtained only 10% average male frequency in crosses between Tuscan or Ligurian parthenogenetic females and males of Latium and Campania populations. However, the males that were obtained were normal. Such a result clearly indicated an initial degree of reproductive isolation.

At the same time, an unexpected finding came from the Tuscan amphigonic females when they were mated to their own males or males of neighbouring populations. It was observed that along with couples that had normal offspring, a few others that copulated normally gave an entirely or almost entirely female offspring, thus behaving as obligatory parthenogenetic [23]. This peculiar aptitude was inherited by the daughters of those couples, and it was followed for several generations in the females in one strain (Table I). The initial development

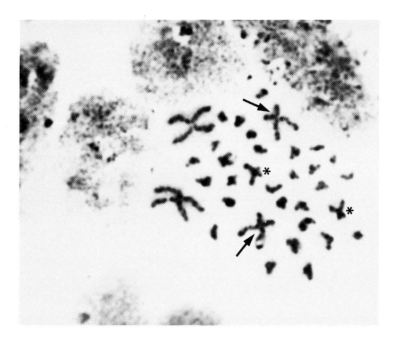

Fig. 9. The 34 chromosome set of B atticus female. Beside the two X chromosomes (asterisks), two pairs of metacentrics are outstanding; the smaller couple (arrows) appears to be new, if we compare the B atticus karyotype with that of B rossius (× 2000).

of embryos of the strain showed the same cytological events described for facultative parthenogenetic embryos. I had to conclude that the all-female families were due to a lack of fertilization. Eggs are fertilized at the end of the common oviduct, just before laying, and several spermatozoa can penetrate into the egg through the micropyle canals. It was ascertained that the lack of fertilization is due in this case to the failure of sperm penetration [37]. This could possibly be due to an unduly early obliteration of the micropylar canals [30]. This kind of obligatory parthenogenesis therefore is not supported by a different diploidization process, but rather by the peculiar characteristics of the egg membranes.

Recently, to further test the reproductive isolation of the Tuscan population, a series of crosses and back-crosses between Tuscan and Sardinian specimens has been performed (Table II). Because females greatly outnumber the males, it is apparent that crosses and back-crosses to females of Tuscan origin give rise to offspring with anomalous sex ratios. On the other hand, reciprocal matings give normally bisexual offspring. This clearly indicates that in Tuscan eggs a normal fertilization with Sardinian sperm does not occur.

TABLE I. Strain "671": Breeding Results for Five Generations

Generation	I	II	III	IV	V

$$\text{♀ } 67I(3 : II3) \rightarrow \begin{cases} \rightarrow A(I : 259) \begin{cases} A_1(5 : 298) \rightarrow A_{1,1}(0 : 40I) \rightarrow A_{1,1,1}(0 : 222) \\ A_2(0 : 28I) \rightarrow A_{2,1}(0 : 340) \rightarrow A_{2,1,1}(I : 290) \end{cases} \\ \rightarrow B(0 : 26I) \rightarrow B_1(0 : 297) \rightarrow B_{1,1}(2 : 306) \rightarrow B_{1,1,1}(0 : 38I) \\ \rightarrow C(5 : 204) \begin{cases} C_1(0 : 244) \rightarrow C_{1,1}(0 : 332) \rightarrow C_{1,1,1}(0 : 298) \\ \quad C_{2,1}(0 : 274) \rightarrow C_{2,1,1}(0 : 252) \\ C_2(0 : 304) \\ \quad C_{2,1}(0 : 295) \rightarrow C_{2,2,1}(0 : 256) \end{cases} \end{cases}$$

The numbers of males and females obtained from each mated female of the strain are given in parentheses.

TABLE II. Results of Crosses and Back–Crosses Between Three Tuscan and Three Sardinian Populations

♀ × ♂	Laid eggs	% Hatching	Larvae sexed at birth ♀	♂
Sard × Sard	628	54.94	126	109
	373	70.24	131	119
	373	48.25	91	79
Tusc × Tusc	486	88.68	205	226
	289	98.61	150	135
	173	79.19	79	58
Sard × Tusc	623	47.83	125	109
	448	65.88	161	134
	316	37.97	57	62
	303	97.69	131	139
Tusc × Sard	134	64.32	73	6
	112	74.11	73	9
	98	76.53	66	7
	93	100.00	82	11
	70	100.00	60	10
	62	66.13	38	2
	58	94.82	52	0
	47	89.36	37	5
Sard/Tusc × Sard	712	81.88	186	195
Sard/Tusc × Tusc	874	78.49	76	70
Sard × Sard/Tusc	112	69.94	31	33
Sard × Tusc/Sard	122	62.88	231	210
Tusc/Sard × Sard	376	78.46	159	3
	79	86.07	61	0
	78	94.94	72	0

GENIC DIFFERENTIATION

Electrophoretic techniques allow us to evaluate the differentiation of several structural genes and therefore to assess the phylogenetic distance among taxa. Analyzing about 20 gene-enzyme systems, Nascetti and Bullini [38] have recently developed this topic in the family Bacillidae. Pertinent data may be summarized as follows:

1) Three genetically differentiated populations have been evidenced. The three groups have been referred to as Tirrenian (A), Southern Italian (B), and Adriatic (C). Type A is mainly found along the whole Tirrenian coast, in Sardinia (where, however, it is slightly differentiated), in France, and in Spain. Type C spreads over the Adriatic coasts, Yugoslavia, and Greece. 2) A and C populations show a genetic distance of about 0.2, according to Nei's method [45]. 3) A and C populations include both bisexual and parthenogenetic demes. 4) B rossius is well differentiated from B atticus and much more from C gallica.

DISCUSSION

The analysis of the reproductive mechanisms and of the morphological, chromosomal, and genic differentiations gives us some information on the evolutionary strategies of B rossius and related species. On morphological grounds no population group including those of the northern border range, seems to approach the speciation level. Subspeciation, too, does not seem to be fully attained, although some indication of this is now being seen. With regard to the external morphology, B rossius seems to belong to that kind of organism whose external phenotype changes very slowly. In my opinion the morphological model realized in B rossius responds well to the ecological, ethological, and physiological requirements of the species. As a result of a selective control B rossius is almost an invariant. Also supporting this position are fossil records having modern phenotypes. Furthermore the numerous instances of close resemblances among species appear to substantiate the idea that parallel evolution is a common event in past and recent Phasmatodean speciation.

B rossius does not seem to have realized alternative karyotypes. Embryonic findings would paradoxically suggest a great possibility of numerical and structural changes, but spermatogonia and spermatocytes have a much more regular chromosomal condition. It is therefore apparent that germ line differentiation and meiosis have an important role in eliminating most of the chromosomal variability. The only finding in favour of a karyotype evolution is that concerning the pericentric inversion detected on the Giglio Island, although a careful check for rearrangements of the small chromosomes is needed.

We should not think that the chromosomal uniformity ascertained at this point in B rossius is the general rule for the Bacillus complex. It is important to remember the 34 chromosome karyotype that I reported for B atticus. This finding led us to the speculation that chromosomal variations will most likely be found in some other species of the genus Bacillus.

Furthermore, the situation found in the Australian species Didymuria violescens, which was thoroughly investigated by Craddock [6,39], clearly illustrates the multiplicity of chromosomal rearrangements that can be realized in the Phasmatodea. In that species (or superspecies) she found no less than 10 karyotypically different races or incipient species, all of which had arisen from the 39-chromosome race. Chromosome fusions and pericentric inversions have produced karyotypes with fewer chromosomes, which decreased gradually down to the 26-chromosome race.

The diagnostic resolution of the gene-enzyme systems, although underestimating genic variability, has evidenced a clear but complex pattern of differentiation among groups of populations of B rossius. Genetic diversity of such groups reaches the sibling species level, as is most convincingly illustrated in the anopheline moquitoes analogs.

The genic differentiation does not separate parthenogenetic versus amphigonic populations, but instead separates geographically isolated groups. Because parthenogenetic populations of both the Adriatic and Tirrenian coast are genetically closer to the amphigonic neighbouring populations, it is clear that parthenogenesis must have arisen independently and repeatedly in the two population groups [38]. Even on theoretical grounds, in this rather unspecialized parthenogenesis, a multiple origin is more likely to occur than in the more specialized ameiotic type. Actually, the well-analyzed instances of the latter type—which is very often associated to hybridization and/or polyploidy—appear to be monophyletic in origin. This is the case of many parthenogenetic weevils [17] and of the two distinct races of the parthenogenetic grasshopper Warramaba (Moraba) virgo [White, personal communication].

The cytological analysis of the parthenogenesis of B rossius perfectly explains its thelytoky and gives us an indication as to the origin of the impaternate but normally functional males. One must remember that aneuploid cells are commonly produced during early embryonic development. It is sufficient to postulate that one X chromosome is lost in the few cells that give origin to the embryo proper. These rare males may play a major role in the reversion of a parthenogenetic population to the amphigony. Perhaps this occurs via an intermediate deme, biased in favour of females.

The witnessed instances of obligatory parthengenesis due to the failure of fertilization do not seem, per se, destined to evolve a more specialized and effective mechanism of diploidization, but I think it is extremely relevant that a basis for the spreading of obligatory parthenogenesis already exists.

One important question that one must ask is: Which is the cellular basis that allows the stick insect egg to develop parthenogenetically? In answer to this question I would like to emphasize the fact that all instances of obligatory parthenogenesis in orthopteroid insects—such as Pycnoscelus surinamensis [40], the already mentioned Saga pedo and Warramaba virgo, and all the obligatory parthenogenetic stick insects—have a common cellular basis, that is to say an extra duplication of DNA during meiosis. In Carausius morosus and Sipyloidea sipylus this duplication has been analyzed by Pijnacker and co-workers [14,41–43]. They found that during the peculiar meiotic prophase a doubling of the DNA takes place so that a somatic number of tetrapartite chromosomes is formed. These divide twice by mitotic divisions, and the final result is an egg with the somatic number of monochromatidic chromosomes. During the first division, chromosome figures are very similar to tetrads. Even if real chiasmata occurred they would be ineffective, because such tetrads are identical copies of the same chromatid. Incidentally, this cytological feature, besides keeping the somatic number of chromosomes in the egg, also maintains heterozygosity in the offspring. Because of this common feature one could think that in amictic parthenogenesis the extra synthesis of DNA is related to the egg's ability to escape the metabolic block. On the contrary, this is the usual fate of the amphimictic egg, prior to fertilization.

The cytological substrate of the facultative parthenogenesis, allowing a haploid egg to divide, could also be related to a similar DNA doubling. With regard to this point, I would like to mention that in B rossius, haploid cells of the very early embryonic stages show thick anaphase chromosomes. Undoubtedly this is due to their contraction, but also to the fact that they are still bichromatidic [26]. With such a chromosome structure haploid eggs would have the diploid amount of DNA, and this could perhaps give us a clue to their developmental capacity.

I would like to state that in Southern France and Northern Italy B rossius is not in its most suited climate. The species is certainly native to much hotter climates. The development of a diapause clearly witnesses a difficult situation. A certain fragility in adult males, which have a much shorter life span than females, could be important in producing advantages to the parthenogenetic reproduction versus amphigony in the northern range. A greater advantage, however, seems to be associated with the possibility that isolated single females give progeny and colonize unoccupied districts [18,20].

In conclusion we would like to state that B rossius has evolved a genetic differentiation according to a geographic pattern and in a gradual mode. The same conclusion seems to hold even for the related Bacillidae. The process of genic changes has proceeded more than the morphological differentiation would suggest and has, at a minimum, brought B rossius to the subspeciation. This event could have a correspondence in the egg capsule sculpturing, which appears to be sensitive to speciation.

Karyotype repatterning does not play a major role within the species, but it is likely to be of importance between species.

The genic pool of B rossius affords a special way of differentiation, because in one generation parthenogenesis produces the fixation of different genotypes with all alleles in a homozygous condition. At the same time it is a starting point for genetic divergence. However, in B rossius genic differentiation along this line does not appear to have gone far, owing to the relatively recent origin of the parthenogenesis and to its diffusely facultative character. An irreversible jump to constant parthenogenesis would make a great difference at both genic and chromosomal levels. This has actually been found to occur in B atticus and C gallica.

I hope that the manifold aspects and levels of B rossius' evolutionary biology which have been discussed will give an idea of the real multiplicity and interdependence of the mechanisms at work in the adaptive and speciation processes.

REFERENCES

1. Günther K: Ueber die tassonomische Gliederung und die geographische Verbreitung der Insectenordnung der Phasmatodea. Beitr Ent 3: 541, 1953.
2. Beier M: Ordnung: Cheleutoptera Crampton 1915. In Bronns HG(ed): Klassen und Ordnungen des Tierreichs. Leipzig: Geest and Portig, 1957, Buch 6, Lief. 2, p 305.
3. Bradley JC, Galil BS: The taxonomic arrangement of the Phasmatodea with keys to the subfamilies and tribes. Proc Entomol Soc Wash 79: 176, 1977.
4. Piza SDT: Primeiras observaçòes sòbre a citologia de fàsmidas brasileiros. Sci Genet 3: 227, 1950.
5. Hughes-Schrader S: On the cytotaxonomy of phasmids (Phasmatodea). Chromosoma(Berl) 10: 268, 1959.
6. Craddock E: Chromosomal diversity in the Australian Phasmatodea. Aust J Zool 20: 445, 1972.
7. Scali V: Problemi di citotassonomia nei Cheleutoptera Crampton 1915. In Atti IX Congr Naz Ital Entomol Siena. 1972, p 285. Tipografia Bertelli & Piccardi-Firenze.
8. White M: Insecta 2. In John B(ed): Animal Cytogenetics. Berlin: Gebruder Borntraeger, 1976, Vol 3, p 1.
9. Cappe de Baillon P, de Vichêt G: Le màle du Clonopsis gallica Charp (Orthoptera, Phasmidae). Ann Soc Entomol Fr 104: 259, 1935.
10. Cappe de Baillon P, de Vichêt G: La parthénogenèse des espèces du genre Leptynia Pant (Orthoptera,Phasmidae). Bull Biol Fr Belg 74: 43, 1940.
11. Chopard L: La parthénogenèse chez les Orthoptéroïdes. Ann Biol 52: 15, 1948.
12. White MJD: Animal Cytology and Evolution. Third Edition, Cambridge: Cambridge University Press, 1973, p 961.
13. Bergerard J: Etude de la parthénogenèse facultative de Clitumnus extradentatus Br (Phasmidae). Bull Biol Fr Belg 92: 87, 1958.
14. Pijnacker LP: The Cytology, Sex-determination and Parthenogenesis of Carausius morosus (Br). Thesis, Groningen, 1964, p 99.
15. Pijnacker LP: Oogenesis in the parthenogenetic stick insect Sipyloidea sipylus Westwood (Orthoptera Phasmidae). Genetica 38: 504, 1967.
16. Bullini L, Bianchi-Bullini P: Ricerche sulla riproduzione e sul corredo cromosomico del fasmide Clonopsis gallica (Cheleutoptera, Bacillidae). Atti Acc Naz Lincei Rend Cl Sci Fis Mat Nat 51: 563, 1971.

17. Lokki J, Saura A: Polyploidy in Insect Evolution. In Lewis WH (ed): Polyploidy Biological Relevance., New York: Plenum Press, 1980, p 277.

18. Benazzi M, Scali V: Modalità riproduttiva della popolazione di Bacillus rossius (Rossi) dei dintorni di Pisa. Atti Acc Naz Lincei Rend Cl Sci Fis Mat Nat 36: 311, 1964.

19. Bullini L: Ricerche sul rapporto sessi in Bacillus rossii (Fab.). Atti Acc Naz Lincei Rend Cl Sci Fis Mat Nat 37: 897, 1964.

20. Bullini L: Spanandria e partenogenesi geografica in Bacillus rossius (Rossi). Atti Acc Naz Lincei Rend Cl Sci Fis Mat Nat 40: 926, 1966.

21. Bullini L: Osservazioni su alcune popolazioni partenogenetiche italiane del fasmide Bacillus rossius (Rossi). La Ric Sci 12: 1270, 1968.

22. Bullini L: Osservazioni su alcuni casi di partenogenesi geografica nei Fasmidi. Arch Bot Biog It 14: 292, 1969.

23. Scali V: Biologia riproduttiva del Bacillus rossius (Rossi) nei dintorni di Pisa con particolare riferimento all'influenza del fotoperiodo. Atti Soc Tosc Sci Nat Mem Ser B 75: 108, 1968.

24. Favrelle M, deVichêt G: Résultats de la fecondation, par un male d'Algerie, de femelles parthénogénétiques françaises du Bacillus rossius (Phasmidae). C R Ac Sci Paris 204: 1899, 1937.

25. Favrelle M, de Vichêt G: Etude de la descendance de Bacillus rossii F issue de croisements entre les lignées bisexuée nord-africaine et parthénogénétique française. C R Ac Sci Paris 226: 1108, 1948.

26. Scali V: Osservazioni citologiche sullo sviluppo embrionale di Bacillus rossius (Insecta Phasmoidea). Atti Acc Naz Lincei Rend Cl Sci Fis Mat Nat 46: 110, 1969.

27. Pijnacker LP: Automictic parthenogenesis in the stick insect Bacillus rossius Rossi (Cheleutoptera,Phasmidae). Genetica 40: 393, 1969.

28. Mosti P, Scali V: Osservazioni sul corredo cromosomico di Bacillus rossius (Insecta Cheleutoptera). Atti Acc Naz Lincei Rend Cl Sci Fis Mat Nat 59: 537, 1975.

29. Scali V, Mosti P: Riarrangiamenti cromosomici in Bacillus rossius (Insecta Cheleutoptera) dell'Isola del Giglio. Atti Acc Naz Lincei Rend Cl Sci Fis Mat Nat 59: 493, 1975.

30. Mazzini M, Scali V: Fine structure of the insect micropyle. VI. Scanning electron microscope investigations on the egg of the stick insect Bacillus rossius (Insecta, Cheleutoptera). Monit Zool It 11: 71, 1977.

31. Mazzini M, Scali V: Ultrastructure and amino acid analysis of the eggs of the stick insects, Lonchodes pterodactylus Gray and Carausius morosus Br (Phasmatodea:Heteronemiidae). Int J Morphol Embryol 9: 369, 1980.

32. Mazzini M, Scali V: Le uova dei Phasmatodea al microscopio elettronico a scansione: loro valore tassonomico. Atti XII Congr Naz It Entomol Roma, 1980 (in press).

33. Scali V, Mazzini M: Fine morphology and amino acid analysis of the egg capsule of the stick insect Clonopsis gallica (Charp.) (Cheleutoptera Bacillinae). Int J Insect Morphol Embryol 6: 255, 1977.

34. Scali V, Mazzini M: The eggs of the stick insects, Sipyloidea sipylus (Westwood) and Orxines macklotti de Haan (Phasmatodea, Heteronemiidae): A scanning electron microscopic study. Int J Invert Reprod 4: 25, 1981.

35. Cappe de Baillon P, Favrelle M, de Vichêt G: Parthénogenèse et variation chez les Phasmes. III. Bull Biol Fr Belg 71: 29, 1937.
36. Montalenti G, Fratini L: Observations on the spermatogenesis of Bacillus rossius (Phasmoidea). In: Proc XV Intern Congr Zool London. 1959, p 749.
37. Scali V: Obligatory parthenogenesis in the stick insect Bacillus rossius (Rossi). Atti Acc Naz Lincei Rend Cl Sci Fis Mat Nat 49: 307, 1970.
38. Nascetti G, Bullini L: Differenziamento genetico e speciazione in fasmidi dei generi Bacillus e Clonopsis (Cheleutoptera, Bacillidae). Atti XII Congr Naz It Entomol Roma, 1980 (in press).
39. Craddock E: Chromosomal evolution and speciation in Didymuria. In White MJD (ed): Genetic Mechanisms of Speciation in Insects. Sydney: Australia and New Zealand Book Co., 1974, p 24.
40. Mattey R: Cytologie de la parthénogenèse chez Pycnoscelus surinamensis L. (Blattaridae-Blaberidae-Panchlorinae). Rev Suisse Zool 52: 1, 1945.
41. Pijnacker LP: The maturation division of the parthenogenetic stick insect Carausius morosus Br (Orthoptera, Phasmidae). Chromosoma (Berl) 19: 99, 1966.
42. Koch P, Pijnacker LP, Kreke J: DNA reduplication during meiotic prophase in the oocytes of Carausius morosus Br. Chromosoma (Berl) 36: 313, 1972.
43. Pijnacker LP, Ferwerda MA: Additional chromosome duplication in female meiotic prophase of Sipyloidea sipylus Westwood (Insecta, Phasmida), and its absence in male meiosis. Experientia 34: 1558, 1978.
44. Scali V: La citologia della partenogenesi di Bacillus rossius. Boll Zool 39: 567, 1972.
45. Nei M: Genetic distance between populations. Amer Natur 106: 283, 1972.

Mechanisms of Speciation, pages 411–433
© 1982 Alan R. Liss, Inc., 150 Fifth Avenue, New York, NY 10011

Speciation as a Major Reorganization of Polygenic Balances

Hampton L. Carson

WHAT DO SPECIATION MODES HAVE IN COMMON?

Both direct and indirect evidence strongly supports the notion that new species arise primarily from population units that are subdivisions of older populations. To understand cladogenesis as a dynamic process, the main task of the population biologist is to study either incipient splits or splits that have occurred only very recently. From the facts obtained, we may then infer the pivotal genetic events. Although few biologists deny the identification of cladogenesis as the key process of species formation, the nature of the genetic changes that transpire during the crucial generations has not been clearly identified. White [1] has recently written a broad review of the numerous discussions and speculations on this topic. *Modes of Speciation* is an apt title, since White accepts the idea that the mechanisms of cladistic speciation are both numerous and varied. Indeed, by ". . . . attributing a much larger role to structural chromosomal rearrangements. . . ." the author has forced serious consideration of a broad new series of cytogenetic dimensions in relation to speciation.

Rather than sift through the numerous modes (allopatric, sympatric, parapatric, stasipatric, peripatric, clinal, area-effect, polyploid, asexual) and select one or more as my favorites, I am attempting a reductionist approach. Thus, I explore to what extent genetic events common to any and all modes can be identified. I will begin by reviewing the evidence indicating that the gene pool of each diploid, cross-fertilizing species is genetically unique and that it comes, over time, to be characterized by a complex and highly-organized polygenic system underlying the observed phenotypes. The evidence further indicates that its many genetic elements are balanced in its gene pool, inserted and held there by an all-pervading stabilizing selection. For such elements, Templeton (this volume) has employed the useful term, "segregating units", which includes not only single loci, but inversions and epistatic blocks as well. The crux of the argument will be that formation of a new species from an older one occurs only following two events that must occur successively. First, the old polygenically balanced gene pool, or segment of it, suffers *disorganization* by forces that are essentially stochastic. This can occur in a relatively small number of generations. Following this, the concept calls for *reorganization* of a new balanced polygenic system over hundreds or thousands of generations under natural stabilizing selection.

This "organization theory" states that speciation requires the development of a new and major organization of polygenic balances. As such, it is a modified and broader application of a theory I presented earlier when stressing the founder effect [2,3]. In the earlier paper, I pointed out that the new theory is not compatible with the classical idea of allopatric speciation in which a large population is supposed to become divided by extrinsic forces into two or more large, geographically-defined sections. Without any disorganization-reorganization cycle, these two large populations (or one of them) are theorized to gradually diverge genetically to form separate species. In contrast, the organization theory requires drastic reduction in population size. This implies a high probability that only one new species (the daughter species) is formed at any one time. After the disorganization phase is over, reorganization of the gene pool of the population and the polygenic synthesis of new adaptations that emerge is accomplished anagenetically by the well-known neodarwinian processes of mutation, recombination, and selection.

In accordance with current trends [4] I am using the term polygene [5] to refer to any gene of very small effect individually that contributes to continuous or quantitative phenotypic variation. As Ayala and MacDonald [6] have suggested, there may be considerable overlap between polygenes as so defined and genes which regulate the activity of structural genes. For the purposes of discussion of balanced gene systems, I will adopt this broad view of the polygene category. Thus, I include regulatory genes as well as all other genes that affect the development of the phenotype by modification of genes which have larger individual effects.

The discussion that follows deals almost exclusively with the origin of diploid cross-fertilizing daughter species from populations which are themselves diploid and cross-fertilizing. As has been pointed out by others (eg [7]) and more recently by me [8], the diploid system is an exceedingly powerful one from the point of view of its capacity to generate recombinational genetic variability and to integrate new mutations of small effect. When exposed to stabilizing selection, this variability can then be built, as suggested above, gene by gene, into a new polygenic balanced system. The diploid gene pool is emphasized since the freedom of the recombination process is reduced in new polyploids so that cladistic speciation will be retarded in such species. The situation, however, is complicated by the "diploidization" of older autopolyploids and wide-cross amphidiploids in which a restored and enhanced recombination potential may be realized. Agamic complexes, "parthenospecies," "asexual species," "microspecies," and obligatory selfers are in a somewhat special category. Such entities may themselves gain considerable individual biological success through fixed heterozygosity and luxuriant heterosis. Nevertheless, such units are essentially clonal. A new clonal variant may arise from one of these; nevertheless, like its parent, the new entity is essentially at an evolutionary standstill. All such forms lack a true gene pool and thus the capacity for speciation of the normal bisexual diploid type. Although conspicuous in the contemporary natural world, agamospecies appear to have only a very restricted cladogenetic evolutionary future. Indeed, these restrictions serve to emphasize the fact that the diploid cross-fertilizing state is characteristically permissive of exhuberant speciation in both animal and plants. Despite the frequency of agamic complexes and polyploids in plants, one is impressed with the extensive cladogenetic speciation in such predominantly diploid genera as *Clarkia, Crepis, Epilobium, Eucalyptus, Gilia,* and *Vicia* to name only a few.

THE BALANCED GENETIC STRUCTURE OF DIPLOID POPULATIONS

How much and what kind of genetic variability does a normal out-crossing diploid species carry in the individuals that make up its gene pool? In 1955, Dobzhansky wrote ". . . the adaptive norm is an array of genotypes, heterozygous for more or less numerous gene alleles, gene complexes and chromosomal structures" [9]. At the time this statement was made, the extent of electrophoretically-detected variability was unknown. Its discovery a decade later sealed the fate of the alternative classical view, that, as Dobzhansky put it, "most individuals should . . . be homozygous for most genes." Probably the only change one might make in Dobzhansky's dictum would be to omit the words "more or less." Most species carry enormous stores of genetic variability in the balanced state.

The emergence of the balance view of population structure has affected the term "gene pool." To some, the term may carry the connotation that the genetic architecture of a species is essentially an unorganized "bean bag" into which the parents pool their genes and from which progeny combinations are drawn by chance [10]. With Haldane [11] I defend this view if it is applied to a single locus with two alleles in a largely homozygous background. In this context, it serves as a useful and dramatic model of the basic attributes of the recombinational aspect of the meiotic divisions and the Hardy-Weinberg equilibrium. These are the basic theorems of population genetics; imbedded within them is the active growing point of the evolutionary process. As Mayr [10] pointed out, however, the analogy to the bean bag is misleading in any larger sense since it fails to include the key feature of organization. The term gene pool is used in this paper as a shorthand designation for the entire, highly organized genic and chromosomal system as it exists in the population of a species.

Evidence for the balance view comes from many sources and has been widely discussed and reviewed in the recent literature. For this reason, I will be brief and will simply recount some examples and cite a few relevant references. Major chromosomal polymorphisms are a striking feature of the populations of a very large number of animals and plants. Some of these observed polymorphisms are transient, either being in the process of elimination by normalizing selection or moving towards fixation. Of great significance, however, is the observation that the most conspicuous elements of this polymorphism appear to be normal balanced attributes of the species, segregating as major blocks of genes in natural populations. Most evidence indicates that the various component elements are not only held in the population but are continually improved and revised by selection favoring multiple, interacting heterozygous states. This interpretation appears to apply not only to the extensive inversion polymorphism within many species of *Drosophila* but also to translocation polymorphism in other animals and in many plants.

Substantial heterochromatic chromosomal blocks are found segregating in many species of both animals and plants (eg [12], [13]). Such blocks, however, may not themselves have intrinsic genic properties which are favored by stabilizing selection. Rather, their distribution in the genome may reveal a condition wherein linked euchromatic blocks are the true object of selection. Accordingly, these blocks may simply be carried along due to linkage. That they may represent a type of very efficiently replicating DNA unrelated to the fitness of the organism is considered a distinct possibility (see p 425).

Another feature of sectional chromosomal polymorphism that needs emphasis in this regard is the fact that most individual cases of polymorphism are species-specific, that is, they appear as characteristic ingredients of specific gene pools. Furthermore, very clear cases are known (eg [14]) in which there are "organization effects." Thus, there is clearly a strong interaction between separate, substantial lengths of the genome.

Artificial selection which attempts to remove these natural polymorphisms and reduce the genome to a structurally monomorphic state frequently meets with considerable resistance in the form of declining vigor and reproductive fitness. This result suggests that there is a pervasive set of interactions linking the action of the separate polymorphisms. That these polymorphisms are integral elements of an organized gene pool, is also manifested by their persistence in laboratory stocks and experimental populations despite the moderate inbreeding that prevails [15–17].

Chromosomal polymorphisms give every evidence of having complex mutigene contents; indeed, even a very short inversion embraces a large amount of the chromosome in molecular terms. There is also evidence [18] that identical gene arrangements by no means have identical genetic contents as measured by electrophoretic methods. Furthermore, the same gene arrangement, originating from different geographical regions displays different behavior in experimental crosses, suggesting that some coadaptation occurs in local subdivisions of the gene pool [19–22]. Emphasis in this paper, however, centers on more fundamental balances that are species-wide in their distribution.

That the gene pool has a complex interactive organization is also indicated by the extensive literature on concealed lethals and other detrimentals in natural populations [23]. Although most lethals tend to be eliminated by normalizing selection, quite a number persist, apparently held in the population because of association with segregating units conferring a high fitness. This behavior is similar to that observed for heterochromatic blocks. The persistence of lethal genes can be ascribed to stabilizing selection favoring some neighboring polygene blocks.

Recessive visible mutant genes exist well above the mutation rate in many populations; they may be demonstrated by simple inbreeding methods [24]. Single gene heterosis cannot be easily demonstrated for these and it seems more likely that they mark blocks of genes contributing to high fitness as heterozygotes [25,16,26,27].

When the astonishing high frequency of allozyme polymorphism was discovered during the 1960's, the notion that each locus imposed its own genetic load, bean bag style, had to be abandoned in favor of the idea that allozyme loci (indeed probably all loci) are part of intricately balanced polygenic blocks. The unit of selection came to be viewed as an interacting system of genes. The separate ultimate loci are not treated by selection as if they had no genetic environment [28]. Recently Hedrick [29] has developed what is essentially the same idea. Some genes which encode quasi-neutral biochemical effects may persist by "hitchhiking" on sections of the genome which contribute to high fitness. When their inheritance is followed individually they are found to label blocks of genes rather than be responsible for a primary effect themselves. This and similar theories tend now to be favored over simpler bean bag models. In consequence, single gene "selectionism" vs. single gene "neutralism" has declined as a controversey.

The extent of allozymic polymorphisms was underestimated at first. More recently, the development of electrophoretic "fine tuning" methods through experimental adjustment of temperature, gel pore size and pH, have revealed even more variability, particularly multiple alleles [30]. This variability is integrated into the genome and is not trivial or transient.

Some methods now exist for recognizing genetic factors which regulate the expression of the enzyme variability in eukaryotes. For example, Dickinson [31] has found that such genes may map in cis, trans or independent positions relative to the structural gene. This adds still another level of variability which is clearly integrated into the gene pool.

Another line of evidence stems from the results of artificial selection experiments. Indeed, experiments on Drosophila such as the classical experiments of Mather and Harrison [32] served as the basis for the building of the multiple factor hypothesis into a sophisticated theory of polygenic balance [33,28] or the shifting balance theory [34,35].

The approach to homozygosity which is theoretically expected under inbreeding seems not always to be realized. For example, the work of Allard and his co-workers [36] has revealed that certain inbreeding plants such as Avena and Hordeum retain heterozygosity for considerable sections of the genome. This provides further evidence for a persistent coadaptation and balance based on strong epistatic interactions.

Basic population genetic theory is able to deal in an elegant fashion with single-locus, two-allele systems. Directional, stabilizing and normalizing selection systems may be modeled and predictions made based on fitnesses assigned to the small number of zygotes involved. Indeed, some sophistication has been gained in the study of two-locus systems [37]. Experiments provide data that confirm the theoretical soundness of this approach.

In the large, variable populations of the real world, however, it is not possible with existing methods to separate the effects of a single gene on fitness from that of other genes. As Hartl [38] puts it

> . . . the genetic underpinning of fitness is extremely complex, involving alleles at many loci that influence fitness in two ways—one through the main effect of each locus (which typically may be an effect on some quantitative trait such as body size, growth rate, or developmental time) and the other through pleiotropic effects on other traits related to fitness.

The result of this is to make such genes of minor interactive effect extraordinarily difficult to identify individually, although it is possible is some organisms [39]. Accordingly, the theory of organization and adjustment by selection of polygenic systems is poorly developed. Gene interaction of many sorts has been demonstrated. For example, we can recognize, in the terminology originally used by Mather [5], *internal balance* or interactions of genes linked in a single homologue and *relational balance,* gene interaction between different arrays of genes on two homologues.

FITNESS AS THE OUTCOME OF GENE POOL ORGANIZATION

Natural selection maximizes Darwinian fitness, the composite trait that relates to the future of the genes of the individual. It compares the number of offspring left by an individual to the average performance of the other individuals present in the same population. The diploid, bisexually-produced individual is pivotal in evolution since it embodies a single complex recombinational genotype generated from an enormous field of variability. Fitness appears to be maximized by stabilizing selection that embraces many loci, resulting in a complex balanced polymorphism whether revealed by chromosome inversions or not. The components of an individual's fitness are notoriously difficult to measure. What data do exist deal mostly with a selected oligogene or chromosomal arrangement and one of the more tractable components of fitness, such as survival, fecundity, or longevity. More difficult to measure, but a component of overriding importance in many cases is the differential sexual activity of the individual.

Partitioning of fitness into components is a somewhat misleading exercise, since it is the composite that in the end determines the relative number of offspring that the individual will leave. Clearly any gene, no matter how minute its effect, which contributes to fitness in any direct or interactive way will find its way into the system by selection. Rather than reach fixation, the facts suggest that such a gene is likely to become a part of the all-pervading balanced system.

The above describes the origin and refinement of what are generally referred to as adaptations, elements which serve Darwinian fitness. Adaptive features of a species, when considered as separate components, virtually never show single-factor inheritance in crosses either within or between species, although there are some valuable exceptions [40,41]. In most species, phenotypic variance is low, suggesting that natural selection maintains a fine-tuned polygenic modifying balance over the genetic basis of adaptations. There is no reason to believe adaptations primarily result from the simple fixation of alleles at specific gene loci. Haldane's dictum [42] "the principal unit process in evolution is the substitution of one gene for another at the same locus" had an appealing simplicity 25 years ago. In view of modern data on population structure it now appears to be misleading, as is dependence on apparently simple models, ie industrial melanism.

As a species in a relatively stable environment becomes older and larger, its gene pool probably becomes organized in a more and more complex manner. It has already sponged up and incorporated many new mutants of minor effect and is holding them in the heterozygous state (see [43] and [44] for both experimental and theoretical treatment of the kinetics of this process). This may reach a saturation point, and give a general impression of stasis.

This apparently static state presents a curious paradox, in that the gene pool itself, when analyzed, manifests a very large genetic variability in the presence of phenotypic stability. It is the organization and storage of this variability that has come to a point of stasis. Such highly heterozygous dynamic equilibria appear to be characteristic of many species as observed in contemporary populations. Accordingly, a sexual population maintains a condition analogous to kinetic energy in physics. The genetic variability held there can be put to work by selection once the environmental and genetic conditions imposing the stasis are altered or removed.

The term "equilibrium" has been used to describe closely similar phenotypes observed over long periods of time in the fossil record [45,46]. Both the paleontologist's data and his concept of the species are difficult for the population biologist and geneticist to deal with. Nevertheless, the equilibria of phenotypes observed in the fossil record might bear some relationship to the dynamic genetic equilibria described above. I will return to this point later (see p 428).

OPEN AND CLOSED SYSTEMS OF BALANCE

In the preceding paragraphs, I have given an account of balanced gene pool structure which may be considered characteristic of a species having a large, outbred, geographically widespread and undivided population. Following Mayr's terminology [47], such species may be described as monotypic, that is, lacking geographically localized subspecies. Monotypic species that have been critically examined appear to have gene pools fully as complex as comparable polytypic species, that is, those that display subspecies. The main difference between such species is probably that gene flow from one subdivision to the next is somewhat more restricted in the case of the polytypic species. In most cases, subspecies appear to retain, even in the face of their divergence in some characters, the essential balanced gene pool characteristic of the species as a whole.

In view of this, I suggested [8] that the gene pool of a mature, widespread species may include two kinds of organized elements. These are 1) a *closed system* of genes in which the balances that have been built up over time cannot normally be broken under the prevailing conditions of strong stabilizing selection and 2) an *open system* consisting of a set of genes not tied in so tightly with the balances.

Genes of the open system are capable of free recombination and fixation without major viability effects. They will be able to respond to directional selection imposed under specific local conditions. Artificial selection can produce extraordinarily divergent morphotypes (eg dog breeds; Drosophila mutants; morning glories) but few investigators take these as serious approaches to speciation. I am in accord with this view and consider such changes as evidence for the existence of an open system of variability that can be easily moulded locally. The subspecies is an open system phenomenon characteristic of species that have widespread continental or archipelago distributions in which gene flow from the central body of the species is only moderately restricted. Under such circumstances, the closed system is unaltered but the open system is capable of adjustments to requirements for certain relatively minor local adaptations. Like polytypic species, monotypic species retain the rigidly balanced closed system but gene flow prevents local adjustment from being pronounced or renders it unnecessary. The difference between monotypic and polytypic species, accordingly, is not a profound one.

Chromosome inversions provide excellent examples of closed systems. Their contents can vary but basically they represent supergenes, that is, large sections of the genome that are effectively cut off from recombination. The result is that a tendency is established that can culminate in fixed heterozygosity: homokaryotypes are less fit than heterokaryotes [48]. In terms of the present theory, accumulation of inversions means that increasing amounts of the genome are insulated from recombination and the balance becomes increasingly obligatory. Disorganization of the balance will be very difficult in central, highly polymorphic populations but considerably easier in marginal populations which carry less chromosomal polymorphism [49].

Certain widespread species (eg the African butterfly Papilio dardanus) display an extreme phenotypic but genetically-based polymorphism which appears to be maintained under a frequency-dependent selection that varies in different geographical areas. Despite its overtly striking character, I would ascribe such a polymorphism to the open rather than the closed system. Thus, the essential ancestral gene pool balances that are characteristic of Papilio dardanus are not affected by the local morph differentiations. As will be developed later, these polymorphisms, as well as the more usual subspecific characters, cannot under most prevailing circumstances, be regarded as leading in the direction of new species. This down-grading of the importance of subspecies is in accord with the notion expressed many years ago by Goldschmidt [50]. Small subspecies, however, may be especially vulnerable to disorganization events, and therefore may acquire great importance in speciation.

THE ORGANIZATION THEORY OF SPECIATION
The Phase of Disorganization

The cladistic view of the origin of species implies that a species, like an individual, goes through life history stages, such as birth, youth, maturity, and old age. The analogy, however, probably should not include senescence and death, as there is no intrisic reason why the gene pool of a species, if it remains large, should not retain a balanced condition indefinitely. In the framework of the present discussion, the attributes of interest are the genetic balances present in the gene pool. Indeed, it may be possible that "closed system" balances increase in direct proportion to the age of the species since its time and place of cladistic origin.

Let us consider a parent species having already a highly balanced and "mature" gene pool. Since its genetic architecture is already complex, ordinary subdivision (into subspecies, for example) might occur but this would not be easily conducive to a break with the old system. The theory requires, then, that before a new integrated gene pool can be established, some type of destructive event would have to ensue that would have the effect of breaking up the old system, especially the closed system described above.

Accordingly, I am suggesting here that there is one basic underlying reductionist feature of all speciation events, namely, disorganization of the old gene pool. This is considered a key initial step leading ultimately to a recognizable cladistic event. Since the disorganizing event may logically be conceived, as mentioned earlier, as being accomplished in a relatively small number of generations, this phase of the speciation phenomenon proposed here is open to experimental attack by population research carried out at the present time level. This exciting possibility may be used as justification for the admittedly rather speculative accounts of disorganization possibilities which follow. I consider the experiment of Powell [51] on the founder-flush model in Drosophila and to some extent the work of Grant [52] on Gilia to represent very promising types of research in experimental population biology relating to the speciation process. Much attention has been given to organization of gene pools under different conditions but relatively little has been given to purposely setting out to study to what extent the gene pool can be disorganized experimentally. There are some strikingly favorable diploid materials which suggest that such experiments might yield data of great interest. Examples of materials that might be experimentally manipulated with this in mind are the semispecies of Drosophila paulistorum [53], Drosophila heteroneura/silvestris [54], Clarkia biloba/lingulata [55], Stephanomeria malheurensis [56], Vandiemenella viatica group [1], Rhagoletis pomonella [57], and Mus poshiavinus [58].

Disorganization by founder events. By its very nature, a disorganization process would, of necessity, be mediated by some sort of strong stochastic event in the history of the population. Considerable discussion has been devoted to the founder effect [59,2,60]. The argument for the importance of such a sudden and chance sundering of the gene pool of an older species is based largely on the specific effects of starting an entire new population from (in extreme cases) a single sexually-reproducing propagule derived from the parent species. Templeton [60] has considered the various population conditions of the parent species that would predispose the immediate descendants of a founder to start what he refers to as a process of transilience. The relevant changes will be not so much simple loss or fixation of alleles through random drift leading to homozygosity. Rather, Templeton's calculations show that the combined effects of inbreeding and gametic disequilibrium on balanced polygenic systems will be especially important. He suggests that the highest probability for a genetic transilience will occur when the parent population is large and panmictic, having a large variance effective size and inbreeding effective size. Founder events will thus be expected to upset selection processes which impinge upon multi-locus systems controlling integrated developmental, physiological, and behavioral traits [61].

What may occur at a founder event and during the generations that immediately follow it are well covered by Wright's well-known shifting balance theory of evolutionary change [34]. Simple attenuation of a population through a single individual will not necessarily in itself result in a breakup of balanced systems. For example, a very large amount of work has been done on many Drosophila species using isofemale strains [62]. These are laboratory strains stemming from a single inseminated female collected in nature. Laboratory strains of many other animals and plants are frequently used in the same manner for laboratory analysis of wild populations. Most such strains, even though they may acquire some special characteristics, are clearly not recognizable as deviants in the direction of new species.

The reason for this, in the framework of my theory, is that the closed system of balances characteristic of the species have not been broken or altered. Thus, a flush-crash cycle has been suggested as an important adjunct of the founder event [63]. By releasing the gene pool from the constraints of natural selection, recombinants from the closed system may appear that are quite outside the norm of the ancestral population. If a small population is started from such a propagule, the key events leading to disorganization may have already been set in motion.

Disorganization in situ by reduction of population size to a vestige. When a population is subjected to a drastic reduction in size by disease, drought, flood, vulcanism, or other natural disaster, a once large continuous population may be reduced to one or more disjunct remnants. Reduction in size may not be as drastic as in the case of the founder event, but within certain isolates the population may reach critical low numbers and high inbreeding coefficients so as to destabilize the balances in the gene pool. Stabilizing selection extracts a price; when selection favors blocks of genes held in the heterozygous state under a recombining system, it generates many genotypes on both sides of the fitness mean. When a certain minimum size is reached, Wright has shown that the population will then wander off its adaptive peak. If this happens to a population with a gene pool structure of the type discussed above, a key disorganization might ensue. Although such a population might contain dysgenic elements, various circumstances of chance might permit it to survive in isolation.

Disorganization by hybridization. As emphasized in the founder-flush model, one of the pivotal elements conducive to disorganization would be a temporary release from selection. This could include a sudden shift from K selection or r selection. A way in which this could be accomplished would be through the chance hybridization between a K-selected species and an r-selected one or between two r-selected species. Under such circumstances, major adaptational adjustments based on balances held tightly to different modes in the parent species might suffer recombinational disorganization when a hybrid swarm between the species is formed over three or four generations of hybridization. As in the founder model, the key event would be the release of the hybrid population from the balancing effect of stabilizing selection. The potential for disorganization is greatest when the hybridization occurs in an area peripheral to the area of abundance of the parent species, as in a "hybridized habitat" [64], which might permit the survival of novel recombinants. As data on the composition of the gene pools of certain species-pairs accumulates, the stage will be set for the reinvestigation of hybrids and hybrid swarms from the perspective of the arguments in this paper. The role of hybridization in the evolutionary process has always been a controversial one; new evidence documenting the degree of gene pool disorganization during hybridization would be very welcome. As mentioned previously, Grant [52] has made a good start on this.

As in the case of the founder model, disorganization by hybridization is easily amenable to experimental study. In recent years, furthermore, the phenomenon of hybrid dysgenesis (see [65]) has attracted a lot of attention. Whatever the basic cause or causes of the observed facts of dysgenesis, such phenomena could make a potential contribution to understanding the disorganization process. Introgressive hybridization [66] has little bearing on organization theory. Leakiness, involving the passage of small amounts of germ plasm from one species into another is akin to gene flow and would not be expected to have any disorganizing results.

Disorganization by shifting balance in small semi-isolated populations. The theme which runs through the present argument is the necessity for some stochastic force to perturb the species-specific balances existing in the gene pool. The founder principle has been invoked as the most extreme kind of drastic change that can occur and still maintain continuity in a population. In 1932, Wright [67] presented his famous model of the adaptive landscape demonstrating mathematically that the greatest possibility for a shift in an adaptive peak occurs when the population structure of a species is broken up into small semi-isolated populations. This model formed the basis for Wright's shifting balance theory of evolution.

Wright's arguments are most powerful when applied to intraspecific differentiation and thus one should be cautious in applying them in the construction of models of shifts in gene pool organization that are great enough to break down what I have called the "closed system." Under some circumstances, however, events comparable to those occuring in founder or vestige populations would be expected to occur. In any case, the reduction of the population to semi-isolated demes is a key event. Any one of these may be vulnerable to having its polygenic balances disorganized.

The effect of disorganization on sexual activity and adaptation. In the earlier discussion relating gene pool organization to fitness, reference was made to the difficulties of separating and measuring fitness components, especially when this composite trait is affected by many genes existing in a complex balance. Techniques for measuring fitness components in individuals in natural populations are unfortunately virtually non-existent and this situation has led to the necessity of instituting tests of individuals from captive, laboratory or field-plot populations. The special difficulties encountered in measuring differences in sexual activity and ultimately relative reproductive success has resulted in very little attention being paid to the evolutionary importance of genetically-determined attributes of sexual activity. Exceptions are the theoretical work pioneered by O'Donald [68] and such observations as those of Ehrman and Spiess [69] and Anderson et al [70].

The result of this neglect has been that the evolutionary aspects of sexual behavior and its morphological underpinnings have had insufficient attention from experimental and theoretical evolutionary biologists. A descriptive science of ethology of sexual behavior has become established (eg [71]). The approach, however, is comparative and, while dealing with relevant behavioral problems, it has lacked the necessary input from analytical genetics to be useful in the understanding of the dynamics of the evolutionary process.

In considering the problem as to which of the many balanced systems of a species are most vulnerable to the proposed cycle of disorganization and reorganization, the delicate adjustments of the sexes by the process of sexual selection stand out (see [72,73,61]. The sexual coadaptation system may be the primary testing ground for fitness.

THE ORGANIZATION THEORY OF SPECIATION.

The Phase of Reorganization

One of the major effects of the disorganization described above is that it often may bring the relevant population close to extinction. Numbers become small; adaptations are impaired by stochastic effects. The mean fitness of the population is lowered as the various balanced genetic components of the gene pool are destabilized. If the population is to survive the threatened extinction, then, the generations that immediately follow the disorganization phase become crucial. Under these circumstances, a change in ambient environment is not a necessary prerequisite for genetic change. It is not a matter of the details of the genotype slavishly tracking the environment. What has happened is that the former genetic organizations of the gene pool, its old epistases and balances, are suddenly in disarray. Accoridngly, selection begins to actively form new balances, using the remnant genetic elements segregating in the depauperate gene pool, which may continue to have a small effective size.

The ensuing one hundred to one thousand generations are considered crucial in the building of the organization of the new gene pool, and the synthesis of the new adaptations. In fact, this stage in the life history of the species, in this reductionist view, is the most important one from the point of view of progressive, significant genetic change per unit time. It is during this time that the adaptations characteristic of the species as a whole are forged by mutation, selection, and recombination along with other correlated morphological, behavioral, and physiological novelties of the new species. Basically, it is a gradual, anagenetic intrapopulational process; there is nothing saltational, rectangular, punctuated, concerted, or instantaneous about it. Macromutations and mutations profoundly affecting development are not required. As the gene pool expands in size and gradually equilibrates, the rate of genetic change is gradually reduced. In most diploid organisms, what has been achieved is considered to be a new complex dynamic balance, not a new fixed homozygous state. The biggest change may well be a change in internal genetic environment and interaction between the many component genes.

The balances sought and maintained by stabilizing selection are dynamic ones. Fitness is maintained as a property of a complex heterozygous state. Since few genes are fixed, it is principally the rate of change that becomes dampened down to a basic equilibrated state. Nevertheless, the gene pool and the fitnesses of its component zygotes must be continually maintained; this is a fundamental attribute of natural selection. As pointed out by Muller [74], relaxation of selection with regard to any character would be expected to lead that character to decay down to the level at which selection does indeed operate. Accordingly, although the capacity for the production of new species may decline with age, dynamic equilibrium with the environment would be expected to remain as a property of the gene pool over many millions of generations, indeed, throughout the life of the species. In certain extremely speciose groups such as the Hawaiian Drosophila, it appears that the principal system that is affected by the disorganization-reorganization cycle in this manner is the very complex coadaptational system involved in mate recognition, sexual behavior, and sexual selection. Courtship patterns and secondary sexual characters of males are developed to an extraordinary degree in most of these species [75,71,54,73]. Indeed, acquiring an altered system of mating can serve as an isolating mechanism if and when it is put to the test in a later population context. As Paterson [76] has stressed, isolation seems to be principally a secondary effect emerging from the fine-tuning of the sexual reproduction system. These secondary effects tend to be much more pronounced in animals than in plants, although some of the latter become co-evolved with, and thus dependent on, insect pollinators. In effect, a plant species may acquire specific behavior from association with specific insect or other animal pollinators.

The complex sexual behavior and recognition systems of many animal groups such as mice and Drosophila are of course basically adaptational in nature, in that they are strongly selected to maximize the efficiency of the reproductive effort. Animals that lack complex sexual behavior and wind-pollinated plants, however, are also subject to intense balancing selecting leading to efficiency of reproduction. The sessile nature of plants puts a premium on the maintenance of ecotypic adaptations. Animals choose their habitats and mates and are genetically endowed for this [77]. As habitat choice is absent in plants, it may be postulated that plant speciation is particularly promoted by disorganization of ecotypic adaptations, especially where ancient habitats have been destroyed, as in the modern world [64].

White [1] sees a strong correlation between chromosomal change and speciation; indeed, he considers that such changes are often pivotal in the formation of species. I contest this view and consider that in most cases their fixation is likely to be merely an incidental accompaniment of small population effects and forced selection for reorganization as the species is formed. For example, among 102 species of picture-winged Hawaiian Drosophila, no Robertsonian changes involving chromosome number are known and 67 full species are involved in a homosequential grouping with at least one other species. There is one group of ten homosequential species. Although fixed inversions between these species are absent by definition, 22 show fixed heterochromatin differences [78]. Until such time as precise, biologically meaningful genetic effects can be demonstrated for such differences in repetitive DNA, I will continue to consider this material as "parasitic" or "selfish" [79,80].

Redistribution of this material could be simply a stochastic accompaniment of the organizational changes. Indeed, organizational theory goes far to explain the many curious stochastic effects that often accompany the evolution of new species. As Wright has repeatedly pointed out, random drift of allele frequencies is only one easily-modelled aspect of the stochastic element in evolution. Organizational change forced under shifting balance systems can account for many kinds of curious morphological, behavioral and ecological shifts as new species are formed. These changes do not always follow in lock-step the changes in the ambient environment, a fact that embarrasses the simplistic and poorly-supported theory that every environmental change must be tracked by direct, fixed changes in the genetic structure of the species.

Earlier, mention was made of the life history of the species. It may well be that an old mature species becomes so locked into obligatory balances that this condition is not conducive to the formation of new species, since the genetic system is resistant to the disorganization phase. Such old species thus may not be competent for the budding off of new ones; they may be looked upon as having essentially become inert from the evolutionary point of view.

Conversely, a fairly young species that has perhaps been through only several thousand generations of organizational balance may be capable of early budding off populations capable of disorganization and reorganization. This may account for the repeated observation, in the contemporary fauna and flora, of clusters of very closely related species ("explosive speciation"). I refer to species clusters found in some freshwater lakes (eg Lake Baikal) or species in clusters such as are found in Hawaiian drosophilids.

Relationship of Organization Theory to Other Theories.

Shifting balance theory. My formulation presented here draws much from Wright's ideas, as he himself has pointed out in the discussion at the end of [3]. I find myself in accord with most of the recent emphases he made in an address given to the Chicago Macroevolution Conference in October 1980 [35]. Herein he places, as I do, special stress on developmental interactions of many genes of small individual effect. He explores what may happen as their balance shifts and the population wanders genetically away from a former adaptive peak under conditions imposed in small populations. Wright's theory, while not specifically dealing with the formation of species, contains elements that are especially conducive to the disorganization-reorganization cycle I have stressed.

Wright also includes in his shifting balance theory a phase that he calls "interdemic selection" (eg [34]). Although a number of authors (eg [81]) have assumed that Wright intended to suggest a form of group selection, careful examination suggests otherwise. As I read it, he visualizes a situation in which many small demes exist; these remain in minor genetic contact with one another. One deme among these many may then come under the control of a superior fitness peak. The evolutionary event that makes this possible is essentially the result of an intrademic process and may involve the reorganization of the genome along the lines that I have been stressing. What Wright calls interdemic selection is said to occur following the favored deme's production of surplus population and its subsequent dispersion into the gene pools of neighboring demes causing these also to come under the control of the superior fitness peak. This process is more in the nature of a gene flow or swamping process than of group selection.

I see no role for group selection in the reorganization theory that I have presented. Group selection is not a powerful force in evolution since groups do not present the selection process with a very broad field of variability. Variation *among* groups is miniscule compared with the field of genotypes generated *within* a group by the meiotic sexual system. For group selection to work as well as individual selection, there would have to be as many groups as there are individual genotypes in a sexual population. This seems very unlikely. In 1932 [67] Wright stated the case powerfully:

> . . . with 10 allelomorphs in each of 1000 loci, the number of possible combinations is 10^{1000} which is a very large number. It has been estimated that the total number of electrons and protons in the whole universe is much less than 10^{100}.

Selection among the discrete zygotes which are generated in a sexual population has a power that utterly eclipses any kind of selection among groups, including "species selection" [81].

Phyletic species formation. Frequent references may be found, especially in the paleontological literature (eg [82]) to the hypothesis that species may arise either cladistically or phyletically (anagenetically). The former is a splitting process in which both products of the split are retained. Anagenesis is often conceived of as a process of transformational phyletic evolution in which an older species, without significant reduction in population size, comes gradually to be replaced in toto by a large population that succeeds the older one. It is eventually recognized as a species different from the earlier one. Such judgments require allochronic comparisons of a kind that the population biologist cannot make, since gene pool comparisons are not possible. Although this kind of speciation is a common assumption of paleontologists, I see no overwhelming evidence that forces me to accept it.

I prefer to consider that phyletic evolution is best viewed as the manner in which a small stochastically disorganized gene pool is rebuilt over time. To repeat what has been said earlier, I consider such a process to be a rather long phase in the early part of the life history of the species during which balanced, complex combinations of genes are formed in populations. The process involves mutation, recombination, and selection in large populations. The rules of this game are well-known; they are embodied in the science of population genetics. It can build new gene pools but it cannot accomplish this without a prior extreme attenuation or destruction of at least a portion of the ancestral gene pool.

Punctuated equilibrium. This theory is based on paleontological data. The data are interpreted as showing that evolutionary change is concentrated in brief metastatic periods ("punctuations") that occur as new species are formed. For most of its existence, however, the species remains in stasis ("equilibrium") [45]. In a rough way, this formulation displays analogies with the organization theory presented in this paper. Accordingly, it will be important to explore similarities and differences.

The most general difference is that organizational-shift theory is built on facts gleaned from the genetics and biology of contemporary populations. Accordingly, it can draw on a vast number of facts of types not available to the paleontologist. It can deal with descent and change in experimental populations and thus can claim in some measure to speak about the kinetics of the evolutionary process. Punctuated equilibrium says nothing about process; it is a statement that is simply descriptive of data taken from the fossil record.

Another confusion is the use of the word "punctuation" to refer to what is clearly considered to be the most active and important phase of evolutionary change. Quite understandably, most readers have considered the authors to be indicating events that are occurring virtually instantaneously in time. Stanley [81], for example, seems to think in terms of phylogenetic diagrams, not populations of organisms. He has described the cladistic origin of a species as "rectangular." How more instantaneous can you get? This easily leads to the promulgation of ideas such as saltation, macromutation, hopeful monsters, group selection, and species selection. When pressed, polemics seem to be substituted for reasoned scientific exchange: for example, Gould has referred to his ideas as "a new and general theory of evolution" [46]. "Neodarwinian gradualism" is set up as a straw man and then demolished. That rates of evolution vary in time was not only made explicit by Darwin himself [83] but also was characteristic of much of the basic neodarwinian literature from the beginning [84].

These exaggerated claims are often repeated in dramatic fashion in press reports, even in scientific journals [85]. This manner of presentation seems to have obfuscated what may indeed be an idea useful for the integration of paleontology and contemporary population biology. Gould and Eldredge [45], for example, admit that a "punctuation" is not rectangular but can be ". . . tolerably continuous in . . . time." In an example given by these authors (Kellogg's data, Figure 3 in their paper), the length of time of the shortest "punctuation" I can measure is about 55,000 years. This allows the population geneticist 1833 generations for the formation of a slow-breeding species (like man), with a generation time of 30 years. For Drosophila melanogaster, which can complete 32 generations a year, the number of generations available during the "punctuation" would be 1,760,000. A great deal can occur genetically in a hundred generations let alone a thousand or a million. Indeed, these numbers are sufficient to accomodate the entire cycle of disorganization, reorganization, and perfection stages proposed here by me. Negation of neodarwinism does not follow; macromutation and sudden developmental shifts, of a sort not observable in contemporary populations, need not be assumed.

I have no quarrel with the Gould-Eldredge equilibrium phase. The concept is based on the stasis they observe in the limited morphological phenotypes found in the fossil record over long geological periods. The morphological criteria used by the paleontologist cannot be equated to the gene pool equilibria discussed by me in this paper. Although comparisons between the two concepts of equilibrium are problematical, it is conceivable that there is a relationship.

Simple genetic origin of isolation. Certain authors place emphasis on a small number of genes that may interact with each other and the environment to produce isolation. This results in a very simplistic theory of species origin. Richardson [86] and Tauber and Tauber [40], for example, take the view that speciation may be accomplished if a daughter population succeeds in fixing two or three such genes, the rest of the genome remaining much the same. Following this supposed simple speciation event, of course, other differences could slowly accumulate. Many stasipatric and sympatric schemes appear to rest on similar arguments. I believe that this interpretation is based on an erroneous view that follows from adoption, implicitly or otherwise, of what I have referred to earlier as the classical view of population structure. As such, it is a great oversimplification of the multitudious genetic adjustment that the genome of a species is continually called upon to make.

Throckmorton [87], for example, is at pains to argue strongly against the view that the species must have a complex, organized gene pool. He offers no crucial data to support his view. That electrophoretically detected variability is sometimes minimal between good full species may suggest to some that the genetic structures of the two species are also similar. In my opinion, electrophoretic similarity between species is less a measure of overall genetic similarity than it is of the irrelevancy of this type of variation to adaptive evolution. Indeed, much of it may be neutral to selection. For example, Drosophila heteroneura and Drosophila silvestris show no fixed electrophoretic differences yet are widely divergent in morphology, behavior, and chromosomal inversion polymorphism [88]. Morphology is clearly under polygenic control [89]; *D silvestris* has 11 inversions segregating in its populations, only one of which is present in D. heteroneura.

As has been mentioned earlier, I believe the evidence indicates that direct selection for isolation and isolating "devices" plays only a small role in the dynamics of the speciation process itself. As Paterson [76] has argued, mate recognition systems give every evidence of being perfected under the kind of stabilizing selection invoked in this article. That polygenic systems exist seems an inescapable conclusion. These systems are important to darwinian fitness of the individuals of the species per se; they are frequently observed to evolve great complexities in complete allopatry. Thus what is observed as "premating isolation" between individuals drawn from separate populations appears most often to be an incidental adjunct of the differential adjustment of the genome within each of the two separate populations.

SUMMARY

Evidence that a diploid bisexual species has a complexly-organized, balanced genetic system is reviewed. Stabilizing selection is the main mechanism which maintains the highly variable but coadapted and stabilized gene pool. Speciation is exclusively cladistic and occurs only after a drastic reduction in effective population size. A small isolate of the ancestral population, often under relaxed selection and free genetic recombination, is subject to stochastic forces that effectively destroy the ancestral balanced equilibrium. There is a break in adaptation and in many cases this involves the genetic coadaptation of the sexes. From this sundered and shattered gene pool, natural selection then rebuilds and reorganizes a genetic system which contains unique elements and combinations. This may take several hundred or several thousand generations and is not instantaneous in any sense. Eventually the new system achieves a stabilized and balanced polygenic polymorphism of its own. Genetic isolation from other groups is considered to arise largely as an incidental by-product of the shift in organizational balance. The theory draws much from Wright's shifting balance theory.

ACKNOWLEDGMENTS

I owe a great intellectual debt to the seminal paper of Ernst Mayr (1954) and to discussions, held some years ago, with the late Wilson Stone. Supported by NSF Grant DEB 79-26692.

REFERENCES

1. White MJD: "Modes of Speciation." San Francisco: Freeman, 1978.
2. Carson HL: Speciation and the Founder Principle. Stadler Genet Symp 3:51, 1971.
3. Carson HL: Reorganization of the gene pool during speciation. In Morton NE (ed): "Genetic Structure of Populations." Honolulu: Univ Press of Hawaii, 1973, p 274.
4. Thompson JN, and Thoday JM (eds): "Quantitative Genetic Variation." London: Academic 1979, p 305.
5. Mather K: Variational selection of polygenic characters. J Genet 41:159, 1941.
6. Ayala FJ, McDonald JF: Continuous variation; possible role of regulatory genes. Genetica 52/53:1, 1980.
7. Wagner WH: Biosystematics and evolutionary noise. Taxon 19:146, 1970.
8. Carson HL: The genetics of speciation at the diploid level. Amer Nat 109:83, 1975.
9. Dobzhansky T: A review of some fundamental concepts and problems of population genetics. Cold Spr Harb Symp Quant Biol 20:1, 1955.
10. Mayr E: Where are we? Cold Spr Harb Symp Quant Biol 24:1, 1959.
11. Haldane JBS: A defense of bean bag genetics. Perspect Biol Med 7:343, 1964.
12. John B, King M: Heterochromatin Variation in Cryptobothrus chrysophorus. II. Patterns of C-Banding. Chromosoma 65:59, 1977.
13. Fukuda I, Channell RB: Distribution and evolutionary significance of chromosome variation in Trillium ovatum. Evol 29:257, 1975.
14. Levitan M: Studies of linkage in populations. VII. Temporal variation and X chromosome linkage disequilibrium. Evol 27:476, 1973.
15. Dobzhansky T, Spassky B: Evolutionary changes in laboratory cultures of Drosophila pseudoobscura. Evol 1:191, 1947.
16. Carson HL: Response to selection under different conditions of recombination in Drosophila. Cold Spr Harb Symp Quant Biol 23:291, 1958.
17. Carson HL: Relative fitness of genetically open and closed experimental populations of Drosophila melanogaster. Evol 15:456, 1961.
18. Prakash S, Levitan M: Associations of alleles of the esterase-1 locus with gene arrangements of the left arm of the second chromosome in Drosophila robusta. Genetics 75:371, 1973.
19. Dobzhansky T, Pavlovsky O: Indeterminate outcome of certain experiments on Drosophila populations. Evol 7:198, 1953.
20. Brncic D: Heterosis and the integration of the genotype in geographic populations of Drosophila pseudoobscura. Genetics 39:77, 1954.
21. Vetukhiv M: Integration of the genotype in local populations of three species of Drosophila. Evol 8:241, 1954.

22. Wallace B: "Topics in population genetics." New York: Norton, 1968, p. 481.
23. Dobzhansky T: Genetics of the Evolutionary Process. New York: Columbia, 1970, p. 505.
24. Spencer WP: Mutations in wild populations of Drosophila. Adv Genet 1:359, 1947.
25. Susman M, Carson HL: Development of balanced polymorphism in laboratory populations of Drosophila melanogaster. Amer Nat 92:359, 1958.
26. Smathers KM: The contribution of heterozygosity at certain loci to fitness of laboratory populations of Drosophila melanogaster. Am Nat 95:359, 1961.
27. Cannon GB: The effects of natural selection on linkage disequilibrium and relative fitness in experimental populations of D. melanogaster. Genetics 48:1201, 1963.
28. Lewontin RC: The genetic basis of evolutionary change. New York: Columbia, 1974, p 346.
29. Hedrick PW: Hitchhiking: a comparison of linkage and partial selfing. Genetics 94:791, 1980.
30. Singh RS, Lewontin RC, Felton AA: Genetic heterogeneity within electrophoretic "alleles" of xanthine dehydrogenase in Drosophila pseudoobscura. Genetics 84:602, 1976.
31. Dickinson WJ: Evolution of Patterns of Gene Expression in Hawaiian Picture-winged Drosophila. J Mol Evol 16:73, 1980.
32. Mather K, Harrison BJ: The manifold effect of selection. Heredity 3:1, 1949.
33. Mather K, Jinks JL: Biometrical Genetics. Ithaca: Cornell, 1971, p 382.
34. Wright S: Evolution and the Genetics of Populations, Vol 3. Chicago: University of Chicago Press, 1977, Ch 13.
35. Wright S: Genic and Organismic Selection Evol 34:825, 1980.
36. Allard RW, Kahler AL, Clegg MT: Isozymes in plant population genetics. In Markert CL (ed): "Isozymes IV. Genetics and Evolution." New York: Academic Press, 1975, p. 261.
37. Barker JSF: Inter-locus interactions: A Review of Experimental Evidence. Theor Pop Biol 16:323, 1979.
38. Hartl DE: "Principles of Population Genetics", Sunderland Mass. Sinaner 1980, p 488.
39. Thompson JN: Polygenic influences among development in a model character. In Thompson JN, Thoday JM: "Quantitative Genetic Variation." London: Academic Press, 1979, p. 243.
40. Tauber CA, Tauber MJ: Sympatric Speciation Based on Allelic Changes at Three Loci: Evidence from Natural Populations in Two Habitats. Science 197:1298, 1977.
41. Carson HL, Ohta AT: Origin of the genetic basis of colonizing ability. In Scudder GGE, Reveal JL (eds): "Evolution Today." Proc 2nd Inter Congr Syst Evol Biol Pittsburgh: Hunt Institute, Carnegie-Mellon Univ p 365, 1981.
42. Haldane JBS: The cost of natural selection. J Genet 55:511, 1957.
43. Mukai T: The genetic structure of natural populations of Drosophila melanogaster. I. Spontaneous mutation rate of polygenes controlling viability. Genetics 50:1, 1964.
44. Mukai T: Maintenance of polygenic and isoallelic variation in populations. Proc 12th Inter Congr Genet 3:293, 1969.
45. Gould SJ, Eldredge N: Punctuated equilibria: The tempo and mode of evolution

reconsidered. Paleobiology 3:115, 1977.

46. Gould SJ: Is a new and general theory of evolution emerging? Paleobiology 6:119, 1980.

47. Mayr E: "Systematics and the origin of species." New York: Columbia, 1942, p. 334.

48. Carson HL: Permanent Heterozygosity. Evol Biol 1:143, 1967.

49. Carson HL: Genetic characteristics of marginal populations of Drosophila. Cold Spr Harb Symp Quant Biol 20:276, 1955.

50. Goldschmidt R: "The Material Basis of Evolution." New Haven: Yale, 1940, p. 436.

51. Powell JR: The founder-flush speciation theory: an experimental approach. Evol 32:465, 1978.

52. Grant V: "Plant speciation." New York: Columbia University Press, 1971, Ch 13 and 14.

53. Dobzhansky T, Pavlovsky O: Experiments on the incipient species of the Drosophila paulistorum complex. Genetics 55:141, 1967.

54. Carson HL: Speciation and Sexual Selection in Hawaiian Drosophila. In Brussard PF: "Ecological Genetics: The Interface." New York: Springer-Verlag, 1978, p. 93.

55. Lewis H, Raven P: Rapid evolution in Clarkia. Evol 12:319, 1958.

56. Gottleib LD: Genetic differentiation, sympatric speciation, and the origin of a diploid species of Stephanomeria. Amer J Bot 60:545, 1973.

57. Bush GL: Sympatric host race formation and speciation in frugivorous fflies of the genus Rhagoletis. Evol 23:237, 1969.

58. Capanna E, Gropp A, Winking H, Noack G, Civitelli MV: Robertsonian metacentrics in the mouse. Chromosoma 58:341, 1976.

59. Mayr E: Change of genetic environment and evolution. In Huxley J, Hardy AC, Ford EB (eds): "Evolution as a process." London: Allen and Unwin, 1954, p. 157.

60. Templeton AR: The theory of speciation via the founder principle. Genetics 94:1011, 1980.

61. Lande R: Models of speciation by sexual selection of polygenic traits. Proc Nat Acad Sci USA 78:3721, 1981.

62. Parsons PA: Isofemale strains and quantitative traits in natural populations of Drosophila. Amer Nat 111:613, 1977.

63. Carson HL: The Population Flush and Its Genetic Consequences. In Lewontin RC (ed): "Population Biology and Evolution". Syracuse: Syracuse Press, Syracuse, New York, pp. 123–137.

64. Anderson E: Hybridization of the habitat. Evol 2:1, 1948.

65. Kidwell MG, Nory JB, Feeley SM: Rapid unidirectional change of hybrid dysgenesis potential in Drosophila. Hered 72:32, 1981.

66. Anderson E: "Introgressive hybridization." New York: Wiley, 1949, p. 109.

67. Wright S: The roles of mutation, inbreeding, crossbreeding and selection in evolution. Proc VI Int Congr Genet 1:356, 1932.

68. O'Donald P: "Genetic models of sexual selection." Cambridge: Cambridge University Press, 1980, p. 250.

69. Ehrman L, Spiess EB: Rare-type mating advantage in Drosophila. Amer Nat 103:675, 1969.

70. Anderson WW, Levine L, Olvera O, Powell JR, dela Rosa ME, Salceda VM, Gaso MI, Guzman J: Evidence for selection by male mating success in natural populations of Drosophila pseudoobscura. Proc Nat Acad Sci USA 76:1519, 1979.
71. Spieth HT: Courtship behavior in Drosophila. Ann Rev Entomol 19:385, 1974.
72. Carson HL, Bryant PJ: Change in a secondary sexual character as evidence of incipient speciation in Drosophila. Proc Nat Acad Sci USA 76:1929, 1979.
73. Spiess EB, Carson HL: Sexual selection in Drosophila silvestris of Hawaii. Proc Nat Acad Sci USA 78:3088, 1981.
74. Muller HJ: The Darwinian and Modern Conceptions of Natural Selection. Proc Amer Phil Soc 93:459, 1949.
75. Hardy DE: "Insects of Hawaii, Vol. 12. Diptera: Cyclorrhapha III. Series Schizophora, Section Acalypterae I. Family Drosophilidae." Honoloulu: University of Hawaii Press, 1965, p. 814.
76. Paterson HEH: More evidence against speciation by reinforcement. S Afr J Sci 74:369, 1978.
77. Taylor CE, Powell JR: Habitat Choice in Natural Populations of Drosophila. Oecologia 37:69, 1978.
78. Carson HL: Homosequential species of Hawaiian Drosophila. Chromosomes Today 7:150, 1981.
79. Doolittle WF, Sapienza C: Selfish genes, the phenotype paradigm and genome evolution. Nature 284:601, 1980.
80. Orgel LE, Crick FHC: Selfish DNA: The ultimate parasite. Nature 284:604, 1980.
81. Stanley SM: A theory of evolution above the species level. Proc Nat Acad Sci USA 72:646, 1975.
82. Simpson GG: "The Major Features of Evolution." New York: Columbia University Press, 1953, p. 434.
83. Templeton AR, Giddings LV: Macroevolution Conference. Science 211:770, 1981.
84. Haldane JBS: The effect of variation on fitness. Amer Nat 67:5, 1937.
85. Lewin R: Evolutionary Theory Under Fire. Science 210:883, 1981.
86. Richardson RH: Effects of dispersal, habitat selection, and competition on a speciation pattern of Drosophila endemic to Hawaii. In White MJD: "Genetic Mechanisms of Speciation in Insects." Sydney: Australia and New Zealand Book Co, 1974, p. 140.
87. Throckmorton LH: Drosophila systematics and biochemical evolution. Ann Rev Ecol Syst 8:235, 1977.
88. Sene FM, Carson HL: Genetic variation in Hawaiian Drosophila IV. Close allozymic similarity between D. silvestris and D. heteroneura from the Island of Hawaii. Genetics 86:187, 1977.
89. Val FC: Genetic analysis of the morphological differences between two interfertile species of Hawaiian Drosophila. Evol 31:611, 1977.

Mechanisms of Speciation, pages 435–459
© 1982 Alan R. Liss, Inc., 150 Fifth Avenue, New York, NY 10011

A Role for the Genome in the Origin of Species?

Gabriel Dover

THE ASCENDANCY OF THE INDIVIDUAL

In a perceptive comparison of the English and German schools of biology in the 19th Century, Ernst Mayr [1] underlined what might have been the subconscious philosophical strategies which led, in the former, to a ready, almost inevitable, explanation for the origin of the species based on natural selection whilst in the latter, despite a plethora of natural scientists, no scientific mechanism was forthcoming. The contrast in outlook is neatly encapsulated by the difference between typological and populational thinking. The typological mode of thought, having its roots in the platonic concept of the eidos, considers the "idea" (type) of a species as the true reality with the observed variation between individuals as the uninteresting shadows on the wall. In contrast to this, British naturalists were pragmatically concerned with individual variation that alone constituted living reality leaving the "type" of the species as a statistical average of the group. Montalenti [2] has drawn attention to a similar pervading divide in the natural philosophy of biology between the atoms (individuals) of Democritus and an Aristotelian holistic ideology in which individual differences are dismissed as unessential and not formally part of the grand final cause of species.

Given such polarisation in the appreciation of individual variation, it is not surprising, with the ascendancy of the theory of the origin of species by means of natural selection of individual differences [3], that "typological" and holistic modes were dismissed with a flurry of scepticism. The observed differences in the abundance and distribution of characters could only be due to differential fitness of individuals (adaptiveness). The solution to the key problem of understanding the mechanism by which variation is fragmented into species (ie, the biological and reproductive distinctiveness of one cluster of individuals from another) is seen as an outcome of fluctuations in frequency of individual characters through the generations. The theory, by means of natural selection, for the origin of species and for the origin of polymorphism within species became so persuasively compelling, with scant intellectual imposition, that no evolutionary phenomenon could escape its mode of action. Sewall Wright, in his contribution to the New Systematics [4] pointedly remarked, "The publication of Darwin's Origin of Species was followed by intensive and ingenious attempts to interpret all sorts of species-differences as adaptive under the belief that natural selection was the sole controlling principle of evolution (Wallace). In recent years Fisher has maintained on theoretical grounds that evolution of organisms is completely subject to the net selection-pressures on the separate genes as the history of physical systems is subject to the increase of entropy." Wright went on to demonstrate that this is not necessarily a correct theoretical conclusion. At the same time Lancelot Hogben [5] insisted "that the mechanical view of evolution as the interplay between external selective agencies and mutation rate raised to the status of a universal physical constant in the mathematical formulation of the theory, does not offer the only explanation of the facts."

Both these men were arguing from different perspectives to which I shall return; however, their early grumbles and most decidedly those of Richard Goldschmidt [6] were soon submerged by an avalanche of conjectures (mathematical and pictorial) that were passed on from one generation of biologists to another. With few exceptions there was no concern for the real methodological weakness of the evolutionary theory in its facility to express too much. "It has such enormous experimental facility that one could hardly imagine anything it could not explain. Now the danger of this is that it rules out any incentive to inquire about any other possible mechanism that could explain the observed facts." (Medawar in conversation with Monod, Popper and others [7]).

OTHER MECHANISMS AND THEIR RELEGATION TO SPECIAL CASES

Historically there have been other mechanisms invoked to explain the observed distribution of variation between species and although they have always operated from within the general framework of mendelian populations, their appropriateness is often relegated to special cases. Interestingly, they all have at their operational centre, a partial reliance on the accidental fragmentation of variation by drift. The most formally rigorous of these is the "shifting balance" theory of Sewall Wright [8] in which the spatial fragmentation of populations into smaller subpopulations causes a random clustering of gene variants that could lead to the establishment of a co-adapted complex of epistatically interacting genes. Differences between subpopulations, arising as a consequence of adaptive and nonadaptive differentiation, could subsequently become the basis for selection between them according to their fitness. "The combination of partial isolation of subgroups with intergroup selection seems to provide the most favourable condition for evolutionary advance" (Sewall Wright [4]).

Several other modes of speciation similarly rely on the accidental distribution of variation (whether genic or chromosomal) by accidents of spatial isolation, eg, Mayr [9], White [10], Carson [11], and Templeton [12] (see contributions to this volume). Despite the differences in emphasis between these proposals, species differences and their reproductive isolation are considered to be the result of genetic or karyotypic transformations within isolated populations in which a random sample of genetic elements become established and fixed.

It is interesting that Michael White's model of stasipatric speciation and Alan Templeton's model of genetic transilience can be grouped together as examples of the founder-drift concept. Both require isolates of a larger population to act as founders for the establishment of a new variant population. In Michael White's model a spontaneous karyotypic rearrangement relies on the reproductive biology of small relatively inbreeding demes to establish homozygotes in sufficient numbers for the rearrangement to function successfully as a barrier between the new and old populations. In Alan Templeton's model a genetic transilience (revolution) is a consequence of inbreeding amongst isolated founders which, under a relatively soft selection regime (flush), permits a new isolating intergenic "complex" to increase in frequency. Describing the models in this manner is not to deny their appropriateness for particular populations with particular biologies and geography (see for example Wilson and colleagues [13,14]). The description is structured to highlight the manner in which an element involved with the biological incompatibility between populations is considered to increase in frequency in the first instance. Other factors may or may not influence the rate at which some appropriate number of individuals are of the variant type but essentially this number is related to the number of selected descendants from variant founders that are spatially isolated from the parental populations.

In a similar manner classical allopatric models of speciation require an initial spatial separation of populations in order for differences in genetic organisation, as adaptive responses to different environments, to become the basis for reproductive incompatibility [9]. Although the reasoning behind these earlier models has not the sharpness of Templeton's theories, the models are formally similar. The one remaining consideration is whether reproductive isolation is an accidental byproduct of adaptive divergence as Muller [15] maintained and Mayr implies, or whether it arises as a subsequent reinforcement by natural selection of existing variation in the reproductive biologies of divergent populations [16].

RESURRECTING THE TYPOLOGICAL CONCEPT

It is clear from the above that there are two basic evolutionary forces considered to be responsible for the observed distribution of variation into species and for the reproductive incompatibilities between them. The one is natural selection operating on adaptive differences and the other is nonadaptive accidents of isolation and genetic drift coupled to natural selection. Both concepts depend on the prior existence of either genic or chromosomal variation between individuals. There has been no recourse to typological thinking. The process of distributing variation into species, to give them their biological and reproductive distinction, involves the selective and/or accidental shifts in frequency of alleles that are segregating in strict mendelian proportions at each and every meiosis. So long as mendelian populations are the arena in which evolutionary changes in frequency are considered to take place then explanations for the clustering of genetic elements into species need to provide a mechanism responsible for the changes in question.

In view of this, it would appear tantamount to heresy to resurrect the typological concept and to question the all too solid assumptions of the nature of mutations and the forces of selection and drift responsible for their distribution and abundance. However, a discrete question can be raised, against the background of the evolutionary behaviour of DNA sequences in the genomes of sexual organisms, if there might not be a need to reintroduce a "typological" form of mutation. This is to say that there might exist a means of typing populations into species as a consequence of molecular mechanisms in the genome that are capable of distributing novel types of mutation to all individuals of a breeding population by a process not dependent on natural selection or drift. The molecular mechanisms responsible for genomic alterations confer on them certain dynamic properties vis-a-vis their distribution within a population that sharply contrasts with individual point mutations and chromosomal rearrangements. Genomic alterations, unlike point and chromosomal mutations, are not simply the raw static inputs into processes of selection or drift, for the mechanisms producing them are formally capable, willy-nilly, of spreading the alterations within a population rather than letting them reside between individuals. The consequence of such events is a more rapid fixation of a variant within a population over and above that which occurs (or not) by the processes of selection and drift. A sexually reproducing population may become "typed" (given time) by a genomic mutation that would distinguish it from another population. This process has been termed molecular drive [43,75].

I want to present the data for the above statements and to explore the possibility that the genome, as a constantly changing population of sequences, is uniquely and dynamically capable of featuring in a process of speciation in a way that is inapplicable to single mendelian genes and to gross karyotypic changes. This is not to deny that there are ways in which differences in karyotypes or genes might underlie premating and postmating barriers or general metabolic incompatability between species. However, an important consideration for all speciation models is to elucidate the manner in which a mutational variant, whether of the karyotype, gene complex, or genome, which is supposedly involved with isolating barriers, increases in frequency in a population to the point that the population in question is no longer compatible with another.

There are several current attempts to relate evolutionary changes in genomes to broader issues concerning biological novelty and its distribution. In particular, the genome is beginning to replace the karyotype as the focus of attention for matters of speciation. Although the switch in emphasis is welcome it would be unfortunate if the word genome were simply to replace the word karyotype in the classical literature, and inadvertently inherit the difficulties of genetic fixation that are associated with karyotype models alluded to above.

INTRODUCING THE EUKARYOTE GENOME

During the past decade sufficient data have accumulated on the organisation of genomes of diverse eukaryote organisms to permit some generalities concerning the mode and behaviour of DNA sequences in evolution. Although there are differences between major taxa concerning the extent and rates at which different types of genomic mutation are occurring, it is clear that they are not separately confined to one or other major form of life. The types of evolutionary changes that have taken place in the genomes of sea-urchin species adequately reflect those that have occurred in genomes of wheat and of the insect Drosophila. It is not possible in a cursory account to cite all the important primary data. These have been reviewed and discussed by Davidson and Britten [17], Flavell [18,19], Dover [20,21], and Peacock [22; and see this volume].

Most higher organisms contain nuclear DNA that is several times in excess of the gene coding requirements of the organism. The degree of redundancy and sequence composition varies extensively. A proportion of the excess DNA is composed of families of repetitive sequences in which the length of repeats (2–5000 base pairs) and the size of families (few to several million members) are highly variable between species. Each species genome can contain several hundreds of different families that are distributed in highly diverse ways relative to the chromosomes and relative to each other. Some families are exclusively confined to centromeres and telomeres. Other families, usually but not necessarily less abundant, occur in complex interspersion patterns with each other and with single-copy DNA.

From studies of size and composition of families that are shared between closely related taxa it is clear that there is a continuous process of sequence amplification that has the effect of boosting individual members of families into new subfamilies as well as producing new families from single-copy DNA.

In the first figure some of the genomic patterns that can arise by recurrent amplification are symbolised. Along the top of the figure a tandem array of sequences (family R1) is depicted which is homogeneous in sequence except for point mutations that have accumulated with time. A process of unequal (non-reciprocal) recombination between chromatids within the array may arbitrarily amplify one or more units to produce a variety of subfamilies. If the units are mutant members then variant subfamilies arise (eg, R1' of family R1). Thus, recurrent amplification ensures that there is a greater sequence homogeneity within subfamilies than there is between them. Many families within a species have a subfamily structure. Families that are shared between species show a greater within-species than between-species homogeneity. Measurements of differences between individuals in the abundance and lengths of tandem arrays of ribosomal DNA confirm that unequal exchange is a continuous process producing a turnover of sequences [23].

Occasionally a length of sequence can become trapped in a round of amplification that is composed of a mixture of the original family and an unrelated nonhomologous sequence. This has been defined as a "compound" unit by Flavell and his colleagues [18,24] and can lead to a variety of complex interspersion patterns symbolised in the first figure. The complex patterns of interspersion of different families (R1, R2, R3, etc) is, in part, a reflection of the particular pathway of amplifications that gave rise to it. A consequence of overlapping amplifications is that a family can be generated of a longer length of DNA containing sub-elements (members of different families) that may be present in a different order from other similarly complex families found elsewhere in the genome (see Fig. 1). Such complex interspersion patterns have been extensively characterised in plant genomes [18,19,24], in D melanogaster [25] and in sea urchins [26]. An interesting feature of "compound" unit amplification is that a length of single-copy DNA can become a repetitive family that is interspersed with another. Recurrent amplification in the evolutionary history of repetitive families, shared between species, of both "single" and "compound" units have been described in plants [18], Drosophila [27], sea urchins [26], calf [28], crabs [29], redneck wallabies [30], and primates [31].

There are alternative ways of creating complex interspersion patterns in addition to amplification of sequences based on equal chromatid exchange. Some DNA sequences (analogous to transposable elements) can physically move between chromosomes and extensive and repeated transposition of sequences may also produce the observed patterns of dispersion [32]. Elimination and inversion of sequences will also contribute to the final outcome. The relative contributions of these alternative mechanisms to the overall evolution of genomes cannot be ascertained for the moment and are a matter of debate (see Dover and Doolittle [33]).

THE GENOME IS IN A DYNAMIC STATE OF TURNOVER

From the events symbolised in Figure 1 it would appear that a process of recurrent amplification of sequences should lead to the gradual accretion of new

RECURRENT AMPLIFICATION

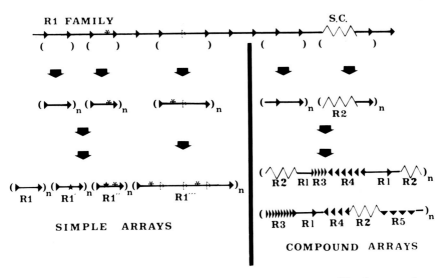

Fig. 1. Schematic outline of a continuous process of sequence amplification to produce from an original simple tandem array of repeats (family R1) a variety of variant but related subfamilies (R1', R1'', R1'''); and two examples of repetitive "compound arrays" consisting of longer units containing scrambled subunits belonging to unrelated families (R2, R3, R4, R5). These latter are produced by recurrent amplifications of units whose lengths bear no simple relationship to the repeating units of R1 and which contain sequences that are not homologous to R1, eg, single-copy (SC) DNA that upon amplification gives rise to the R2 family. On the left symbols represent mutational changes in an R1 sequence. On the right symbols represent nonhomologous families.

families or subfamilies and hence to the steady growth of the genome at a rate dependent on the rate of unequal exchange (or other mechanisms). Different genomic regions clearly do have different facilities for expansion, for new additional chromosome arms and regional size polymorphisms are known to occur. Centromeres and telomeres are regions of high rates of sequence amplification for reasons that are entirely unknown. For example, six of the seven known sibling species of the melanogaster species subgroup of Drosophila have a standard acrocentric X chromosome whereas a seventh, D orena, has a metacentric X in which the new arm is composed of an additional family of sequences unique to the species [27,34,35]. However, it is apparent from a wide survey of species that genomes and karyotypes do not grow indefinitely by the addition of new families, despite the activities of a constant process of sequence amplification. On balance the genomes of related species are in a steady-state condition that is probably achieved by an equal and counterbalancing process of sequence elimination. Although there are exceptions [36,37], the steady state is a reflection of an apparent concomitant addition and elimination of family members by one and the same molecular processes. The processes are often random in direction in that there is no preferred replacement of old for new sequences. Families of sequences are continuously being homogenised, in a stochastic manner, by one variant member or another. This phenomenon is known as concerted evolution. There are observations on the distribution of sequence variation within and between species that are difficult to rationalise on any other grounds (see below).

One of the important features of family homogenisation, and one that has critical consequences for the spread of variants in a population, is that it may proceed irrespective of the chromosome distribution of a family of sequences [38,39,75]. A process of replacement which transgresses chromosomes ensures that, in a sexually reproducing population, a new variant becomes fixed throughout the family of sequences in a population (discussed later). Through sexual reproduction a population can be considered as a pool of chromosomes between which free exchange of sequences is able to take effect at some definable rate. The rate at which a particular family of sequences becomes homogeneous in a population depends on several parameters that are affecting the total biology of the process. The nature of these parameters are described by the following two case-histories of homogenisation and fixation.

FAMILY HOMOGENISATION AND THE EVOLUTION IN CONCERT OF A TANDEM ARRAY OF RIBOSOMAL GENES

The first indication that families of sequences are subject to unusual and unexpected homogenisation processes came from the studies of Wellauer, Reeder, and Dawid on the family of sequences that serve as regular spacers between tandemly repeating genes coding for ribosomal RNAs, in species of Xenopus [40,41]. The multiple-copy genes are highly homologous, both within a species and between species. However, in marked contrast to this the spacers are nonhomologous between the species and yet are homogeneous within a species. A mechanism is operating that homogenises the spacers for new mutational variants in each species so that over time no between-species homology remains. The process is a result of a nonreciprocal transmission of sequence information between the chromosomes of an individual, resulting in a spread of a variant sequence to other individuals within a sexual population (see below).

The concerted evolution of coding and noncoding, tandem and dispersed families suggests that homogenisation may be the fate of large proportions of the genome [38,39,42,43,75].

The family of sequences that comprise the ribosomal RNA genes in Drosophila consists of a regular interspersion of the major two genes (18S RNA and 28S

RNA genes) with spacer DNA such that there is a large spacer separating the end of an 28S RNA gene and the beginning of an 18S RNA gene and a smaller spacer between the end of an 18S RNA gene and the beginning of an 28S RNA gene. The large spacer is for the most part not transcribed, and the smaller spacer is transcribed and eventually spliced from the functional RNAs. In D melanogaster there are approximately 200 copies of such a repeating unit on each of the X and Y chromosomes. About 60% of the 200 units on the X chromosome contain an insertion sequence (type I–INS) within the 28S gene and about 15% of the X and Y units contain a different insertion (type II–INS). Within each nontranscribed longer spacer (NTS) there is a region of short repeating sequences. A full description of the ribosomal units in Drosophila is to be found in Long and Dawid [41]. Given the complexity of the unit that comprises the basic repeat it is ideally suited for an examination of the relative rates at which different regions are evolving. This type of analysis has been made on rDNA of seven sibling species of the melanogaster species subgroup [23,42,44].

All the components within the compound repeating unit show high levels of structural similarity and sequence homology between the species, including the NTSs [42,44]. However, it is possible to identify an NT spacer by the presence of a mutation that is diagnostic for a species. The identification is unambiguous because all the NT spacers of the X and Y clusters within a species are identical with respect to the mutation. The extensive homogenisation for such mutations within and between individuals of a species reflects the process of homogenisation that is occurring within each genome. This is to say that there exists a mechanism which is constantly adding and subtracting units of the family leading to the fixation of a variant unit within a population of repeats, and subsequently within a population of individuals. There is evidence that for arrays of ribosomal genes the mechanism is one of an unequal exchange of sequences between chromatids [23,42,45,46]. Unequal chromatid exchange varies the lengths of repeating units and measurements of differences in lengths of repeats within and between species and suggests that the short internal repetition within the NTS of Drosophila [23] and of other species [41,47] is the "hot spot" of unequal exchange.

In order to assess the rates at which family turnover and unequal exchange might be proceeding, it is necessary to examine the differences between individuals whose genetic relatedness is understood. Our analysis [23] of the organisation of rDNA of individual X and Y chromosomes taken from isofemale lines of one small population of D melanogaster reveals a very extensive polymorphism for lengths of spacers and for the abundances of repeats of each length. Changes in these parameters reflect the continuous nature of unequal exchange. However, the investigations have not revealed any new lengths or abundances during the 1000 generations of laboratory conditions. Furthermore, no differences have been observed between the isofemale lines and between other world-distributed populations for the key diagnostic sequence mutation that char-

HETEROGENEITY ⟨ LENGTH.   HOMOGENEITY →SEQUENCE.
(POLYMORPHISM)   COPY NUMBER.

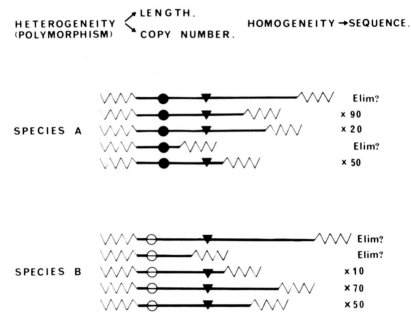

Fig. 2. Symbolic representation of sequence homogeneity and length/ abundance heterogeneity of a repetitive family that is shared between species A and B. Solid line represents the sequence in question; symbols represent mutational changes: ● is diagnostic for A; ○ is diagnostic for B; ▼ is diagnostic for A + B. Numbers on the right represent the copy number of a particular length. Differences in copy number and length exist both within a tandem array of an individual and between individuals. Both types of heterogeneity are considered to be the result of "unequal exchange" within the arrays. Lengths that are too long or too short may be selectively eliminated. References [23,38,41–43] can be checked for the particulars of rDNA polymorphisms within and between species of Drosophila from which the figure is derived.

acterises the NT spacers of D melanogaster [23]. The absence of sequence heterogeneity, in contrast to length and abundance heterogeneity, is a reflection of a low mutation rate relative to the rate of family turnover and homogenisation.

In Figure 2 the species distribution of length, abundance, and sequence variation is depicted. Lengths and abundances of repeating units can differ within a tandem array on a chromosome, between chromosomes of an individual, and between individuals of a population [23]. However, species A can be diagnosed for a sequence mutation in all its rDNA repeats, of whatever length and abundance, which is absent from species B (see legend, Fig. 2). There is no particular favoured mutant that becomes fixed in a species. The process is not one of rectification in a master-slave manner. It is stochastic and relies on the accidental fixation due to repeated cycles of expanding and contracting families of sequences by unequal exchange. Some of the parameters that might be involved in the rates of fixation of a variant sequence in multigene families have been described mathematically [48,49]. An easy analogy for understanding sequence turnover in multiple copy families is the constant replacement of old banknotes by new. The identical nature of the notes obscures the fact that the process is continuous. A new design is a recognisable (mutational) difference that does, in time, replace the old.

FAMILY HOMOGENISATION AND THE EVOLUTION IN CONCERT OF A DISPERSED FAMILY

Unequal exchange between chromosomes probably underlies the sequence co-identity of tandem clusters. For example, it is conceivable that the extensive sequence identity between five clusters of ribosomal genes, scattered on non-homologous chromosomes of primate species, is the result of a low but significant level of unequal exchange between all such regions [50]. However, extensive homogenisation has been found in families that are dispersed in thousands of separate locations on the chromosomes, and alternative molecular mechanisms are considered to be responsible. A family of interspersed sequences that is common to several rodent species and for which there is a high degree of interspecific homology, contains some key diagnostic mutations indicating a process of homogenisation within each species [39]. The family in question is composed of approximately 20,000 members interspersed within each species genome and the distribution of the variation clearly suggests that a turnover of sequences is taking place independently in each genome. There are no large differences in family size between the species and this fact, coupled with the extensive distribution of the units amongst nonhomologous sequences would mitigate against unequal exchange as the mechanism of homogenisation. As yet, there are no indications which particular molecular events are responsible for the observations. There are several mechanisms, formally analogous to "gene conversion," which are capable of homogenising two disparate sequences located on either homologous or nonhomologous chromosomes [39,51–53]. A serious problem, however, is the extent to which homogenisation can proceed within a large interspersed family by simple reliance on a stochastic bidirectional process of conversion. However, a preference for the direction of conversion, no matter how small, would eventually fix the preferred sequence within the family, and subsequently to all the chromosome locations within a sexually reproducing population. There is evidence for a degree of uni-directional "biased" gene conversion during meiosis [54,43,75].

It is possible that very large families, tandem or interspersed, are insufficiently homogenised throughout their membership before a new variant begins to spread in another part of the family. This would have the effect of subdividing large families into segments each of which could be characterised by a mutational variant. Segments have been observed in the abundant satellite DNAs of several species (see [21] for review), and interestingly, in the dispersed rodent family, referred to above [39]. The data suggest that a segment is not confined to a particular chromosome and presumably does not owe its existence to localised amplifications [55] which would lead to regional increases in family size. If the rates of homogenisation are slow then segments would remain a permanent feature of large families in the sense that they would be following one on another, like ripples in a pool. A full discussion of these phenomena are discussed in [43,75].

In Figure 3 a family RI is indicated in which new mutational variants R1′ and R1″ arise. The R1 family is either tandem (upper) or interspersed (lower). For large families a stage is reached at which the RI′ and RI″ variants have replaced part of RI to the extent that three segments co-exist. On the right of the figures RI″ has become fixed in most of the original array. It is feasible that in tandem arrays an entirely extraneous unrelated sequence R2 becomes trapped in cycles of unequal exchange and replaces the RI family. The existence of nonhomologous sequences in the NT spacers of the ribosomal genes of D hydei, D virilis, and D melanogaster, and between species of Xenopus [41] supports this contention. The genomes of related species of Gramineae similarly differ by numerous additions and eliminations of separate families that are suggestive of a process of attrition of one family by another [19,56].

Although the phenomenon of gradual replacement is known to have occurred within a relatively few families of coding and noncoding sequences [38], the number simply reflects those families in which it has been specifically investigated. Many families that are shared between species show a greater within-species than between-species homogeneity and also differences in abundance in the separate genomes [17,18,21,57,58]. Although a sequential process of replacement has not been proven for many of these families the pattern of distribution of the variation is very suggestive of analogous systems. Families in which relatively more rapid changes in position are indicative of a mechanism of active mobility show no evidence, for the few that have been looked at in detail, of independent concerted evolution in separate species [32,59]. The molecular mechanisms responsible for the distribution and rates of change of such nomadic sequences are probably different in kind from those involved with continual family homogenisation [33].

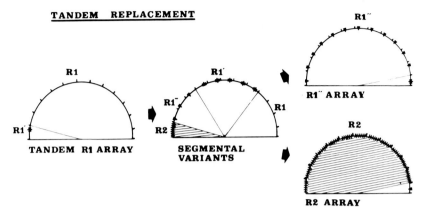

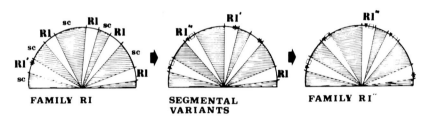

Fig. 3. Symbolic representation of sequence turnover and replacement, (underlying the phenomenon of concerted evolution) for tandem and interspersed repetitive families. In each an original family (R1) is gradually replaced by variant sequences (R1', R1"). If the families are large then a transitional phase consisting of segmental variants exists: The so-called "type B" segments of repetitive families (see text for references). The molecular processes underlying replacement are considered to be different: "Unequal exchange" for tandem arrays; mechanisms analogous to biased "gene conversion" for dispersed families. For tandem arrays, unequal exchange can generate length and abundance heterogeneity (see Fig. 2); and a nonhomologous sequence (R2) may replace R1. The replacement phenomenon occurs within and between individuals of sexual populations.

AN ACCIDENTAL PROCESS OF FAMILY HOMOGENISATION WITHIN POPULATIONS: MOLECULAR DRIVE

The fixation of a variant member within a population of sexually reproducing individuals is in part a consequence of intragenomic homogenisation, whether

INTRAGENOMIC SPREADING

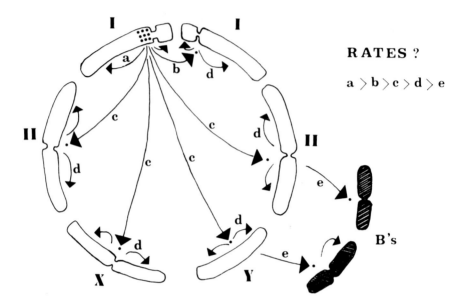

Fig. 4. Symbolic representation of intragenomic replacement of a mutational variant, arising on chromosome I, of a repetitive family dispersed over regular and supernumerary (B) chromosomes. The rate of replacement within the chromosome of origin is probably faster than the rate between homologues. Rates between nonhomologues and between A and B chromosomes are probably slower. Overall rates are primarily determined by the separate rates of diverse mechanisms underlying asymmetric sequence exchange.

by unequal exchange, "gene conversion," or direct transposition. In order to understand the manner in which a variant member may spread in a population it is necessary to examine the spreading processes within a genome.

Figure 4 illustrates the kinds of homogenisation occurring within a karyotype containing two pairs of autosomes, a pair of sex chromosomes and supernumerary "B" chromosomes. It is probable that the rate of homogenisation of a variant is faster in the chromosome of origin (a in figure) than between homologues (b), which in turn is faster than the rate between nonhomologous chromosomes (c). The rates are basically determined by the rates of unequal exchange, "gene conversion," or direct transposition, which differ not only between themselves but also between different parts of the karyotype.

From an analysis of minor differences in X and Y chromosome rDNA in D. melanogaster it would appear that the rate of homogenisation within a chromosome is faster than that between chromosomes [60]. Additionally, some transposable elements of D melanogaster move more frequently to the homologous than to the nonhomologous chromosomes [61]. However, differences in the rate of spread to different chromosomes of a complement might be small relative to the overall rate at which homogenisation can proceed throughout a karyotype, including B chromosomes. For example, it has been noted above that the segmental variants of large families are neither confined to a single chromosome [55] nor to a single species [62], and presumably there is a sufficient rate of unequal exchange or conversion between most chromosomes to ensure a wide distribution of the variants. Sequence homology between A and B chromosomes in species of tsetse [63] suggests that the spread of sequences can take place irrespective of the biological status of the chromosome complements.

The effect of spreading variant sequences to several chromosomes within the lifetime of an individual is that an unexpected non-mendelian segregation of the variant ensues after meiosis. Any facility for homogenising two originally heterozygous sequences on a pair of chromosomes ensures that there is some proportion of the gametes, in excess of one half, which carry the variant in question. The process is not analogous to "meiotic drive." The numbers of chromosomes in a population of gametes carrying a variant is directly proportional to the rate of homogenisation between the chromosomes. Intragenomic homogenisation of tandem arrays is a stochastic outcome of cycles of expansion and contraction of arrays by unequal exchange [23,42,49]. Intragenomic homogenization of families that are finely interspersed require a degree of biased conversion (see above).

The proportion (p) of gametes in excess of one half that carry a variant should result, on the assumption that the sequence is neutral in effect, in the fixation of the variant within a population. If the proportion p is small and the family in question is large and distributed over many chromosomes, then the time taken for complete homogenisation throughout a population of chromosomes is long. Many of the parameters assumed to influence the rates of fixation, within and between individuals have been described in depth by Ohta [49]. Should the variant undergoing homogenisation prove to be detrimental to the fitness of the individuals that carry it, then the rate of fixation will depend on the net balance between the rate of spread between chromosomes (ie, the magnitude of p) and the size of the reduction in fitness. Recent models based on both constant and fluctuating effects on fitness by an element spreading in a population have shown that quite considerable detrimental effects on fitness need to take place to prevent its fixation to the population (John Barrett and Donal Hickey, personal communications). It is interesting to note that a process of homogenisation both within and between chromosomes would, in any event, allow selection to operate more expediently in deciding the fate of the variant in question [23,75].

In summary, the fixation of a variant within a sexually reproducing population, and indeed the observation that considerable homogenisation has occurred within families of sequences in many species, is a consequence of the spread of a variant sequence, by a variety of mechanisms, from one chromosome to another. The unusual molecular behaviour of DNA sequences is such that it could, in principle, cause an increase in the abundance and wide distribution of a variant member of a multiple-copy family, in the patterns in which it is observed. Natural selection due to differences in fitness plus random fixation due to genetic drift may also play their part in contributing to the phenomenon of concerted evolution. However, enough is known about the nonreciprocal transfer of information between chromosomes at the molecular level to raise the possibility of a third evolutionary force involved with the distribution of genetic variation. The process of driving a variant in this manner relies neither on fitness differences nor on the contraction and expansion of populations. In theory a variant should spread to the effective limits of a reproductive population, under its own steam [For details of molecular drive see 43,75].

It is for this reason that the typological concept has been raised. A mechanism which can self-propel a mutation throughout a sexual population is in a sense typing the population with the variant in question. It is not instantaneous; nevertheless there are potentially rapid mechanisms to increase the frequency of a variant in a manner not relying on natural selection or drift. Homogeneity in sequence both within and between individuals has been observed for many families of repeated sequences from single pairs of globin genes to more abundant and dispersed families (see [38,39,42,43,49,50–53]). It is possible that populations are typed by diagnostic variants within a great proportion of their genomes. If changes in some of these genic and nongenic repetitive families were to affect the reproductive or general biology of organisms then it could be that the concerted evolution of a population of individuals has raised the population in question to the status of a species.

It is interesting to note that the extent of genome differentiation between populations as described is a simple reflection of the rates of mutation and homogenisation. The processes responsible for the homogenisation of families are continuous and promote a constant turnover of sequences. This will largely go unrecognised unless a variant sequence becomes fixed within the family or a segment of it. The number of variants that distinguish a family in one population from its counterpart in another would be small, over a given time span, given the low mutation rate and the apparently faster rate of homogenisation [23,43]. The net effect of these contrasting rates of change is that genomes would appear to undergo long periods of stasis followed by episodic periods of change. The stasis is one of illusion, given a constant turnover of nonvariant sequences; and the qualitative "jumps" would reflect the appearance and spread of new mutations.

MOLECULAR DRIVE AND AN ORIGIN OF SPECIES BY ACCIDENT

There are several different routes by which populations may differentiate into species. These are generally classified as pre- and post-mating reproductive barriers. There is no particular reason, given the paucity of information on the molecular biology of eukaryote genomes, for preferring one over the other regarding the possible outcome of continuously homogenising families of coding and noncoding sequences. Transcriptional controls are unknown and the genetics of chromosome behaviour are only partially understood [73,74]. There are many ways by which one could envisage the role of much of the repetitive nature of the genome in either of these fundamental processes [17,43,56,64–66,76]. It is unlikely that the continuously changing nature of sequence composition, abundance, and position of such families would be of zero effect on the behaviour of the chromosomes and their transcription. This is not to say that such effects are the function of repetitive families: There is a distinction between effect and function which needs to be emphasised [67].

Many families are finely interspersed with genes (see [17,20]) and concerted changes in their composition might affect the transcription of the genes possibly by their effects on chromatin accessibility to polymerases. Some such changes might affect the genes controlling sexual behaviour or the time of flowering, and pre-mating barriers could ensue. If the genes affecting the reproductive biology of organisms were themselves members of a multiple-copy gene family, then the molecular mechanisms responsible for within-family homogeneity would act directly on the phenotype.

Notwithstanding the different routes by which the reproductive biologies of organisms might be affected, the phenomenon of molecular drive could ensure that a population of individuals is more or less coincidental in its behaviour as well as its sequences (Fig. 5). Molecular drive imparts a degree of cohesiveness to a population. The problem of sexual isolation of one or a few individuals by mutation of a single gene or by a karyotype rearrangement with the attendant difficulties of increasing the frequency of such isolating elements does not arise. Within any one group of concertedly evolving individuals there would not necessarily be any within-group reproductive incompatibilities unless in extremis, the rapidity of homogenisation is such that all members of a family of a *single* individual acquire a variant identity. If this were the case, and if the reproductive biology of the individual were affected, then there would be a "species" of one member. The rapidity of the process is, in such an event, reducing a family of sequences effectively to one genetic element and the whole mutant family suffers the fate of a lethal point mutation of a single gene. It is possible that the families

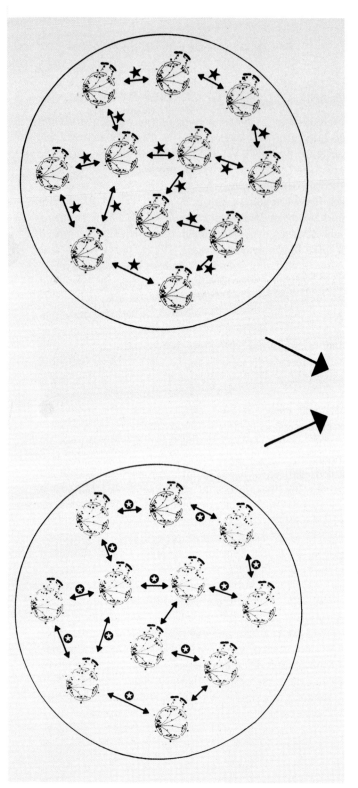

HYBRID

I   II   X/Y   B

MAXIMUM GENOME DISSIMILARITY.
BIOLOGICAL INCOMPATIBILITY ?

of sequences in which concerted evolution has been observed are precisely those in which the homogenisation within an individual is slow relative to the rate of fixation in a population. The size of the family, the number of chromosomes over which it is spread, and the generation time of the organism would contribute to the relative balance between homogenisation within an individual and homogenisation across individuals. The final figure illustrates an extreme possibility in which biological incompatibility takes effect only when maximum genome dissimilarity arises in an interpopulational hybrid. The incompatibility need not be restricted, as drawn, to a post-mating dysgenesis between individuals taken from populations in which the genomes have evolved in concert in their separate ways. The incompatibility could be pre-mating in type, as discussed above. Whatever the nature of the barriers, the important point is that the incompatibility can be envisaged as taking effect only between two individuals that are maximally dissimilar in their repetitive families. Maximality would be achieved in the case of post-mating dysgenesis when two dissimilar haploid complements of chromosomes arrive in the same individual. There is only a low probability, the magnitude of which depends on the size of a family, its chromosome distribution, and the rates of homogenisation, for two maximally dissimilar haploid genomes to arise in a newly formed zygote from *within* a population. As discussed above if the rate of fixation through a sexual population is not inordinately slow relative to the rate within an individual, then only a few diploid individuals would inherit two completely contrasting haploid sets: One with the original members and the other with the variant members. During the homogenisation process the majority of individuals would consist of random mixtures of chromosomes exhibiting a spectrum of new and old variants of a family both within and between the chromosomes. On this basis, there would not be too severe a reduction in individual fitness and the majority of individuals from within a population would remain intercompatible. All the above statements may be transcribed into mathematical language; however, no increased precision (in contrast to clarity) would ensue given our ignorance of the all important relative rates of homogenisation within and between chromosomes, ie, within and between individuals for sexual populations [43,75].

Fig. 5. Two sexually reproducing populations in which a molecular drive of variants of repetitive families (symbolised by stars) is occurring independently in each. Within each individual a process of intragenomic spread of variants (homogenisation) is occurring at a rate dependent on the rates of unequal exchange, gene conversion or direct transposition (see Fig. 4). The rate of intragenomic spread determines the proportion (p) of gametes of an heterozygote, carrying the variant sequence, which are in excess of one half. The magnitude of p determines the rate of fixation of the variant in the population assuming neutrality. Maximum genomic dissimilarity arises in a hybrid between the populations that have concertedly evolved in sequence. This could result in post-mating biological incompatibility. Pre-mating incompatibility is equally as consequential on the process of molecular drive.

A post-mating dysgenesis in hybrids between individuals of different populations of D melanogaster is understood to be the result of stochastic fluctuations in the numbers and, probably, positions of a class of mobile elements [68,69]. Accidental gain and loss of such elements is analogous to the processes of unequal exchange and biased gene conversion that control the fate of variant members of other types of repetitive families. The rates of change in number and distribution of mobile elements are apparently higher than those of other repetitive components [32,42,59,61,69].

Many repetitive elements, whatever their function and dispersion, exhibit a common mode of change in frequency in populations as a consequence of the molecular behaviour of DNA, and not solely as a result of either natural selection due to differences in individual fitness or accidents of genetic drift in fluctuating populations. On this basis, changes in frequency and distribution by molecular drive can be considered as nonadaptive. Any changes in phenotype resultant on homogenisation of variants of repetitive families can be diagnosed, similarly, as nonadaptive. There might be, in the phenomenon of molecular drive, an explanation for much of the phenotypic differences between species considered to be nonadaptive [70,71]. Long periods of morphological stasis punctuated by more rapid periods of change observed in the fossil record of some organisms ([72] and Stanley, this volume), could also be consequential on the molecular processes of continuous genomic turnover followed by the rare fixation of variants, as described above.

The "typing" of populations as a consequence of molecular drive could be considered to be a mode of "accidental speciation" if some of the reproductive or other biological effects described above follow suit. Species by accident is simply to suggest that there are molecular processes capable of spreading mutational variants through populations in essentially a nonadaptive mode. An accidental mutational basis of species differentiation is an old idea reaching back to Muller [15], Goldschmidt [6], and Hogben [5]. An important critical difference is that there are no inherent problems in the present theory in explaining the increase in frequency of mutations possibly involved in the biological isolation of species. This is not to deny that variation between individuals, the result of point mutations in single genes, relies on natural selection in order to increase in frequency. However, the process of adaptation through natural selection should not be confounded with the process of speciation unless it can be shown that the two are the result of one and the same mechanism. If some species differentiation arises in the manner suggested then it is a nonadaptive mode of speciation.

Testing models of speciation is notoriously problematic, and evolutionary theory has been beset with such a difficulty ever since Darwin. However, in lieu of our experimental inability to homogenise variant sequences at will, there are some observations which can be made to refute the theory. For example, it is interesting that the perimeters of molecular drive are delimited by the boundaries between species. If molecular drive is inconsequential to speciation, then one would expect to find in widely scattered species, geographical regions in which different variants have been independently fixed. If it were to be shown that no reproductive incompatibility ensued from genomic differences between such regions, then the proposed dynamic relationship between genomes and species would be seriously weakened. However, in the absence of such observations, there are no grounds as yet for dismissing the contribution of molecular drive to the distinctiveness and cohesiveness of species.

The phenomenon of molecular drive does not appear to be restricted to particular organisms with particular population structures. The molecular behaviour not negate the existence of alternative modes of speciation; however, it is possible that much of the natural history of species differences, which furnish the plurality of proposed speciation mechanisms, are simply the accidental phenotypic results of independent sequence homogenisation. Without wishing to unduly force the caricaturisation, I would suggest that the acceptance of Plato's concept of the eidos of the species (with the concomitant relegation of individual variation to the process of adaptation alone) was conceptually right for reasons that have only recently come to light.

ACKNOWLEDGMENTS

These studies have been supported by grants GR/A 65379 and GR/B 6810.7 from Science Research Council, (GB).

NOTE ADDED IN PROOF

Since writing this chapter, the workings and genetic consequences of molecular drive have been clarified and examined in detail, elsewhere (see [43,75]).

REFERENCES

1. Mayr E: Evolution and the Diversity of Life. Cambridge: Harvard University Press, 1976, p 26.
2. Montalenti G: From Aristotle to Democritus via Darwin: a short survey of a long historical and logical journey. In Ayala F, Dobzhansky T (eds): Studies in the Philosophy of Biology. Berkeley: University of California Press, 1974, p 3.
3. Darwin C: The Origin of Species by Means of Natural Selection. London: John Murray, 1859.
4. Wright S: The statistical consequences of mendelian heredity in relation to speciation. In Huxley J (ed): The New Systematics. London: The Systematics Assoc., 1940, p 161.
5. Hogben L: Problems on the origin of species. In Huxley J (ed): The New Systematics. London: The Systematic Assoc., 1940, p 269.

6. Goldschmidt R: The Material Basis of Evolution. New Haven: Yale University Press, 1940.
7. Medawar P: In Ayala J (ed): Studies in the Philosophy of Biology. Berkeley: University of California Press, 1974, p 357.
8. Wright S: Genic and organismal selection. Evolution 34:825, 1980.
9. Mayr E: Animal Species and Evolution. Cambridge: Harvard University Press, 1963.
10. White MJD: Modes of Speciation. San Francisco: Freeman, 1978.
11. Carson HL: Speciation and the founder principle. Stadler Genetics Symposia. 3:51, 1971.
12. Templeton A: The theory of speciation via the founder principle. Genetics 94:1011, 1980.
13. Wilson AC, Sarich VM, Maxon LR: The importance of gene rearrangement in evolution: evidence from studies on rates of chromosomal, protein and anatomical evolution. Proc Natl Acad Sci USA 71:3028, 1974.
14. Wilson AC, Bush GL, Case SM, King M-C: Social structuring of mammalian populations and rate of chromosomal evolution. Proc Natl Acad Sci USA 72:5061, 1975.
15. Muller HJ: Isolating mechanism, evolution and temperature. Biol Symp 6:71, 1942.
16. Fisher RA: The Genetical Theory of Natural Selection. New York: Dover, 1958.
17. Davidson EH, Britten RJ: Regulation of gene expression: possible role of repetitive sequences. Science 204:1052, 1979.
18. Flavell RB, Rimpau J, Smith DM, O'Dell M, Bedbrook JR: The evolution of plant genome structure. In Leaver C (ed): Genome Organization & Expression in Plants. New York: Plenum Press, 1980.
19. Flavell RB: Molecular changes in chromosomal DNA organisation and origins of phenotypic variation. In Bennett MD, Bobrow M, Hewitt GM (eds): Chromosomes Today. London: Allen, 1981, p 42.
20. Dover GA: Ignorant DNA? Nature 285:618, 1980.
21. Dover GA, Strachan T, Brown SDM: The evolution of genomes in closely related species. In Scudder GG, Reveal JL (eds): Evolution Today. Proc 2nd Int Congr Syst & Evol Biol. Pittsburgh: Hunt Institute, 1981, p 337.
22. Peacock, J: Satellite DNA—change and stability. In Bennett MD, Bobrow M, Hewitt GM (eds): Chromosomes Today. London: Allen, 1981, p 30.
23. Coen ES, Thoday JM, Dover GA: Rate of turnover of structural variants in the rDNA multigene family of D melanogaster. Nature 295:564, 1982.
24. Bedbrook JR, O'Dell M., Flavell RB: Amplifcation of rearranged repeated DNA sequences in cereal plants. Nature 288:133, 1980.
25. Wensink PC, Tabata S, Pachl C: The clustered and scrambled arrangement of moderately repetitive elements in Drosophila DNA. Cell 8:1231, 1979.
26. Anderson DM, Scheller RH, Posakony JW, McAllister LB, Trabert SG, Beall C, Britten RJ, Davidson EH: Repetitive sequences of the sea urchin genome. Distribution of members of specific repetitive families. J Mol Biol 145:5, 1981.
27. Strachan T, Coen ES, Webb DA, Dover GA: Modes and rates of change of complex repetitive DNA families in species of the melanogaster species subgroup. J Mol Biol (in press), 1982.
28. Pech M, Streeck RE, Zachau HG: Patchwork structure of a bovine satellite DNA. Cell 18:883, 1979.

29. Christie NT, Skinner DM: Selective amplification of variants of a complex repeating unit in DNA of a crustacean. Proc Natl Acad Sci USA 77:2786, 1980.

30. Dennis ES, Dunsmuir P, Peacock WJ: Segmental amplification in a satellite DNA: restriction enzyme analysis of the major satellite of Macropus rufogriseus. Chromosoma 79:179, 1980.

31. Donehower L, Furlong C, Gillespie D, Kurnit D: DNA sequence of baboon highly repeated DNA: evidence for evolution by nonrandom unequal crossovers. Proc Natl Acad Sci USA 77:2129, 1980.

32. Young MW: Middle repetitive DNA: a fluid component of the Drosophila genome. Proc Natl Acad Sci USA 76:6274, 1979.

33. Dover GA, Doolittle RF: Modes of genome evolution. Nature 288:645, 1980.

34. Lemeunier F, Dutrillaux B, Ashburner M: Relationships within the melanogaster species subgroup of the genus Drosophila. III. The mitotic chromosomes and quinacrine fluorescent patterns of the polytene chromosomes. Chromosoma 69:349, 1978.

35. Barnes SR, Webb DA, Dover GA: The distribution of satellite and main-band DNA components in the melanogaster species subgroup of Drosophila. Chromosoma 67:341, 1978.

36. Macgregor HC: Some trends in the evolution of very large chromosomes. Phil Trans R Soc Ser B 283:309, 1978.

37. Rees H: DNA in higher organisms. Brookhaven Symp Biol 23:394, 1972.

38. Dover GA, Coen ES: Springcleaning ribosomal DNA: a model for multigene evolution? Nature 290:731, 1981.

39. Brown SDM, Dover GA: The organisation and evolutionary progress of a dispersed repetitive family of sequences in widely separated rodent genomes. J Mol Biol 150:441, 1981.

40. Wellauer PK, Reeder RH: A comparison of the structural organisation of amplified DNA from Xenopus mulleri and Xenopus laevis. J Mol Biol 94:151, 1975.

41. Long EO, Dawid IB: Repeated genes in eukaryotes. Ann Rev Biochem 49:727, 1980.

42. Coen ES, Strachan T, Dover GA: The dynamics of concerted evolution in the ribosomal and histone gene families within the melanogaster species subgroup of Drosophila. J. Mol. Biol. (in press), 1982.

43. Dover GA, Brown SDM, Coen ES, Dallas J, Strachan T, Trick M: The dynamics of genome evolution and species differentiation. In Dover GA, Flavell RB (eds): Genome Evolution. London: Academic Press, 1982, p 343.

44. Tartof KD: Evolution of transcribed and spacer sequences in the ribosomal RNA genes of Drosophila. Cell 17:607, 1979.

45. Petes TD: Unequal meiotic recombination within tandem arrays of yeast ribosomal DNA genes. Cell 19:765, 1980.

46. Szostak JW, Wu R: Unequal crossing over in the ribosomal DNA of Saccharomyces cerevisiae. Nature 284:426, 1980.

47. Fedoroff NV: On spacers. Cell 16:697, 1979.

48. Kimura M, Ohta T: Population genetics of multigene family with special reference to decrease of genetic correlation with distance between gene members of a chromosome. Proc Nat Acad Sci USA 76:4001, 1979.

49. Ohta T: Evolution and Variation of Multigene Families. Berlin: Springer-Verlag, 1980.

50. Arnheim N, Krystal M, Schmickel R, Wilson G, Ryder O, Zimmer E: Molecular evidence for genetic exchanges among ribosomal genes on non-homologous chromosomes in man and apes. Proc Nat Acad Sci USA 77:7323, 1980.

51. Scherer S, Davis RW: Recombination of dispersed repeated DNA sequences in yeast. Science 209:1380, 1980.

52. Jackson JA, Fink GR: Gene conversion between duplicated genetic elements in yeast. Nature 292:306, 1981.

53. Baltimore D: Gene conversion: some implications for immunoglobulin genes. Cell 24:592, 1981.

54. Fincham JRS, Day PR, Radford A: Fungal Genetics. Fourth ed, Oxford: Blackwell, 1979.

55. Brown SDM, Dover GA: The specific organisation of satellite DNA sequences on the X-chromosome of Mus musculus: partial independence of chromosome evolution. Nucl Acid Res 8:781, 1980.

56. Flavell RB: Sequence rearrangement and amplification: major contributors to chromosome evolution. In Dover GA, Flavell RB (eds): Genome Evolution. London: Academic Press, 1982, p 301.

57. Dover GA: The evolution of DNA sequences common to closely related insect genomes. In Blackman RL, Hewitt GM, Ashburner M (eds): Insect Cytogenetics. Symp R Ent Soc London 10. Oxford: Blackwell, 1980, p 13.

58. Moore GP, Scheller EH, Davidson EH, Britten RJ: Evolutionary change in the repetition frequency of sea urchin DNA sequences. Cell 15:649, 1978.

59. Young MW, Schwartz HE: Nomadic gene families in Drosophila. Cold Spring Harbor Symp Quant Biol 45:629, 1981.

60. Yagura T, Yagura M, Muramatsu M: Drosophila melanogaster has different ribosomal RNA sequences on X and Y chromosomes. J Mol Biol 13:533, 1979.

61. Ising G, Block K: Derivation-dependent distribution of insertion sites for a Drosophila transposon. Cold Spring Harbor Symp Quant Biol 45:527, 1981.

62. Brown SDM, Dover GA: Conservation of segmental variants of satellite DNA of Mus musculus in a related species: Mus spretus. Nature 285:47, 1980.

63. Amos A, Dover GA: The distribution of repetitive DNA, between regular and supernumerary chromosomes in species of Glossina (tsetse): a two-step process in the origin of supernumeraries. Chromosoma 81:673, 1981.

64. Miklos GLG: Sequencing and manipulating highly repeated DNA. In Dover GA, Flavel RB (eds): Genome Evolution. London: Academic Press, 1982, p 41.

65. Bennett MD: Nucleotypic basis of the spatial ordering of chromosomes in eukaryotes and the implications of the order for genome evolution and phenotypic variation. In Dover GA, Flavell RB (eds): Genome Evolution. London: Academic Press, 1982, p 239.

66. Finnegan DJ, Will BM, Bayer AA, Bowcock AM, Brown L: Transposable DNA sequences in eukaryotes. In Dover GA, Flavell RB (eds): Genome Evolution. London: Academic Press, 1982, p 29.

67. Maynard Smith J: Overview: unsolved evolutionary problems. In Dover GA, Flavel RB (eds): Genome Evolution. London: Academic Press, 1982, p 375.

68. Bregliano JC, Picard G, Bucheton A, Pelisson A, Lavige JM, L'Heritier PL: Hybrid dysgenesis in Drosophila melanogaster. Science 207:606, 1980.
69. Engels WR: Hybrid dysgenesis in Drosophila and the stochastic loss hypothesis. Cold Spring Harbor Symp Quant Biol 45:561, 1981.
70. Gould SJ, Lewontin RC: The spandrels of San Marco and the Panglossian paradigm: a critique of the adaptationist programme. In: The Evolution of Adaptation by Natural Selection. London: The Royal Society, 1979, p 147.
71. Lewontin RC: Adaptation. Sci Am 239:156, 1978.
72. Stanley, SM: Macroevolution. San Francisco: Freeman, 1979.
73. Dover GA, Riley R: Inferences from genetic evidence on the course of meiotic chromosome pairing in plants. In Riley R, Bennett, MD, Flavell RB (eds): The Meiotic Process. Phil Trans R Soc Lond B 277:313, 1977.
74. Baker BS, Carpenter ATC, Esposito MS, Esposito RE, Sandler L: The genetic control of meiosis. Ann Rev Genet 10:53, 1976.
75. Dover GA: Molecular drive: a non-Darwinian mode of evolution. Nature: (submitted).
76. Stern H: Chromosome organisation and DNA metabolism in meiotic cells. In Bennett MD, Bobrow M, Hewitt G (eds): Chromosomes Today. 1981, p 94.

Mechanisms of Speciation, pages 461–470
© 1982 Alan R. Liss, Inc., 150 Fifth Avenue, New York, NY 10011

Coadaptation of the Genetic System and the Evolution of Isolation Among Populations of Western Australian Native Plants

S.H. James

INTRODUCTION

Many Western Australian native plant species set reduced numbers of seeds following self pollination relative to cross pollination. This effect is very pronounced in Anigozanthos (the Kangaroo paws) [1], in many species of Stylidium (the trigger plants) [2–6], in many species and forms of Laxmannia [Keighery, personal communication], and it occurs commonly within Myrtaceae [7].

In these groups, "self-incompatibility" is not due to an inhibitory interaction between participatory tissues following pollination and prior to fertilization, but is due to postzygotic abortion. Thus, following self pollination in these groups, normal numbers of ovules are fertilized, but only restricted numbers develop into formed seeds. Partly developed aborted products may be discerned within such ovaries. This response indicates that the individual plants of these groups are heterozygous for recessive lethal factors (or for factors capable of generating zygotic lethals through recombination).

Recessive lethal factors have been demonstrated in Isotoma petraea complex hybrids [8–10] where permanent hybridity is maintained by a balanced lethal system operating postzygotically in the almost totally autogamic lineages of this species.

Anigozanthos and Stylidium, however, are not inbreeders. The kangaroo paws are bird pollinated [11], while the trigger plants provide a classic example of floral adaptation to promote allogamy. In groups such as these, any hybridity maintained by recessive lethality must be impermanent and therefore variable in space and time.

Few studies have been made to assess the importance and consequences of recessive lethals in the genetic architecture of plant population systems.

In this paper, relevant observations on some of these groups are presented as a series of "case histories." They are then used as a basis for interpreting aspects of population differentiation and for generating speculations about the process of speciation in these groups and, perhaps, more generally among plants. They invariably support C.D. Darlington's contention [12] that the evolution of genetic systems may be interpreted in terms concerning the pursuit of hybridity and the control of recombination.

CASE HISTORIES
Isotoma petraea (Lobeliaceae)

This species occurs in small isolated populations on granites and other rocky areas throughout the dry, central Eremean Province of Australia. It is largely self pollinating. It has a haploid chromosome number of $n = 7$, and usually two terminally localized chiasmata per bivalent at meiosis. The populations carry a small proportion of "floating" interchange heterozygotes, but there is a primitive chromosome end sequence which is constant over the bulk of the species' distributional range and which predates the speciation of its eastern relatives, I. axillaris and I. anaethifolia. The plants are herbaceous perennials, quite floriferous, and contain 800–1500 ovules per ovary; they produce enormous numbers of very small seeds. Interpopulational crosses yield heterotic progeny.

Hybridity levels may be generated, in this system, by crosses, including rare interpopulational crosses. This hybridity will be dissipated by inbreeding until limiting levels are achieved. At this point, selection for hybridity levels comparable to that of the parent becomes intense. The generation of adequate numbers of sufficiently heterozygous seeds is facilitated by the terminally localized chiasmata, $n = 7$, high ovule number and small seed size, and the high reliability of the self pollination habit. Such a genetic system might discard 127 of every 128 seeds, and this implies a genetic load greater than 99%.

Strong evidence that this interpretation is correct arises from the fact that complex hybridity has arisen in the southwestern corner of the species, distributional range. Complex hybridity evidently arose as a zygotically balanced ring of six ($\odot_6$) on Pigeon Rock, about 100 km north of Southern Cross in Western Australia. Once initiated, the genetic system spread throughout populations in a southwesterly direction, picking up interchanges, improving its lethal system, and evolving into complete $\odot14$ complex heterozygotes [8,9].

The complex hybrids of Isotoma are meiotically very irregular, with levels of disjunction as low as 20%, in both anthers and ovules. Coupled with the zygotic lethal system, it would appear that the complex hybrids carry a genetic load of 90% relative to the seemingly fertile structural homozygotes they have displaced. But this merely highlights the inefficiency of the primitive genetic system, and supports the proposal that the primitive system is burdened with an even higher genetic load, whereas the blatantly irregular complex hybrids are much more efficient in conserving heterozygosity.

Analysis of the lethal system in ⊙12 Bencubbin and its synthetic twin hybrid progeny [10] provides evidence of chromosomal segment transpositions in the complexes. Transpositions have two important properties. First, segregation from transposition heterozygotes may result in deficiencies, which may act as lethal genes, and duplications. Secondly, transpositions and duplications provide enhanced recombinational capabilities for their carriers. Utilizing these properties, a model accounting for the origin of complex hybridity and its evolutionary elaboration has been proposed [9]. It is probable that complex hybridity arose in the selfed progeny of a plant heterozygous for two transpositions involving three chromosomes, possibly following a rare interpopulational crossing event.

The evolutionary elaboration of complex hybridity in Isotoma has been accompanied by several other changes of biological significance. In the first place, the self-pollination habit has become more intense, the flowers have become smaller, and the numbers of ovules per ovary have increased. But more importantly there appears to have been a redistribution of variation in the species. Preliminary studies which require extension and consolidation indicate that the complex hybrids exhibit less variation within plants, between plants in selfed progenies, and between plants in populations than do the primitive structural homozygotes, but they exhibit greater levels of variation between populations. Also, crosses between populations yield negatively heterotic progeny; the coadaptation which existed between populations of primitive forms has been replaced by a coadaptation between the two complexes constituting the hybrids.

In summary, the cytoevolution of Isotoma petraea suggests the following: 1) Very high genetic loads may exist in nature. Indeed the 90% reduction in reproductive capacity of the complex hybrids must represent a quite substantial improvement over the genetic load carried by the structural homozygotes they replace. 2) Transposition heterozygosity enables the generation of lethals (deficiencies) and chromosome structural change. This capability, in Isotoma petraea (suitably preadapted through its high levels of autogamy, terminally localized chiasmata, and low chromosome number), has resulted in the assembly of functional complex heterozygotes at one stroke. 3) Once initiated, the evolutionary elaboration of complex hybridity is orthogenic. 4) The evolution of this conservative genetic system is associated with a loss of interpopulational coadaptation with respect to interpopulational hybrid vigour, and with increased interpopulational divergence with respect to morphological attributes.

Laxmannia (Liliaceae)

In some forms of Laxmannia sessiliflora, the organization of the genetic system is very simple, and may be taken to represent the primitive condition. Here, the small fly-pollinated white flowers are open pollinated and although self compatible, will not set seed unless pollinated by vectors. The haploid chromosome number is 4, chiasmata are randomly located in the bivalents, and there is one ovule in each of the three ovarian loculi. Full complements of seed are set on self pollination.

In other forms of Laxmannia sessiliflora, and in other species of Laxmannia, departures from this primitive genetic system occur. Importantly, these different forms exhibit automatic self pollination on flower closure when the anthers are brought into contact with the stigma. Self pollination appears to be the usual pollination method in these forms, and it is associated with a variety of devices which can best be interpreted as facilitating the pursuit of hybridity. These devices include zygotic lethal systems, localization of chiasmata, interchange hybridity and dysploidy, increased numbers of ovules, and polyploidy. These investigations are being assembled into thesis form by Mr. G.K. Keighery, and should be published in the near future.

In summary, the studies on Laxmannia strongly support the contention that the evolutionary pursuit of hybridity is orthogenic and towards increasingly conservative genetic systems. The ubiquity of conservative elements within the genetic systems of Laxmannia suggests that perhaps a majority of its taxa may be conservative derivatives of a relatively few more open forms.

Stylidium (Stylidiaceae)

Trigger plants are unique with respect to their pollination mechanism which is clearly adapted to utilize insect vectors, but which cannot exclude geitonogamy. The plants are self compatible in that no stigma or stylar inhibition of the male gametophyte after self pollination is in evidence. Little information on the genetic systems of Stylidium species outside Western Australia is available, but the few analyses we have made indicate that free seed set occurs in these forms following self pollination. The primitive chromosome number in Stylidiaceae $n = 15$ and this number, or its euploid derivatives, is characteristic of all its species occurring outside Western Australia.

Within Western Australia, however, chromosome numbers in Stylidium are extremely varied, with $n = 5 \leftarrow 15 \rightarrow 16$, and polyploidy occurring on $x = 13, 14$, and 15. Numerous dysploid series occur, and many closely related species pairs differ in chromosome number. The abundant and dysploid speciation of Stylidium in Western Australia is associated with recessive lethal polymorphisms which discriminate between selfing and cross-pollination events by eliminating most of the self-pollination products postzygotically. Recessive lethal polymorphisms are characteristic of most Western Australian species we have studied, the major exceptions being the ephemeral annuals of Section Despectae [2–5, and Burbidge, personal communication].

The patterns of chromosome variation and the occurrence of lethal systems in Stylidium crossocephalum have been studied by D.J. Coates [5,6,13]. This species exhibits remarkable levels of karyotypic variation. Some 80 different haploid karyotypes were identified in a sample of about 250 plants taken from throughout the species, distributional range (coastal sandplains, north Perth, Western Australia). Both polymorphic and polytypic variation in karyotype form could be discerned. The species is divided into a mosaic. Within any unit of the mosaic, one or a very few chromosome arrangements occurred as homozygotes, while several other arrangements were observed in heterozygous condition. Some chromosome arrangements could be related to others as simple recombinants while other quite distinctive forms also occurred.

Hybrids between plants from different localities, and having different karyotypes, were made. These were structurally heterozygous, exhibiting multiple chromosome associations at meiosis, and were almost completely sterile. However, pollen-sterile plants were not found in nature, even in transects connecting distinctly different chromosomal forms.

Seed set following self pollination in S crossocephalum is much less than that following cross pollination, and considerable variation with respect to seed set and the level of discrimination between cross and self pollination exist within and between populations. It would appear that each population is composed of its own uniquely coadapted array of lethal factors. Interpopulational cross pollinations yield interesting seed sets when viewed in relation to the boundaries defining karyotypic mosaic segments; interpopulational crosses within mosaic elements tend to result in seed set levels below those of intrapopulational cross pollinations, while crosses involving populations from different mosaic segments tend to result in seed-set levels comparable with or better than intrapopulational crosses.

It has been suggested [6] that transposition heterozygosity may underly the lethal systems, the chromosome repatterning, and the dysploid evolution within Stylidium. This suggestion is strongly supported by the multiple chromosome associations and segregational sterility observed in synthetic heterogenomic interpopulational hybrids in S. crossocephalum.

The hypothesis is that each chromosome acts essentially as a single locus in heredity, and each chromosome may carry from one to many recessive lethals, which may be deficiencies, or it may be alethal. Thus, if n different lethals may be carried on a single chromosome, 2^n different chromosome alleles could be constructed by recombination; one of these is alethal, the remainder carrying from 1 to n lethal factors. Selection would operate against chromosome alleles carrying the greatest number of lethal factors so that a coadapted subset of chromosomes would be sifted out. The remaining lethal chromosome alleles would be subjected to negative frequency-dependent selection. The multiple allelic array of chromosome alleles so established would provide a basis for discrimination between cross and self pollination. The efficiency of this discrimination would depend upon the number of chromosome alleles, the relative frequency of the alethal chromosome allele, and the number of chromosomes showing this type of polymorphism. Realistic levels of discrimination can be achieved with short multiple allelic series at many chromosome loci, or longer series at fewer loci. It is probable that the different populations vary with respect to these parameters.

Also, it is probable that different isolates from the same ancestral population carry differently coadapted subsets of the same set of chromosome alleles. These subsets may not be mutually coadapted, so that a chromosome allele from one population may overlap, in lethal content, with two or more of the chromosome alleles from the second population. Crossing between such populations would be characterised by reduced seed set. The implication here is that the karyotypic mosaic units in S. crossocephalum may represent collections of populations whose internally coadapted but relationally uncoadapted lethal arrays derive from a common source population.

The boundaries between these mosaic segments would thus represent interfaces between more ancient isolates. Although cross pollinations involving plants from different sides of such boundaries may set higher numbers of seeds, suggesting perhaps, that different chromosomes may support the lethal arrays in different mosaic segments, the products of such crosses are grossly structurally heterozygous and highly sterile.

In summary, evolutionary adjustment of local gene pools in Stylidium crossocephalum (and other Stylidium species) leading to coadaptation of their lethal gene arrays may lead to the genetic divergence of neighbouring populations to a greater or lesser degree. Population interaction must result in the generation of new arrays of chromosome alleles, from which new coadapted gene pools will be selected. If this dynamic coadaptive process fails, population interaction would provide a situation in which barriers to genetic communication would be adaptive; speciation through reinforcement would be promoted.

Dampiera linearis (Goodeniaceae)

This species contains diploid, tetraploid, and hexaploid forms. Clearly the diploid condition is primitive. The diploids are confined to two areas, one the lateritic plateau and scarp slopes of the Whicher Range, south of Busselton, the other around Albany, both in south Western Australia. All euploid forms are self-incompatible in that pollen germination and pollen tube penetration is inhibited on selfing. The flowers of this species are uniformly uniovulate. The diploids have a chromosome number of $n = 9$, usually with a single chiasma per bivalent [14].

Clearly, the tetraploids and hexaploids are derived from the diploids. However, the significant genetic system I wish to review is the diploid B chromosome-carrying system of populations on the ecotonal slopes of the Whicher Range

[14]. One to six B chromosomes may be found in some of the diploid plants occurring in these populations. They exhibit three significant properties.

1. Plants without B chromosomes may have pollen fertility levels of more than 90%, or less than 80%. The higher levels of sterility indicate the plants to be heterozygous for conditions resulting in synthetic gametic lethals through segregation at meiosis. Plants with B chromosomes tend to exhibit only the higher levels of fertility. It appears that the B chromosomes restore fertility to the otherwise more sterile hybrids. (Increasing sterility also appears to accompany increasing numbers of B chromosomes).

2. B chromosomes may form physical associations with one of the A chromosomes in bivalents at meiosis. This is especially apparent at diplotene where the B chromosomes themselves form "pseudo-multivalent" associations which may attach to three or more chromosomes in three or more separate bivalents. This anchoring of one chromosome of several bivalents into a pseudomultivalent B chromosome association may well coordinate bivalent congression onto the M-1 plate and serve as a conservative device restricting segregational capabilities at meiosis. This capability could account for the restoration of fertility to certain hybrid genotypes.

3. Some of the B chromosome-containing plants exhibit nondisjunction of A chromosome bivalents at A-1, leading to aneuploid meiotic products. The nondisjoining bivalents are often seen to be physically associated with B chromosomes, and the nondisjunction is polarised in that the A bivalents nondisjoin to the A-1 pole having the greater number of B chromosomes. It is conceivable that extreme events involving nondisjunction could result in nonreduction and the production of diploid gametes to give rise to tetraploid plants.

These observations, deductions, and speculations have been incorporated into a proposal concerning the evolutionary origin of tetraploidy in D. linearis as follows.

The primitive diploid D. linearis was adapted to the laterite conditions of the Whicher Range and possessed genetic endowments rendering it incapable of penetrating the newer environments of the coastal plain. However, normal evolutionary processes associated with diploid sexuality resulted in the assembly of coastal plain-adapted genomes in the ecotonal periphery of the laterite-adapted population.

These novel genomes, however, would be disassembled by sexual processes following gene flow from the continuously adjacent parental population unless conserved. B chromosomes, curiously enough, provided the conservative device in this instance in a system reducing segregation in genomically hybrid individuals. A byproduct of B chromosome-A chromosome interaction, however, was the production of unreduced gametes which combined to yield tetraploids, now fully isolated from their diploid ancestors, and these tetraploids won the coastal plain.

In summary, it is proposed, with some evidence, that B chromosomes mediated the tetraploid speciation event in Dampiera linearis; the B chromosomes found their initial adaptive utility as a conservative device in heterogenomes, but the speciational event derived from their serendipitous property of inducing polarized nondisjunction at meiosis.

Anigozanthos (Haemodoraceae)

The kangaroo paws of Western Australia have been studied in considerable biosystematic detail by S.D. Hopper [1, 16, 17]. The flowers of these plants are of quite remarkable colours, texture, and conformation, and they are bird pollinated. Seed set after self pollination is much lower than that following intrapopulational cross pollination. No stigmatic or stylar incompatibility is in evidence following self pollination, while the many partly developed but aborted seeds in selfed ovaries attest to the operation of postzygotic lethal systems.

All species of Anigozanthos are diploid, with n = 6. No evidence for intraspecific cytological variation is presently available, but interspecific hybrids commonly exhibit one or two interchange or inversion configurations at meiosis, and high levels of pollen sterility. Interspecific hybridization is common within the genus, and Hopper has shown that the components of reproductive isolation separating species are multiple, and include pollinator behaviour, mechanical attributes of the flowers, ecological preferences, hybrid infertility, and temporal differences in flowering.

The genus provides examples of all the major stages postulated in the classical pattern of geographical speciation, ranging from allopatric geographical races with weak reproductive isolation to sympatric fully isolated biological species. The kangaroo paws appear to be much more ''conventional'' than the case histories reviewed earlier.

Nevertheless, the kangaroo paws exploit recessive lethal systems to discriminate between cross and self pollination. In view of the chromosomal uniformity of the genus, it is unlikely that the lethals, in most cases at least, derive from transposition heterozygosity.

Interpopulational crosses within kangaroo paw species are quite variable in their results, and few consistent patterns can be detected. One situation of interest, however, is that concerning the dwarf form of A viridis; interpopulational crosses within this form, and within the tall normal form, yield much higher seed set than interpopulational crosses between the tall and dwarf form. This suggests that the morphological boundary here, which may well be associated with pollinator discrimination, may also mark a discontinuity in lethal system coadaptation. Morphological boundaries of taxonomic significance in Anigozanthos are frequently associated with pollinator discrimination, and they may also mark chromosome structural discontinuities.

CONCLUSIONS AND SPECULATIONS

The case histories reviewed above encourage the following conclusions:

1. The pursuit of hybridity is a potent and ubiquitous force in the cyto-evolutionary dynamics of populations.
2. The evolution of conservative genetic systems, once initiated, is orthogenetic, and a large proportion of existing taxa are conservative derivatives of more open systems.
3. The pursuit of hybridity is facilitated in many plant groups by arrays of multiple alleleic recessive lethals.
4. Such populations are coadapted with respect to the frequency distributions of their multiple allelic lethal arrays so as to maximize the discrimination between cross and self pollination.
5. Recessive lethals may be deficiencies of chromosome segments arising from segregation in transposition heterozygotes.
6. Barriers to genetic communication between populations will arise merely as consequences of the coadaptation of recessive lethal systems (especially deficiencies) within populations.
7. Recessive lethal systems arising from transposition heterozygosity are associated with dramatically enhanced chromosome repatterning capabilities, ie, the capacity to engender genetic, or cytogenetic, revolution.

In addition to these conclusions, the following observations may be made:

1. Coadaptation and genetic conservation may be promoted by i) restricting the recombinational capabilities of hybrids, and ii) avoiding hybrid formation. In this latter sense, devices promoting reproductive isolation between differentiated gene pools, devices both prezygotic and postzygotic, are clearly conservative.
2. The strength of the reproductive isolation established by intrapopulational coadaptation will determine the outcome of population interaction.

Following the interaction of divergent populations, either a new coadapted population will be generated (population fusion) or the two populations may diverge further (reinforcement), and in either case, stabilized serendipitous isolates may be generated.

Thus, in certain instances, speciation may be initiated as reproductive discontinuities emanating from the forces of coadaptation operating within populations, and fashioned where the coadaptive abilities of the genetic system are incapable of achieving population fusion. This mode of speciation is most likely among groups of organisms disposed to deme formation and inbreeding (especially hermaphroditic plants) but is unlikely in groups whose enhanced vagility and enforced biparental sexuality diminish deme formation. The population genetic architecture in these two contrasting types of species is likely to be quite different. In the deme-forming inbreeder, conservative devices built into the genetic system in response to a pursuit of hybridity may predispose the population to sharp episodes of extensive chromosomal reorganization, leading to new coadaptations, to increasingly significant barriers between populations, and to speciation. The vagile outbreeder, on the other hand, having avoided the need to encumber its gene pool with devices inserted by a pursuit of hybridity, will be that much more resistant to this type of speciation.

REFERENCES

1. Hopper SD: A biosystematic study of the kangaroo paws, Anigozanthos and Macropidia (Haemodoraceae). Aust J Bot 28: 659, 1980.
2. James SH: Chromosome numbers and genetic systems in the triggerplants of Western Australia (Stylidium, Stylidiaceae). Aust J Bot 27: 17, 1979.
3. Banyard BJ, James SH: Biosystematic studies in the Stylidium crassifolium species complex (Stylidiaceae). Aust J Bot 27: 27, 1979.
4. Farrell PG, James SH: Stylidium ecorne (F.Muell. ex. Erickson and Willis) Farrell and James comb. et stat. nov. (Stylidiaceae). Aust J Bot 27: 39, 1979.
5. Coates DJ: Cytoevolution and Speciation in the Scale Leaved Triggerplants (Stylidium Section Squamosae). PhD Thesis, University of Western Australia, 1979.
6. Coates DJ, James SH: Chromosome variation in Stylidium crossocephalum (Angiospermae:Stylidiaceae) and the dynamic coadaptation of its lethal system. Chromosoma (Berl) 72: 357, 1979.
7. Rye B: Chromosome Numbers, Reproductive Biology and Evolution in the Myrtaceae. PhD Thesis, University of Western Australia, 1980.
8. James SH: Complex hybridity in Isotoma petraea I. The occurrence of interchange heterozygosity, autogamy and a balanced lethal system. Heredity (Lond) 20: 241, 1965.
9. James SH: Complex hybridity in Isotoma petraea II. Components and operation of a possible evolutionary mechanism. Heredity (Lond) 25: 53, 1970.
10. Beltran IC, James SH: Complex hybridity in Isotoma petraea III. Lethal system in $\odot_{12}$ Bencubbin. Aust J Bot 18: 223, 1970.
11. Hopper SD, Burbidge AH: Assortative pollination by red wattlebirds in a hybrid population of Anigozanthos Labill. (Haemodoraceae). Aust J Bot 26: 339, 1978.
12. Darlington CD: The evolution of genetic systems. Cambridge, London, 1939.
13. Coates DJ: B chromosomes in Stylidium crossocephalum (Angiospermae: Stylidiaceae). Chromosoma (Berl) 77: 347, 1980.
14. Bousfield LR: Chromosome Races in Dampiera linearis R.Br. PhD Thesis, University of Western Australia, 1970.
15. Bousfield LR, James SH: The behaviour and possible cytoevolutionary significance of B chromosomes in Dampiera linearis (Angiospermae: Goodeniaceae). Chromosoma (Berl) 55: 309, 1976.
16. Hopper SD: The structure and dynamics of a hybrid population of Anigozanthos manglesii D. Don and A. humilis Lindl. (Haemodoraceae) Aust J Bot 25: 413, 1977.
17. Hopper SD: Speciation in the Kangaroo Paws of South-Western Australia (Anigozanthos and Macropidia: Haemodoraceae). PhD Thesis, University of Western Australia, 1978.

Mechanisms of Speciation, pages 471–478
© 1982 Alan R. Liss, Inc., 150 Fifth Avenue, New York, NY 10011

Cytogenetic Evidence on the Nature of Speciation in Wheat and Its Relatives

Ralph Riley

INTRODUCTION

Extensive studies have been made of the cytogenetic relationships of species of wheat (Triticum) and of related genera including Secale (rye), Aegilops, and Agropyron because of the economic importance of wheat and rye. Clearly, the practical objectives of this research have been to enable the controlled transfer of useful genes into the crop species from related species and genera. However, examination of factors affecting gene flow between species is clearly relevant to the understanding of reproductive isolations and therefore of speciation.

The purpose of this paper will be to draw attention to evidence showing that reproductive isolation, effected by the limitation of chromosome pairing and recombination at meiosis, can be enhanced by one-step changes in the nuclear environment. The possibility will be raised that such enhanced limitations of pairing depends upon the interaction of the changed nuclear environment with the different reiterated DNA sequences possessed by the potential pairing-partner chromosomes.

As a preliminary it may be useful to indicate that a large proportion of the genome of wheat and its relatives is made up of other than structural genes. Appropriate calculations are set out in Riley [1], which estimate that the 1C DNA content of any one of the three genomes combined in the hexaploid wheat Triticum aestivum is about 5.8 pg. If there are 10,000 structural genes, averaging about 1 kb in length, then 0.185% of the genome constitutes conventional informational content. There may be three times as many structural genes, but even so it appears that less than 1% of the wheat genome is structural gene DNA, and certainly a very high proportion of the genome consists of high copy number repeat sequences [2].

MODIFICATION OF MEIOTIC CHROMOSOME PAIRING
Aegilops speltoides × Triticum boeoticum

Ae speltoides and T boeoticum are both diploid species with 2n = 14. T boeoticum is an inbreeding species and a close relative of the donor of the A genome to hexaploid wheat. Ae speltoides is an outbreeder—widely distributed in land areas at the eastern end of the Mediterranean—and while it may not be precisely the species which contributed the B genome to hexaploid wheat it has many characters of that parent. Some forms of Ae speltoides carry supernumerary chromosomes.

While the supernumerary chromosomes have no apparent influence on the pairing at meiosis of other chromosomes of the Ae speltoides complement, when the Ae speltoides × T boeoticum hybrid is made the presence of a supernumerary chromosome results in a sharp reduction in pairing (Table I). The level of meiotic pairing in the absence of a supernumerary is sufficiently low for hybrids to be sterile, but seed may be set on backcrossing. However, in the presence of supernumeraries the likelihood of seed fertility on backcrossing is much reduced, and because chiasma frequency in the hybrid is considerably lower the probability is much reduced of gene introgression from one parental species to the other via the hybrid [3].

The presence of a supernumerary chromosome thus enhances the reproductive isolation of the species by reducing hybrid meitoic chromosome pairing. Presumably this results from the interaction of the supernumerary with some property which differentiates potentially homologous chromosomes derived from different parents of the hybrid. Flavell, O'Dell, and Smith [4] have compared high copy number DNA sequences of Ae speltoides with those of T boeoticum. Ae speltoides possessed highly repeated sequences—constituting 2–3% of the genome—which are not detectably repeated in T boeoticum. Also, 0.5% of the genome of diploid wheat consists of highly repeat sequences of DNA not present in Ae speltoides. These species-specific sequences differentiating the genome of Ae speltoides and T boeoticum may be sufficient to enable the enhancement of pairing failure to be triggered by the presence of supernumeraries. However, the species-specific repeats constitute a rather small percentage of the total genome, and if the supernumerary were to interact solely with these differentiating parts of the hybrid genome it seems likely that the response would derive from the position of the repeats in the chromosomes, rather than from their frequencies.

Triticum aestivum

Triticum aestivum (2n = 6x = 42) combines within its genomes sets of chromosomes assembled, during the course of allopolyploid evolution, from three distinct 14-chromosome diploid species. There is marked homology between the structural genes incorporated from the three diploid parents of T aestivum, yet at meiosis each chromosome pairs only with its fully homologous

TABLE I. Mean Chromosome Pairing at M1 of Meiosis in F1 Hybrids from the Cross Ae. speltoides × T.boeoticum with and Without a Supernumerary Chromosome Derived from Ae. speltoides*

Supernumerary	Chromosome no.	Number of hybrids	Range of plant means			
			univalents	bivalents	triv. + quad.	chiasmata
0	14	8	0.3– 7.8	3.0–5.9	0.1–0.5	3.7–11.2
1	15	5	12.5–14.7	0.2–1.3	0.0	0.2– 1.3

*Thirty cells/plant.

partner and never, or extremely rarely, with its homoeologues derived from a different diploid parent—for references see Riley and Dover [5]. A genetic system, of which the principal component is the Ph gene on chromosome 5B, is responsible for the limitation of meiotic pairing to full homologues—with the resulting regularity of chromosomal segregation and a disomic pattern of inheritance [3].

The origins of the polyploid wheat species were thus dependent on the occurrence of a sequence of abrupt changes. First interspecies hybridisation and chromosome doubling in the hybrid occurred to produce the allotetraploid forms. Because the integrity of the chromosome structures of the parental species has been maintained at the tetraploid level, it is likely the Ph allele found in the present polyploid wheats also determined the course of the first tetraploid meioses in wheat. Since this form of the allele is not now present in diploid relatives of polyploid wheat it may be speculated that mutation, to the form of Ph present on chromosome 5B today, occurred in the first polyploid or in one of the parents or gametes from which it was derived. Thus, the origin of polyploid wheat at the tetraploid level was probably an abrupt event dependent on the coincidence of hybridisation, the doubling of the chromosome number, and mutation of a gene affecting meiotic processes.

Subsequently further hybridisation and chromosome number doubling events took place to give rise to the allohexaploid forms from their tetraploid and diploid parents. The genetic determinants of meiotic diploidisation were as effective at the hexaploid as at the tetraploid level.

The mechanism is not understood by which meiotic pairing discrimination occurs in ployploid wheat despite much study (for references see Riley and Dover [5]). Clearly, however, a component of the system must be an interaction between the pairing control genes, especially Ph, and the differential structure of homoeologous chromosomes which have been derived in evolution from a common chromosome of a remote ancestor of all the diploid species whose chromosome sets are assembled in the polyploids. This must be true even if a third factor—such as relative spatial positioning and orientation of potential partners—is involved in the determination of pairing specificity.

Flavell, O'Dell, and Smith [4] have compared repeat sequences in the DNA of T aestivum and in diploid species likely to have contributed chromosome sets in its evolution. Apparently, about 11% of the hexaploid wheat genome consists of repeated sequences not present in the DNA of T boeoticum or present infrequently. However, most of the families of repeat sequences in T aestivum occur in Ae speltoides and in Ae squarrosa, which have affinity with the B and D genome donors, respectively.

Additionally some of the most highly repeated sequences in T aestivum are concentrated in the B genome. Thus there already seems to be evidence of differentiation between the DNA assembled in hexaploid and tetraploid wheat from different diploid donors. Since there is no evidence of differentiation due to major structural changes, such as translocations or inversions, and since the structural gene contents are very similar, it seems probable that the Ph gene system confines meiotic pairing to full homologues by a recognition system that discriminates, either directly or indirectly, on the basis of differences in the possession or distribution of repeat sequences.

Crucially, however, is the existence of a trigger system—the Ph gene—the abrupt origin of which may be presumed to have caused the stabilisation of a polyploid genetic system and the origin of the highly successful polyploid wheats.

Secale cereale and Secale montanum

Secale cereale and S montanum are diploid species of rye with 2n = 14. S cereale is the cultivated tough-eared, free-threshing annual species of agriculture, while S montanum is the brittle-eared perennial, wild-growing species of the Balkans. Hybrids between the species are of low fertility, principally because they differ by three interchanges involving four pairs of chromosomes so that, rarely, a ring of eight chromosomes occurs at meiosis [6]. Despite this structural heterozygosity meiotic chromosome pairing is high in diploid hybrids with rarely more than a single univalent (Table II).

Both species can be crossed artificially to polyploid wheat, either at the hexaploid or the tetraploid level, and from the hybrids synthetic amphiploids can be produced at the octoploid or hexaploid level. Such synthetic species are called Triticale, and they have reasonably normal meiotic chromosome pairing with no multivalents but with occasional univalents due to the failure of bivalent formation. Seeds are set in many florets.

Hybrids between different octoploid forms of Triticale can be made, including those between Triticale parents derived from S cereale and S montanum, respectively (for convenience these will be referred to as Triticale cereale or as Triticale montanum). At meiosis in such hybrids the frequency of univalents is very high [7], nor is there the frequency of multivalents that is to be expected from the structural heterozygosity of the S cereale and S montanum chromosomes present in the hybrid (Table II). This leads to the conclusion that although in

TABLE II. Mean Chromosome Pairing at M1 of Meiosis in Secale cereale, S montanum, and Hybrids Between Them and in Triticale cereale and Triticale montanum Which Comine the Chromosome Complements of the Secale Species and that of Triticum aestivum and Hybrids Between the Triticale Forms

Species or hybrid	Chromosome no.	Univalents	Bivalents	Triv.	Quad.	Quin.	Sex.
T aestivum	42	0.0	21.0	—	—	—	—
			(21)				
S cereale	14	0.0	7.0	—	—	—	—
			(7)				
S. montanum	14	0.0	7.0	—	—	—	—
			(7)				
S cereale × S montanum	14	0.4	4.4	0.2	0.2	0.3	0.4
		(0–1)	(4–7)	(0–2)	(0–1)	(0–1)	(0–1)
Triticale cereale	56	1.8	27.1	—	—	—	—
		(0–4)	(20–28)				
Triticale montanum	56	2.4	26.8	—	—	—	—
		(0–8)	(24–28)				
Triticale cereale × Triticale montanum	56	12.0	21.8	0.03	0.07	—	—
		(8–20)	(16–24)	(0–1)	(0–1)		

Ranges are in parentheses.

diploid hybrids the chromosomes of S cereale and S montanum pair at meiosis they do not pair with the same frequency when the wheat genome is also present in the disomic condition. An analogous phenomenon occurs when hexaploid Triticale forms are compared in the same way [7].

Apparently the state of differentiation of the chromosomes of the two Secale species is such that interaction with another process affecting meiotic chromosome pairing—perhaps comparable with the Ph activity of T aestivum—would largely preclude pairing and hence gene flow between the species. It is as though these species are poised for the reinforcement of their reproductive isolation by the acquisition, on one side or the other, of a genetic activity that, in interaction with interspecies heterozygosity, would grossly disrupt meiosis in hybrids.

It is not possible to describe the form of interspecies chromosome differentiation that interacts with the wheat chromosome complement to limit pairing of the rye chromosomes. However, Jones and Flavell [8] have compared the frequency of occurrence and the distribution of several highly reiterated sequences of DNA in S cereale and S montanum. The sequences concerned are largely concentrated in the telomeric heterochromatin, and in S cereale they constitute 11.8% of the genome [9]. By contrast, in S montanum a maximum of 8.16% of the genome was comprised of these sequences and, as Table III shows, there are striking differences between the species in the contents of the

TABLE III. Number of Copies and Percentage of the Genome-Taken-Up Families of DNA Repeat Sequences in Secale cereale and S montanum

Repeat sequence family in base pairs	S. cereale	S. montanum
480	1×10^6 copies	$2-7 \times 10^5$
	6.1% of genome	1.2–4.3%
610	3.5×10^6	6.5×10^4
	2.7%	0.5%
120	1.5×10^6	$1-2 \times 10^6$
	2.4%	1.6–3.2%
630	7.5×10^4	2×10^4
	0.6%	0.16%

reiterated families with 480 and 610 base pairs (bp) in the sequences. Moreover, Jones and Flavell [8] report big differences between S cereale and S montanum in the chromosomal distribution of the 480-bp sequence and indeed in the structure of this sequence in S montanum compared with other rye species. Heterozygosity or hemizygosity in hybrid Triticale for these largely telomeric repeats may provide one part of the mechanism by which meiotic pairing is interrupted and by which species isolation could be completed were the chromosomal distinctions to be triggered by an appropriate nuclear environment.

DISCUSSION

The Meiotic Trigger

Among the numerous routes to speciation there are some which have long been accepted as permitting the abrupt origin of new species. These include the occurrence of polyploidy—with the instantaneous creation of reproductive isolation between forms with the old and the new chromosome numbers. Similarly, one-step changes to modified chromosome number segregation must permit the rapid erection of barriers to reproduction, as must certain forms of gross chromosomal structure change.

The present account has the purpose of suggesting another means by which reproductive isolation might be created abruptly between populations between which gene flow had previously been possible. It draws together evidence from plant cytogenetics and molecular biology, but the prime evidence comes from meiosis wherein it is shown that single changes in the nuclear environment can give rise to marked changes in the pairing of chromosomes. It is clear that changes like the introduction of a supernumerary chromosome or the presence of an activity like that performed by the Ph gene in T aestivum can result in the modification of meiosis in a way that can lead to hybrid sterility or allopolyploid stability, with consequences for speciation.

Clearly it can only be speculation that changes in meiosis, which can be abruptly triggered, relate to the possession of different repeat sequences (which would henceforth be called species-specific sequences) in the hitherto partner chromosomes. However, in the examples described here there is some evidence for differences in repeat sequences in the sets of chromosomes which pair differently in changed nuclear environments. There is no evidence for any other difference between the chromosomes which could offer an alternative explanation for the means by which the meiotic trigger could become effective.

Thus the differential amplification and chromosomal distribution of repeat sequences in separate parts of the geographical range of a species might constitute the fortuitous preliminaries to the creation of reproductive isolation. Its effectiveness would depend upon the emergence of an appropriate meiotic trigger. The trigger would constitute the interactive element that enabled meiotic pairing discrimination to occur between potential partner chromosomes carrying repeat sequences that are different in kind, in quantity, or in chromosomal distribution. Alternatively if no trigger develops, and the previous limitations to gene flow were removed, differences will be lost in the DNA repeat sequences and so the potentiality for speciation by this route would be lost.

This paper has attempted, in a speculative manner, to consider whether a relationship may exist between differentiation between populations in terms of DNA repeat sequences and fortuitous preparation for speciation. It is an idea which gains strength if it is accepted that the consequences of such differentiation may on occasion depend upon other nuclear components which have the potentiality for change. The idea may be more acceptable in terms of an effect through meiotic behaviour, which affects every chromosome pair of the complement, when population divergence in repeat sequences equally affects many, if not all, chromosomes.

REFERENCES

1. Riley R: Prospects. In Evans LT, Peacock WJ (eds): Wheat Science—Today and Tomorrow. Cambridge: Cambridge University Press, 1981, p 271.
2. Flavell RB, Smith DB: Nucleotide sequence organisation in the wheat genome. Heredity 37: 231, 1976.
3. Riley R, Chapman V, Miller TE: The determination of meiotic pairing. In: Proceedings of the Fourth International Wheat Genetics Symposium. Columbia: Missouri Agricultural Experimental Station, 1973, p 731.
4. Flavell R, O'Dell M, Smith D: Repeat sequence DNA comparisons between Triticum and Aegilops species. Heredity 42: 309, 1979.
5. Riley R, Dover G: Inferences from genetical evidence on the course of meiotic chromosome pairing in plants. Phil Trans R Soc Lond B 277: 313, 1977.

6. Riley R: The cytogenetics of the differences between some Secale species. J Agricult Sci 46: 377, 1955.

7. Miller TE, Riley R: Meiotic chromosome pairing in wheat-rye combinations. Genetica Iberica 24: 241, 1972.

8. Jones JDG, Flavell RB: The structure, amount and chromosomal localisation of defined repeat DNA sequences in species of the genus *Secale* (unpublished).

9. Bedbrook JR, Jones J, O'Dell M, Thompson RD, Flavell RB: A molecular description of telomeric heterochromation in Secale species. Cell 19: 545, 1980.

Mechanisms of Speciation, pages 479–509

Speciation Patterns in Woody Angiosperms of Tropical Origin

Friedrich Ehrendorfer

INTRODUCTION

A remarkable diversity of speciation patterns has become established in different groups of plants and animals, at different phases of their evolution, and under different environmental regimes. This has become more and more apparent during the last decades and is obvious from several relevant examples presented during this symposium. Why have I chosen woody and predominantly tropical Angiosperms for my contribution?

1) Tropical rain forests are among the most complex and species-rich in our biosphere. This has stimulated the formulation of a number of strongly controversial hypotheses on speciation in trees of the humid tropics during the latter sixties [1–4]. Too few facts were available then to make well-founded statements on the problem. 2) Many woody and tropical groups of Angiosperms exhibit "primitive" character states and are regarded as remnants of progenitors of temperate and often herbaceous groups with "advanced" features (eg, Mimosaceae + Caesalpiniaceae → Fabaceae, Capparaceae → Brassicaceae, Araliaceae → Apiaceae). One may therefore ask: Are extant woody Angiosperms of the humid tropics stasigenetic relics on the way to extinction or do they continue to evolve and radiate successfully [5]? More and more relevant observations on the taxonomy and population structure of such groups have become available during the last decade [eg, 6,7]. 3) Speciation patterns strongly depend on the reproductive systems of the plant groups in question. The last years have brought out the first relevant studies on the breeding system of woody tropical Angiosperms [eg, 8–12]. 4) Very little cytogenetic data have been available for tropical tree groups until 1970. This is understandable in view of the difficulties involved (numerous and small chromosomes, long generation sequences, etc). Yet, already the few early contributions [eg, 13–16] have suggested speciation patterns remarkably different from those of temperate woody Angiosperms, for instance in regard to the widespread occurrence of paleopolyploidy [cf, 17,18]. The

development of new methods for the study of chromosomes, nucleotypes, and DNA [cf, 19,20, several papers in this symposium] has greatly enhanced such work on tropical woody Angiosperms as well, eg, in our own laboratory.

Points 1–4 substantiate the development and importance of the problems surrounding speciation and evolution in woody tropical Angiosperms. But time and space allotted to my contribution do not allow for a complete survey or a general discussion. Instead, I propose to present a number of selected cases which have recently been studied and range from the level of populations and races to that of taxonomic tribes and families. The results will be used to illustrate a number of salient aspects of our problem in a summary fashion.

POPULATIONS, RACES, AND EXTRINSIC ISOLATION MECHANISMS

The Winteraceae are an isolated and depauperate relic family of the Angiosperms (Magnoliales) with eight genera, and exhibit numerous "primitive" morphological and anatomical characters [21]. Their ancient nature is documented by fossils since the Upper Cretaceous and by a presently disjunct Gondwana land distribution, ranging from Australasia and a center of diversity in New Caledonia to Madagascar, New Zealand, South and Central America, with Tertiary remains in Antarctica. With the exception of the aberrant dioecious genus Tasmannia (which is paleotetraploid and has a secondary chromosome base number of $x_2 = 13$), all of the five other general studied cytologically so far have uniformly preserved the highly paleopolyploid (probably dodekaploid: 12x) base number $x_5 = 43$, ie, they have $2n = 86$; Zygogynum even has reached the 24x level with $2n = 172$ [14]. All lower polyploids or diploids apparently have become extinct during the long history of the family.

One would expect morphologically uniform, stenoecious species and stagnant evolution in a relic family like the Winteraceae. Such expectations are at least partly refuted by detailed studies on very polymorphic and ecologically radiating groups, eg, Tasmannia (= Drimys s. lat.) in New Guinea [22] or Drimys s. str. in Central and South America [21,23]. The latter genus consists of six closely related geographically vicarious and ± disjunct species of which observations on the E Brazilian D brasiliensis, and D angustifolia are relevant for our discussion. The plants form shrubs or trees. Their rather conspicuous white flowers are diurnal and attract a variety of insect pollinators. Apart from outcrossing, inbreeding is possible because there is no self-sterility mechanism. Seeds are mainly dispersed by fruit-eating birds. In addition, some vegetative reproduction occurs by way of rooting branches. Chromosome numbers ($2n = 86$) and structures appear to be uniform throughout the genus [21,24].

A detailed analysis of a gallery forest population of D brasiliensis subspecies brasiliensis from São Paulo, near Botucatu (Pardinho), extending over an area of ca 150 × 30m and with about 67 individuals (map:Fig. 1) has revealed a remarkable amount of morphological variation in regard to leaf size and shape, nerve angle, flower number per inflorescence, peduncle length, anthocyan coloration, petal length, carpel number, etcetera (scatter diagram: Fig. 2). These characters vary between neighboring plants but are quite constant between the different branches of one individual. One can therefore assume a genetic basis for the polymorphism of the Pardinho population. There is even a suggestion

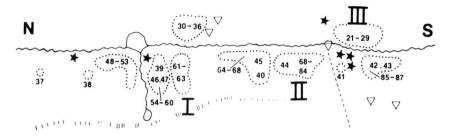

Fig. 1. Map of a gallery forest population of Drimys brasiliensis subsp. brasiliensis: Pardinho near Botucatu (São Paulo), ca. 150 × 30 m, with 67 individuals (numbered 21–87) and indications for part populations I, II, and III also apparent in Figure 2 (modified from Ehrendorfer et al [21]).

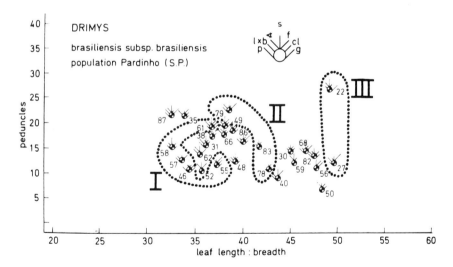

Fig. 2. Scatter diagram illustrating the variation of selected individuals from the Pardinho population of Drimys brasiliensis subsp. brasiliensis (Fig. 1) with reference to the individual numbers and part populations I, II, and III. Characters analysed are peduncle length (in mm), leaf shape (length:breadth), petiole length (p), leaf area (length × breadth × 0.66), nerve angle (∢), and color of lower side (s), flower number (f), petal length (cl) and carpel number (g) (modified from Ehrendorfer et al [21]).

for a microgeographical differentiation within the population: Examine the morphological position in the scatter diagram (Fig. 2) of the neighbour individuals forming groups I, II, and III (Fig. 1). Microgeographical and morphological proximity are obvious and attest to a very early phase of divergence among partial populations probably due to limited gene flow over distances of 150 m or less.

The trends for allopatric racial differentiation within D brasiliensis becomes conspicuous when we consider the variation of the species over its total area in E Brazil (Figs. 3,4). While subspecies brasiliensis with shrubs or small trees is limited to the gallery forests of the somewhat drier interior, subsp. sylvatica attains the size of large trees and ranges widely throughout the more humid coastal and mountainous areas. The populations of this subspecies vary considerably and partly in a clinal fashion from north to south. Near timberline in the high mountains of Itatiaia and Campos do Jordão a third race has differentiated, subspecies subalpina.

All these subspecies of D brasiliensis are basically vicarious and allopatric, and they are connected by transitional forms in zones of contact. Their individuality obviously is maintained primarily by spatial isolation; but there is also some variation in flowering periods. Subspecies sylvatica flowers from August

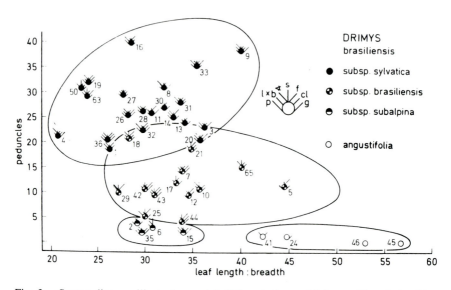

Fig. 3.   Scatter diagram illustrating racial differentiation and infraspecific relationships within Drimys brasiliensis and D. angustifolia in E. Brazil. Individual symbols as in Figure 2; numbers refer to provenances in Figure 4 (from Ehrendorfer et al [21]).

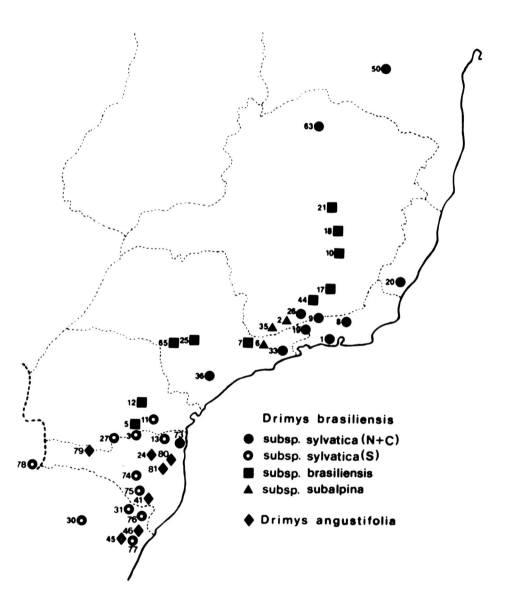

Fig. 4. Distribution of Drimys brasiliensis (with infraspecific races) and D angustifolia in E. Brazil. Numbers of provenances correspond to those of the scatter diagram, Figure 3 (from Ehrendorfer et al [21]).

to September in the north, throughout from August to February in the center, but only during the summer (January and February) in the south. This separates it from the geographically close and ecologically only slightly different D angustifolia which develops its flowers in the spring (August to October): A remarkable case where the integrity of two intimately related taxa is maintained by the divergence and stabilization of flowering times, ie, by a temporal isolation mechanism.

The morphological, ecological, geographical, and cytogenetic relationships of the taxa of Drimys in Central and South America suggest differentiation by allopatric segregation of wide-ranging taxa (eg, northern, central, and southern populations of D brasiliensis subspecies sylvatica) or by allopatric "budding" into new habitats (eg, D brasiliensis subsp. sylvatica → subsp subalpina), both backed by spatial, ie, geographical and/or ecological isolation. In D angustifolia parapatric divergence seems to have been supported by temporal isolation. But there are also reasons to postulate hybridization phenomena: D brasiliensis subsp. brasiliensis, eg, appears to be the product of former gene exchange between subspecies sylvatica and D angustifolia (compare its morphological position and correlated variation in Fig. 4). Speciation seems to be reversible in groups like Drimys where cycles of differentiation and hybridization [25] are promoted by stagnant chromosome structural divergence and the very delayed origin intrinsic crossing barriers. This and the genetic storage and recombination potential of high polyploidy apparently have helped to maintain evolutionary capacity and adaptability even in very ancient groups like the Winteraceae.

As a contrast, our next example, the genus Jacaranda, comes from the Bignoniaceae, a relatively "advanced" sympetalous family (Scrophulariales), mostly woody and distributed throughout (and beyond) the tropics. Jacaranda, with about 50 species [26] has radiated from a great variety of humid to dry tropical habitats in South and Central America. Its basic evolutionary strategy is quite similar to that outlined for Drimys: The genus is uniform in its paleotetraploid chromosome number $x_2 = 18$ ($2n = 36$), and even with the refined modern technique of Giemsa C-banding only minimal structural differences can be detected between the karyotypes of distantly related species (Fig. 5). The conspicuous flowers have a rather uniform appearance throughout the genus and are pollinated by larger Hymenoptera, resulting in out- and inbreeding. Seeds are winged and wind-dispersed.

A geographical–ecological transect from E Brazil (Fig. 6) illustrates that closely related species of Jacaranda tend to occupy different vegetation zones (eg, J puberula agg., J montana, J pulcherrima, J subalpina). Species which occur in the same area are either more distantly related (eg, J decurrens and J caroba belong to different sections, J macrantha and J crassifolia or J micrantha and J jasminoides to different species groups within one section) and/or have different flowering periods (eg, the closely related J puberula agg. in August

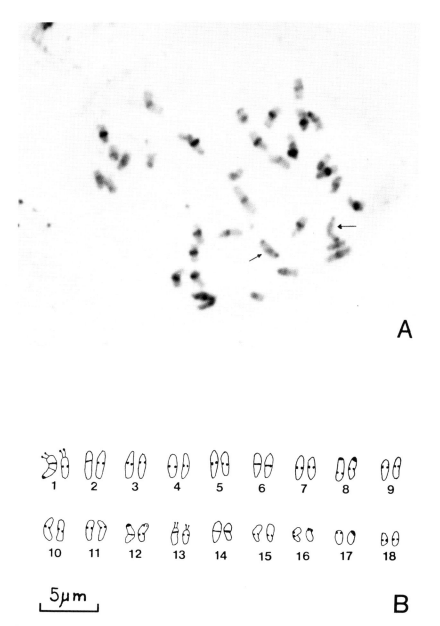

Fig. 5. Giemsa C-banded karyotypes of Jacaranda mimosifolia (A, sect. Monolobos) and J. macrantha (B, sect. Dilobos), differing only very slightly in one chromosome pair with two terminal bands (instead of one; arrows) (from Morawetz [38]).

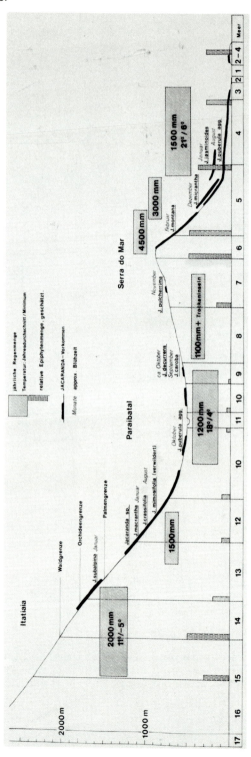

and J micrantha in December; cf, also the species pairs mentioned above). Figure 7 illustrates the divergence of flowering periods as a mean of temporal isolation and better pollinator exploitation for another neotropic genus of Bignoniaceae: Arrabidaea [27].

Do the species of Jacaranda compete with each other in zones of contact, or in cases of sympatric occurrence? The numerous field observations by Morawetz result in the remarkable conclusion that they usually "avoid" direct competition: Allopatric species hardly come into immediate contact, and even sympatric species tend do form locally separated populations due to their scattered distribution or slightly divergent ecological preferences. Hybridization obviously is possible in Jacaranda too, but appears to be a rare phenomenon of localized significance—at least under the present conditions.

Observations on the polymorphic J puberula agg. have led to a model of parapatric evolutionary divergence at steep ecotones (Fig. 8): Under the influence of different environmental regimes part populations on either side of rain forest/savannah or rain forest/restinga contact zones reach different stem heights and have different flowering peaks with different flower quantities and different pollinators. This semi-isolation will obviously lead to the selection of different sets of genes among part population and the fixation of adaptive traits comparable to those differentiating mature species in Jacaranda.

To complement our data about infraspecific evolution from the neotropics, reference is made to examples from the paleotropical Dipterocarpaceae [7]. Here too we encounter the pattern of remarkable chromosomal stability so characteristic for rain forest trees [29]. Figure 9 illustrates cases of clear-cut and secondarily swamped allopatric geographical, and of sympatric (or rather parapatric), more ecologically orientated differentiation in Dipterocarpus, Shorea, and Vatica. Furthermore, studies on the isozyme polymorphism in populations and species of Shorea and Xerospermum (Sapindaceae) by Gan et al [29] substantiate the assumption of high levels of outbreeding and genic variability but only restricted ranges of pollen and fruit dispersal, as suggested by our own observations on Drimys and Jacaranda.

Fig. 6.   Geographical-ecological transect from the coast to the Itatiaia Mountains in E.Brazil, with the distributions of 12 species of Jacaranda and their flowering peaks. Numbers at the bottom refer to different vegetation types and columns to relative amounts of epiphytes; frames include mean and minimum temperatures, and annual precipitation (from Morawetz [26]).

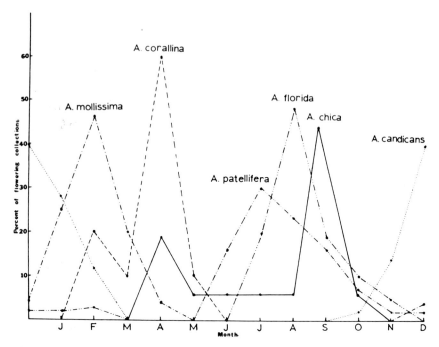

Fig. 7. Different flowering periods in species of Arrabidaea (Bignoniaceae) from Costa Rica and Panama (from Gentry [27]).

ORIGINS OF POLYPLOIDS

Recent studies on African Ebenaceae [30] have contributed to the problem of how polyploidy makes its appearance in the evolution of tropical woody Angiosperms. Again, the chromosome numbers (2n = 30) and karyotypes are remarkably uniform throughout this pantropical family. Polyploidy is rare and mainly has been ascertained for groups expanding beyond the tropics, eg, in the complexes of Diospyros roxburghii–D kaki–D virginiana (2x–4x–6x) in Asia and North America, and of D lycioides in Africa (Fig. 10). Within the latter complex there is a center of variation which extends from Transvaal to South Rhodesia and Natal. The 2x subspecies guerkei and the 4x subspecies nitens are limited to this area, and it is only here that 2x cytotypes have been found within subsp. lycioides and subsp. serica. In the large areas of the two latter subspecies extending to the S and N, only 4x cytotypes have been found so far. We can conclude from these findings that polyploidy must have originated several times in the D lycioides complex, and that it is primarily of the autopolyploid type, but probably also has been involved in interracial hybridization. From the much

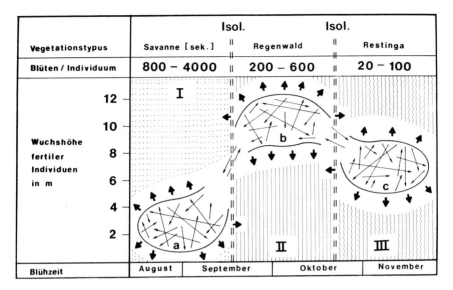

Fig. 8. Model of parapatric evolutionary divergence based on data from Jacaranda puberula agg. Part populations in adjacent plots of savannah (I), rain forest (II), and restinga (III) with different stem heights (2–12 m), flowering periods, and flower quantities per individuum; heavy and thin arrows suggest trends of seed dispersal and gene flow (from Morawetz [26]).

wider distribution of 4x as compared with 2x cytotypes, one has to assume that polyploidy has promoted adaptability and geographical expansion in this subtropical woody group just as in so many temperate herbaceous Angiosperms.

For Malayan Dipterocarpaceae, eg, taxa of Shorea and Hopea, Kaur et al [31] have shown that the origin of polyploid cytotypes and polyploid species (2x, 3x, 4x) sometimes is linked to apomictic reproduction: This is a phenomenon also well known from numerous partly agamic complexes of woody and herbaceous extratropical Angiosperms.

CHANGES IN CHROMOSOME STRUCTURE AND NUMBER

Let us now turn to the role of changes in chromosome structure and number, both dysploidy and polyploidy, in the long-term evolution of tropical Angiosperms, as we compare related species, genera tribes, and subfamilies. Among the relatively "primitive" families of Magnoliales only the diverse pantropical Annonaceae [ca. 120 genera and 1100 species; cf 32,33] are still well represented on the diploid level and thus give us insights into the basic processes of their

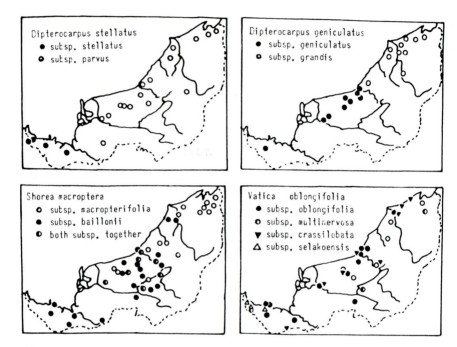

Fig. 9. Patterns of infraspecific differentiations in Dipterocarpaceae from northwestern Borneo: allopatric subspeciation in Dipterocarpus stellatus and D geniculatus, secondary swamping of allopatric differentiation by migration in Shorea macropterifolia and parapatric ecotypic divergence in Vatica oblongifolia (from Ashton [73]).

chromosomal evolution. Haploid chromosome numbers within the family appear to be linked by dysploidy and polyploidy according to Figure 11. No genus is known to include species with different chromosome base numbers, and groups of related genera are often characterized by a single base number, eg, the pantropical group of Fusaea, Xylopia, Neostenanthera, Cananga, and Goniothalamus with x = 8. Dysploid changes of base number brought about by major chromosome mutations apparently have been very rare phenomena in the long history of the family dating back the Cretaceous. Considering certain parallels with morphological and palynological change [33,34], we have to postulate x = 7 → 8 at least twice for the Malmea and once for the Fusaea subfamily, one possible reversal x = 8 → 7 for the Annona subfamily, and furthermore x = 8 → 9 twice for the Malmea and once for the Annona subfamily.

The slow rate of chromosomal divergence among Annonaeae is also brought out by a comparison of karyotypes of distantly related genera with the same base number, eg, Duguetia (Malmea subfamily) and Xylopia (Fusaea subfamily) (Fig.

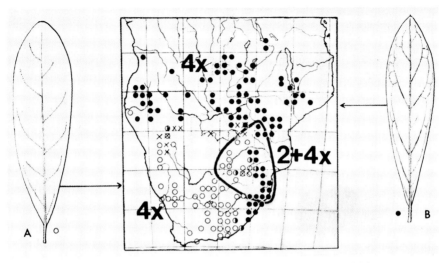

Fig. 10. Leaves and distribution of Diospyros lycioides subsp. lycioides (A,o) and subsp. sericea (B, ●); reference to sympatric occurrence is by half-filled symbols, to intermediates by x (from White and Vosa [30]).

Changes of chromosome numbers (n)
in Annonaceae

6x	21	24	
4x	14	16	
2x	7	8	9

Fig. 11. Haploid chromosome numbers recorded for members of the Annonaceae. The arrows indicate postulated main directions of change in regard to polyploidy and dysploidy.

12): both have $2n = 16$ ($x = 8$) with six subtelocentric chromosome pairs of decreasing size, one larger submetacentric pair (with satellites in Duguetia), and one very large satellite pair (with different centromer position in the two genera). Distinct species within Xylopia appear to be identical judging by the standards of classical karyotype analysis. The more refined method of comparing chromosomes during mid-prophase, prometaphase, and metaphase allows us to differentiate between early, later, and very late condensing chromatin [35] (Fig. 13). With this method microstructural differences between species of one genus have been detected, eg, between Annona cherimolia and A muricata (belonging to different sections). Our own Giemsa C-banding has shown that very early

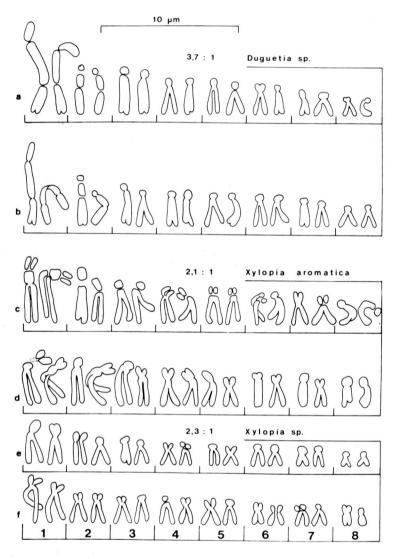

Fig. 12. Giemsa karyotypes (non-banded) of neotropical Annonaceae: One species of Duguetia and two distinct species of Xylopia (from Morawetz [24]).

condensing chromatin does not always correspond to constitutive heterochromatin, as has been supposed. Banding patterns due to the insertion of constitutive heterochromatin segments differ widely among Annonaceae and sometimes even reveal structural heterozygosity (eg, in Porcelia cf macrocarpa [24]). Constitutive

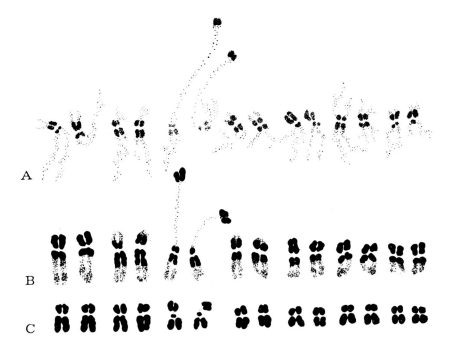

Fig. 13. Somatic chromosomes of Annona cherimolia (2n = 14) after 8-hydroxyqui-noline pretreatment, Carnoy fixation and aceto-orcein staining. A) Mid-prophase, B) pro-metaphase, C) metaphase (× 3200) (from Tanaka and Okada [35]).

heterochromatin therefore appears to be relatively the most versatile component of chromosome structure in tropical trees just as in many herbaceous Angiosperms.

Infrageneric polyploidy so far has been found only in few Annonaceae, eg, in the large genus Annona with x = 7. From 12 species checked nine are diploids. From the two tetraploids one is the isolated A glabra, single representative of a monotypic section. As a component of alluvial forests it has expanded from the Neotropics across the Atlantic to West Africa. The only hexaploid also has radiated into rather extreme habitats, it is A coriacea, a widespread member of the xeric Brazilian Cerrado formation. Apart from such cases of relatively more recent infrageneric polyploidy there are others where genera appear to be polyploid throughout and where we have to postulate more ancient paleoployploidy and extinction of diploids; examples are some of the morphologically most primitive genera, like Unonopsis and Guatteria in the Malmea subfamily and Anaxagorea in the Fusaea subfamily, all 4x on x = 7, ie, $x_2 = 14$.

How further cycles of chromosome evolution are brought into motion among ancient woody paleopolyploid groups is best demonstrated by the Magnoliaceae family (12 genera, ca 210 species). Also of Cretaceous age, it has a disjunct distribution in (sub)tropical to warm temperate forests of East Asia and the Americas today. The highly derived chromosome base number $x_3 = 19$ can be regarded as paleohexaploid [14]. The related monotypic relic families Degeneriaceae and Himantandraceae with $x_2 = 12$ appear to be paleotetraploid [36]. Okada [37] has elaborated karyotypes and idiograms for the Magnoliaceae, again considering early, later, and very late condensing chromatin (Fig. 14). These idiograms are compatible with the thesis of ancient paleohexaploidy. In spite of obvious overall similarities (nine or more chromosome paris a, 2–3 b, 1 satellite pair ℓ etc), they demonstrate small amounts of further chromosome structural divergence within the family. The dynamics of this process is underlined by the occasional observation of structural heterozygosity (cf the satellite chromosome pairs ℓ and r in Fig. 14 C,E,G, and H).

The obvious systematic isolation of Liriodendron (in a separate tribe) is hardly apparent from its idiogram which is quite similar, eg, to that of Magnolia hypoleuca (Fig. 14A,C). The three species of Michelia (I–K) are relatively uniform and differ by the prominence of f- and j-type chromosomes from the more heterogeneous species of Magnolia. Its section Gwillima (H) has 14 pairs of a, only two of c, and none of k. Within this genus the different sections exhibit some change and coherence in idiograms. From the evergreen sect. Gwillima (H) and section Magnolia (B) to the summergreen sect. Buergeria (F-G) and section Rhytidospermum (C–E) an overall trend is apparent from one to two satellite chromosome pairs (ℓ–t), from 14 to 13, 12, 10, and 9 pairs of $\pm$ metacentrics a with proximal early condensing chromatin and a corresponding increase of $\pm$ submetacentric or subtelocentric pairs of b to k with early condensing chromatin expanding to whole arms and terminal positions. In this karyological trend the morphologically most derived genus Liriodendron also takes a most advanced position. The recently established Giemsa C-banded karyotype of this genus (Figs. 14, 15) underlines this suggestion by its highly diverse banding pattern and a high proportion (74%) of chromosomes with terminal heterochromatin [38].

Within Magnolia the amount of morphological and chromosome structural divergence apparently is well correlated with the possibilities and ease to produce artificial hybrids and further progency [39]. Such hybridization has been carried out successfully between species of the same section (eg, $\pm$ fertile hybrids within section Rhytidospermum), but also between different sections (eg, $\pm$ sterile hybrids between section Magnolia with Ma virginiana and members of section Rhytidospermum or section Oyama, all belonging to subgenus Magnolia, or between the sections Buergeria, Tulipastrum, and Yulania, all belonging to

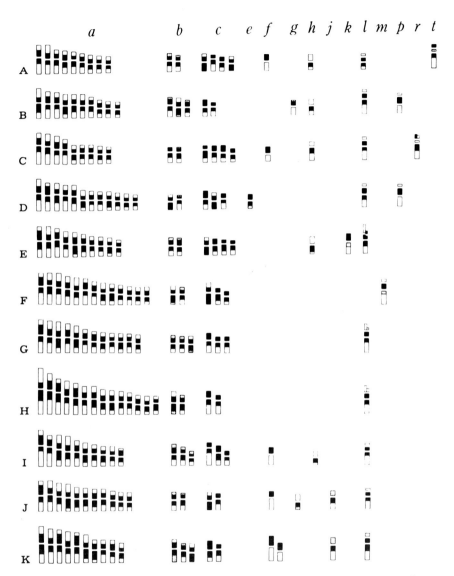

Fig. 14. Idiograms of haploid somatic chromosomes in prometaphase of Magnoliaceae: Early, later, and very late condensing chromatin marked black, stippled, and white, respectively. Small letters refer to chromosome types, capital letters to taxa: A) Liriodendron tulipifera (tribe Liriodendreae), all others are tribe Magnolieae: B–H) Magnolia; I–K) Michelia. B, Magnolia virginiana (sect. Magnolia); C, Ma. hypoleuca; D, Ma tripetala; E, Ma macrophylla (all sect. Rhytidospermum); F, Ma salicifolia; G, Ma stellata (all sect. Buergeria); H, Ma coco (sect. Gwillima); I, Michelia champaca; J, Mi compressa, K, Mi figo (from Okada [37]).

subgenus Yulania). Significantly, no hybrids have been obtained between the two subgenera of Magnolia. One may ask why this strongly delayed origin of intrinsic crossing barriers has not resulted in more extensive hybridization between sympatric Magnolia species. Again, the reasons for this apparently are to be sought in the efficiency of extrinsic barriers, ie, divergent flower biological specialization [40].

Within Magnolia polyploids have arisen on top of the paleohexaploid level of $x_3 = 19$ (2n = 38), ie, teraploids (2n = 76) and hexaploids (2n = 114). These additional polyploidization processes must date back a long time, as no cases of closely related diploids and polyploids are known. Only within section Theorhodon, a species group with diploids (M. hamori), and one with hexaploids (M. grandiflora, M. schiediana), are known. For all the other sections only one ploidy level has been recorded, eg, within subgenus Yulania 2x for section Buergeria, 4x for section Tulipastrum, and 6x for section Yulania.

CHROMOSOMES, ENVIRONMENT, AND EVOLUTION OF ANGIOSPERM FAMILIES

The last two examples represent the broader frame of taxonomic families and should illustrate the channeling effect of environmental conditions on patterns of evolution. Figures 16 and 17 summarize our present knowledge about relationships, dysploidy, polyploidy, and distribution for the pantropical Dilleniaceae (11 genera and about 325 species: [18,41–45]). The chromosome base number appears to be x = 7; it has survived only among paleopolyploid members from mesic forest habitats of the humid paleotropics and is found in both subfamilies, Dillenioideae (eg, Dillenia: 8x, 16x) and Tetraceroideae (eg, Tetracera: 4x). It is remarkable that the greatest accumulation of character states considered "primitive" in regard to morphology, anatomy, phytochemistry, etc also is found among these paleopolyploid representatives. We can accept this correlation as a guideline for interpreting the main directions of ecological and geographical radiation of the Dilleniaceae: from humid-mesic to xeric, and from an ancient center in Australasia (or western Gondwanaland) to the rest of the tropics (and some extratropical areas).

The evolution of the neotropical Dilleniaceae-Tetraceroideae has proceeded on the 4x level with descending dysploidy from n = 14 and 13 in Tetracera to n = 13 in Doliocarpus and Curatella, and to n = 10 in Davilla. Linked to the origin of some diversity (close to 100 species), the Tetraceroideae have radiated into dry and open woodlands or even savannahs. In the paleotropical Dillenioideae the genus Dillenia (with ca 65 species) takes a central position. Only paleopolyploidy on 8x (n = 27, 26, 24) and 16x (n = 56) has been established so far. Dillenia includes rainforest trees (rarely shrubs) with unspecialized flowers. The woody Didesmandra and Schumacheria, and the herbaceous Acrotrema

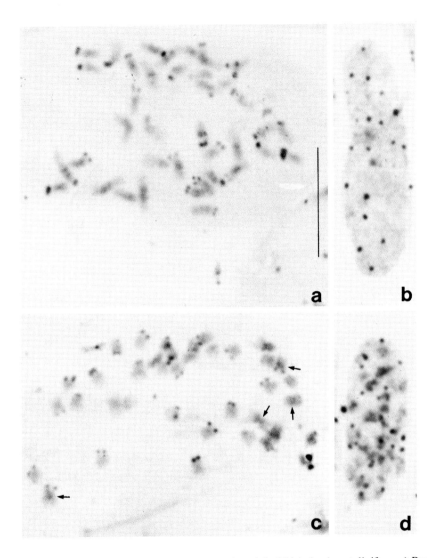

Fig. 15. Giemsa C-banded chromosomes and nuclei of Liriodendron tulipifera. a) Prometaphase and c) mataphase chromosomes; b) interphase and d) telophase nucleus. Bar = 5 μm (from Morawetz [38]).

(with n = 27) all have few species (1, 3, and ca 10, respectively); also limited to mesic habitats, they are morphologically clearly more advanced than Dillenia. Hibbertia, on the other hand, has diversified extensively with the majority of its ca 150 species in Australia [46,47]. The origin of the astonishing spectrum of

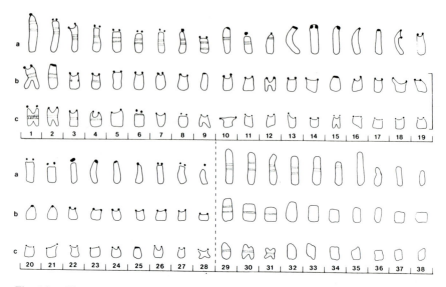

Fig. 16. Giemsa C-banded karyotypes of Liriodendron tulipifera. During prometaphase (a) more heterochromatic bands appear than in metaphase (b,c). Chromosomes 1–28 have terminal heterochromatin, 29–38 do not. Bar = 5 μm (from Morawetz [38]).

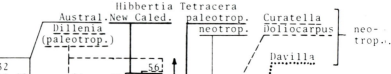

Fig. 17. Haploid chromosome numbers recorded for members of the Dilleniaceae. Numbers in square brackets apparently have become extinct. The horizontal arrows indicate postulated dysploid and polyploid lineages, respectively. Frames around chromosome numbers correspond to genera and distribution areas (from Ehrendorfer [18]).

chromosome numbers within the genus can be interpreted as a combination of gross structural change in chromosome structures, descending dysploidy (with n = 5 and 4, but n = 6 still missing), and ascending polyploidy (4x–6x–8x–16x). This daramatic chromosome evolution has sustained a not less dramatic ecological radiation, ranging from small trees of mesic habitats to climbers or under-shrubs of dry woodlands, and finally to pygmy xerophytes of semideserts. It is significant for the evolutionary history of Hibbertia that its morphologically most primitive taxa occupy mesic habitats, belong to the extremely ancient relic flora of New Caledonia, and again are paleopolyploids on the 8x level (with n = 27 and 28; [45]).

The different patterns of chromosomal differentiation within extant Dilleniaceae thus reflect different phases of the family's historical evolution: The extensively dysploid and diploid-polyploid pattern of Australian Hibbertias obviously corresponds to a phase of rapid diversification and radiation into a multitude of ± dry and floristically unsaturated habitats of this continent since the Mid-Tertiary. The less dramatic diversification of Tetraceroideae in a variety of neotropical habitats has left a series of related genera on the paleo-4x level and may date back at least to the early Tertiary. Finally, the extremely disjunct members of Dillenia, of related depauperate genera, and of New Caledonian Hibbertias, all represented by paleo-8x and -16x only, apparently are relics from the early differentiation of the family which may well have survived in stable mesic habitats of the humid Paleotropics since the late Cretaceous.

The Rutaceae are a large, nearly exclusively woody family with more than 150 genera and ca 1800 species [48] which has radiated widely from humid-mesic into dry habitats in the tropics but also extends into the subtropical and temperate zones. In regard to the karyological and nucleotypic differentiation of the family my former collaborator M. Guerra [49] has recently made very substantial contributions, and has summarized our present knowledge. Figure 19 gives a survey of changes in chromosome numbers as related to major tribes and their phylogenetic advancement. The dramatic differences of nuclear and chromosome size within the family are illustrated by Figure 20. The basis for this is a very wide range of DNA values which has been established for 17 Rutaceae species by Guerra [49]. For 4C interphase nuclei there is a range from 0.78 to 34.84 pg DNA (about 1:45). Values listed in Table I represent 10^{-2}pg of DNA calculated for single 1C chromosomes. Finally, in Figure 21 these data have been plotted for each species against ploidy level; furthermore, they are combined with information on relative amounts of euchromatin vs heterochromatin or condensed chromatin in interphase nuclei, on stability of chromosome number in the groups involved, and on their tropical or extra-tropical occurrence.

The core of the tribe Zanthoxyleae (= Evodiinae) includes about 20 diverse and ± isolated genera which range throughout the humid tropics. They are characterized by the largest number of primitive morphological, anatomical, and phytochemical traits within the family [50], and are today only represented by paleopolyploids (mostly 4x and 8x) with some dysploid chromosome number oscillation. The nucleotypic parameters of the Zanthoxyleae (species 1–4 in Table I and Fig. 21) mostly are near or slightly above the mean for the family. Similarly, the monotypic and depauperate Flindersieae of humid Australasia are mostly

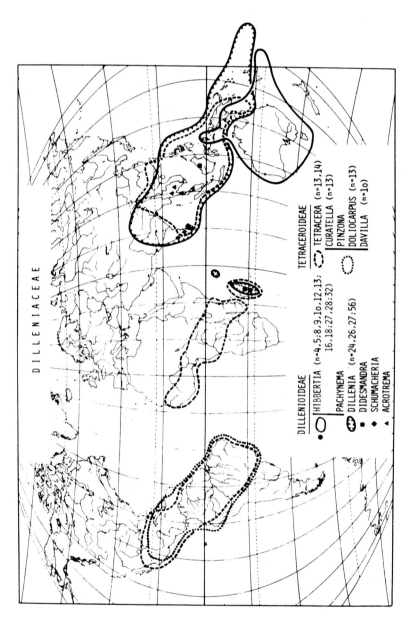

Fig. 18. Distribution map of Dilleniaceae, with reference to haploid chromosome numbers within genera (from Ehrendorfer [18]; data partly from Kubitzki [41–43]).

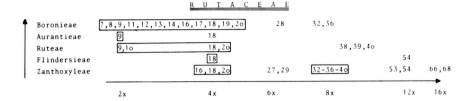

Fig. 19.   Haploid chromosome numbers recorded for major tribes of the Rutaceae. More widespread and typical numbers are framed. Dominance of primitive towards derived character states within the taxa is expressed by the vertical arrow, while the horizontal arrow indicates postulated main directions of dysploid and polyploid changes.

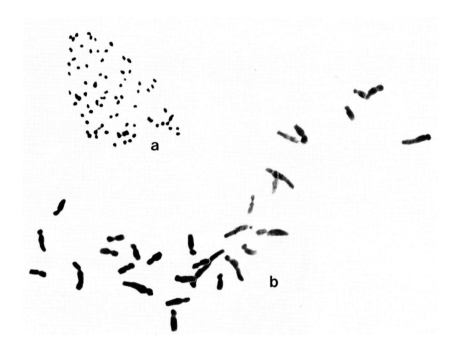

Fig. 20.   Mitotic metaphase chromosomes of Rutaceae stained with Feulgen + aceto-carmine. A) Ruta graveolens (2n = 78); B) Skimmia japonica (2n = 30) (from Guerra [49]).

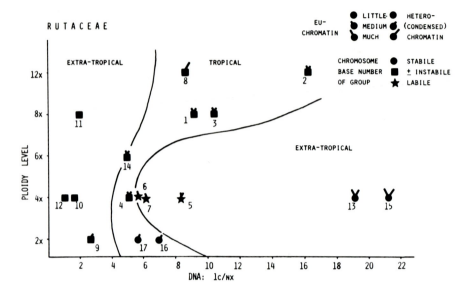

Fig. 21. Nucleotype parameters and evolution in the Rutaceae. Species numbers and data are from Table I and from Guerra [49]. Coordinates: DNA in 10^{-2} pg per chromosome and ploidy level. Symbols indicate relative amounts of chromatin types in interphase nuclei, and stability of chromosome number in the groups involved; lines separate tropical and extra-tropical groups.

paleo-4x, but exclusively on x = 9. There is little doubt that the majority of the Zanthoxyleae as well as the Flindersieae are stasigenetic remnants of an early phase of evolutionary differentiation of the Rutaceae stock, dating back to the Cretaceous.

Evolutionary patterns are very different in the three remaining tribes of the Rutaceae to be discussed here. The woody Aurantieae (with species 16, 17 in Table I and Fig. 21) have a much more advanced character spectrum and have differentiated into an array of about 30 genera which are closely related and difficult to separate. They exhibit little ecological radiation beyond the humid (sub)tropics of the Old World. Evolution has proceeded nearly exclusively on the diploid level (x = 9, 2n = 18) without any substantial change in chromosome and chromatin structure or in other nucleotypic parameters. Hybridization, therefore, is possible between many species and sometimes even genera. The tribe thus represents a rather juvenile pattern of cladogenetic differentiation by small steps, ie, predominantly by gene mutations.

TABLE I. DNA and Ploidy Level (2x-4x-6x-8x-12x) for 17 Species of Rutaceae Representing Seven Major Tribes: Zanthoxyleae (1–4), Boronieae (5), Diosmeae (6,7), Cusparieae (8), Ruteae-Rutinae (9–12), Ruteae-Dictamninae (13), Toddalieae (14,15) and Aurantieae (16–27) [49].

	Rutaceae	Ploidy	DNA:1c/nx $(10^{-2}pg)$	
1	Fagara zanthoxyloides	8x	9.10	
2	Zanthoxylum alatum	12x	16.43	
3	piperitum	8x	10.26	
4	Melicope ternata	4x	5.17	
5	Correa virens	4x	8.27	
6	Coleonema album	4x	5.65	
7	pulchrum	4x	6.01	
8	Erytrochiton brasiliensis	12x	8.57	
9	Boenninghausenia albiflora	2x	2.55	
10	Ruta chalepensis	4x	1.63	
11	graveolens	8x	1.94	
12	montana	4x	0.98	←
13	Dictamnus albus	4x	19.10	
14	Ptelea baldwinii	6x	4.94	
15	Skimmia japonica	4x	21.43	←
16	Citrus sinensis	2x	6.86	
17	Murraya paniculata	2x	5.56	

Arrows indicate minimum and maximum values.
From [49].

Examples for the early expansion of the Rutaceae with small shrubs or perennial herbs into extratropical zones are afforded by the few genera of the Ruteae (species 9–12: Rutinae and 13: Dictamninae in Table I and Fig. 21). Nucleotypic differentiation is extraordinary: While the Mediterranean and Near East members of the Rutinae exhibit extremely reduced DNA values and consequently very small chromosomes (with some dysploid and polyploid variation: 2x–4x–8x), the Dictamninae, only with the Eurasian genus Dictamnus, have stabilized at a very high DNA level and paleotetraploidy. The two groups are probably not closely related. Parallel cases of development from tropical to temperate Rutaceae are found in the (unnatural) tribe of Toddalieae of which Ptelea from North America (with little DNA) and Skimmia from East Asia (with very much DNA) are incorporated into Table I and Figure 21 (species 14 and 15).

Finally, we have to mention the predominantly Australian Boronieae and the South African Diosmeae (with species 4, 5, 6, respectively, in Table I and Fig. 21). Both tribes include numerous, partly quite aberrant and extremely specialized genera and species, and have radiated extensively into a multitude of extratropical and ± xeric mediterranoid habitats. This radiation probably is not much older than the middle Tertiary and has been accompanied by extraordinary changes in chromosome structure, dysploidy, and polyploidy (cf, Fig. 19). This has been originally elaborated by Smith-White [51,52] for the Boronieae and is now becoming established as a parallel phenomenon for the not directly related Diosmeae as well. In regard to the nuclear parameters the members of the two tribes again are quite close to the mean values for the family. Boronieae and Diosmeae are relatively juvenile groups and exemplify a pattern of anagenetic evolution which is accentuated by major steps, ie, chromosome and genome mutations and thus leads to an accelerated rate of intrinsic barrier formation and divergence. It is obvious that this evolutionary pattern is very similar to that of Hibbertia (Dilleniaceae) in Australia, and differs greatly from those demonstrated for woody groups of the humid tropics. We, therefore, feel safe to interpret these different patterns as conditioned by different environmental regimes of the past and present under which the various groups have differentiated.

Can we recognize correlations between various nucleotypic parameters and patterns of evolution within the Rutaceae from Figure 21? In regard to DNA values per chromosome and levels of polyploidy we can only say that relatively high DNA values apparently limit the origin of high polyploids. Which of the chromatin components are involved in the excessive increase or decrease of DNA values? Obviously both, as a comparison of species 13 + 15 and 9–12 clearly indicates. Still, the highly condensed or heterochromatic component appears to be the more labile. This is in line with the general suggestion that it includes primarily highly repetitive and not transcribed nucleotide sequences. But there is no clear indication that an increase of highly condensed chromatin or of heterochromatin also leads to an increase in structural or numerical instability of chromosomes. It is medium amounts of this chromatin component and medium amounts of DNA as a whole which we find in the groups excessively unstable in this respect. Both very low and very high amounts of DNA apparently have a retarding or buffering effect on structural or dysploid changes. Furthermore, it is obvious that the DNA values of the majority of (sub)tropical Rutaceae, both more primitive and more advanced, fall into a medium range, while extra-tropical genera are either very low or very high. This clearly contradicts the thesis by Levin and Funderberg [53] that there is a general trend towards increased DNA values from tropical to temperate Angiosperms. Obviously much more relevant data are needed on the nucleotypic parameters of other plant groups and their ecological radiation before we can hope to gain a better understanding of how they influence patterns of speciation and evolution.

GENERAL CONCLUSIONS

In retrospect, we can draw a number of general conclusions from the examples presented and from literature data on comparable groups.

A) In the humid (sub)tropics the mesic rain forest ecosystem is characterized by extreme species diversity, ecological complexity, and relative historical stability. The high degree of "species saturation" has led to an extreme partitioning of ecological niches correlated with much specialization, intensive biotic interference, and selective pressures (particularly in juvenile stages). Under such an environmental regime we usually encounter the following syndrome of evolutionary strategies among woody Angiosperms:

1. Individuals within populations and populations within a species are normally widely dispersed and often spatially ± isolated.
2. Outbreeding in most groups is enhanced by flower biological devices and/or self-incompatibility; still, at least some selfing and inbreeding often is possible. In addition, there are also dioecious, obligately autogamous, and even apomictic groups.
3. The range of pollen and seed dispersal mostly is quite limited. This results in limited gene flow and small effective population size. Founder phenomena are frequent and an important aspect of spatial expansion of populations.
4. Intra-population variability and individual heterozygosity often are high because of unspecialized flowers, outbreeding, high chromosome numbers, polyploidy, lack of intrinsic barriers to gene flow etc, all contributing to high recombination rates.
5. Initial steps of speciation mostly are on the basis of allopatric (often peripatric) geographical or parapatric ecological differentiation. Underlying genetic changes predominantly include small scale and often adaptive gene mutations and microstructural chromosome rearrangements.
6. This initial differentiation is gradual and mostly backed by extrinsic pre-mating barriers, such as spatial distance, different flowering times and/or different pollinators. Among closely related taxa true sympatry is rare, and direct competition is thus avoided.
7. Intrinsic, post-mating crossing barriers, eg, in the form of accumulated major chromosome rearrangements or dysploid change, usually appear slowly and are retarded during the course of evolution. Therefore, the potentials for hybridization are maintained beyond considerable morphological and ecological divergence among biological species. While hybridization is apparently rare under ± stable conditions, periods of environmental instability

have resulted in hybrid swamping, "reversible speciation," and differentiation-hybridization cycles. Ancient stasigenetic lineages thus have remained variable and well adapted, but basically conservative.

8. The establishment of neopolyploids is rare; mostly they are of autopolyploid nature and/or appear to be linked to the hybridization of closely related races and species with very similar karyotypes. Because of their great age many tree groups of the humid tropics have passed through one or several cycles of polyploidization and further genic and some chromosome structural differentiation. With the extinction of diploid or lower polyploid ancestors, today they usually have become paleopolyploids.

B) In geologically more recent and less stable habitats of the dry (sub)tropics ecological niches often are less "species saturated" and there is less biotic interference and competition. Under such an environmental scenario evolutionary strategies of woody Angiosperms tend to differ in several respects:

1. Individuals within populations are more aggregated, but small size and spatial isolation of populations, importance of founder phenomena, trends to specialized flowers, lower chromosome numbers, establishment of intrinsic crossing barriers, etc reduce the gene flow and lower the recombination rates.
2. Initial differentiation often is based on major chromosome structural rearrangements and dysploidy on diploid or lower polyploid levels. This also includes instances of obvious allopolyploidy. The bottleneck situations linked to such changes are obviously more easily overcome here as compared to tropical humid environments.
3. The fast and early origin of intrinsic post-mating crossing barriers greatly limit hybridization and back a "rectangular" or "punctuational" pattern of evolution, often characterized by revolutionary and anagenetic changes in morphology and the penetration into new ecological niches.

It is obvious, then, that the different evolutionary strategies in woody Angiosperms of (sub)tropical origin are channeled by different environmental regimes as exemplified under (A) and (B) above. But this must not obscure the basic role of the different potentials of the various groups as laid down in their cytogenetic organisation and nucleotype (eg, DNA, eu- and heterochromatin differentiation, etc). We are only beginning to grasp the fascinating interplay of the many parameters involved in these evolutionary processes.

ACKNOWLEDGMENTS

Much of this presentation would not have been possible without the great help and stimulation of my collaborators, particularly Dr. W. Morawetz, Vienna, and Dr. M. Guerra Filho, Recife. We are all grateful to the "Fonds zur Förderung der Wissenschaftlichen Forschung in Österreich" (Projekt-No. 4052) for generous support of our research work.

REFERENCES

1. Fedorow AA: The structure of the tropical rain forest and speciation in the humid tropics. J Ecol 54:1–11, 1966.
2. Richards PW: Speciation in the tropical rain forest and the concept of the niche. Biol J Linn Soc 1:149–153, 1969.
3. Steenis van CGGJ: Plant speciation in Malesia, with special reference to the theory of non-adaptive saltatory evolution. Biol J Linn Soc 1:97–133, 1969.
4. Ashton PS: Speciation among tropical forest trees: Some deductions in the light of recent evidence. Biol J Linn Soc 1:155–196, 1969.
5. Ehrendorfer F: Chromosomen, Verwandtschaft und Evolution tropischer Holzpflanzen, I.Allgemeine Hinweise. Österr Bot Z 118:30–37, 1970.
6. Ashton PS: An approach to the study of breeding systems, population structure and taxonomy of tropical trees. In Burley J, Styles BT (eds): Tropical Trees, Variation, Breeding and Conservation. New York: Academic Press, 1976, pp 35–42.
7. Ashton PS: A contribution of rain forest research to evolutionary theory. Ann Mo Bot Garden 64:694–705, 1978.
8. Bawa KS: Breeding systems of tree species of a lowland tropical community. Evolution 28:85–92, 1974.
9. Bawa KS: Breeding systems of trees in a tropical wet forest. New Zealand J Bot 17:521–524, 1979.
10. Bawa KS, Opler PA: Dioecism in tropical forest trees. Evolution 29:167–179, 1975.
11. Opler PA, Baker HG, Frankie GW: Plant reproductive characteristics during secondary succession in neotropical lowland forest ecosystems. Biotropica 12 Suppl.:40–46, 1980.
12. Baker HG, Bawa K, Frankie GW, Opler PA: Reproductive biology of plants in tropical forests. In Golley FB, Leith H (eds): Tropical Forest Ecosystems. New York: Springer-Verlag, 1980.
13. Mangenot S, Mangenot G: Enquete sur les nombres chromosomiques dans une collection d'espèces tropicales. Rev Cyt Biol Vég 25:411–447, 1962.
14. Ehrendorfer F, Krendl F, Habeler E, Sauer W: Chromosome numbers and evolution in primitive Angiosperms. Taxon 17:337–468, 1968.
15. Mehra PN, Bawa KS: Chromosomal evolution in tropical hardwoods. Evolution 23:466–481, 1969.
16. Mehra PN: Cytogenetical evolution of hardwoods. Nucleus 15:64–83, 1972.
17. Ehrendorfer F: Evolutionary significance of chromosomal differentiation patterns in Gymnosperms and primitive Angiosperms. In Beck CB (ed): Origin and Early Evolution in Angiosperms. New York: Columbia University, 1976, pp 220–240.
18. Ehrendorfer F: Polyploidy and distribution. In Lewis WH (ed): Polyploidy and distribution. New York: Plenum, 1980, pp 45–60.
19. Stebbins GL: Chromosome, DNA and plant evolution. Evol Biol 9:1–34, 1976.
20. Grant WF: The evolution of karyotype and polyploidy in arboreal plants. Taxon 25:75–84, 1976.
21. Ehrendorfer F, Silberbauer-Gottsberger I, Gottsberger G: Variation on the population, racial and species level in the primitive relic Angiosperm genus Drimys (Winteraceae) in South America. Pl Syst Evol 132:53–83, 1979.
22. Vink W: The Winteraceae of the Old World.-I.Pseudowintera and Drimys—morphology and taxonomy. Blumea 18:225–354, 1970.

23. Gottsberger G, Silberbauer-Gottsberger I, Ehrendorfer F: Reproductive biology in the primitive relic Angiosperm Drimys brasiliensis (Winteraceae). Pl Syst Evol 135:11–39, 1980.
24. Morawetz W: Unpublished observations.
25. Ehrendorfer F: Differentiation-hybridization cycles and polyploidy in Achillea. Cold Spring Harbor Symp Quant Biol 24:141–152, 1959.
26. Morawetz W: Morphologisch-ökologische Differenzierung, Biologie, Systematik und Evolution der neotropischen Gattung Jacaranda (Bignoniaceae). Diss Formal- u Naturwiss Fak Univ Wien, 1980, p 308.
27. Gentry AH: Flowering phenology and diversity in tropical Bignoniaceae. Biotropica 6:64–68, 1974.
28. Jong K: Cytology of the Dipterocarpaceae. In Burley J, Styles BT (eds): Tropical Trees, Variation, Breeding and Conservation. New York: Academic Press, 1976, pp 79–84.
29. Gan YY, Robertson FW, Ashton PS, Soepadmo E, Lee DW: Genetic variation in wild populations of rain-forest trees. Nature 269:324–325, 1977.
30. White F, Vosa CG: The chromosome cytology of African Ebenaceae with special reference to polyploidy. Bol Soc Broter Sér 2,53:275–297, 1980.
31. Kaur A, Ha CO, Jong K, Sands VE, Chan HT, Soepadmo E, Ashton PS: Apomixis may be widespread among trees of the climax rain forest. Nature 271:440–441, 1978.
32. Fries RE: Annonaceae. In Engler A, Prantl K: Die natürlichen Pflanzenfamilien. 2.Aufl., Berlin: Duncker & Humbolt, 1959, 17aII:1–171.
33. Walker JW: Pollen morphology, phytogeography, and phylogeny of the Annonaceae. Contr Gray Herb Harv 202:1–132, 1971.
34. Walker JW: Chromosome numbers, phylogeny, phytogeography of the Annonaceae and their bearing on the (original) basic chromosome number of Angiosperms. Taxon 21:57–65, 1971.
35. Tanaka R, Okada H: Karyological studies in four species of Annonaceae, a primitive Angiosperm. J Sci Hiroshima Univ Ser B2(Bot) 14:85–105, 1972.
36. Sauer W, Ehrendorfer F: Chromosomen, Verwandtschaft und Evolution tropischer Holzpflanzen, II.Himantandraceae. Österr Bot Z 118:38–54, 1970.
37. Okada H: Karyomorphological studies of woody Polycarpicae. J Sci Hiroshima Univ Ser B2(Bot)15:115–200, 1975.
38. Morawetz W: C-Banding in Liriodendron tulipifera (Magnoliaceae): Some karyological and systematic implications. Pl Syst Evol 138:209–216, 1981.
39. Spongberg SA: Magnoliaceae hardy in temperate North America. J Arnold Arb 57:250–312, 1976.
40. Thien LB: Floral biology of Magnolia. Am J Bot 61:1037–1045, 1974.
41. Kubitzki K: Die Gattung Tetracera (Dilleniaceae). Mitt Bot München 8:1–98, 1970.
42. Kubitzki K: Doliocarpus, Davilla und verwandte Gattungen (Dilleniaceae). Mitt Bot München 9:1–105, 1971.
43. Kubitzki K: Relationships between distribution and evolution in some heterobathmic tropical groups. Bot Jahrb Syst 96:212–230, 1975.
44. Stebbins GL: Flowering Plants: Evolution above the Species Level. Cambridge: Belknap, 1974.

45. Ehrendorfer F: Unpublished observations.
46. Stebbins GL, Hoogland RD: Species diversity, ecology and evolution in a primitive Angiosperm genus: Hibbertia (Dilleniaceae). Plant Syst Evol 125:139–154, 1976.
47. Rury PM, Dickison WC: Leaf venation patterns of the genus Hibbertia (Dilleniaceae). J Arn Arb 58:209–256, 1977.
48. Engler A: Rutaceae. In Engler A, Prantl K (eds): Die natürlichen Pflanzenfamilien. 2.Aufl, Leipzig:Engelmann, 1931, 19a:187–359.
49. Guerra FM: Karyosystematik und Evolution der Rutaceae. Diss Formal u Naturwiss Fak Univ Wien, 1980.
50. Waterman PG: Alkaloids of the Rutaceae: Their distribution and systematic significance. Biochem Syst Ecol 3:149–180, 1975.
51. Smith-White S: Chromosome numbers in the Boronieae (Rutaceae) and their bearing on the evolutionary development of the tribe in the Australian flora. Austr J Bot 2:287–303, 1954.
52. Smith-White S: Cytological evolution in the Australian flora. Cold Spring Harbor Symp Quant Biol 24:273–289, 1959.
53. Levin DA, Funderburg SW: Genome size in Angiosperms: Temperate versus tropical species. Am Nat 114:784–795, 1979.

Mechanisms of Speciation, pages 511–538
© 1982 Alan R. Liss, Inc., 150 Fifth Avenue, New York, NY 10011

Speciation and Transspecific Evolution

Georges Pasteur

INTRODUCTION

During this highly rewarding week at the Accademia dei Lincei, we have been looking into how a species, whether by subdividing or by budding off, becomes two sister—congeneric—species. Assuming that every supraspecific taxon begins as a mere species, it seems of interest, for this last contribution to make some inquiry into the possibility of supraspecific taxon genesis through a speciation event. Does transspecific evolution require, at least at times, a somewhat different kind of speciation from the mechanisms that have been dealt with at this symposium or are mentioned in the Bush [1], Endler [2], White [3], or Stanley [4] reviews? The present paper will defend that this can indeed happen, as an alternative to Stanley's [5,6] species selection.

THEORETICAL CONSIDERATIONS

There is little doubt that some supraspecific taxa, up to at least the class category, originated with no noticeable episode of accelerated anagenesis through a combination of slow phyletic evolution in large populations and speciation processes each of which involved relatively little morphological change. Fossil record continuity from reptiles to mammals in Triassic times, as well as between amphibians and reptiles in late Paleozoic, testifies to this. A thick fossil sequence is even known, as has been recalled by Thaler [7], of a genus growing into another genus with no trace of, and apparently no room for, any speciation event whatever [8].

However cases such as those just related are not my concern here. Commonly, higher taxa appear suddenly in geological layers. The exponentially growing paleontological collecting that has been going on in this century has steadfastly substantiated, rather than disproved, this familiar observation [9,10]. It has also long been known that gaps preceding taxon appearances as fossils often go together with a clear episode of accelerated anagenesis [11]. This is a pertinent point to the present study.

The Simpson-Mayr Model of Transspecific Evolution

Simpson (1944) showed that these classical features of the fossil record can be rationally explained if missing links at the origins of taxa were small fast-evolving ("tachytelic") populations rather than macromutant individuals, a viewpoint population geneticists can only concur with. As envisioned by Simpson [12], supraspecific taxa constitute, together with their respective environments, "adaptive zones" of various ranks, each zone corresponding to a unique way of life. A new adaptive zone is generated when one or a few isolated populations from an existing adaptive zone "become preadaptive and evolve . . . at an exceptionally rapid rate to a radically different ecological position", such a rapid ecological shift indicating a "quantum" of evolution.

Weaknesses, related to how quantum evolution can start, undermined Simpson's pioneering model. He proposed that 1) a large population becomes fragmented into "small isolated lines of descent", and 2) prior to becoming preadaptive, any such small line has to enter an inadaptive phase, losing the adaptive equilibrium of its ancestors and collaterals. That a small population can survive unadapted for an appreciable length of time is hardly imaginable; moreover, small populations resulting from the fragmentation of a large one are relicts, more likely to be on their way to extinction than endowed with fresh evolutionary potential.

The 1944 Simpson model was to be completed by the Mayr (1954) hypothesis of a genetic revolution elicited by the change of genetic environment that can take place in one or a few founders of a new population, when this or these individuals have suddenly been cut off from a large species population and plunged into an alien environment. While Mayr [13] made no reference to Simpson and genetic revolution has often been thought of or discussed as just another possible mechanism for speciation, a genetic revolution can actually be seen as a process capable of hoisting a population from one adaptive zone to another [14,15]. Then the theory of quantum evolution is supplied with the founder effect as a definitely more plausible trigger than "inadaptivation" of isolated populations. Since Mayr's theory aptly combines with Simpson's, we may, as an ad hoc convenience, refer to a Simpson-Mayr model.

Substantiation and Review of Genetic Revolution

The last three years have seen developments of utmost importance for the theory of genetic revolution. Experimental work largely inspired by the hypotheses and drosophila findings of Carson [16–18] has beautifully illustrated genetic revolution at the infrageneric level, as summarized in Table I. A distinctive new species—in the Simpson [19] notion of evolutionary species—has been obtained by applying to a strain of Drosophila mercatorum the elements theorized to underlie genetic revolution, all in an extreme form except the change of physical environment [20]. It can therefore be stated that, of the numerous modes of *primary* speciation which are contemplatable, the only one that has been substantiated by direct experimental evidence is speciation by genetic revolution. (See [21] for the notion of primary speciation.)

Recent theoretical contributions to the subject are no less impressive. Speciation via the founder effect has been thoroughly reviewed by Templeton [22]. What is basic to triggering a genetic revolution is the sudden change from high heterozygosity and outbreeding to strong homozygosity and inbreeding, leading, in addition to keen genetic random drift, to selection of alleles fitter in homozygous condition. On other points, however, Templeton finds shortcomings in the theory contended by Mayr (see [23,24]). To stress the divergence, he chooses to refer to genetic transilience rather than revolution.

Genetic transilience is "a rapid shift in a multilocus complex influencing fitness in response to a sudden perturbation in genetic environment" ([25].) The shift necessarily leads to some physiological change, even when this is not reflected in visible morphology, but the shift may sometimes be too weak to generate reproductive isolation. "Multilocus" does not mean all the loci, far from it. Templeton [26] has given clear evidence that the loci involved in a genetic transilience 1) have important regulatory roles, and 2) are not isozyme loci. It is in fact quite clear by now that the genes detected through protein electrophoresis do not evolve in correlation with the speciation process [27], a point actually predictable [28].

The perturbation in genetic environment produced by a founding event is due to the strong inbreeding which takes place, all of a sudden, in the small founding party. Alleles are suddenly selected for their homozygous effects. Therefore, the genetic variability required to respond to the new selective pressure must be brought along by the founder(s). Mutation does not have time to intervene, since a genetic transilience must occur in the initial few generations [29]. It follows that factors capable of favoring higher initial genetic variability, as well as factors that reinforce and maintain inbreeding in the founding population, make a transilience more likely to happen in this population [30]. The alternative to genetic transilience is that either the organism does not significantly change or the founding population is wiped out—very frequent outcomes no doubt.

Further overhaul of genetic revolution theory is provided by Lande [31]. While Templeton [32] was interested in species formation only, Lande is also interested in transitions between adaptive zones and the genesis of substantial phenotypic change. Together, Templeton and Lande have brought vast refinement, and by doing so considerable support, to the Simpson-Mayr model—and have served as an inspiration to the present work.

TABLE I. Summary of Experimental Evidence for Genetic Transilience

Drosophila species	Origin	Main events	reproductive isolation[a]	Observed results		Source
				divergence		
				morphological	physiological	
Serendipitous Laboratory Occurrences						
D adiastola	Hawaiian	repeated breeding bottlenecks	+ +	–	–	[78]
D silvestris	Hawaiian	repeated breeding bottlenecks	+ +	–	–	[79]
Planned experiments						
D pseudoobscura	continental	four 1-couple/flush cycles	+	–	–	[80]
D virilis	lab stocks	drastic change in external environment	?	–	+ +	[81]
D mercatorum	Pacific	1-female founding, utterly drastic change in genetic environment	+ + +	+ +	+ +	[20]

[a] +, Partial premating isolation; + +, complete premating isolation; + + +, premating plus postmating isolation.

Genetic Revolution and Hopeful Transilience

The aim of this paper is to present arguments and observations in favor of the possibility that an ongoing genetic revolution is at the origin of those supraspecific taxa which result from accelerated anagenesis. In other words, a speciation triggered by a founder effect may end up by forming a genus or higher taxon when the mechanisms of genetic transilience continue to work rather than stop once reproductive isolation is achieved. Subdivision into an island-model situation that is supposed to optimize the transilience in the colonizing population [33], but which should soon stabilize it, would be delayed. As perceived by Lande [34], in a process of this kind reproductive isolation is a byproduct of the divergence between the populations. Reproductive isolation can only be potential anyway, since ancestral and founder populations are in essence geographically separated.

In default of hopeful monsters to account for transspecific evolution, we may call "hopeful" a nonstabilized transilience. For it to continue beyond the stage where potential reproductive isolation is reached, hopeful transilience should be launched by a *non random* foundation, resulting from properties of ancestors and founders that are outlined below. For the continuing, more or less far-reaching process that spans the whole transition, the name "genetic revolution" can be preserved, even though the rate of change may be relatively mild when the process is long. The distinction between transilience and revolution is quantitative here: A genetic transilience is a genetic revolution which stops once it has produced a congeneric species. A genetic revolution is a genetic chain reaction, a genetic turnover which can go on and on well beyond the point where potential reproductive isolation has been reached, spanning the entire transition between two adaptive zones.

Time Involved and Population Size

If major adaptive transitions were accomplished by successions of species wherein anagenesis was concentrated on speciation, as illustrated by Stanley [35,36], fossils should be found in droves from the static, large-species-population "vertical" phases, according to a basic argument of Stanley and prior punctuationalists themselves (see [37–40]). Since transitional fossils between higher taxa are actually the exception rather than the rule, transitional populations between adaptive zones *had* to be small and changing throughout. This may seem hard to admit in view of the million years involved in the birth of an order and still more of a phylum or a class. In about 1,500 Myr, however, something like 60 phyla, 250 classes, and one thousand orders have been formed—some of which, as has already been recalled, without tachytely (at least continuous tachytely). We are thus dealing with an unlikely event. On the other hand, "small" is not "very small." With effective breeding size being from a few hundreds to a few thousands, a panmictic population could still evolve very substantially faster than common species populations, and still with a much lower probability of fossilization. Moreover, since all fertile adults can become available for reproduction in case of disaster, a population with such a N_e should not be easily wiped out—the less so if females have big progenies or numerous progenies or both (see below, Predispositions). Finally, the drastic mosaic evolution so typically observed in the rare transitional fossils is to be expected, on theoretical grounds, under circumstances of genetic revolution [41].

Environment and Setting

The fact that the last phylum arose in Paleozoic times and the last class did so during the Mesozoic era also suggests that there was more ecospace left than there is available nowadays for small populations to perform major adaptive shifts. The environment must offer a new channel, a new avenue to the colonizing population, so that it can undergo sustained ecological drift (in the general, non genetic meaning of "drift"—a uni-oriented motion). From this standpoint, the diversification of animals and plants on modal-size islands and volcanic archipelagos, a classical evidence of speciation by founder effects, is misleading insofar as it makes believe that transspecific evolution has to be insular when it requires a genetic revolution. In fact, the anagenesis possible to a population colonizing a small island is conspicuously limited. The usual differentiation on archipelagos is merely specific and does not seem to ever go beyond the generic level. That a population stays in long genetic disequilibrium and drifts steadily through a succession of niches into the adaptive zone of a family inescapably demands, and this is truer for a higher taxon, the ecological wealth of a continent or a very large island—or of an ocean or a very large lake. (The celebrated example of the Drepanididae, as a whole bird family that probably evolved from a single immigrant in the Hawaiian Islands, is no exception. Bird taxonomists tend to "split" categories at all levels except the species. What is defined as a family for ornithologists would be a tribe for specialists of most other animal groups, and the Drepanididae, whose "genera" are mere species groups, are certainly not more diversified than the Hawaiian Drosophila fauna, which does not even constitute a new genus and has at most given rise to one further genus—even this is in doubt.)

Predispositions

For an organism to successfully carry out a quantum leap into a new adaptive zone, this organism has to have predisposition to do so. These can be considered at three levels.

Type of organism within the ancestral taxon. It has been superbly exemplified by Nevo [42] and generally emphasized during this symposium that we ought to be extremely cautious about making generalizations about speciation mechanisms, in view of the profound and immense variety of relevant characteristics among eukaryotic adaptive types. Yet this is vague. At what taxonomic level are adaptive differences to be considered: phyla, classes, or families? Factors leading to reproductive isolation are drastically different among major groups of micro-mammals [43], ie, above the family level. However, in as small a genus as Artemia in terms of number of species, since it has no more than five or six species Beardmore has shown us that widely dissimilar modes of speciation apparently have also been acting [44]. The same is true in the genus Drosophila, some 1,800 species strong, a third of which, interestingly enough, especially with regard to the present topic, are prone to speciate by transilience—namely the Hawaiian group, in sharp contrast to continental flies [45].

A genetic revolution will have a better chance to start and succeed in a colonizing population when the organism, whatever the level of comparison with its relatives may be, possesses the following properties: 1) aptitude for storing sperm a long time; 2) ability to survive in alien ecological conditions; 3) long reproductive life and/or big broods, that is females producing numerous progenies or big progenies or both; 4) overlapping generations; 5) no heterogamy of any sort, including rare male advantage; 6) high chromosome number and maximal possible crossing over acting on euchromatic parts. Properties 5 and 6 have been pointed out by Templeton [46] as allowing consanguinity to have maximal impact. Properties 3 and 4, in addition to their also favoring consanguinity and to their obvious value for plain survival, permit a maximum of the remaining genetic variability to be released during the population growth phase that immediately follows the initial settlement of the founding individual(s).

Ancestral population. As shown by Templeton [47], the most favorable ancestral population is a large and panmictic population. On the one hand, genetic variability is high and heterozygosity is maximal. Therefore, a propagule will take along maximal genetic variability. On the other hand, inbreeding is minimal, giving full force to the change of genetic environment. Suppose on the contrary that the ancestral population is constructed according to Wright's [48] "island-model" structure of semi-isolated local populations. Within each local population homozygosity is high and genetic variability is restricted, with opposite effects to those just described in a propagule detached from such a population.

Individual. At the individual level, the essential property is a lucky constellation of alleles, bestowing upon its bearer a strong drive for long-distance emigration, making it, at least in part, an active rather than passively transported founder. Coluzzi has shown us evidence that, in mosquitoes, migration can be active or passive depending on genotypic control [49], and I know from personal experience that such a drive can exist in individual vertebrates. In 120 days of herpetological exploration in Morocco, I was fortunate enough to chance upon an unmistakably migrating lizard twice, first a Lacerta perspicillata and in the second instance a Chamaeleo chamaeleon, the first a lizard and possibly both being long-distance migrants. The two lizards were in an environment alien to their species, utterly so for the L perspicillata. Accordingly both displayed thoroughly abnormal behavior, even though they were in excellent condition. Distance from normal species range was at the very least many kilometers for the L perspicillata (head-and-trunk length, 7 cm). With the chameleon, distance could not be known but migration was self-evident: On a cool mountain morning, the lizard was walking in a straightforward manner, paying no attention to soil nature, on feet that are the most ill-adapted feet for walking among past and present reptiles; an unusual psychophysiological state was also reflected by the chameleon's being entirely white. Such abnormalities cannot conceivably result from random dispersal, the less so as these individuals were adults. They had to be driven by an internal urge to migrate. So this does exist.

Preadaptation

Preadaptation can be viewed as being at two levels: the ancestral population, and the founding population.

Ancestral population. At this level, preadaptation should be marked in two ways: possession of alleles having extremely unequal selective values in different environments, and possession of alleles having very high fitness in the new environment awaiting colonization.

Founding population. The founding population itself has to keep preadapted, and this as long as its tachytelic evolution persists. Soon the necessary alleles can only be obtained from mutation. Now, ordinary mutability will be quite sufficient to replace eliminated and lost alleles. Two lines of research, one on quantitative mutations (started with, among others, Mukai [50]; see [51]), the other on polymorphism generating further polymorphism (started by the Ohno group [52]; see [53]), have shown that most mutation rates are orders of magnitude higher than the 10^{-5} to 10^{-6} rates still classically cited.

Clearly, the more developed the preadaptation the more circumscribed the new environment is likely to be, and the smaller the population within it. At the same time, the chance of such a preadaptation succeeding becomes progressively unlikely. All this is in accordance with 1) the observed rarity of major adaptive changes and the fact that there are fewer and fewer such changes the higher their rank, and 2) the proposed continental rather than insular origin for higher taxa, if active rather than passive emigration is required.

Other Properties

Distinction between predispositions and preadaptation has been conveniently didactic. Not all possible properties of the organism that has a special ability for genetic revolution are easy to arrange into one or the other group. As a borderline case, should it be mentioned that the gifted organism is supposed to be able to easily resort to r reproductive strategy? Since species that tend to be panmictic would seem to be rather K selected, a gifted one may be richly endowed with those jack-of-all-trades alleles or recombination possibilities which would allow a colonizing offshoot to switch to selection for early reproductive life and small body size. It is a paleontological rule that adaptive pioneers have small individual size as compared with average sizes in the descending *and* the ancestral taxa. This small size may have been shaped by a strenuous r-selection regime. Larger individual size goes along with success within the adaptive zone, hence bringing about some degree of specialization and K strategy which should be incompatible with interzonal evolution (of the same rank).

To summarize, I suggest that preadaptation, conquest of new environments, and long-lasting genetic turnover go hand in hand and are three facets of the same phenomenon; that is: In order to conquer new environments, the tachytelic population has to remain preadapted, and preadaption renewal results from constant allele substitution in a direction that is favorable to the ecological shift. Mathematical treatment of this question [54] seems to provide support to this view.

AN ILLUSTRATION IN PRESENT-DAY REPTILES

Both 1) the probable plurality of animal speciation mechanisms according to adaptive subtype within a single genus, and 2) a synthesis of the Simpson 1944 and Mayr 1954 theories for transspecific evolution, were suggested to this lecturer [55,56] by ecological and geographical situations within a small group of living vertebrates: the lygodactyls, lizards of the family Gekkonidae, or geckos.

Besides an archaic genus which need not concern us here, lygodactyls include the Ethiopian genus Lygodactylus and an isolated Madagascar population which I named Millotisaurus mirabilis in the early 1960's. This population displays every expected characteristic, in present-day nature, of a tachytelic population in the process of passing through a quantum of evolution toward a strikingly novel adaptive zone.

Lygodactylus is one of the smallest types of amniotes. Like most other gekkonid types, it is endowed with adhesive pads on its digits. As a genus-specific peculiarity, Lygodactylus also has an adhesive pad on the ventral tip of its tail; this extra pad is reproduced when the tail regenerates. A general trend in all Lygodactylus phylads is for the mental scale to become entire (Fig. 1c) by losing intramental sutures (Fig. 1b) that are remains of a primitive inner mental scale (Fig. 1a). In a given phylogenic line, the intramental index (mean suture length divided by distance between sutures) decreases from primitive to advanced species, but not one of the more than 50 known Lygodactylus species and semispecies has ever shown entire and so-called "tripartite" mentals together, even as a single exception. This suggests that the ultimate phases in suture shortening are associated with counterselected traits, so that the time when a lygodactyl population includes together entire-mental and tripartite-mental bearers is a phase

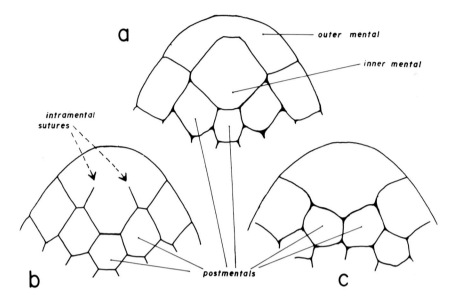

Fig. 1. Mental scale (or plate) area in lygodactyls. a) Plesiomorphic condition, only found in some individuals of a few primitive species—here, in Lygodactylus somalicus. b) "Tripartite" mental in the verticillatus species group—here, in L decaryi. c) Synapomorphic condition—here, in L madagascariensis.

of rapid evolution, as if the "entire" condition were difficult to reach, but overwhelmingly selected for as soon as it is reached by a first individual (Fig. 2). In other words, a genetic transilience is indicated in the end phase of lygodactylian mental plate evolution.

On the African continent and adjacent islands, lygodactyls are represented by four phylads, all belonging to the same subgenus of Lygodactylus. The distributions of the main two phylads, which have three species groups each, are shown on the right-hand side of Figure 3. A mere glance at these maps suggests differences in species structure and speciation between the two phylads. This is confirmed by detailed studies (see [57]).

The Eastern-Austral phylad typically exemplifies, albeit in an advanced state, classical allopatric speciation due to erection of ecological barriers that split species populations and result in their independent evolution. Two species groups that have evolved in close parallelism are represented by the three southern ranges on Figure 3 map B. Taking the Lygodactylus bonsi group as an example, from north-north-east to south-south-west the species are: L bonsi, whose sole population inhabits Mount Mlanje, Malawi, at about 2500–3000 m; L bernardi,

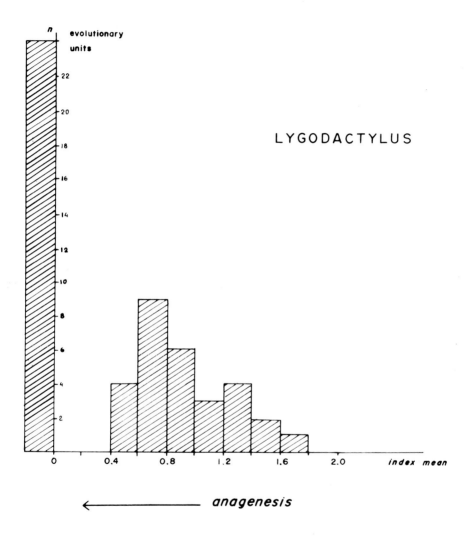

Fig. 2. Evidence of a gap, hence of a threshold and a small quantum of evolution, in the final phase of a general evolutionary trend in lygodactyls: Intramental suture regression. The index (mean of the two sutures divided by their distance) is zero when there are no such sutures. Evolutionary units are species, semispecies, and well isolated subspecies. Several Madagascar species discovered while preparing this study are not included; the picture will only be reinforced when they are.

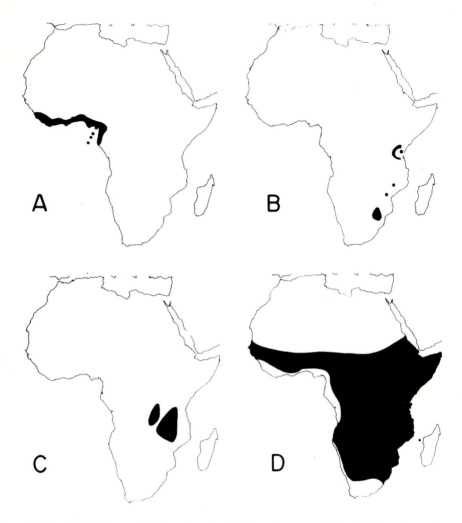

Fig. 3. Distribution of African lygodactyls (after Pasteur, 1977), all belonging to the nominal subgenus of Lygodactylus. A) Guinean phylad; B) Eastern-Austral phylad; C) Malawi-Tanganikese phylad; D) Panafrican phylad [55].

a bit less restricted but still narrowly located in the Inyangana Mountains, Zimbabwe, between about 1500 and 2000 m; and L ocellatus which, around 700–1200 m, occupies a sizable range in Transvaal and Swaziland. Now, this pattern of mountain relicts combines with NNE-SSW gradients in numerous traits. Nevertheless, the three species display strong homogeneity in ecological and morphological peculiarities. Clearly, they are remains of a unique species where numerous clines were similar to present-day gradients in nature and direction, and whose lowland populations have become extinct along with climate warming/drying.

But if we now have a look at the group of Lygodactylus picturatus (Fig. 4), a branch of the Panafrican phylad and the most important species group of all lygodactyls in geographic extension, number of individuals, average individual size, and therefore biomass, the situation is the opposite. Rather than clear-cut and geographically separated, species are all in one homogenous block, indistinguishable from one another except by colors where they are contiguous or overlapping. In this successful group of remarkable climbers and poor walkers, typical characters of the genus Lygodactylus reach extreme developments [58]. The mental plate exists only in its apomorphic condition, ie, entire.

The Figure 4 map suggests that parapatric modes of speciation may have been the major ones at work in the picturatus group. In eastern populations, strongly differentiated species-specific colors in males, unusually lively for geckos, may be a result of selection against hybrids following partial postzygotic reproductive isolation as well as of sexual selection acting as a primary, prezygotic factor of reproductive isolation. Whether stasipatric speciation is involved will have to wait, to be known, until the picturatus group chromosomes are known at the gamma karyology level. (Beta karyology is of limited interest in geckos, where Robertsonian and other large translocations seem quite rare.)

The geographical pattern of the closely related Lygodactylus capensis group, another branch of the Panafrican phylad, offers a strikingly different picture (Fig. 5) from either the picturatus or the Eastern-Austral species groups. It certainly looks typical of what Prof Mayr called peripatric speciation in the inaugural lecture [59], and in all likelihood what is involved is just that. Knowing that all five peripheral units are well differentiated and definitely have more advanced traits than the two main species [60], we are faced in the two subgroups with unmistakable examples of genetic transiliences. An interesting point is that only two of these five daughter units are on islands: The capensis group offers an insular pattern in continental speciation. Just as in peripatric insular speciation proper, one of the continental daughter species, L lawrencei, is different enough to almost deserve to be classified as a subgenus. L lawrencei is also the only African form of the capensis group with an entire mental plate, and this can be seen as further evidence that a genetic revolution took place in its formation. Along with its founder-effect-speciation pattern, the capensis group thus presents deeper diversification than the other two types of species groups we have dealt with. And yet the African species of the capensis group are far from being the whole story, the continuation of which follows.

Two members of the group live in South America, as remnants of a once huge population occupying from north-eastern Brazil to Gran Chaco, Paraguay. The two American species, although easy to distinguish from each other [61],

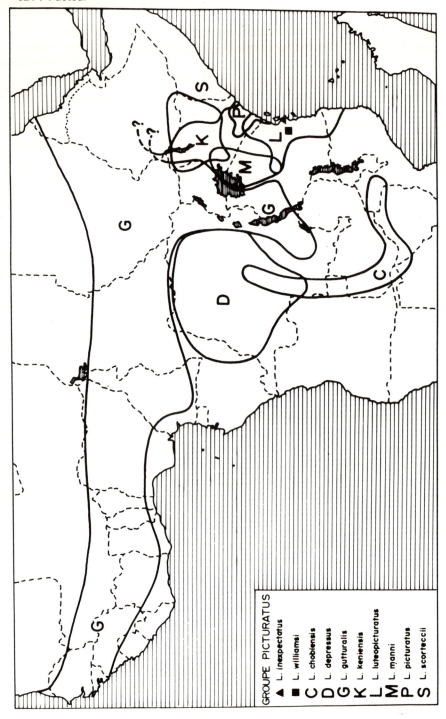

GROUPE PICTURATUS

▲ L. inexpectatus
 L. williamsi
■ L. chabanaudi
C L. depressus
D L. gutturalis
G L. keniensis
K L. luteopicturatus
L L. manni
M L. picturatus
P L. scorteccii
S

are both very similar to L capensis itself, except that they have entire mentals [62]. This trait suggests that at some time a genetic transilience occurred in their ancestry—presumably with the colonization of America. On the Indian side of Africa, the dramatic adaptive radiation that Lygodactylus underwent on Madagascar, which I will refer to later, has been traced back to a propagule of, again, the species L capensis [63]. All told, the descendancy of L capensis today includes no fewer than 26 independent lineages, and these lineages, as we shall see, are more diversified than the 30 or so lineages of all African lygodactyls put together.

In relation to what was theorized about in the first part of this paper, the following points, whether already noted or not, are made evident by lygodactyls: 1) Modes of speciation can differ strikingly according to adaptive subtype and species-group structure within a single lizard genus. 2) An insular pattern of species-forming genetic transiliences can be observed continentally, without even such island-like natural sites as caves or isolated mountains. 3) Lygodactyls reproduce all year long; each fertile female can lay two eggs several times a year; female modal longevity is about three years in natural populations of L verticillatus, a species descended from L capensis (unpublished observations). 4) The trend toward long-distance colonization and transilience existed early in the capensis group, since it is observable in the sibling species L angolensis (Fig. 5, left). But L angolensis is (in bush and gallery forests) strictly arboreal. L capensis is rather loosely so, and often inhabits open spaces where it can commonly be found on boulders or, in association with wood, on the ground. L capensis appears consequently more movable and adaptable, in accordance with its better aptitude to colonize at long distance, including maritime transport. L angolensis population structure conforms to the Wright "island model," whereas L capensis is so dense that the species is probably reticular, if not fully panmictic, in many places. In Transvaal, it is known as "the common gecko," although plenty of larger gecko species exist in the area. These differences between L capensis and L angolensis are reflected in quantitative character ratios, whose within-sample variation is larger in the former, and between-sample variation is larger in the latter [64]. Because selection is necessarily more diversifying in L capensis, the more eurytopic species, its population is necessarily more heterozygous, a factor enhancing the chance of successful transilience.

Fig. 4. Provisional map of the Lygodactylus picturatus species group. Reproductive isolation is only ascertained between the following pairs: L luteopicturatus/L picturatus, L luteopicturatus/L williamsi, and (indirectly) L luteopicturatus/L keniensis. Two color subgroups are apparent: G-D-C-M and L-P-S-K. (After Pasteur, 1964 [14].)

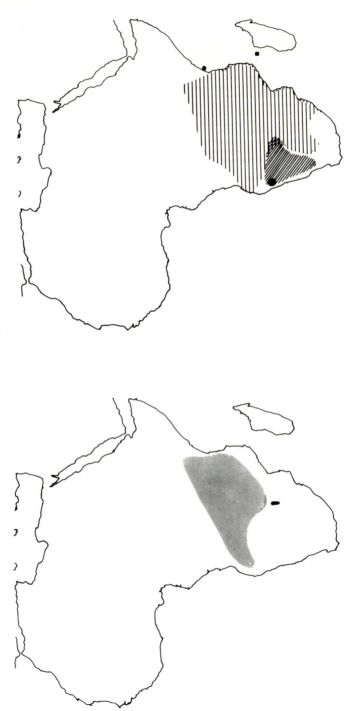

Clearly, in the genus Lygodactylus, L capensis was uniquely predisposed and preadapted to be a successful long-distance colonist, and it is then no surprise that the Madagascar lygodactyl fauna is an offshoot of L capensis rather than from another African lineage. However, during a time when L capensis changed very little (p 530)—nowadays still a rather primitive-looking lygodactyl—the adaptive radiation of its propagule on Madagascar engendered 1) a new Lygodactylus subgenus, 2) three new species groups in the old subgenus, and 3) what could be, with Millotisaurus already referred to, the early stages of a new adaptive zone transcending the genus level—the whole thing totalling 19 living species, versus definitely fewer than twice as many on an area 35 times as large, all within one subgenus, in Africa. As in any adaptive radiation, tachytelic lines and their genetic revolutions are strongly implied in this Malagasy descendance of L capensis.

In spite of this relatively tremendous cladogenesis, the Madagascar lygodactyl fauna shows strictly invariant states in nostril characters that were highly variable in Africa, L capensis population included. In addition, all Madagascar Lygodactylus have in common minor traits which are never observed in African forms. The nostril traits suggest descent from one individual that happened to be homozygous at the controlling loci. Novel uniform characters would result from subsequent homozygosity in new gene combinations of the founding population descended from the first individual. Knowing that Squamata females can long retain sperm [65], that L capensis can eat prey of extremely varied sizes [66], and that eastbound rapidly flowing Zambezi-like streams have always existed as powerful raft-propellants since the Paleogene, a likely scenario for the one propagule that succeeded in colonizing Madagascar is as follows[1].

A female L capensis was thrust off the African coast with a floating tree pack, had progeny on that raft, and a highly inbred population became established in this unprecedented habitat. Reproducing xylophagous insects provided for the colony sustenance, and bark fissures offered sufficient shelter against splashing water and green sea. The sea was never turbulent and when what remained of

---

[1]That other attempts took place is testified by endemic Lygodactylus insularis on atoll Juan de Nova. See Figures 5 and 6.

Fig. 5. Peripatric speciation in the two subgroups of the Lygodactylus capensis species group. Left: superspecies of L angolensis; peripheral form: L stevensoni. Right: superspecies of L capensis; peripheral forms: counterclockwise, L lawrencei, L bradfieldi, L insularis, L [capensis] pakenhami. (L bradfieldi occupies a substantial range but is 1) exceptionally homogenous, and 2) a sibling species of L capensis, suggesting it originated through a recent and rapid rush over its niche by a strongly flushing peripheral isolate of L capensis [14].)

the lucky raft gently drove ashore on Madagascar, there was environmental continuity with such fallen logs and tree trunks as can be seen today on Madagascar's western seaside. This made an easy settlement possible.

Around the first ecological niche, the empty, available ecospace offered by Madagascar to very small climbing lizards was immense. For the initial population, competition would rather come, if at all, from spiders and other rapacious arthropoda than other lizards (all much bigger judging from present-day species). A vast population flush may well have soon followed the initial settlement, allowing important recombinations to take place [67]. Thus a variety of new preadaptations may have been ready around the first major ecological interface right when it was reached. Whatever the particular process may have been, there is no doubt anyway, judging from present-day diversification, that Madagascar provided lygodactyls—as it did lemurs and many less famous groups (see [68])—with environmental diversity of continental scale. This has especially benefited the lineage of the above-mentioned far-isolated lygodactyl species whose differentiation is such that Pasteur [69] gave it another genus name, Millotisaurus. From 2300 m up in the Ankaratra Mountains of central Madagascar (see Fig. 6), the M mirabilis population is perched on the northern summit of the highest mount (Mount Tsiafajavona, 2644 m). Covering about two to three square kilometers, it numbers a few thousand adults. This population, as compared with Lygodactylus geckos, which are tropical tree dwellers or at any rate specialized climbers, has accomplished a huge ecological shift: There are no trees on the Ankaratra summit, only short grass and scattered stones; the average temperature is about 10°C in day time and often below 0°C on winter nights. In this improbable spot, nevertheless, M mirabilis is actively reproducing all year round.

In association with ecological novelty, the Millotisaurus population displays six major morphological traits that were unknown in the genus Lygodactylus, three of which are new for the family Gekkonidae—hence the idea that first steps in the genesis of a suprageneric adaptive zone may be at stake. New traits at the genus level are: 1) existence of keels on all dorsal, lateral, and gular scales, 2) dorsal scales that overlap, and 3) the loss of adhesive pads on the tail. New traits at the family level include 4) keeled scales having two or more keels (up to nine on some animals) on many body parts, 5) hands with four fingers, the thumb having been lost in the process, and 6) the existence of a longitudinally striped, skink-like coloration.

All these characters are adaptively related to the new way of life. Millotisaurus digits have fewer and lesser adhesive scales than the Lygodactylus homologues, which obviously promotes running on the ground rather than climbing. The loss of finger I and the tail adhesive pad have distinctly similar adaptive significance, confirmed by an exceptionally high rate of tail autotomy. Even though the significance of other above-noted traits is not so obvious, that these traits are also adaptive is attested to by the fact that they tend to converge with usual

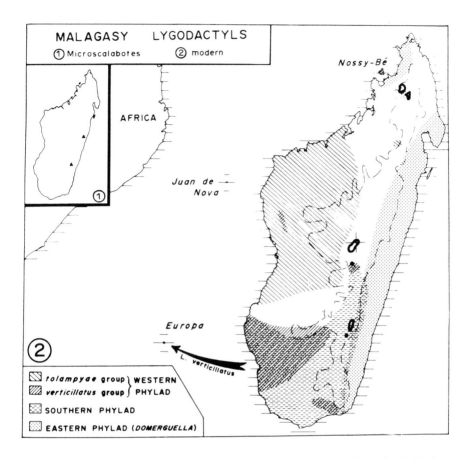

Fig. 6. Provisional map of Madagascar lygodactyls (after Pasteur 1964, updated). Dashed contour line: 1000 m; heavy solid contour line: 2000 m. The cross in the center of the island points to the Millotisaurus mirabilis single deme. (Domerguella is the second subgenus of Lygodactylus, and Microscalabotes is a relict from a previous African invasion.) [14]

characters of skinks. Skinks are typical ground dwellers. Indeed two of the three other lizard species that live with Millotisaurus or nearby are species of the family Scincidae.

A most probable consequence of the new way of life is panmixia in the M mirabilis population. The average male cruising range in a lifetime is likely to be a sizable fraction of the small population range diameter. Male environment must be coarse grained at the extreme. While, in other lygodactyl species, harem-keeping male territoriality is fostered by such heterogeneities as trees [70] or boulder structures [71], on the Ankaratra summit the environment is perhaps so homogenous as to induce random mating. This would not preclude differential reproduction within either sex, of course. At any rate M mirabilis must have an N_e in the hundreds or at most the lower thousands, quite in keeping with a long sustained, if relatively soft, genetic revolution (see [72]).

In spite of many original species characters that come in addition to the major traits mentioned above, the morphology of M mirabilis clearly indicates a relationship with both the Western and the Southern phylads of Madagascar Lygodactylus, the tie being closer with the L montanus group of the Southern phylad. However, some primitive postmental traits indicate a more direct relationship with the capensis species group in Africa than with any Madagascar species. Still more interestingly, some postmental scale numbers are so plesiomorphic as to be *more* primitive than their homologs are in present day L capensis population. In fact, they are at least as primitive as in the somalicus species group, which is the Panafrican phylad's primitive species group in contemporary nature (Table II). This has three important implications: 1) Since the time when L capensis dispatched its successful propagule to Madagascar, there has been significant phyletic evolution, at least for postmentals, in the L capensis lineage; 2) the founders of the Malagasy population still had alleles controlling plesiomorphic chin character states, and these alleles, or the right combination(s) thereof, have been subsequently lost in lygodactyl lines other than M mirabilis; 3) M mirabilis is an odd mixture of advanced and primitive characters, ie, it is a typical example of the drastic mosaic evolution characteristically shown during shifts to new adaptive zones (see review in [73]). This brings about the related question of a further striking attribute of the M mirabilis population: However small, it is far more variable morphologically than any Lygodactylus species, no matter how large that species' population may be. This phenotypic variability manifests itself in two most significant ways:

1) Variation in characters that have no phylogenetic significance. This variation, which reaches here a maximum among lygodactyls in several traits, especially some fluctuating asymmetries, is evidence of a relatively weak developmental homeostasis. Since there is no immigration, this weak homeostasis in turn points to strong directional selection (see [74]) and/or a relatively high level of inbreeding (see a brief review in [75]).

2) Traits whose variation is stretched between old and new character states. That an intense, unfinished anagenetic evolution is in process is strongly suggested in this case. For instance, disappearance of the thumb is not quite complete yet; a spectrum of cases is observed between hands that show no trace of it and hands where a rather marked bump remains. Dorsal scale imbrication, adhesive parts regression, and extended distributions in various traits of phylogenic significance present the same picture, namely variation stretching between a condition being given up and a future condition, as in postmental numbers already

TABLE II. Numbers of Postmental Scales Observed in Related Lygodactyls, From Primitive to Advanced Forms

Taxonomic entity	Postmental plates (%)					n
	5	4	3	2	1	
Africa						
L somalicus species group	—	4.2	95.8	—	—	48
Lygodactylus capensis	—	0.9	91.0	8.1	—	210
Madagascar lower altitudes						
L. verticillatus species group	—	—	99.4	0.6	—	1416
Lygodactylus robustus	—	—	21.7	78.3	—	120
Lygodactylus tuberifer	—	—	5.0	93.7	1.3	382
Madagascar mountain tops						
L. montanus species group	—	—	89.8	10.2	—	127
Intermediate species	—	—	94.8	5.2	—	58
Millotisaurus mirabilis	1.2	2.4	90.4	6.0	—	83

Relationships can be seen in Figure 8.
In these and all other lygodactyl lines, the trend is toward numerical reduction.

mentioned (Table II). A somewhat special but spectacular example is also provided by mental plate variation. Contrary to the absolute rule in present-day species of Lygodactylus (see p 519), both entire and tripartite mentals can be seen in Millotisaurus. The distribution of intramental indexes (Fig. 7) fits the hypothesis formulated earlier that the last stages of intramental suture regression are somehow selected against, and have then to be overcome. With this trait, some disequilibrium, and the necessity for the population to go on through complete mental scale evolution, are indicated.

In conclusion, the characteristics theoretically assumed to exist in a transitional population performing a major adaptive change and in its ancestors have been found in the M mirabilis population and its ancestors. If we suppose that this population has always been small, and has undergone a long sustained genetic revolution which made it shift from arboreal or semi-arboreal to ground-dwelling life while making it drift from warm western lowlands through the Malagasy Plateau to cold central mountains, then we may think that the population is trapped on its mountain peak. However, a climatic change to colder weather would probably suffice for it to spread, opening, as it were, the new adaptive zone. Further information about this evolution can be obtained from an examination of parallel evolutions.

Preadaptation has been insisted upon in the first part of this paper. Ancestor preadaptation implies that several lines could evolve in parallel directions to the main stem of a new adaptive type, as is so repeatedly emphasized in evolutionary literature about the early stages of mammalian evolution. This is exactly what

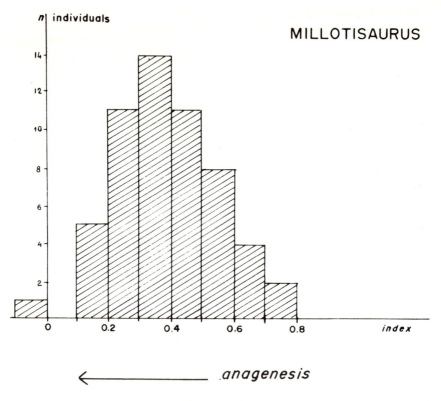

Fig. 7.   Intramental suture index in the Millotisaurus mirabilis population. Compare with Figure 2.

occurred in Madagascar with the Southern phylad of Lygodactylus. No fewer than four species (L montanus, L blanci, Lygodactylus sp., and Intermediate species on Figure 8), all with restricted if not small populations, have followed the M mirabilis path, each on a different mountain. The four species show typical dissimilar mosaic evolution from one to the next (see one comparison in [76]). The first three are but partially modified in the Millotisaurus direction. Ecologically and geographically as well as morphologically, they are not very far removed from a species of lower altitudes, L robustus. But the fourth species, which lives above 2000 m in the Andringitra Mountains, is intermediate enough between Millotisaurus and Lygodactylus to throw doubt, at first, on generic distinction validity, even though M mirabilis represents an extreme gekkonid trend. The trend is already well embodied in the Andringitra species.

Most remarkable is the fact that all four species share with M mirabilis an important color polymorphism, having a striped morph and a nonstriped, usually spotted, one—both morphs being more salient and contrasting beautifully in the intermediate species and M mirabilis [77], where the striped morph is definitely skink-like. Since this polymorphism is not shared by strongly convergent mountain species of other groups (see Fig. 8), common immediate ancestry is clearly pointed out for the montanus group and Millotisaurus. If this is true, then no Southern species did lose the character state "more than 3 postmentals" until the color polymorphism settled in, and then all lost it. Besides deeper mosaic evolution on this ground, what also opposes the M mirabilis lineage to all others is body size. The Lygodactylus species heavily underlined on Figure 8 (Intermediate included) have larger individual sizes than relatives, in conformity with a now cherished rule held by many herpetologists about mountain lizards. Not so with M mirabilis: It is the second shortest lygodactyl, even though it is the highest-living Madagascar species. We might ask the question, has selective pressure for shorter size dominated in it adaptation to cooler climate and thinner air (protection against cosmic radiations)? If so, how can sheer survival be explained? Anyway, smaller size, as noted in the first part of this paper, conforms with intense phyletic interzonal evolution.

The color polymorphism of Millotisaurus and mountain Southern Malagasy species teaches us an important lesson. Since the day when a capensis founding party set foot on Madagascar, nowhere else in the lygodactyls has cladogenesis been more profound than in this ensemble. And yet, the color polymorphism has gone through all of it undamaged. This does show that polymorphisms other than biochemical polymorphisms can be completely independent from the speciation process, albeit anything but neutral. In the present case, the dichromatism signal tenacity in the concerned species, and its complete absence in all related species, are evidence that it has been highly selective since the very day of its inception, knowing that the new morph, ie, the striped one, is the more visible for predators.

The interpretation presented here of the Millotisaurus mirabilis population is admittedly very hypothetical. It certainly cannot have us forget that genera are often nothing else than clusters of species which have become so defined via extinction of intermediate species. What the Millotisaurus question needs now is reinforcement through technical expertise. To this writer's knowledge, only one chromosome preparation has ever been made from a lygodactyl, unfortunately on defective glass. That was from a M mirabilis testis, showing 39 or 40 acrocentric chromosomes. From the biochemical genetics standpoint, lygodactyls are absolute terra incognita. Accordingly, investigations of prime interest can still be conducted along the two lines emphasized in this paper—speciation plurality and subsequent transspecific evolution—a sine-qua-non condition being live samples. Kenya will be foremost in the study of lygodactyl speciations,

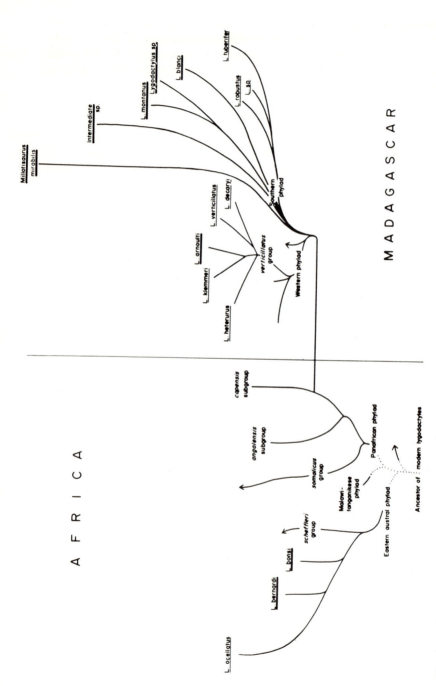

Fig. 8. Kindred of Millotisaurus. Heavily underlined species, which are restricted to upper parts of small isolated mountain ranges or isolated mounts, are convergent in many traits; prominent are an eyelid expansion above the eye and a dark shade all over the dorsal side.

because all major forms of the picturatus group, three of the capensis group, and two other fascinating complexes are all present in that country. In Madagascar, exceedingly stringent rules about the use of any kind of wild animal seem to preclude, in the foreseeable future, the possibility of scientific work on lygodactyls.

ACKNOWLEDGMENTS

This work has benefited from critical comments made upon reading the manuscript by Profs. Edouard R. Brygoo, Joël Mahé, Jacques Michaux, Renaud Paulian, G. Ledyard Stebbins, and Alan R. Templeton. I am most grateful to them all, a recognition of debt which is not intended to be a clue of their opinions.

REFERENCES

1. Bush GL: Modes of animal speciation. Ann Rev Ecol Syst 6:339, 1975.
2. Endler JA: Geographic Variation, Speciation, and Clines. Princeton: Princeton University Press, 1977.
3. White MJD: Modes of Speciation. San Francisco: W.H. Freeman, 1978.
4. Stanley SM: The nature of quantum speciation. In Stanley SM (ed): Macroevolution. San Francisco: W.H. Freeman, 1980, p 143.
5. Stanley SM: A theory of evolution above the species level. Proc Natl Acad Sci USA 72:646, 1975.
6. Stanley SM: Macroevolution. San Francisco: W.H. Freeman, 1980.
7. Thaler L: Paleontological patterns of evolution in rodents and biological processes of speciation. In Barigozzi C, Montalenti G, White MJD (eds): Mechanisms of Speciation. New York: Alan R. Liss, 1982.
8. Chaline J, Mein P: Les Rongeurs et l'Evolution. Paris: Doin, 1979.
9. Valentine JW: Transspecific evolution. In Dobzhansky TH, Ayala FJ, Stebbins GL, Valentine JW: Evolution. San Francisco: W.H. Freeman, 1977, p 233.
10. Valentine JW: The evolution of multicellular plants and animals. Sci Am 239:104, 1978.
11. Valentine JW: The geological record. In Dobzhansky Th, Ayala FJ, Stebbins GL, Valentine JW: Evolution. San Francisco: W.H. Freeman, 1977, p 314.
12. Simpson GG: Tempo and Mode in Evolution. New York: Columbia University Press, 1944.
13. Mayr E: Change of genetic environment and evolution. In Huxley J, Hardy AC, Ford EB (eds): Evolution as a Process. London: George Allen and Unwin, 1954, p 157.
14. Pasteur G: Recherches sur l'Evolution des Lygodactyles. Rabat: Institut Scientifique Chérifien, 1964.
15. Lande R: Genetic variation and phenotypic evolution during allopatric speciation. Am Nat 116:463, 1980.
16. Carson HL: The population flush and its genetic consequences. In Lewontin RC (ed): Population Biology and Evolution. Syracuse: Syracuse University Press, 1968, p 123.

17. Carson HL: The genetics of speciation at the diploid level. Am Nat 109:73, 1975.
18. Carson HL: Speciation and sexual selection in Hawaiian Drosophila. In Brussard PF (ed): Ecological Genetics: The Interface, New York: Springer-Verlag, 1978, p 93.
19. Simpson GG: Principles of Animal Taxonomy. New York: Columbia University Press, 1961.
20. Templeton AR: The unit of selection in Drosophila mercatorum. II. Genetic revolution and the origin of coadapted genomes in parthenogenetic strains. Genetics 92:1283, 1979.
21. Grant V: Plant Speciation. New York: Columbia University Press, 1971.
22. Templeton AR: The theory of speciation via the founder principle. Genetics 94:1011, 1980.
23. Mayr E: Animal Species and Evolution. Cambridge: Belknap Press, 1963.
24. Mayr E: Populations, Species, and Evolution. Cambridge: Belknap Press, 1970.
25. Templeton AR: reference 22.
26. Templeton AR: reference 20.
27. Pasteur G, Pasteur N: Les critères biochimiques et l'espèce animale. In Bocquet C, Génermont J, Lamotte M (eds): Les Problèmes de l'Espèce dans le Règne Animal. Paris: Société Zoologique de France, 1980, vol 3, p 99.
28. Templeton AR: Modes of speciation and inferences based on genetic distances. Evolution 34:719, 1980.
29. Templeton AR: reference 22.
30. Templeton AR: reference 22.
31. Lande R: reference 15.
32. Templeton AR: reference 22.
33. Templeton AR: reference 22.
34. Lande R: reference 15.
35. Stanley SM: reference 6, Fig 7-3.
36. Stanley SM: Speciation and the fossil record. In Barigozzi C, Montalenti G, White MJD (eds): Mechanisms of Speciation. New York: Alan R. Liss, 1982, p 41.
37. Eldredge N, Gould SJ: Punctuated equilibria: an alternative to phyletic gradualism. In Schopf TJM (ed): Models in Paleobiology. San Francisco: Freeman and Cooper, 1972, p 82.
38. Hecht MK, Eldredge N, Gould SJ: Morphological transformation. the fossil record, and the mechanisms of evolution: a debate. Evol Biol 7:295, 1974.
39. Gould SJ, Eldredge N: Punctuated equilibria: the tempo and mode of evolution reconsidered. Paleobiology 3:115, 1977.
40. Stanley SM: reference 6.
41. Lande R: reference 15, equations 12-13.
42. Nevo E: Speciation in subterranean mammals. In Barigozzi C, Montalenti G, White MJD (eds): Mechanisms of Speciation. New York: Alan R. Liss, 1982, p 191.
43. Nevo E: Speciation in subterranean mammals. In Barigozzi C, Montalenti G, White MJD (eds): Mechanisms of Speciation. New York: Alan R. Liss, 1982, p 191.
44. Beardmore JA: Genetic differentiation and speciation in Artemia. In Barigozzi C, Montalenti G, White MJD (eds): Mechanisms of Speciation. New York: Alan R. Liss, 1982, p 345.
45. Templeton AR: reference 22, p 1032.

46. Templeton AR: reference 22.
47. Templeton AR: reference 22.
48. Wright S: Evolution in Mendelian populations. Genetics 16:97, 1931.
49. Coluzzi M: Micro- and macrogeographical distribution of chromosomal inversions and speciation in anopheline mosquitoes. In Barigozzi C, Montalenti G, White MJD (eds): Mechanisms of Speciation. New York: Alan R. Liss, 1982, p 143.
50. Mukai T: The genetic structure of natural populations of Drosophila melanogaster. I. Spontaneous mutation rates of polygenes controlling viability. Genetics 50:1, 1964.
51. Lande R: reference 15.
52. Ohno S, Stenius C, Christian L, Schipmann G: De novo mutation-like events observed at the 6PGD locus of the Japanese quail, and the principle of polymorphism breeding more polymorphism. Biochem Genet 3:417, 1969.
53. Koehn RK, Eanes WF: An analysis of allelic diversity in natural populations of Drosophila: The correlation of rare alleles with heterozygosity. In Karlin S, Nevo E (eds): Population Genetics and Ecology. New York: Academic Press, 1976, p 377.
54. Lande R: reference 15.
55. Pasteur G: Quelques commentaires à propos de l'espèce chez les lygodactyles. In Bocquet C, Génermont J, Lamotte M (eds): Les Problèmes de l'Espèce dans le Règne Animal. Paris: Société Zoologique de France, 1977, vol 2, p 335.
56. Pasteur G: reference 14.
57. Pasteur G: reference 55.
58. Pasteur G: reference 14.
59. Mayr E: Patterns of speciation in animals. In Barigozzi C, Montalenti G, White MJD (eds): Mechanisms of Speciation. New York: Alan R. Liss, 1982, p 1.
60. Pasteur G: reference 14, chaps 10-11.
61. Smith HM, Martin RL, Swain TA: A new genus and two new species of South American geckos. Papéis Avulsos Zool 30:195, 1977.
62. Bons J, Pasteur G: Solution histologique à un problème de taxinomie herpétologique intéressant les rapports paléobiologiques de l'Amérique du Sud et de l'Afrique. Compt Rend Acad Sci D 284:2547, 1977.
63. Pasteur G: reference 14, chap 14.
64. Pasteur G: reference 14, chap 11.
65. Saint-Girons H: Déplacements et survie des spermatozoïdes chez les reptiles. Colloq Instit Natl Santé Rech Médic 26:259, 1973.
66. Pianka ER, Huey RB: Comparative ecology, resource utilization and niche segregation among gekkonid lizards in the southern Kalahari. Copeia:691, 1978.
67. Carson HL: reference 17.
68. Paulian R: La Zoogéographie de Madagascar et des Iles Voisines. Tananarive: Institut de Recherche Scientifique, 1961.
69. Pasteur G: Notes préliminaires sur les lygodactyles (Gekkonidés) III. Diagnose de Millotisaurus gen. nov., de Madagascar. Compt Rend Soc Sci Nat Maroc 28:65, 1962.
70. Greer AE: The ecology and behavior of two sympatric Lygodactylus geckos. Breviora 268:1, 1967.
71. Blanc CP, Pasteur G: Unpublished observations.
72. Lande R: reference 15.

73. Mayr E: reference 23, p 596.
74. Soulé M: Phenetics of natural populations. II. Asymmetry and evolution in a lizard. Am Nat 101:141, 1967.
75. Futuyma DJ: Evolutionary Biology. Sunderland: Sinauer, 1979, p 373.
76. Pasteur G: Note préliminaire sur les geckos du genre Lygodactylus rapportés par Charles Blanc du Mont Ibity (Madagascar). Bull Mus Natl Hist Nat (2) 39:439, 1967.
77. Pasteur G: reference 14, plate I.
78. Arita LH, Kaneshiro KY: Ethological isolation between two stocks of Drosophila adiastola Hardy. Proc Hawai Entom Soc 13:31, 1979.
79. Ahearn JN: Evolution of behavioral reproductive isolation in a laboratory stock of Drosophila silvestris. Experientia 36:5, 1980.
80. Powell JR: The founder-flush speciation theory An experimental approach. Evolution 32:464, 1978.
81. Wallace B: The adaptation of Drosophila virilis to life on an artificial crab. Am Nat 112:971, 1978.

INDEX

PROGRESS IN CLINICAL AND BIOLOGICAL RESEARCH

Series Editors

Nathan Back
George J. Brewer
Vincent P. Eijsvoogel
Robert Grover

Kurt Hirschhorn
Seymour S. Kety
Sidney Udenfriend
Jonathan W. Uhr